World Water Resources

Volume 2

Series Editor
Vijay P. Singh, Texas A&M University, College Station, TX, USA

This series aims to publish books, monographs and contributed volumes on water resources in the world, with particular focus per volume on water resources of a particular country or region. With the freshwater supplies becoming an increasingly important and scarce commodity, it is important to have under one cover up to date literature published on water resources and their management, e.g. lessons learnt or details from one river basin may be quite useful for other basins. Also, it is important that national and international river basins are managed, keeping each country's interest and environment in mind. The need for dialog is being heightened by climate change and global warming. It is hoped that the Series will make a contribution to this dialog. The volumes in the series ideally would follow a "Three Part" approach as outlined below: In the chapters in the first Part *Sources of Freshwater* would be covered, like water resources of river basins; water resources of lake basins, including surface water and under river flow; groundwater; desalination; and snow cover/ice caps. In the second Part the chapters would include topics like: *Water Use and Consumption*, e.g. irrigation, industrial, domestic, recreational etc. In the third Part in different chapters more miscellaneous items can be covered like impacts of anthropogenic effects on water resources; impact of global warning and climate change on water resources; river basin management; river compacts and treaties; lake basin management; national development and water resources management; peace and water resources; economics of water resources development; water resources and civilization; politics and water resources; water-energy-food nexus; water security and sustainability; large water resources projects; ancient water works; and challenges for the future. Authored and edited volumes are welcomed to the series. Editor or co-editors would solicit colleagues to write chapters that make up the edited book. For an edited book, it is anticipated that there would be about 12–15 chapters in a book of about 300 pages. Books in the Series could also be authored by one person or several co-authors without inviting others to prepare separate chapters. The volumes in the Series would tend to follow the "Three Part" approach as outlined above. Topics that are of current interest can be added as well.

Readership

Readers would be university researchers, governmental agencies, NGOs, research institutes, and industry. It is also envisaged that conservation groups and those interested in water resources management would find some of the books of great interest. Comments or suggestions for future volumes are welcomed.

Series Editor:

Vijay P. Singh, Department of Biological and Agricultural Engineering & Zachry Department of Civil Engineering, Texas A & M University, USA, vsingh@tamu.edu

More information about this series at http://www.springer.com/series/15410

Nilgun B. Harmancioglu • Dogan Altinbilek
Editors

Water Resources of Turkey

Springer

Editors
Nilgun B. Harmancioglu (retired)
Civil Engineering, Hydraulics and Hydrology
Dokuz Eylul University
Buca, Izmir, Turkey

Dogan Altinbilek
Civil Engineering Department
Middle East Technical University
Cankaya, Ankara, Turkey

ISSN 2509-7385 ISSN 2509-7393 (electronic)
World Water Resources
ISBN 978-3-030-11728-3 ISBN 978-3-030-11729-0 (eBook)
https://doi.org/10.1007/978-3-030-11729-0

This Springer imprint is published by the registered company Springer Nature Switzerland AG.
The registered company address is: Gewerbestrasse 11, 6330 Cham, Switzerland

The Ataturk Dam on the Euphrates River, was built as the major dam of the Southeastern Anatolia Project with 85 million m^3 embankment, creating a reservoir of 48 km^3, with installed capacity of 2400 MW. It generates hydroelectric energy up to 9 TWh/y. The dam is functional since 1992 and is widely considered to be not only the largest dam in Turkey, but also one of the largest in the world (Photo by Prof. Dr. Unal Ozis)

Foreword

In world geography, Turkey occupies a unique place connecting Europe and the Middle East. It has varied landscape and varied climate, which, in turn, reflect diverse space-time characteristics of its water resources. These diverse characteristics present a challenge and an opportunity to policy makers, the agencies responsible for water resource management, the agricultural sector involved in growing crops, the energy sector involved in generating hydropower, the municipalities responsible for water supply, the departments of tourism involved in providing recreational facilities, as well as those involved in environmental and ecosystem management. Further, because of its situation, Turkey has a number of transboundary river basins, which compound the aforementioned challenge and opportunity. A comprehensive treatise dealing with a broad range of water resource issues pertaining to Turkey has long been desired, and the need for such a treatise has increased under the specter of climate change and global warming and the transformation of the country from a water-rich nation to a not-so-water-rich nation, facing frequent water scarcity because of a growing population, rising standard of living, increasing demand for water, and less-than-proper water management. This book, edited by Professors Harmancioglu and Altinbilek, meets this long-felt need.

Beginning with a discussion of the factors that shape Turkey's water resources and their availability and distribution, the book, encompassing 17 chapters, provides a detailed treatment of water use and consumption, as well as management, legislation, sustainability and security, climate change, ancient water works, and future outlook. The book contains a wealth of information, reflecting the vast and rich experiences of not only the editors but also the chapter contributors.

There is much to learn from this book which more than meets the objectives of the World Water Resources book series. It shows how water management strategies can be developed in concert with legal, institutional, economic, and capacity development requirements. These strategies enhance sustainability and water security and promote the water-food-energy nexus. The material presented in the various chapters is equally relevant to many countries in the world—a point for launching the book series.

Both Professors Harmancioglu and Altinbilek deserve a lot of applause and ought to be congratulated for preparing this treatise, which will be of enormous value to the people of Turkey and will also enrich the water resource literature.

Distinguished Professor, Regents
Professor, Caroline & William N Lehrer
Distinguished Chair in Water
Engineering, Texas A&M University,
College Station, Texas, USA

Vijay P. Singh

Preface

Environmental and water crises dominate the present world we live in. Pollution is observed at varying degrees of intensity in almost all natural resources. The quality of surface and ground waters is continuously degrading to limit their use. Similarly, land resources are exposed to problems of soil erosion, deforestation, urban sprawl, and desertification in many parts of the world. Another problem in particular regions is air pollution, which has already reached life- and health-threatening levels. All these adverse conditions have eventually endangered physical habitat for biodiversity. There are also some recent problems to aggravate the situation, i.e. climate change and its possible effects on various components of the environment. Certainly, environmental degradation not only endangers nature, but it also has serious social and economic implications. An important feature of the above problems is that they are experienced globally; that is, all countries experience one or more of these adverse conditions at different degrees of severity and coverage. The same is valid for water resources so that it is important to work out commonalities and differing aspects of world water resources.

Springer has initiated a new book series entitled World Water Resources, edited by Professor Vijay P. Singh from Texas A&M University, USA. The reason for starting the book series was to have a comprehensive understanding of water resources of different countries and how these countries are planning and managing their water resources. With the freshwater supplies becoming an increasingly important and scarce commodity, it is important to have under one cover up-to-date literature published on world water resources and their management. Lessons learned from one river basin may be quite useful for other basins. Since many countries share their resources with each other, it is deemed useful to have an unbiased assessment of these resources, which will help resolve legal issues arising from mutual conflicts. That means national and international basins must be managed, keeping each country's interest and environment in mind.

This book is proposed as one of the first volumes to contribute to the above book series and provides an in-depth description of water resources in Turkey. Turkey has a unique geographical location in the world, extending in part from Europe to the

Mediterranean and further to the Middle East. This means that its water resources also reflect diverse characteristics under varying geography, topography, hydrology, geology, and climate among its water basins. Furthermore, due to its geographical location, Turkey has a significant number of transboundary river basins as an upstream and downstream country and has to share her water resources with her neighbors, an issue that sometimes may lead to water conflicts.

Turkey is also an interesting example as a developing country that tries to adapt to universal water management strategies but with legal, institutional, economic, and capacity development drawbacks. The country has long remained as a water-rich country, but the situation is now reversed to problems of water scarcity due to increasing population, inefficient use of the resource, impacts of climate change, and environmental degradation.

The book is designed in three parts, as required for the book series. The first part is devoted to physical features shaping the country's water resources, e.g., geography, topography, geology, hydrology, meteorology, climate, and the similar. The second part focuses on water use and consumption comprising the three major uses of water in Turkey, which relate to the domestic (drinking and industrial), agriculture (irrigation), and water power (hydroelectric energy) sectors. The third part of the book addresses the so-called miscellaneous issues, but all of which are highly significant in terms of water management. These issues, in particular, are river basin management, water-related legislation, sustainability and water security issues, impacts of global warming and climate change, ancient water works, and challenges for the future. A special section is devoted to transboundary river basins in Turkey as a significant portion of the water resources is transboundary and deserves special attention with regard to political conflicts and peace in the region.

The above three parts are addressed via 17 chapters, including the introductory chapter (Chap. 1), which provides a summary of each section. Chapters 2 and 3 reflect an interesting feature of water resource systems and structures in Turkey, that is, one may encounter in the country both the world's most ancient water works (Chap. 2) and the modern advanced systems developed during the Republican era (Chap. 3), which are unique in many ways. Chapters 4, 5, 6, 7, and 8 present an in-depth description of the physical features of water resources in the country, including a comprehensive chapter on Turkey's climate. The following three chapters constitute the second part of the book, i.e., water use and consumption. Three areas, agriculture (Chap. 9), water power (Chap. 10), and domestic/industrial water (Chap. 11), are considered in these chapters as the major water-consuming sectors. The third part of the book starts with Chap. 12 on transboundary river basins in Turkey. Transboundary and/or boundary-forming rivers have a significant contribution to the overall water potential of Turkey; they also play an important role with respect to the possible mutual conflicts among riparian countries and peace in the region. Chapter 13 addresses the practices of river basin management in Turkey, the importance of which is recognized only within the last decade. The major issues that influence water management are discussed in Chap. 14 on the expected impacts of climate change and Chap. 15 on the Turkish legislation relevant to water resources. Chapter 16 basically complements Chap. 13 to elaborate further on water management through discussing concepts of sustainability, water security, water allocation, and the water-food-energy nexus. The last chapter, Chap. 17, is basically a wrap-up of current problems regarding water resources and management in Turkey and discusses challenges for the future in the light of these problems.

The reader may find some commonalities between the chapters since many aspects of water resources cannot be separated by clear-cut lines but are components of the same water continuum (e.g., features like hydrology and water potential, groundwater and precipitation, climate and precipitation). Likewise, figures given for the same entities may vary slightly from chapter to chapter, depending on possible discordances between the derivations of these figures by different authors. However, these variations are minor and do not affect the overall level of the entities.

No book has been published as yet in Turkey to cover all aspects of water resources in the country, related problems, and water management issues as comprehensive as in this volume. Thus, the presented work will be, in a way, a useful "reference book" or a guide for all members of water communities in the country, including authorities, institutions, water users, academicians, and all other professionals. It is believed that the international water community will also welcome this volume as it demonstrates a multifaceted example of water developments and associated problems in a typical developing country under water stress. Water professionals, practitioners, managers, governmental authorities, and scientists at the international level are expected to benefit from the book in terms of working out commonalities and differing aspects of world water resources. Information and lessons derived from this example will be useful for basins and their management problems in other countries. In particular, Turkey's unique geographical position linking Europe and Asia and the presence of significant transboundary river basins will be important factors in arousing international interest in the volume.

It must be mentioned here that the authors contributing to this book are eminent professors and government authorities, renowned in their area of expertise both in Turkey and on the international platform. The editors would like to express their deep appreciation to all the contributors of the book, who have generously devoted their efforts and time toward the realization of this volume.

The editors extend their special thanks to the key editor of the book series World Water Resources, Professor Vijay P. Singh of Texas A&M University, USA, and to Mrs. Petra van Steenbergen, executive editor of Springer Earth Sciences, Geography and Environment, for inviting them to prepare this book. Their valuable roles and support in initiating and realizing the present volume are gratefully acknowledged. Special thanks go to our colleagues Drs. Filiz Barbaros and Cem P. Cetinkaya, who committed themselves to scrupulous editing of the book throughout all stages of the work.

Finally, we would like to dedicate this book to the Republic of Turkey and to all those who have contributed to the development of water resources in the country. We expect that the work will shed light to future water developments and help water professionals and authorities recognize both the country's strenghts to build on and the expected water problems that are ever growing.

Buca, Izmir, Turkey — Nilgun B. Harmancioglu
Cankaya, Ankara, Turkey — Dogan Altinbilek

29.10.2018
95th Anniversary of the Foundation of the Republic of Turkey

To the Memory of Prof. Dr. Orhan Uslu

At the time when the production process of this book started, we were deeply saddened to learn that Prof. Dr. Orhan Uslu, one of the principal contributors to this work, unexpectedly passed away. His loss left us with the most heartfelt sorrow but we have found solace in having had the opportunity to have worked with him in his last work. It was an honor to have known such a great person and scientist. His memory will always be with us as we read this book and regret that he did not have time to see it published.

We would like to express our sincere condolences to his family and to the national/international academia. May he rest in peace.

05.12.2018

Nilgun B. Harmancioglu
Dogan Altinbilek
Editors
on behalf of all authors of this book

Contents

Contributors

Hafzullah Aksoy Department of Civil Engineering, Istanbul Technical University, Istanbul, Turkey

Hakan Aksu Faculty of Aeronautics and Astronautics, Department of Meteorological Engineering, Samsun University, Samsun, Turkey

Ahmet Alkan Faculty of Engineering, Civil Engineering Department, Dokuz Eylul University, Izmir, Turkey

Dogan Altinbilek Past Director General, State Hydraulics Work (DSI), Ankara, Turkey

Past Vice President, World Water Council, Marseille, France

Civil Engineering Department, Middle East Technical University, Ankara, Turkey

Nazan An Center for Climate Change and Policy Studies, Bogazici University, Istanbul, Turkey

Institute of Environmental Sciences, Bogazici University, Istanbul, Turkey

Filiz Barbaros Faculty of Engineering, Civil Engineering Department, Dokuz Eylul University, Izmir, Turkey

Ertugrul Benzeden Faculty of Engineering, Civil Engineering Department, Dokuz Eylul University, Izmir, Turkey

Selmin Burak Institute of Marine Sciences and Management, Istanbul University, Istanbul, Turkey

Cem P. Cetinkaya Faculty of Engineering, Civil Engineering Department, Dokuz Eylul University, Izmir, Turkey

Mahmut Cetin Department of Agricultural Structures and Irrigation, Faculty of Agriculture, University of Cukurova, Adana, Turkey

Mehmet Ekmekci Hydrogeological Engineering Program, and International Research Center for Karst Water Resources, Hacettepe University, Ankara, Turkey

Nilgun B. Harmancioglu (retired) Faculty of Engineering, Civil Engineering Department, Dokuz Eylul University, Izmir, Turkey

Murat Ali Hatipoglu Investigation, Planning and Allocation Department, General Directorate of State Hydraulic Works (DSI), Ankara, Turkey

Aysegul Kibaroglu Faculty of Economics, Administrative and Social Sciences, Department of Political Science and International Relations, MEF University, Istanbul, Turkey

M. Levent Kurnaz Center for Climate Change and Policy Studies, Bogazici University, Istanbul, Turkey

Department of Physics, Bogazici University, Istanbul, Turkey

Hakan Mat MAT Consult Ltd. Co., Istanbul, Turkey

Yalcin Ozdemir Faculty of Engineering, Civil Engineering Department, Dokuz Eylul University, Izmir, Turkey

Unal Ozis Faculty of Engineering, Civil Engineering Department, Dokuz Eylul University, Izmir, Turkey

Tugba Ozturk Center for Climate Change and Policy Studies, Bogazici University, Istanbul, Turkey

Department of Physics, Isik University, Istanbul, Turkey

Bülent Selek Ministry of Agriculture and Forestry, Ankara, Turkey

Zeliha Selek Technology Faculty, Gazi University, Ankara, Turkey

M. Tufan Turp Center for Climate Change and Policy Studies, Bogazici University, Istanbul, Turkey

Institute of Environmental Sciences, Bogazici University, Istanbul, Turkey

Murat Turkes Center for Climate Change and Policy Studies, Bogazici University, Istanbul, Turkey

Orhan Uslu Emeritus Professor, Environmental Engineering Department, Dokuz Eylul University, Izmir, Turkey

Hasan Yazicigil Department of Geological Engineering, Middle East Technical University, Ankara, Turkey

About the Editors

Nilgun B. Harmancioglu
Prof. Dr. Nilgun B. Harmancioglu
Emeritus Professor
Dokuz Eylul University
Faculty of Engineering, Civil Engineering Department
Director, EA-TEK
International R&D, Engineering, Software and Consultancy Company
DEU Technopark, Izmir, Turkey

Prof. Dr. Nilgun B. Harmancioglu is a professor of hydrology and water resources, who carried out teaching and research activities at the DEU Faculty of Engineering between 1976 and 2017 (formerly Aegean University between 1976 and 1982). She received her BSc in civil engineering in 1973, her MSc in water resources engineering in 1976, and her PhD in hydrology and water resources in 1981. She was promoted to full professorship in 1992 at the DEU Faculty of Engineering, Civil Engineering Department.

She carried out her postdoctoral studies first at Ecole National Superieure des Mines de Paris, Centre d'Informatique Geologique, Laboratoire d'Hydrogeologie Mathematique Paris, (1980–1981), and then at George Washington University, School of Engineering and Applied Science, International Water Resources Institute (1984–1986), where she also worked as a research associate.

Dr. Harmancioglu's basic areas of research are water resource management, simulation of hydrologic processes, and information theory as applied to water resources. She has about 350 publications in the relevant areas, 70% of which are in English. She published and edited 4 books published by Springer and has contributed to several international conferences and symposia as a codirector, lecturer, and member of organization committees. She conducted various research projects funded by NATO, IWMI, the British Council, EU Framework Programs, the State Planning Agency, DEU, and TUBITAK (Scientific and Technical Research Council of Turkey). She also acted as a member of the Theme Advisory Board for the UNESCO IHP-VI Program between 2002 and 2007.

She was awarded by EWRA (European Water Resources Association) in 2015 as a distinguished member for her sustained and remarkable contributions to the scientific field of water resource management and for her continuous service to society.

Dr. Harmancioglu was the director of SUMER (DEU Water Resources Management Research & Application Center) between 2001 and 2017. She is currently the scientific director of EA-TEK, International R&D, Engineering, Software and Consultancy Company, located at the Technopark of Dokuz Eylul University, to continue her research and project activities.

Dogan Altinbilek
Past Director General, State Hydraulics Work (DSI) of Turkey
Past Vice President, World Water Council
Professor, Middle East Technical University, Civil Engineering Department
Ankara, Turkey

Prof. Dr. Dogan Altinbilek is an adjunct professor of water resource engineering at Middle East Technical University, Ankara, Turkey. After receiving his BSc degree in Civil Engineering from Middle East Technical University (METU) in Ankara, Turkey, in 1965, he received his MSc and PhD degrees in Civil Engineering from the Georgia Institute of Technology, USA. Since 1970, he has been involved in academic profession holding posts at Middle East Technical University of Turkey and King Abdulaziz University of Saudi Arabia. During 1996–2001, he was the Director General of State

Hydraulic Works (DSI), which is the major government agency in Turkey responsible for water resource development, including water supply to large cities. During 2005–2009, Prof. Altinbilek was appointed as president of Energy Group at IC Holding, Ankara, Turkey, and also as chairman of the board of AES-IC Ictas Energy Production Company.

He served as a governor (2006–2018) and Vice President (2012–2018) of World Water Council (WWC). Prof. Dogan Altinbilek was also an honorary member and the President (2013–2015) of International Water Resources Association (IWRA). Prof. Altinbilek also served as the President (2004–2008) and Honorary Member of International Hydropower Association (IHA).

Chapter 1
Introduction

Nilgun B. Harmancioglu

Abstract Springer has initiated a new book series entitled World Water Resources, which aims to develop a comprehensive understanding of water resources of different countries and how these countries are planning and managing their water resources. This book is proposed as one of the first volumes to contribute to the above book series and provides an in-depth description of water resources in Turkey. The book is designed in three parts. The first part is devoted to physical features shaping the country's water resources; the second part focuses on water use and consumption, including domestic, agricultural (irrigation), and hydroelectric energy production uses. The third part of the book addresses the so-called miscellaneous issues, e.g., river basin management, water-related legislation, sustainability and water security issues, impacts of global warming and climate change, ancient water works, transboundary basins, and challenges for the future. The three basic parts of the book mentioned above are covered via 16 chapters, each of which gives a comprehensive description of the various features of water resources in Turkey. This introductory chapter is intended to shortly summarize the contents of these chapters so that it will be easier for the reader to select sections of interest.

Keywords Turkey · Water resources · Water resources systems · Basin management · Water security · Transboundary river basins

We live in an age of environmental alertness. Almost all natural resources are attacked by pollution at varying degrees of intensity. The quality of surface and ground waters is continuously degrading. The situation is similar for land resources with problems of soil erosion, deforestation, and desertification in many parts of the world. Air pollution has already reached life and health threatening levels in particular regions. These problems have eventually endangered physical habitat for biodiversity. Further difficulties are expected due to the possible effects of climate

N. B. Harmancioglu (retired) (✉)
Faculty of Engineering, Civil Engineering Department, Dokuz Eylul University, Izmir, Turkey
e-mail: nilgun.harmancioglu@deu.edu.tr

N. B. Harmancioglu, D. Altinbilek (eds.), *Water Resources of Turkey*, World Water Resources 2, https://doi.org/10.1007/978-3-030-11729-0_1

change on various components of the environment. All these adverse developments are induced by diverse human activities as well as by natural occurrences. The result is that environmental degradation not only endangers nature, but it also has serious social and economic implications. An important feature of the above problems is that we live them globally; that is, all countries experience one or more of these adverse conditions at different degrees of severity and coverage.

The global nature of the present environmental crisis, including water and its central role in the environment, has essentially been the major motive in initiating the new book series on "World Water Resources" by Springer. The basic idea behind this initiative is to work out commonalities and differing aspects of world water resources by dedicating each volume in the series to a different country. Regarding water resources, the book series will be a valuable source of information sharing at basin, national and international scales. Information and lessons learned on a river basin from a particular country may be useful for basins in other countries. At the national scale, the series foresees a comprehensive understanding of water resources of different countries and of how these countries are planning and managing their water resources. At the international scale, since many countries share their resources with each other in transboundary or international basins, it is deemed useful to have an unbiased assessment of these resources, which will also help resolve legal issues arising from mutual conflicts. There are other recent global issues to be addressed under the same cover, such as global climate change, water management, sustainability and water security, water-energy-food nexus, politics and water resources, and economic development.

This book is proposed as a volume to contribute to the above book series and provides an in-depth description of water resources in Turkey. Turkey has a unique geographical location in the world, extending in-part from Europe to the Mediterranean and further to the Middle East (Fig. 1.1). This means that its water resources

Fig. 1.1 Geographical map of Turkey and her neighbors

also reflect diverse characteristics under varying geography, topography, hydrology, geology, and climate among its water basins. Furthermore, due to its geographical location, Turkey has a significant number of transboundary river basins and has to share her water resources with her neighbors, an issue that sometimes may lead to water conflicts.

In view of the above-mentioned global environmental (water) crisis most countries experience today, it must be stated that Turkey is also a country currently under water stress although it was considered as being water-rich in the past. Countries regarded as being rich in water resources have 8–10 thousand m^3 water per capita per year; and, on the basis of international standards and indices, the annual threshold value for water scarcity is stated as 1700 m^3 per capita per year. The recent reports by state authorities reveal that the available water per capita in Turkey is in the order of 1400 m^3. It is predicted that the population of Turkey will increase to 100 million in 2040 so that this value will reduce to 1120 m^3. Thus, Turkey will experience a more enhanced water stress in the future, coupled with increases in relevant environmental problems. At this stage, it is important that prompt and sound actions are taken to reassess the prevailing water problems and to develop more effective policies for the planning and management of available water resources in the country. These issues are discussed in depth in the following chapters of this book.

The overall picture of Turkey's water resources structure, including the sources, developments and management of water, is highly complex and again unique in a number of ways, ranging from historical to present day features. Thus, only a book as comprehensive as this volume can "almost fully" cover the diverse characteristics of water resources in the country. The book is designed in three parts. The first part is devoted to physical features shaping the country's water resources, e.g. geography, topography, geology, hydrology, meteorology, climate, and the similar. Thus, sources of freshwater, surface water, groundwater, river basins, lakes, coastal waters, water quality, and wetlands are described in this first part. Given that the source is identified with its many features, the second part focuses on water use and consumption. Three major water uses of water in Turkey relate to domestic (domestic and industrial), agriculture (irrigation), and water power (hydroelectric energy). Within the last decade, preservation of environmental (ecological) integrity has assumed the second priority among the above water uses after provision of domestic water. The third part of the book addresses the so-called miscellaneous issues, but all of which are highly significant in terms of water management. These issues, in particular, are river basin management, water-related legislation, sustainability and water security issues, impacts of global warming and climate change, ancient water works, and challenges for the future. A special section is devoted to transboundary river basins in Turkey as a significant portion of the water resources are transboundary and deserve special attention with regard to political conflicts and peace in the region. The three basic parts of the book mentioned above are covered via 16 chapters, each of which gives a comprehensive description of the various features of water resources in Turkey. This introductory chapter is intended to shortly summarize the contents of these chapters so that it will be easier for the reader to select sections of interest.

An interesting feature of water resources systems and structures in Turkey is that one may encounter both the world's most ancient water works and the modern advanced systems, which are unique in many ways. Turkey is particularly known for her historical attractions in the tourism sector. The same is true for the water sector; that is, the country is renowned for her outstanding remnants of numerous ancient water works from a four-millenia-long period, some of which are still in operation. These historical works are rich in kind to encompass various types of water structures. Geographically, they extend all over the country, indicating the various civilizations who realized them. Chapter 2 of this book presents many examples of these ancient structures in colorful photos.

After the foundation of the Turkish Republic in 1923, Turkey started both to assess the availability of her water resources for various purposes and to harness the potential through concentrating on development plans and projects. The first projects, mostly dams, date back to the 1930s and were realized under financial constraints and significant data limitations. These projects were certainly coupled with the establishment of water-related institutions and authorities. In the following years, water resources systems and structures flourished in Turkish river basins for purposes of irrigation, domestic water supply, power generation, flood control, and other purposes. The last three to four decades have witnessed large scale and unique development plans and systems, including water transfer among basins and even transport to neighboring water scarce regions. These new developments are covered in Chap. 3, providing interesting examples of the advanced modern projects in Turkish river basins.

The next five chapters relate to the first part of the book, i.e., the description of the physical features (geography, topography, geology, hydrology, meteorology, climate, and the similar) that constitute the driving forces or sources producing the overall picture of water resources in Turkey. It will be observed in all these chapters that the above mentioned physical features are marked by significant diversity, leading also to diverse characteristics of water resources in Turkey. The final output of these chapters will be the delineation of the water potential in the country, derived as the most recent figures presented so far.

Chapter 4 focuses on the comprehensive analysis of Turkey's climate, including the current aridity and drought conditions, by implementing a contemporary physical geographical, hydro-climatological and meteorological approach. The chapter covers a description of variable climatic conditions across all the geographical regions of the country. It discloses that "*the southern and western portions (Marmara, Aegean, Mediterranean and South-eastern Anatolia regions) of Turkey are characterized mainly with the dry summer sub-tropical Mediterranean climate*", and further that "*All of these regions are also influenced by mid-to-high degree drought probability and drought risk, while the Black Sea Region is characterized with a mid-latitude temperate climate with a low-level drought probability and risk.*" The climate of Turkey is marked by diversity since it is affected by the transition zone of various atmospheric disturbances and weather types. The complexity of the topography, land-sea interactions and many other influences aggrevate this diversity. The chapter also discusses climate-induced soil properties and land degradation and, thus, relates to the topic of desertification as one of the significant problems in Turkey.

The diversity in Turkey's climate is directly reflected in the hydrologic characteristics of freshwaters as detailed in Chap. 5 on surface water. Surface water resources of Turkey are divided between 25 river basins, each with its own hydrological and hydro-meteorological character. The share of surface waters in these basins contribute to 80% of the total water potential of the country. Chapter 5 presents in detail the characteristics of each river basin, including lakes, by disclosing their drainage areas, input precipitation, and average flow conditions. This information is derived from data provided by the prevailing hydrometric gauging network in Turkey, which is also described as to its assets and limitations. The chapter further focuses on streamflow maxima and minima in the river basins since, due to the heterogeneity of hydro-meteorological conditions in the country, some regions are flood-prone while others can be affected by extreme droughts. The trend analyses mentioned in the chapter finally conclude that extremes in surface flows will become more pronounced with larger maxima and lower minima than those observed in the past.

A comprehensive evaluation of groundwater in Turkey is presented in Chap. 6. A significant portion of the streamflow of major rivers is supplied by groundwater through springs and base flow. The aquifers in the country are grouped into four categories as the alluvial deposits/gravel aquifers, karstic aquifers, volcanic rock aquifers, and the fractured bedrock aquifers. The available groundwater potential provided by these aquifers is about 15 km^3, accounting approximately to 18% of the total water resources potential of Turkey. Groundwater is basically used in irrigation, domestic and industrial water supply, where irrigation consumes about approximately two-thirds of the total amount of groundwater used. It is depicted in the chapter that "*the share of the groundwater resources in sectoral water allocation has increased from 8.5 billion m^3/year in 1995 to about 15 billion m^3/year in 2014*" and further "*that the groundwater resources supply about 25% of the total water use in 2004 and about 30% in 2008*". The apparent overexploitation of groundwater resources and the accruing decrease of springs and/or spring flows show the inevitable result that the groundwater resources are under severe natural and/or anthropogenic stresses in Turkey.

Following the description of freshwaters and groundwater in quantitative terms in the preceding chapters, Chap. 7 presents an assessment of water quality in Turkish waters and the related pollution problems. It is stressed in the chapter that increases in production and consumption have caused serious pressures on water resources and the environment in Turkey. These problems are due to population increase, inland migration from rural to urban areas, industrialization, agriculture, expansion of tourism, and increases in economic activities and resource depletion. The work discusses the current status of water quality of inland and coastal waters (river, lake, groundwater and marine pollution) in Turkey, along with the associated legislation. Current wastewater management schemes and efforts towards the improvement of the pollution situation (surface water and groundwater pollution, lake pollution, marine pollution and removal of sludge), using available treatment and disposal technologies, are also considered in depth. The chapter concludes that, although environmental loads have increased several hundred times in the last 50–60 years, Turkey has shown a fast and consistent response to the situation as sufficient expertise, funds, regulations, and responsible governmental agencies have developed.

The first part of the book concludes with Chap. 8, which presents the overall results for the water potential of Turkey. The chapter is based on a very recent study carried out by the former Ministry of Forestry and Water and the related institutions (State Hydraulic Works, Meteorological Services, and the Directorate General of Water Management). The work comprises a compilation and review of the available data on precipitation, evapotranspiration, groundwater and surface flows between 1981 and 2010, and the main factors affecting Turkey's water potential. The computed values for the water resources potential in each of the 25 river basins in the country are disclosed together with the country's total water potential, which was derived by a water balance approach. The chapter further presents recent information on maximum and minimum values of the hydrologic and meteorological components of the water cycle, using the period between 1981 and 2010 as the reference years.

The following three chapters constitute the second part of the book, i.e. water use and consumption. Three sectors, agriculture, domestic/industrial water and water power, are considered in these chapters as the major water consuming sectors. Chapter 9 focuses on irrigated agriculture and states that irrigation is a must in almost all basins since the spatial and temporal distribution of input precipitation is highly variable across the country. The total irrigable area in Turkey is 25.85 million hectares (Mha), of which 22.6 Mha land can be economically and technically irrigable under present conditions. However, only 8.5 Mha area is expected to be equipped with irrigation schemes by DSI (the State Hydraulic Works) by the year 2023, which is the centennial of the foundation of the Turkish Republic. Irrigation is the largest water consuming sector in Turkey and uses about 74% of the total water potential of Turkey. However, the share of water use for irrigation is projected to be reduced to 64% by 2023. The chapter gives information on the institutional framework in irrigation developments and discusses irrigation practices in each of the 25 river basins of the country. It also introduces large scale regional development projects including irrigation investments. Among these, the *Southeastern Anatolia Project (GAP)* deserves special attention as it is considered as one of the biggest integrated development projects in the world, comprising several aspects such as irrigation, flood control, environment, hydropower generation, education, health, industry, etc.

Chapter 10 is devoted to water power and hydroelectric potential of Turkey. The primary energy resources of Turkey are quite limited; therefore Turkey has to develop her water power potential to produce electrical energy. The hydroelectric energy production of Turkey was only 1 TWh/year in 1960 but increased to about 75 TWh/year in 2017, which is equal to half of the ecomically feasible hydroelectric potential of the country. The total installed capacity of hydroelectric power plants was 0.4 GW in 1960 and increased to 27.3 GW in 2017, thus nearly 70 times in

57 years. The economically feasible hydroelectric potential of Turkey is in the order of 150 TWh/year. Turkey anticipates to harness the remaining part of this potential so that several hundreds of major hydroelectric power plants have still to be developed in the near future. The total installed capacity of all types of electricity generating plants in Turkey was around 73 GW in 2015. The total installed capacity of hydroelectric power plants is 26 GW, and its share is in the total power generation is 35%. The water power potential of Turkey is basically harnessed by power plants located at the toe of dams along the middle stretches of the rivers and by high-head diversion plants at the higher stretches of the rivers. The chapter presents several examples of the existing water power schemes in the country, detailing the various structures (dams, conduits, diversions, etc.) involved in the systems in colorful photographs.

Urban and industrial water use and management in Turkey is considered in Chap. 11, using available data related to sectoral uses, constraints and a realization of envisaged targets up to the present. Water supply and sanitation sector is discussed on the basis of the existing regulatory system, particularly in relation to the EU legislation as part of Turkey's accession process to the EU. The sectoral analysis presented in the chapter covers such issues as drinking water production and its use by households and the public sector, industries connected to the public water supply system, self-supplied industries, and pollution generated by municipal and industrial water consumption and its disposal. The first part of the chapter focuses on water uses in the urban and industrial sector, water supply and treatment, sewerage, wastewater treatment and disposal. A general description of the existing institutional structure related to water and sewerage administrations in Turkey is covered in the second section.

The third part of the book starts with Chap. 12 on transboundary river basins in Turkey. Transboundary and/or boundary-forming rivers contribute about 70 km^3/year or 40% to the gross surface water potential originating in Turkey. Their basins cover an area of 250,000 km^2 or roughly one third of the land surface of Turkey. The largest is the Euphrates-Tigris Basin, which represents about four fifths of the water potential of transboundary rivers in Turkey. The rest is contributed by the basins Orontes, Kura-Araks, Chorokhi, Maritza, and a few other quite small basins. The chapter presents detailed information on these basins as to their potential, development, existing water structures, and the amounts of water potential shared by each riparian country. Often, downstream riparian countries are concerned about the development of land and water resources in Turkey as they anticipate decreases in the quantity of waters that flows into their territories in addition to possible deterioration in quality of the water. These concerns are particularly related to the implementation of the Southeastern Anatolia Project in the Euphrates-Tigris Basin, where Turkey is the upstream riparian in both main subbasins. Concerns over water quantity also stem from water scarcity in the Middle East, which is expected to get worse through the impacts of climate change. However, Turkey claims that water resources developments in the country comply with the UN principles of equitable and reasonable use and further that dams in Turkey provide significant benefits to downstream countries, such as flood mitigation, sediment retention, and temporary low flow augmentation.

Chapter 13 addresses the issue of river basin management in Turkey. The water community recognized the significance of the concept of river basin management only within the last 2 decades. However, as a European Union (EU) candidate country, Turkey has taken significant steps towards the implementation of EU norms and requirements on river basin management. Still, there is yet no management plan that is practiced fully in Turkish river basins due to a number of problems. These problems relate basically to institutional structures, water management based on administrative boundaries, insufficient databases, poor monitoring and surveillance, capacity building, and insufficient sanctions and policies. In particular, there is no national Water Law yet finalized in the country. It still remains at the draft scale. Despite these deficiencies, River Basin Management Plans (RBMPs) are prepared in accordance with the EU Water Frame Directive (WFD) for 25 river basins in the country to achieve "good status" by 2036 in all water environments. To this end, the previously completed Basin Protection Action Plans (BPAPs) are transformed into RBMPs.

Chapter 14 essentially follows Chap. 4 on Climate of Turkey but, this time, focuses on the expected impacts of climate change over the country. The work covers projections of future climate conditions by using regional climate model simulations. To this end, changes in seasonal precipitation climatology, extreme weather conditions, and aridity conditions of Turkey are assessed for the 30 years between 2021 and 2050, based on the reference period of 1971–2000. The majority of geographical regions in Turkey are characterized by semi-arid climate conditions. Only the northeastern and the western parts (some parts) of the Black Sea region appear to be hyper-humid areas. The climate projections realized indicate a strong decrease in precipitation for almost all areas, but an increase in the intensity of drought conditions. Accordingly, more arid conditions are expected in the region for the near future. This means that Turkey will become significantly vulnerable to climate change, particularly to increased droughts.

Chapter 15 addresses one of the crucial issues in Turkey, i.e. legislation relevant to water resources. Institutional and legal problems have long hindered the proper management of these resources since institutions and legislation have gone through many changes and amendments in time. Laws governing the use and management of water failed to keep up with the increasing water demand and decreasing water supply. At present, the national Water Law is still in draft form, but many new primary and secondary water legislations in the domestic water, irrigation, hydropower, and the environment sectors have been issued. This chapter describes the principal water legislation in Turkey, along with institutional changes that have taken place due to domestic and regional political issues. Turkey's transboundary water policy is also discussed on the basis of related basic principles and prevailing practices.

Sustainability and water security issues considered in Chapter 16 basically complement Chapter 13 on river basin management. Sustainability is a long renowned concept in Turkey, but it is not fully understood as to its linkages with developmental targets and water management. The same is true for the fundamental basis of water security. The problem has been that every sector, every authority, or every individual evaluated sustainability through his/her own window. Mostly, the term has been associated with efforts on preventing long term adverse effects of development on the environment. In the water sector, this definition was directly associated with water quality. As noted above, the difficulty has been to comprehend the link between sustainability, water resources, and development. Water security is a newer concept in Turkey, and it has been only 4–5 years since water authorities came to recognize it as a requirement. Studies on these issues long remained at academic level, and the early institutional responses were quite slow. The chapter discusses the above problems and presents examples of the few studies carried out on sustainability and water security.

The last chapter, Chap. 17, is basically a wrap-up of current problems regarding water resources and management in Turkey and discusses challenges for the future in the light of these problems. It also discloses how water communities in the country react to these challenges and plan new targets for the future. The most challenging issue appears to be water scarcity resulting from fast increases in population, urbanization, industrialization, agriculture, expansion of tourism, and increases in economic activities, and resource depletion. Turkey is also an interesting example as a developing country that tries to adapt to universal water management strategies but with legal, institutional, economic and capacity development problems. The country has long remained as a water-rich country but the situation is now reversed to problems of water scarcity due to increasing population, inefficient use of the resource, impacts of climate change and environmental degradation.

In Turkey, no book has been published as yet to cover all aspects of water resources in the country, related problems, and water management issues as comprehensive as this volume. Thus, the presented work will be, in a way, a useful "handbook" or a guide for all members of water communities in the country, including authorities, institutions, water users, academicians, and all other professionals. It is believed that the international water community will also welcome this volume as it demonstrates a multifaceted example of water developments and associated problems in a typical developing country. Information and lessons derived from this example will be useful for basins and their management problems in other countries. In particular, Turkey's unique geographical position linking Europe and Asia and the presence of significant transboundary river basins will be important factors in arousing international interest in the volume.

Chapter 2
Water Works of Ancient Civilizations

Unal Ozis, Ahmet Alkan, and Yalcin Ozdemir

Abstract Turkey is renowned for her outstanding remnants of ancient water works and is thus one of the foremost open-air museums of the world with regard to hydraulic structures in the world. Numerous ancient water works from a four-millenia-long period are still in operation after several centuries or even several millennia. These historical works are rich in kind to encompass dams, irrigation canals, masonry conduits, aqueduct-bridges, tunnels, water collection works, water conveyance systems, pipes, inverted siphons and water mills. Geographically, they extend all over the country, indicating the various civilizations who realized them. They date back to the second millenium BC, the Hittite civilization in Central Anatolia; to the first half of the first millenium BC, the Urartu civilization in Eastern Anatolia; to the second half of the first millenium BC and the first millenium AD, the Hellenistic, Roman and Byzantine civilizations in Western and Southern Anatolia; to the eleventh up to the fourteenth centuries, the Seljukide civilization in Central and Eastern Anatolia; to the fourteenth up to the early twentieth centuries, the Ottoman civilization in Turkey. Some of these ancient water works were given as interesting examples in relevant books; several of them were dealt with more detail in other specific publications, journals and proceedings.

Keywords Turkey · Civilization · Ancient · Water · Hydraulic · Dam · Conduit · Tunnel · Aqueduct · Bridge · Canal · Pipe

2.1 Introduction

Turkey is one of the foremost open-air museums of the world with regard to historical water works for the last 4000 years (Fig. 2.1). Numerous ancient water works from this four-millenia-long period are still in operation after several centuries or even several millennia.

U. Ozis · A. Alkan (✉) · Y. Ozdemir
Faculty of Engineering, Civil Engineering Department, Dokuz Eylul University, Izmir, Turkey
e-mail: ahmet.alkan@deu.edu.tr; yalcin.ozdemir@deu.edu.tr

N. B. Harmancioglu, D. Altinbilek (eds.), *Water Resources of Turkey*, World Water Resources 2, https://doi.org/10.1007/978-3-030-11729-0_2

Fig. 2.1 Location of major historical water schemes in Turkey (Ozis & Arisoy)

The ancient hydraulic works in Turkey encompass dams, irrigation canals, masonry conduits, aqueduct-bridges, tunnels, water collection works, water conveyance systems, pipes, inverted siphons, and water mills.

They date back to the second millenium BC, the Hittite civilization in Central Anatolia; to the first half of the first millenium BC, the Urartu civilization in Eastern Anatolia; to the second half of the first millenium BC and the first millenium AD, the Hellenistic, Roman and Byzantine civilizations in Western and Southern Anatolia; to the eleventh up to the fourteenth centuries, the Seljukide civilization in Central and Eastern Anatolia; and to the fourteenth up to the early twentieth centuries, the Ottoman civilization in Turkey (Ozis and Benzeden 1977; Ozis 1981b, 1984a, 1994a, b, c, 1995, 1996, 1998, 1999, 2015a, b, 2017; Bildirici 1994, 2004; Buyukyildirim 1994a, 2017; Ozis et al. 2006, 2009, 2014a; DSI 2008; Unutmaz 2013, Baykan and Baykan 2015).

Some of these ancient water works were presented as interesting examples in relevant international books (Garbrecht 1985, 1987a, 1991, 1995; Fahlbusch 1987a; Tolle-Kastenbein 1990; Hodge 1992; Grewe 1998, 2010a; Viollet 2000; Crouch 2004; Wiplinger 2006a, 2013a, 2014, 2016; Nikolic 2008; Mays 2010; IWA 2012; Wiplinger and Letzner 2018). Several of them were dealt with more detail in other international specific publications, journals, and proceedings, as cited in the subsequent sections.

This chapter summarizes the principal sites and main features of ancient water works in Turkey. Further details can be found in more specific literature given in the lengthy list of references at the end of the chapter.

2.2 Hittite Water Works

2.2.1 Dams

The most ancient hydraulic works in Turkey date back to the second millenium BC, remaining from the Hittite civilization in Central Anatolia. Certain indices indicate that some wells and canals may even date back to the sixth millenium BC (Emre 1993; Bildirici 1994, 2004; Ozis 1994a, 1999; Cinaroglu 2007; DSI 2008; Inal 2009; Grewe 2010b; Strobel 2013).

The 8 m high Karakuyu dam in Uzunyayla, with a total crest length of 400 m, is a remarkable example. The restituted Golpinar dam near Alacahoyuk is thus the most ancient dam actually in use. Eflatunpinar dam near Beysehir lake, Koylutolu and Yalburt dams near Ilgin, and Guneykale dam near Bogazkale are other dam remains from the Hittite Period.

2.2.2 Water Supply

The spring-water collection chamber in Hattusha (Bogazkale) (Fig. 2.2), formed by rubble stones, has an almost triangular cross-section of about 1.4 m width and 2.6 m height (Neve 1969/70).

Fig. 2.2 The springwater collection chamber in the Hittite capital Hattusha (Bogazkale). (Photo by U. Ozis)

2.2.3 *Irrigation*

There are remains of several water conduits in ancient Hittite settlements and irrigation systems in Central Anatolia (Bildirici 1994, 2004; Bildirici and Bildirici 1996; DSI 2008).

2.3 Urartu Water Works

2.3.1 *Dams*

There are several small dams from the first half of the first millenium BC, belonging to the Urartu civilization in Eastern Anatolia. These are located mostly on various watercourses flowing to Lake Van. Some of them, like the 7 m high dam raising the level of the Kesis Lake and those on Doni creek, are still in use. Kircagol dam near Adilcevaz, Suphan and Argit dams near Muradiye, and several small dams between Van and Hakkari are other remains (Garbrecht 1987a; Belli 1996; Ozis 1999; Bildirici 2004; DSI 2008; Hepbostanci et al. 2015).

2.3.2 *Water Supply & Irrigation*

2.3.2.1 Samram Canal

The 56 km long Samram canal (Fig. 2.3) supplied the Urartu capital Tushpa (Vankale). This canal dates back to 800 BC and is one of the oldest canals still in use. The Samram canal irrigates, with very few modifications, about 2000 ha of land (Ogun 1970; Burney 1972; Garbrecht 1975; Ozis 1994a; Belli 1997; Bildirici 2004; DSI 2008; Grewe 2010b; Hepbostanci et al. 2015).

2.3.2.2 Ferhat Canal

Remains of the Ferhat canal, conveying water from Lake Balikli towards the western edge of the Igdir plain, and some other irrigation systems in the Van area, might also date back to the Urartu period (Bildirici 2004; DSI 2008).

Fig. 2.3 The Samram Canal near the Urartu capital Tushpa (Van). (Photo by U.Ozis)

2.3.2.3 Qanats Near Van

Some of the underground water conduits of Van, still partly in use, might also date back to this period and can be considered as predecessors of the later qanats. Certain cisterns in this region appear to be of Urartu origin (Bildirici 2004; DSI 2008).

2.4 Hellenistic, Roman and Byzantine Water Works

2.4.1 Water Supply & Conveyance Schemes

2.4.1.1 General Remarks

Long-distance water supply schemes in Western and Southern Turkey, from the Hellenistic, Roman and Byzantine civilizations, date back to the second half of the first millenium BC and the first half of the first millenium AD.

The long-distance water conveyances in the Aegean and Mediterranean regions of Turkey are very numerous and new discoveries add to the rich variety of them (Weber 1904, 1905; Ozis 1981b, 1994a, 1995, 1996, 1998, 2015a, b; Fahlbusch 1982, 1987a; Garbrecht 1985, 1995; Tolle-Kastenbein 1990; Hodge 1992; Buyukyildirim 1994a; Cecen 1996a, b; Grewe 1998, 2010a, 2014; Viollet 2000; Wiplinger 2006a, 2013a, 2016; Ozis et al. 2007, 2009, 2014a, 2018a; DSI 2008; Turk et al. 2010; IWA 2012; Baykan and Baykan 2015).

These systems include spring-water collection chambers; lead-, stone-, clay-pipes of various sizes; rock-cut and masonry canals; tunnels of over 2 m height; inverted siphons under up to 190 m water pressure with lead-pipes and to 155 m water pressure with stone-pipes; and aqueduct-bridges of up to 40 m height.

The water conveyance to Constantinopolis (Istanbul) with a length of 242 km is the longest Roman water conveyance in the world, and that to Phoceia (Foca) with 100 km length ranks among the longest Roman conveyance systems. The multiple water conveyance systems to Pergamon (Bergama), Smyrna (Izmir), and to Ephesus (Efes) make these cities, besides Rome, Lyon and few others, among the most interesting examples of multiple urban water supply schemes in the antique world. The water conveyance systems to Perge, Hierapolis, Tralleis, Antiochia/Orontes are other, somewhat shorter multiple water conveyance systems in Anatolia.

The lead-pipe inverted siphon of the Madradag water conveyance to Pergamon resists up to 190 m water pressure, and the 3.3 km long stone-pipe inverted siphon of the Karapinar water conveyance to Smyrna resists up to 155 m water pressure. These were siphons operating under respective largest pressures in the antique world, dating back to the late centuries of the first millenium BC, the Hellenistic period.

With a length of about 1.7 km, the stone-pipe inverted siphon of Aspendos is the longest one on arches in Turkey. The ruined aqueduct-bridge over Karkassos (Ilyas), a tributary of Kaikos (Bakircay), on the Soma conveyance to Pergamon, had probably a height of 40 m, being the second highest Roman aqueduct after the Pont-du-Gard of Nîmes.

Various important sites are summarized below in an approximately counterclock-wise geographical sequence.

2.4.1.2 Constantinopolis (Istanbul)

This city served as the capital of three empires in the course of 16 centuries: of the Roman Empire from 330 to 395, of the Byzantine Empire from the year 395 to 1453, and of the Ottoman Empire from 1453 to 1922.

Istanbul was supplied by water during the Roman and early Byzantine times through important long-distance conveyance systems (Forchheimer 1890; Forchheimer and Strzygowski 1893; Dalman 1933; Eyice 1979; Fahlbusch 1982; ISKI 1983; Tolle-Kastenbein 1990; Cecen 1994, 1996a, b; Ozis 1987, 1994a, 1995, 1996, 2001; Unutmaz 2013; Ozis et al. 2018a). These systems were heavily damaged in the course of numerous sieges of the city, during the second half of the first millenium and the first half of the second millenium, until her fall in 1453 to the Ottoman Empire.

The first major water conveyance system to Istanbul dates probably back to the Roman emperor Hadrian's time (117–138 AD). The remains of a water conduit, found roughly 220 m south of the Mihrimah Mosque in Edirnekapi, indicate that this

Fig. 2.4 Valens (Bozdogan) Aqueduct-bridge in Constantinople (Istanbul). (Photo by U.Ozis)

conduit transported water from the westwards Halkali area to the city. The conveyance continued over the 23 m high and 970 m long Bozdogan aqueduct-bridge (Fig. 2.4), which has been earlier called as the Hadrian aqueduct-bridge. It is later called Valens-aqueduct-bridge, who reigned during 364–378 AD and thoroughly repaired it in the context of the very long water conveyance from Thrace.

The 242 km long water conveyance system to Istanbul, along the southwestern slopes of the Istranca mountains in Thrace, was constructed for the new capital of the late Roman Empire in the fourth century AD, when emperor Constantinus I shifted the capital from Rome to Istanbul in 330 AD. This was by far the longest water conveyance system of ancient times (Cecen 1996a, b). The conveyance passes over 40 aqueduct-bridges, whereby the 33 m high Kursunlugerme is the highest.

Two aqueduct-bridges of this system, the already mentioned Bozdogan and the 19 m high and 110 m long Mazul, were later used by several Ottoman systems (some of the Halkali water conveyances, including the Suleymaniye water conveyance built by the great Architect Sinan in the sixteenth century).

A long-distance water conveyance from the north, the Belgrade Forest area close to the Blacksea, date probably back to the reign of Valens or of Theodosius I, the latter of whom reigned in the period of 379–395 AD, being the last emperor of the entire empire before the split in 395 AD. This conveyance was laid at an altitude significantly higher than that of the later Ottoman Kirkcesme water conveyance to Istanbul, which is one of the masterworks of the Architect Sinan in the middle of the sixteenth century. Only the lowest row (four arches) of the three-storey Egri (Kovuk) aqueduct-bridge, with a broken alignment, might belong to another Roman system.

2.4.1.3 Parion (Kemerkoy)

A ruined aqueduct-bridge, along with various baked clay and marmor pipe elements, exist at the ancient city of Parion on the southern shore of the Marmara Sea, near the village Kemerkoy northwest of Biga. It is presumed that water was conveyed from the vicinity of the village Cataltepe, southwest of Kemerkoy, so that the length of the water conveyance may be estimated to be in the order of 20 km (Wiplinger 2016).

2.4.1.4 Troia – Ilion (Truva)

A water conveyance of probably 25 km length brought the water of the springs near the village of Camlica to Troia-Illion (Truva), located to the east of the city. The most interesting element of this conveyance is the aqueduct-bridge at the village of Kemerkoy, with an arch of 16 m span width, located 27.5 m higher than the creek underneath (Kayan 2000; Aylward et al. 2002; Wiplinger 2006a; Unutmaz 2013; Ozis et al. 2018a).

2.4.1.5 Pergamon (Bergama)

Pergamon was supplied by several long-distance water conveyances dating back to Hellenistic and Roman periods, from the north-western Selinus (Bergama) valley, from the northern Madradag area, and from the eastern Kaikos (Bakircay) basin (Garbrecht and Holtorff 1973; Garbrecht and Fahlbusch 1975, 1978; Fahlbusch 1982, 1987a, e; Hecht 1975, 1976, 1978, 1979, 1983; Garbrecht 1976, 1985, 1987b, 1995; Tolle-Kastenbein 1990; Ozis 1994a, 1995, 1996; Grewe 1998; Viollet 2000; Nikolic 2008; Mays 2010; Alkan et al. 2014; Ozis et al. 2018a).

The 44 km long Madradag water conveyance system, consisting of three baked clay pipe conduits, crossed the last valley by means of a lead pipe inverted siphon under a maximal water pressure of 190 m, the highest pressure in antiquity, to reach the Akropolis area. This system was later replaced by a masonry conduit in the Roman period and crossed the same valley over an arched aqueduct-bridge (Fig. 2.5), to supply the blooming middle-level part of Pergamon.

Also in Roman times, a 53 km long conveyance brought the water of Turgutalp springs, located to the east of Pergamon in the Kaikos (Bakircay) river valley, through six tunnel stretches and over 40 aqueducts-bridges. The completely ruined aqueduct-bridge over the Karkassos (Ilyas) tributary likely had a length of 550 m and a height of 40 m. This conveyance is later extended by 10 km eastwards in order to harness the water of the Aksu springs.

Fig. 2.5 The Roman Aqueduct-bridge of the Madradag Water Conveyance crossing the last valley before Pergamon. (Photo by U.Ozis)

Fig. 2.6 Rock-cut canal of the water conveyance to Phocea. (Photo by U.Ozis)

2.4.1.6 Phoceia (Foca)

The roughly 100 km long water conveyance to Phocea was most likely fed from the Goksu springs near Manisa. It has an alignment similar to the modern first stage water supply project of Izmir along its first 19 km. The alignment is quite close to that of the modern Menemen right bank irrigation canal for the next 48 km. It follows most probably with a gentle gradient the contour lines along the coastal slopes for the last 33 km in order to reach Phocea (Ozis 1994a, 1995, 1996; DSI 2008; Wiplinger 2016; Ozis et al. 2018a).

The conduit displays stretches of open channels (Fig. 2.6), simple horse-shoe shaped masonry galleries, rock-cut canals, and rock-cut tunnels. The bottom widths are in the order of 0.4–0.7 m, and the longitudinal slopes vary between 0.15 and 0.3‰. Remains of any significant aqueduct-bridges have not been encountered.

2.4.1.7 Sardis (Sart)

The almost 16 km long water conveyance system to Sardis was fed from the Kocapinar springs issuing on the slopes of Bozdag to the south of the city. Another spring is connected at km 6. The conveyance includes stretches of masonry galleries, tunnels, clay pipes, and small aqueduct-bridges up to 8 m height. The conduit bifurcates in Teknetas at km 10; one branch reaches the Artemis temple after 1.5 km, and the other continues for 5 km until it reaches the baths and the gymnasium (DSI 2008; Wiplinger 2016; Ozis et al. 2018a).

Fig. 2.7 Aqueduct-bridges over the Melas (Melez) creek in Izmir. (Photo by Y.Ozdemir)

2.4.1.8 Smyrna (Izmir)

The 30 km long Karapinar water conveyance from the east is the longest of the seven water conveyance systems supplying Izmir in the past (Weber 1899; Fahlbusch 1982; Tolle-Kastenbein 1990; Hodge 1992; Ozis 1994a, 1995, 1996; Ozis et al. 1999, 2018a; DSI 2008; Nikolic 2008; Pinar 2011; Unutmaz 2013; Alkan et al. 2014; Wiplinger 2016). This clay pipe conduit may date back to the Hellenistic period. It includes a 3.3 km long stone-pipe inverted siphon crossing the Melas (Melez) river under a remarkable water pressure of 155 m.

The 27 km long Akpinar water conveyance from the south is the second longest one and may date back to the Roman period. The conduit is basically a covered masonry canal, crossing the small creeks along the alignment over heavily damaged modest aqueduct-bridges.

The probably 20 km long conveyance from the vicinity of Buca in the southwest crosses the Melas (Melez) river by means of two aqueduct-bridges (Fig. 2.7). The downstream one with masonry channel is 21 m high and dates probably back to the Roman period. The other aqueduct-bridge, roughly 100 m upstream of it, with a conduit of clay pipes, appear to be constructed later as a bypass to the former one and dates probably back to Byzantine or Ottoman periods.

2.4.1.9 Metropolis (Yenikoy)

Metropolis (Yenikoy) was supplied by a 21 km long water conveyance which carried the water of a spring northwest of the city. The elevation of the spring limited the water supply to only the lower half of the city. The conduit is a masonry channel and crosses a valley over an aqueduct-bridge (Weber 1904; DSI 2008; Ozis et al. 2018a).

2.4.1.10 Ephesus (Efes)

Ephesus was supplied by at least four water conveyances in antiquity: the 8 km long Selenus (Sirince) system from the east; the 7 km long Marnas (Derbentdere) system from the southeast; the 37 km long Kenchrios (Degirmendere) system from the south; and the 42 km long Kaystros (Kayapinar) system from the northeast (Forchheimer 1923; Wilberg 1923; Ozis and Harmancioglu 1980; Fahlbusch 1982; Alzinger 1987; Tolle-Kastenbein 1990; Hodge 1992; Ozis 1994a, 1995, 1996; Ozis and Atalay 1999; Ozis et al. 1998, 2005a, b, 2014b, 2018a; Ortloff and Crouch 2001; Crouch 2004; Wiplinger 2006a, b, 2008, 2010, 2013b, c, 2016; DSI 2008; IWA 2012; Unutmaz 2013; Alkan et al. 2014).

The Sirince conveyance, fed by groundwater collected near the village, supplied water to the area around the Artemis temple and may date back to the fifth century BC. The main conduit consists of conical baked clay pipes and ends with lead pipes under the altar of the temple, joined by marmor elements. This conveyance was probably diverted to the Ayasuluk hill in Selcuk in the sixth century AD, over the 625 m long Selcuk aqueduct-bridge of up to 15 m height, and served as the venter of a stone-pipe inverted siphon. The remains of water-balance towers parallel to this aqueduct-bridge are elements of a baked clay pipe conduit supplying the fourteenth century Isabey mosque and baths in its vicinity.

The Derbentdere system displays several baked clay pipe conduits, which probably had been subject to various repairs and bypasses. The system lay partly on rock-cut terraces and reaches the castellum near the Magnesia gate of Ephesus. The conveyance passes over the 15 m high two-storey Sextilius Pollio aqueduct-bridge (Fig. 2.8), replacing probably a clay- or stone-pipe inverted siphon. It then extends over two smaller, single-arch aqueduct-bridges (Becerik I & II).

The Degirmendere water conveyance carries the water of the Degirmendere and Keltepe springs by means of a cut-and-cover masonry conduit, passes through half a dozen tunnels, crosses the creeks by two dozens aqueduct-bridges, and reaches Ephesus at the southwestern hill slopes. The longest aqueduct-bridge was Arvalya with 400 m and the longest tunnel Atalay with 815 m. The system was initially constructed in Hadrian's time, early decades of the second century AD. However, an earthquake in the year 159 AD caused a sink of 3 m at the fault by km 18 and disrupted the function of the conveyance. A second larger and almost parallel channel had to be built in order to reach the city. This conveyance, with almost parallel aqueduct-bridges, was established during the later decades of the second century and reached Ephesus at an 8 m higher elevation.

The Kayapinar conveyance carried the springwater of the village Kursak near Belevi to Ephesus, receiving also the water of the Pranga springs at km 23. The conduit is basically a cut-and-cover masonry channel, crossing the creeks over small aqueduct-bridges. It reaches the antique stadium at km 40, passes then under the central seat row of the theater's upper tier, and dicharges into the Traian fountain, constructed in 114 AD by Claudius Aristion. The conduit continues under the marble road and supplies water also to the slope-houses quarter.

Fig. 2.8 The Sextilius Pollio Aqueduct-bridge on the Derbentdere Water Conveyance to Ephesus. (Photo by U.Ozis)

2.4.1.11 Magnesia/Meandros (Ortaklar)

Three relatively short water conveyances supplied water to Magnesia ad Meandros. The 7.5 km long Naipli conveyance from the northwest passes over the almost 1 km long Ortaklar aqueduct-bridge. The 6 km long Tekinkoy conveyance from the north consists mainly of a masonry channel. The 8 km long Arguvanli conveyance consists of baked clay pipes (Baykan et al. 2001a; DSI 2008; Ozis et al. 2018a).

2.4.1.12 Tralleis (Aydin)

The northwestern branch of the Tralleis water conveyance extends from the Caykavustugu springs to the Taskemer aqueduct-bridge and is 8.6 km long. The northern branch from the Olmez spring to the same bridge is 2.5 km long. The conveyance continues for 4.5 km to reach the Kizlarkulesi tower. A third, north-eastern branch from the Kocabag spring to the same tower is 10.4 km long. The final stretch, from the tower to the Ucgozler aqueduct-bridge in Tralleis, is 1.5 km long (Weber 1904; Wiplinger 2006a; DSI 2008; Ozis et al. 2018a).

2.4.1.13 Nysa (Sultanhisar)

Remains of an eventual diversion work and of some aqueduct-bridges led to the assumption that water was brought to Nysa, near Sultanhisar, from the Tekkecik (Malkoc) creek, 1 km to the north. Moreover, a very interesting water reservoir exists to the north of the city, covering an area of 40 m × 50 m (Ozis 1984a, 1994a). More recent investigations indicated a 5.5 km long water conveyance from Malgacemir area in the north, consisting of clay pipes (DSI 2008). Baked clay pipes for water distribution were also found in the city.

2.4.1.14 Tripolis (Yenicekent)

The 22 km long water conveyance carried the water of the Karsipinar spring, in the northeast, to Tripolis between Buldan and Yenicekent. The interesting springwater collection work still serves to supply water to the city Guney (DSI 2008; Ozis et al. 2018a).

2.4.1.15 Eumenia (Isikli)

A roughly 6 km long conveyance carried the water of Ortadag springs from the north to Eumenia, at the village Isikli of the county Civril. The conduit consists of baked clay pipes. The remains of a water tower exist in Eumenia (DSI 2008).

2.4.1.16 Hierapolis (Pamukkale)

Three relatively short water conveyances supplied Hierapolis: the 5.4 km long Karahayit conveyance from the north, the 6.4 km long Kocapinar conveyance from the northeast, and the 4.8 km long Mustak conveyance from the east (Weber 1904; Ozis and Harmancioglu 1980; Fahlbusch 1982; Ozis 1994a, 1995, 1996; Baykan 1999; Baykan et al. 2001b; Wiplinger 2006a; DSI 2008; Grewe 2014; Ozis et al. 2018a).

Remains of masonry channels and aqueduct-bridges indicate that the Karahayit conveyance dates back to the Roman period. Remains of baked clay pipes are encountered elsewhere. The most interesting water work is the water reservoir (Fig. 2.9), a 'castellum aquae' close to the city, with 11 × 12 m inner area and roughly 2.5 m height.

2.4.1.17 Laodicea (Eskihisar)

Water was brought to Laodicea by a 7 km long conveyance system from Baspinar springs, located southwards near Denizli. The initial conduit consisted of two parallel rows of baked clay pipes, dating back to the Hellenistic period. Later, in the Roman period, it was replaced by a masonry channel which extended partly over aqueduct-bridges, reaching the stone-pipe inverted siphon (Weber 1898; Ozis 1994a; Baykan 1999; Baykan et al. 2001b; Wiplinger 2006a; DSI 2008; Unutmaz 2013).

Fig. 2.9 Water reservoir at Hierapolis. (Photo by U.Ozis)

The inverted siphon is 820 m long, crosses the 40 m deep valley by two parallel rows of stone blocks (Fig. 2.10) and reaches the city's first distribution tower. From this tower, which has an actual height of around 5 m, but which probably had a height of 8–9 m, water was conveyed for 430 m to a second distribution chamber to supply other parts of the city.

The archeological excavation team under the leadership of Celal Simsek discovered in 2015 at Laodicea a water-related inscription from 114 AD, reflecting the ancient regulations of water use in the city.

2.4.1.18 Attuda (Hisarkoy) and Trapezopolis (Bekirler)

A roughly 5 km long conveyance from the southwest supplied Attuda, at Hisarkoy near Babadag, with the water of the Ikizce spring. A roughly 10 km long conveyance from the south supplied Trapezopolis, at Bekirler near Babadag, with the water of the springs on western slopes of the Salnakos mountain. The conduits consisted probably of baked clay pipes in both cases (Weber 1904; DSI 2008).

2.4.1.19 Sebastapolis (Kizilca)

A roughly 6 km long conveyance from the northeast supplied Sebastapolis, near Kizilca to the southeast of Tavas, with the water of the Caylak spring (DSI 2008).

Fig. 2.10 Stone-pipe inverted Siphon of the water conveyance to Laodicea. (Photo by U.Ozis)

2.4.1.20 Aphrodisias (Geyre)

Several water conveyance systems supplied Aphrodisias in ancient times (Weber 1904; Ozis 1994a; DSI 2008; Commito and Rojas 2010; Wiplinger 2016; Ozis et al. 2018a). The 25 km long Timeles (Yenidere) system from the southeast carries the water of two springs near the Guzelkoy village, located in the adjacent Tavas closed basin. The conduit is basically a cut-and-cover masonry channel (Fig. 2.11), passing through tunnels of several km length before the villlage Karapinar. The springs and the initial part of the conveyance is actually submerged in the reservoir of the Yenidere dam.

A 10 km long water conveyance from the north carried the water of a spring, located near the village Isiklar, to Aphrodisias. There are traces of a probably 9 km long water conveyance near the village Seki. An aqueduct-bridge of this conveyance was used as a bridge in Ottoman times.

2.4.1.21 Alinda (Karpuzlu)

An up to 7 m high and roughly 85 m long aqueduct-bridge (Fig. 2.12) crosses the last valley before Alinda near Karpuzlu. Parts of a canal formed by stone plates are still in situ on top of this bridge. Remains of a ruined masonry canal exist southwest of the bridge. A 12 km long water conveyance, with baked clay pipes, carry water from of a spring located to the northeast of the city towards the aqueduct (DSI 2008; Grewe 2014; Ozis et al. 2018a).

Fig. 2.11 Remains of the Timeles water conveyance to Aphrodisias. (Photo by A.Alkan)

2.4.1.22 Alabanda (Doganyurt)

Alabanda was supplied with water by an at least 22 km long conveyance system from the south (Ozis et al. 1979a, 2018a; Ozis 1991, 1994a, 1995, 1996; DSI 2008). The conveyance was fed from five springwater collection chambers (Fig. 2.13) at hill slopes along the left bank of Cine river. The conduit is basically a masonry channel with rock-cut stretches, passes over seven aqueduct-bridges, and ends in the city's water reservoir, which is actually totally silted.

About 5 km further to the south of the first water collection chamber is the Incekemer aqueduct-bridge over the river Cine, which served as a 'venter' for a stone-pipe inverted siphon. Road construction activities left no trace of any water conveyance between these two locations. Remains of a much smaller inverted siphon were found on a creek at the right bank of Cine river.

Initially, it was believed that Incekemer belonged to another water conveyance in the direction of Gerga on the right bank. However, the large elevation difference between Incekemer and the several hundred meters higher Gerga led later to the assumption that Incekemer should belong to a conveyance to Alabanda to eventually constitute an upstream element of the actual system, increasing its length to about 30 km. This part is actually submerged in the reservoir of Cine Adnan-Menderes dam.

Fig. 2.12 The last Aqueduct-bridge of the water conveyance to Alinda. (Photo by U.Ozis)

Fig. 2.13 A springwater collection chamber of the water conveyance to Alabanda. (Photo by U. Ozis)

2.4.1.23 Priene (Gullubahce)

The water of the springs on hill slopes to the north of the city is carried to two locations of Priene. The conveyance consists of baked clay pipes, laid on the ground or in masonry canals. The conveyance bifurcates after 125 m: a 290 m long branch discharges in the water reservoir at the acropolis northwest of Priene, and the other 1140 m long branch into water reservoirs to the east of the city (Crouch 1993, 1996,

2004; Ozis 1994a, 1995, 1996; Crouch et al. 1997; Ortloff and Crouch 1998; Alkan et al. 1999; Wiplinger 2006a; DSI 2008).

2.4.1.24 Miletus (Balat)

The roughly 4 km long conveyance carried water from the spring at the Stefania-Plateau to the south between the villages Akkoy and Yenikoy. It supplied the nympheum and the northeastern part of Miletus. The conduit consists of baked clay pipes at the beginning, then continues as a masonry channel, and reaches the nympheum with an aqueduct-bridge, of which only two arches remain. The almost 2 km long other conveyance carried water from the spring at the Jeralex area in the south between Balat and Akkoy, reached the Holy Gate, and supplied the smaller southwestern part of Miletus (Ozis 1994a; Tuttahs 1998, 2001, 2007; Crouch 2004; Ozis et al. 2018a).

2.4.1.25 Iasos (Gulluk)

The remains of an almost 490 m long aqueduct-bridge exist in Iasos (Fig. 2.14). It begins near a well, and it is assumed that water was taken out from this well and heightened at least 3 m to supply the conveyance system. Furthermore, there are several clay pipe lines and masonry canals in the city (Tomasello 1991; Ozis 1994a; Ozis et al. 2018a).

2.4.1.26 Mylasa (Milas)

The 7.4 km long water conveyance from the northeast was fed from a spring near the creek Saricay. It passes over six modest aqueduct-bridges upstream of the Akgedik dam and reaches Mylasa on a highly interesting, but partly damaged, 2.3 km long, so far the longest aqueduct-bridge in Turkey (Fig. 2.15) (DSI 2008; Mays 2010; Alkan 2015; Wiplinger 2016; Ozis et al. 2018a).

2.4.1.27 Keramos (Oren)

The roughly 6 km long water conveyance from the northeast carried the water of five springs to Keramos at northern shore of the Gokova bay. One of these springs, Sucikan, still supplies Oren. The conduit passes over several partly ruined aqueduct-bridges, discharges into pools, which might have served as silting basins, and then to a cistern (Wiplinger 2006a; DSI 2008; Ozis et al. 2018a).

Fig. 2.14 The Aqueduct-bridge in Iasos. (Phot by U.Ozis)

Fig. 2.15 The 2.3 km long Aqueduct-bridge in Mylasa. (Photo by A.Alkan)

2.4.1.28 Knidos (Datca)

A roughly 7 km long conveyance carried the water of the springs in cape Kalamis, from the east to Knidos at the western end of the Datca peninsula between the Gokova and the Hisaronu bays. The conduit consisted of baked clay pipes (DSI 2008).

2.4.1.29 Kibyra (Golhisar) and Oenoanda (Incealiler)

The water conveyance system to Kibyra near Golhisar is 2.3 km long and includes a 400 m long stone-pipe inverted siphon. The water conveyance system to Oenoanda, near Incealiler, is 3.5 km long and includes a 500 m long stone-pipe inverted siphon (Stenton and Coulton 1986; Coulton 1987; Baykan and Cantilav 1997; DSI 2008; Turk et al. 2010; Alkan et al. 2014). These two antique cities, together with Bubon and Balbura in between, formed the Kibyratis Union.

2.4.1.30 Xanthos (Kinik)

The water of the Inpinar spring was brought to Xanthos near Kinik by a roughly 9 km long conveyance from the east. The conduit is basically a masonry channel, rock-cut at some stretches, and passes over modest aqueduct-bridges (Buyukyildirim 1994a; Burdy and Lebouteiller 1998; Ozis et al. 2018a).

2.4.1.31 Patara (Gelemis)

The 22 km long conveyance carries the water of the Islamlar spring from the northeast. The conduit is basically a masonry channel with some rock-cut stretches and even displays a baked clay pipe stretch. It passes over several modest aqueduct-bridges and through an interesting stone-pipe inverted siphon. The length of the siphon is about 500 m and that of the masonry venter 190 m, curved in alignment and in elevation; stone pipes are under a maximal water pressure of about 20 m (Buyukyildirim 1994a; Ozis 1994a, 1995, 1996; Baykan et al. 1997; Baykan and Iskan 2011; Iskan and Baykan 2013; Unutmaz 2013; Sahin 2016; Wiplinger 2016; Ozis et al. 2018a).

2.4.1.32 Pinara (Minare)

A roughly 7 km long water conveyance from the west brought the water of the spring Ericek to the upper part of Pinara near the village Minare southwest of Fethiye. A shorter conveyance from the south brought the water of the spring Muargoz to the lower part of the city (DSI 2008; Turk et al. 2010).

2.4.1.33 Arykanda (Arif)

An almost 4 km long conveyance from the northwest carried the water of Baskoz springs to Arykanda, near the Arif village north of Finike. The conduit was mainly a rock-cut channel along the mountain slopes. A short conveyance from the southeast carrying the water of the Badil spring also supplied the city (Buyukyildirim 1994a; DSI 2008; Turk et al. 2010).

2.4.1.34 Phaselis (Tekirova)

A long, partly damaged aqueduct-bridge (Fig. 2.16) exists in Phaselis near Tekirova. It appears probable that an almost 10 km long conveyance from the north brought the water of a spring at the hill slopes (Buyukyildirim 1994a; Kurkcu 2015b; Ozis et al. 2018a).

2.4.1.35 Perge (Aksu)

Besides some cisterns and five springwater collection works close to the city, water was brought to Perge by two conveyance systems of large capacities (Fahlbusch 1987c; Baykan and Dag 1994; Buyukyildirim 1994a, 1997; Ozis 1994a, 1995, 1996; Ozis et al. 2018a).

The 11 km long conveyance from the north carried the water of the Gelindusen spring, 2.5 km to the north of the Kursunlu waterfall, to supply Perge. The conduit is partly rock-cut and partly cut-and-cover masonry gallery. It passes over a few aqueduct-bridges, such as the 11 m high Egridere and 16 m high Ahmetali.

The 22 km long conveyance from the northwest carried the water of Dudenbasi springs, which originate, in turn, from the karstic Kirkgozler springs. The conduit begins as a tunnel, then continues as a masonry channel, and passes over some aqueduct-bridges, like the 10 m high Sogucaksu.

Fig. 2.16 The Aqueduct-bridge in Phaselis. (Photo by U.Ozis)

2.4.1.36 Selge (Zerk)

The ancient city Selge (Zerk), located in the mountainous upper Koprucay basin, northwest of Beskonak, was supplied by two rather short water conveyances. The 5 km long conveyance from northwest passes over a 10 m high, partly damaged aqueduct-bridge; the conduit consists of a stone plate canal on the ground, but a baked clay pipe conduit exists in the ground underneath. The 6 km long conveyance from southwest consists of semi-cylindrical or U-shaped baked clay elements (Buyukyildirim 1994a, b; Ozis et al. 2018a).

2.4.1.37 Aspendos (Belkis)

The 17 km long water conveyance to Aspendos is fed by springwater and crosses the Koprucay valley, the last one before the city, by a 1.7 km long stone-pipe inverted siphon (Fig. 2.17) (Ward-Perkins 1955; Garbrecht 1977; Ozis and Harmancioglu 1980; Fahlbusch 1982, 1987a, b; Garbrecht 1985, 1995; Tolle-Kastenbein 1990; Hodge 1992; Buyukyildirim 1994a; Ozis 1994a, 1995, 1996; Kessener and Piras 1997; Grewe 1998, 2014; Kessener 2001; Piras 2001; Ortloff and Kassinos 2003; Nikolic 2008; Unutmaz 2013; Alkan et al. 2014; Wiplinger 2016; Ozis et al. 2018a).

The venter of the inverted siphon is an up to 18 m high aqueduct-bridge and displays two partly damaged water balance towers, which probably had 38 m and 30 m heights. This structure can be considered as three consecutive inverted siphons. The length of the upstream part between the inlet surge chamber and the upstream tower is about 600 m; the length of the middle part between the two towers 950 m; and the length of the downstream part, between the second tower and the outlet

Fig. 2.17 The Stone-pipe inverted Siphon of Aspendos, on Arched bridge as Venter. (Photo by U. Ozis)

chamber is 150 m. The alignment of the aqueduct-bridge displays two changes of direction, 16° at the first tower and 55° at the second one.

The hydraulic capacity of the siphon is calculated as 65 l/s. The two towers assure the safe functioning of the system under air displacements and dynamic pressure changes (Kessener 2001; Ortloff and Kassinos 2003; Nikolic 2008; Wiplinger 2016). Moreover, the second tower avoids the horizontal bending forces due to change in direction, which otherwise might damage the pipe system and interrupt the water flow (Garbrecht 1977; Ozis 1994a, 1995, 1996; Alkan et al. 2014; Ozis et al. 2018a).

2.4.1.38 Side (Side)

The 25 km long conveyance to Side diverts a small part of the Manavgat river flow on the right bank, at the location where the Dumanli spring discharges into Manavgat on the left bank. The karstic Dumanli spring has an average discharge around 50 m^3/s flowing out from a single orifice. The spring and the initial 5 km of the water conveyance to Side are submerged in the reservoir of Oymapinar dam.

The conduit passes through several rock-cut channels (Fig. 2.18), tunnels and over 24 aqueduct-bridges. The longest is the 340 m long Homa (or Kirkgozler) aqueduct-bridge, situated 2 km downstream of the Oymapinar dam (Izmirligil 1979; Ozis and Harmancioglu 1980; Fahlbusch 1982, 1987d; Tolle-Kastenbein 1990; Hodge 1992; Buyukyildirim 1994a; Grewe 1994, 1998, 2010a, 2014; Ozis 1994a, 1995, 1996; Ozis et al. 2018a).

2.4.1.39 Sagalassos (Aglasun)

A 24 km long conveyance collected the water of several springs, from the east to the west, along the slopes of the mountain in the north, and supplied Sagalassos near Aglasun with water. Some remains of rock-cut or masonry conduits are encountered (Wiplinger 2006a, 2016; Ozis et al. 2018a).

2.4.1.40 Antiochia/Pisidia (Yalvac)

A probably 11 km long conveyance from the north supplied water to Antiochia ad Pisidia. The conduit is basically a masonry channel, ends over a 300 m long partly ruined aqueduct-bridge of curved alignment and crosses the last valley with a stone-pipe inverted siphon (Weber 1904; Bildirici 1994; Burdy and Taslialan 1997; Ozis et al. 2018a).

Fig. 2.18 Rock-cut channel of the water conveyance to Side and, in the background, the Modern Oymapinar Arch Dam on Manavgat river. (Photo by U.Ozis)

2.4.1.41 Antiochia/Cragum (Guney)

Several drainage channels, a ruined part of a masonry gallery, and the remains of a large Roman bath were found at Antiochia ad Cragum, located near the village Guney, southeast of Gazipasa at the Mediterranean sea. These elements indicate a relevant water conveyance from the eastern area (Wiplinger 2016).

2.4.1.42 Anemurium (Anamur)

Remains of a significant aqueduct-bridge indicate an interesting water supply for the antique city of Anemurium (Weber 1904; DSI 2008).

2.4.1.43 Sbede (Yukari Caglar)

A 4 km long water conveyance carries the water of the spring Boncukcayiri to Sbede, near the village Yukari Caglar, northeast of Ermenek. Besides a short baked clay pipe stretch, the conduit is a rock-carved channel, partly in the karstic underground. A stretch of this conduit is obtained by an horizontal qanat system with horizontal lateral tunnels instead of vertical shafts (Bildirici 2014).

2.4.1.44 Diocaeserea (Uzuncaburc)

Three antique settlements, Diocaeserea (Uzuncaburc), Olba (Ugra), and Elaiussa Sebaste (Ayas), were supplied by the water of the Lamas (Limonlu) river, which flows into the Mediterranean sea near Erdemli. The 36 km long water conveyance to Diocaeserea consists of masonry and rock-cut channels and some tunnel stretches, the longest of them constructed by the qanat technique. The water conveyance to Diocaeserea dates probably back to the first century AD (Arisoy et al. 1987; Tolle-Kastenbein 1990; Bildirici 1994; Ozis 1994a, 1995, 1996; Ozbay 1998; DSI 2008; Unutmaz 2013; Wiplinger 2016; Ozis et al. 2018a).

2.4.1.45 Olba (Ugra)

The 18 km long water conveyance to Olba consists of masonry and rock-cut channels and numerous tunnel stretches; the total length of the tunnel stretches is 3.7 km. Furthermore, there is also an almost 18 m high aqueduct-bridge in Olba. The water conveyance to Olba dates probably back to the middle of the first century AD or latest to the end of the second century (Arisoy et al. 1987; Tolle-Kastenbein 1990; Bildirici 1994; Cangiri and Akpinar 1994; Ozis 1994a, 1995, 1996; Ozbay 1998; DSI 2008; Unutmaz 2013; Ozis et al. 2018a).

2.4.1.46 Elaiussa Sebaste (Ayas)

The water conveyance to Elaiussa Sebaste (Ayas) is 25 km long and extended by 3 km to Korykos (Kizkalesi). The conduit consists of rock-cut or masonry channels and passes over eight aqueduct-bridges (Fig. 2.19) (Limonlu, Tirtar, Kumkuyu, Yemiskumu A to D, Ayas). The extension to Korykos ends in a large cistern. The water conveyance to Elaiussa Sebaste was probably constructed between 140 and 260 AD and extended to Korykos at the end of the fifth or early sixth century AD (Arisoy et al. 1987; Bildirici 1994; Ozis 1994a, 1995, 1996; Grewe 1998; Ozbay 2001; DSI 2008; Murphy 2013; Unutmaz 2013; Ozis et al. 2018a).

2.4.1.47 Antiochia/Orontes (Antakya)

Four water conveyance systems supplied Antiochia ad Orontes in ancient times. The 4 km long Kuruyer conveyance from the east and the 5 km long Dursunlu conveyance from the south are the shorter ones. The longer conveyances of 12 km and

Fig. 2.19 Kumkuyu Aqueduct-bridge of the water conveyance to Elaiussa Sebaste. (Photo by U. Ozis)

14 km are fed from the Defne springs farther to the south (Lassus 1983; Doring 2012; Pamir and Yamac 2012; Ozis et al. 2018a).

The quite ruined Kantara aqueduct-bridge, 35 m high and 160 m long, appears to be the largest of the system. The Demirkapi aqueduct-bridge was very close to the city gate; it was heightened by an arched gravity dam in the sixth century and transformed into a dam, but collapsed later.

2.4.1.48 Edessa (Sanliurfa)

The Karakoyun water conveyance supplying Edessa (Sanliurfa), evidenced by the remains between Samsat and Millet bridges over the Karakoyun creek, is also called Justinian's aqueduct-bridge, so that it might date back to the early Byzantine period (Kurkcuoglu 1992; Gerger and Kurkcuoglu 1997; IWA 2012; Yenigun et al. 2013).

2.4.1.49 Samosata (Samsat)

Samosata was supplied by a 40 km long water conveyance from the northeast. The conduit was partly rock-cut and partly masonry channel and passed over 15 aqueduct-bridges, which are ruined to a great extent. The city and the entire conveyance are submerged by the reservoir of the Ataturk dam (Izmirligil 1983; Ozis 1994a, 1995, 1996; Ozis et al. 2018a).

Fig. 2.20 Rock-cut Canal of the water conveyance to Amaseia. (Photo by U.Ozis)

2.4.1.50 Amaseia (Amasya)

A 24 km long water conveyance from the south supplied Amaseia. It had an alignment mostly parallel to the river Yesilirmak along the foot of the mountain range on the right bank. The conduit includes impressive rock-cut stretches (Fig. 2.20) (DSI 1994, 2008; Unutmaz 2013; Ozis et al. 2018a).

2.4.1.51 Ankyra (Ankara)

Considering the alignment of the last stretch, a water conveyance from the Elmadag area in the southwest is assumed to supply Ankyra (Firatli 1951). Such a conveyance probably had a length of around 30 km. The conduit consisted probably of baked clay pipes. Numerous remains of stone pipes are encountered in the area, and a large number of stone pipes are used in the construction of the castle's walls. A significant stone-pipe inverted siphon, eventually in double rows, probably have crossed the last large valley near the city.

2.4.2 Water Distribution and Sewerage

Water distribution and wastewater collection systems of certain Hellenistic-Roman-Byzantine cities in Anatolia deserve also special attention, such as Priene, Miletus, Ephesus, Hierapolis, and Istanbul (Bildirici 2002; Crouch 2004; Ortloff and Crouch 1998, 2001; Wiplinger 2006b; Tuttahs 2007; Strobel 2013; Uytterhoeven 2013).

2.4.3 Cisterns

Covered and open cisterns in Istanbul, dating back to the fourth and up to the sixth centuries AD, are extraordinary examples of antique cisterns, totalling a volume of roughly 1,000,000 m^3. They were the largest of their kinds, with side lengths up to 150–250 m. Noteworthy are the covered cisterns Yerebatan with 336 (Fig. 2.21) and Binbirdirek with 234 columns (Forchheimer and Strzygowski 1893; Eyice 1979; Ozis 1982; IWA 2012).

There are also other sites with important cisterns, like Termessos to the northwest of Antalya (Kurkcu 2014), Assos near Behramkale, Pergamon (IWA 2012), Aigai near Yuntdag (DSI 2008), Keramos (Wiplinger 2006a), Sagalassos (Wiplinger 2006a), Patara (Wiplinger 2016), Arykanda (DSI 2008), Rhodiapolis to the north of Kumluca (Wiplinger 2006a), and Ariassos near Bucak (Wiplinger 2006a; Kurkcu 2015a).

Fig. 2.21 The Yerebatan Cistern in Istanbul. (Photo by Y.Ozdemir)

2.4.4 *Tunnels*

2.4.4.1 Seleuceia Piera (Cevlik)

The tunnel and river diversion system at Cevlik dates back to the Roman period. The construction began in the first and ended in the second century AD. It was aimed to prevent the silting of the harbor of Seleucia Pieria to the northwest of Samandag near Antakya.

The 875 m long system has a capacity of 70 m^3/s. It encompasses two tunnel stretches of 90 m and 30 m in length. The cross sections are of simple horseshoe or trapezoidal, with dimensions in the order of 6–7 m, being the largest of its time (Fig. 2.22) (Alkan and Ozis 1991a, b, 2013; Garbrecht 1991; Grewe 1998, 2010a; DSI 2008; Grewe et al. 2010; Ozis et al. 2010; Baykan et al. 2011).

2.4.4.2 Bezirgan Near Kalkan

The 250 m long Bezirgan tunnel near Kalkan, with 1.1 m width and 2.2 m height, serves as flood water emissary of the karst polje (Genc et al. 2010; Ozis et al. 2010; Baykan et al. 2011, 2013; Wiplinger 2016).

Fig. 2.22 Inlet of the first tunnel stretch in Cevlik. (Photo by U.Ozis)

2.4.4.3 Underground Conduits of Amaseia (Amasya)

There are some water channels, dating back to the fourth and the third centuries BC, at the Amasya castle, located on the mountain overlooking the city (DSI 2008).

2.4.4.4 Tunnels of Water Conveyances

Several water conveyances to ancient settlements include tunnel stretches, as dealt with in Sect. 2.4.1; most of these tunnels have widths in the order of 1 m, heights in the order of 2 m.

2.4.5 Structures Covering Water Courses

2.4.5.1 Pergamon (Bergama)

The tunnel-like twin structures from the Roman period, covering the Bergama creek, date back to the early second century AD. They were the largest of their kind with 7.5 m height and 9 m width each. They have a total capacity of 720 m^3/s and are still in situ (Fig. 2.23) (Ozis et al. 1979b, 2010; Grewe et al. 1994; Ozis 1994a, 1995, 1996; DSI 2008; Baykan et al. 2011).

Fig. 2.23 Inlet of the twin structures covering the Bergama Creek. (Photo by U.Ozis)

2.4.5.2 Nysa (Sultanhisar)

The tunnel-like structure covering the Tekkecik creek in Nysa has dimensions close to 6 m and a capacity of 290 m^3/s (Ozis et al. 1979b, 2010; Grewe et al. 1994; Ozis 1994a, 1995, 1996; DSI 2008; Baykan et al. 2011).

2.4.5.3 Acarlar Near Ephesus

The tunnel-like structure covering the creek in Acarlar near Ephesus has dimensions of 3.3–4.9 m and a capacity of 70 m^3/s (Ozis et al. 2010; Baykan et al. 2011).

2.4.6 Dams

Several dams in Central Anatolia, dating back to the early centuries of the first millenium AD, like the 16 m high Orukaya dam near Corum (Fig. 2.24), 16 m high Cevlik dam near Antakya, 10 m high Cavdarhisar near Kutahya, 4 m high Boget dam near Nigde, and others dating back to the sixth century AD, like the Dara dams near Mardin in Southeastern Anatolia, are interesting remains from these periods. The Ildir dam near Cesme and the Lostugun dam near Amasya date probably back to the second half of the first millenium AD. Sultan and Sihke dams near Van, with some Urartu origins, might also date back to the same period (Schnitter 1979; Garbrecht 1991; Ozis 1999; Hepbostanci et al. 2015). A structure over the creek up in Termessos appear to be a dam (IWA 2012).

2.4.7 Water Power

There are numerous remains of water mills in Turkey; it is hard to conclude that some of them date back to this period. It is believed, however, that the first water wheel was constructed in the fourth century BC in Cabeira (Niksar) (IWA 2012).

Several water mills are encountered in the upstream part of the Degirmendere water conveyance to Ephesus and of the Anaia (Kadikalesi) water conveyance near Kusadasi (Kreiner 2013). Remains of Roman water-powered stone saws were found at certain locations in Anatolia (Wiplinger 2006a; Grewe and Kessener 2007).

Fig. 2.24 The bottom outlet of the Orukaya Dam. (Photo by U.Ozis)

2.5 Seljukide Water Works

2.5.1 Dams

A few dams in Turkey date back to the eleventh up to the fourteenth centuries, the Seljukide period in Central and Eastern Anatolia. The remains of some dams have been submerged in the reservoirs of modern dams like Altinapa and Sille; others are damaged by outside effects.

An interesting example is the 12 m high Faruk dam near Van, with a crest length of 30 m; however, the left half collapsed in 1988. The estimates for the construction date of the Faruk dam varies from the Urartu to the Ottoman periods, but the appropriate dating appears to be Seljukide (Schnitter 1979; Cecen 1987; Garbrecht 1991; Ozis 1999; Bildirici 2004; Ozis et al. 2007; DSI 2008).

2.5.2 *Water Supply & Conveyance Schemes*

Certain remains of the water supply system in Sanliurfa date back to the Seljukide period (Kurkcuoglu 1992; Gerger and Kurkcuoglu 1997; IWA 2012; Yenigun et al. 2013).

2.5.3 *Cisterns*

The cisterns, especially in western and southern regions of Turkey are quite interesting (Ozis 1982; IWA 2012).

2.5.4 *Irrigation*

The Sahip Ata irrigation canals in Konya date back to the thirteenth century. The irrigation systems in Eregli and at other places in Central Anatolia, some of them with probable Hittite origins, date also back to the Seljukide period; a few of them are still in operation (Bildirici 1994, 2004; Bildirici and Bildirici 1996).

2.5.5 *Water Power*

The supply canal of a water-mill at Cermik passes through an asymmetrical opening of the Seljukide Haburman masonry bridge belonging to the twelfth century (Fig. 2.25). This is apparently one of the most ancient water power schemes in Anatolia still in operation (Ozis et al. 2007; DSI 2008).

Some ancient water-mills encountered in the upstream part of the Degirmendere water conveyance to Ephesus and of the Anaia (Kadikalesi) water conveyance near Kusadasi may eventually date back to the Seljukide period (Kreiner 2013).

2.5.6 *Hydromechanics*

The book by Ebul-feyz El Cezeri, who was named after the town Cizre in South-eastern Anatolia, is a twelfth century masterwork on ingenuous hydro-mechanical devices (Cezeri 1196; Hill 1974; Cecen 1979a; IWA 2012).

Fig. 2.25 A Canal, supplying a water Mill, through the Seljukide Haburman Bridge in Cermik. (Photo by U.Ozis)

2.6 Ottoman Water Works

2.6.1 Water Supply & Conveyance Schemes

2.6.1.1 General Remarks

The use of water during the Seljukide and Ottoman periods was relatively more modest, compared to the abundant water use in the Roman period. Water was very appreciated by the Ottomans, and adequate measures were utilized for water safety and wastewater removal (Cecen 1999; IWA 2012).

The basic discharge unit during the Ottoman period was "lule" (a word used also for 'orifice') (Fig. 2.26), equivalent to 36 liters per minute. 1 lule is the discharge flowing through a circular orifice of 26 mm inner diameter, under a water pressure of 96 mm over the center of the orifice. The subunits of lule are: 'kamis' (= 1/4 lule), 'masura' (= 1/8 lule), 'cuvaldiz' (= 1/32 lule), 'hilal' (= 1/64 lule) (Cecen 1988, 1991a, b, 1999, 2000; Ozis and Arisoy 1987, 1996; Ozis 1994a; IWA 2012).

2.6.1.2 Suleymaniye and the Other Halkali Water Conveyances to Istanbul

The Halkali water conveyance systems to Istanbul were constructed in the period of 1450's to 1750's. They consist of 16 systems with a total length of 130 km, including the 50 km long Suleymaniye water conveyance by the great engineer and architect Sinan in the 1550's. The conduits are basically baked clay pipes, with certain tunnel sections, and passing over some aqueduct-bridges. The Suleymaniye system even made use of the fourth century Mazul and Bozdogan aqueduct-bridges (Cecen 1979b, 1984, 1986a, b, 1988, 1990, 1991a, 1999, 2000; Ozis 1984b, 2001; Ozis and Arisoy 1987, 1996, 2000, 2003; Ozis et al. 2007, 2016, 2018b; Acar 2010; IWA 2012).

Fig. 2.26 The Orifices (lule) in the operation chamber of the Yeni Dam. (Photo by Y.Ozdemir)

Fig. 2.27 Yedigoz Aqueduct-bridge on the Taslimüsellim Water Conveyance to Edirne. (Photo by U.Ozis)

2.6.1.3 Taslimusellim Water Conveyance to Edirne

The 50 km long Taslimusellim water conveyance system to Edirne is also considered as a work of Sinan, dating back to the 1530's and expanded some decades later. The conduit is a masonry gallery; the alignment passes over several aqueduct-bridges and includes certain tunnel sections. The system is for the large part still in operation (Fig. 2.27) (Akmandor 1968; Ozis and Arisoy 1986, 1987, 1996, 2000, 2003; Ozis et al. 2007, 2016, 2018b; DSI 2008).

Fig. 2.28 Maglova Aqueduct-bridge on the Kirkcesme Water Conveyance to Istanbul. (Photo by U.Ozis)

2.6.1.4 Kirkcesme Water Conveyance to Istanbul

The 55 km long Kirkcesme water conveyance system to Istanbul in the 1560's is one of the masterworks of Sinan, with four major aqueduct-bridges (Uzun, Egri, Maglova, Guzelce). The system includes more than thirty aqueduct-bridges of various sizes. The conduit is a masonry gallery and is for the large part still in operation (Ozand 1968; Cecen 1979b, 1984, 1986b, 1988, 1990, 1999, 2000; Ozis 1984b, 1987, 2001; Schnitter 1990; Ozis and Arisoy 1987, 1996, 2000, 2003; Ozis et al. 2007, 2016, 2018b; DSI 2008; IWA 2012).

Sinan's schemes are the most important long-distance water conveyance systems since Roman times. The aqueduct-bridges Uzun, Egri, Maglova, Guzelce of the Kirkcesme system, with heights up to 35 m and lengths up to 700 m, rank among the largest of their kinds in all times (Fig. 2.28).

2.6.1.5 Uskudar Water Conveyances to Istanbul's Asian side

The Uskudar water conveyance systems, to the east of Bosphorus in Istanbul, date back to the sixteenth up to the nineteenth centuries. The conduits are mostly baked clay pipes; noteworthy are the water balance towers for pressure control and distribution (Fig. 2.29) (Cecen 1979b, 1991b; Ozis et al. 2007; Dinckal 2001; IWA 2012).

2.6.1.6 Taksim Water Conveyance to Istanbul

The 23 km long Taksim water conveyance to Istanbul dates back to the 1730's and is still for the large part in operation. The conduit is a masonry gallery; it includes a

400 m long aqueduct-bridge over Buyukdere and initially an interesting inverted siphon. A new system for drinking water purpose, the Hamidiye water conveyance, collecting water from the same area, was constructed in the nineteenth century (Yungul 1957; Cecen 1979b, 1984, 1986b, 1992; Ozis 2001; Ozis et al. 2007; IWA 2012).

2.6.1.7 Kilyos Water Conveyance Near Istanbul

The short Kilyos water conveyance near Istanbul is interesting with regard to the water balance towers (Bildirici 2008).

Fig. 2.29 A water balance tower of the Uskudar Water Conveyances. (Photo by U. Ozis)

2.6.1.8 Other Ottoman Water Conveyances

The water supply systems of Sanliurfa, having roots in the Seljukide period (Kurkcuoglu 1992; Gerger and Kurkcuoglu 1997; IWA 2012; Yenigun et al. 2013); of Bursa (IWA 2012), of Corum (IWA 2012), and of Safranbolu (IWA 2012) are some other interesting systems.

Some of the Ottoman water conveyances to Izmir (Ozis et al. 1999; DSI 2008) and the Pasasuyu water conveyance to Izmit (Unal 2001) have roots in ancient periods. The diversion and conveyance of Ephesus' Degirmendere waters to Kusadasi (Ozis and Atalay 1999; Ozis et al. 1998, 2005b) and a water conveyance with aqueduct-bridges to Foca (Ozis 1994a, 1995, 1996) date back to the Ottoman times.

2.6.2 Water Distribution and Sewerage

Beyond the water conveyances, the water distribution and wastewater collection systems of Istanbul deserve also special interest (Sarikaya et al. 2001; Bildirici 2002; Dinckal 2005; IWA 2012). Besides the traditional baked clay pipes and cut-and-cover masonry channels, traditional kharizes are used for clean water collection and transport as well as wastewater collection (IWA 2012).

2.6.3 Cisterns

Several cisterns in the western and southern regions of Turkey date also back to the Ottoman period (Ozis 1982; Commito and Rojas 2010).

2.6.4 Dams

2.6.4.1 Istanbul Dams

In the period of 1620–1839, the Kirkcesme system was supplemented by four (Topuz, Buyuk, Ayvat, and Kirazli) and the Taksim system by three (Topuzlu, Valide, and Yeni) masonry dams, with heights up to 17 m and crest lengths up to 104 m. All these dams, located at the Belgrad forest to the north of Istanbul, are still in operation (Figs. 2.30 and 2.31) (Yungul 1957; Cecen 1968, 1979b, 1984, 1986b, 1987, 1988, 1990; Ozis 1977, 1981a, 1984b, 1999; Tutuncuoglu and Benzeden 1979; Schnitter 1994; Ozis et al. 2007; DSI 2008; Acar 2010).

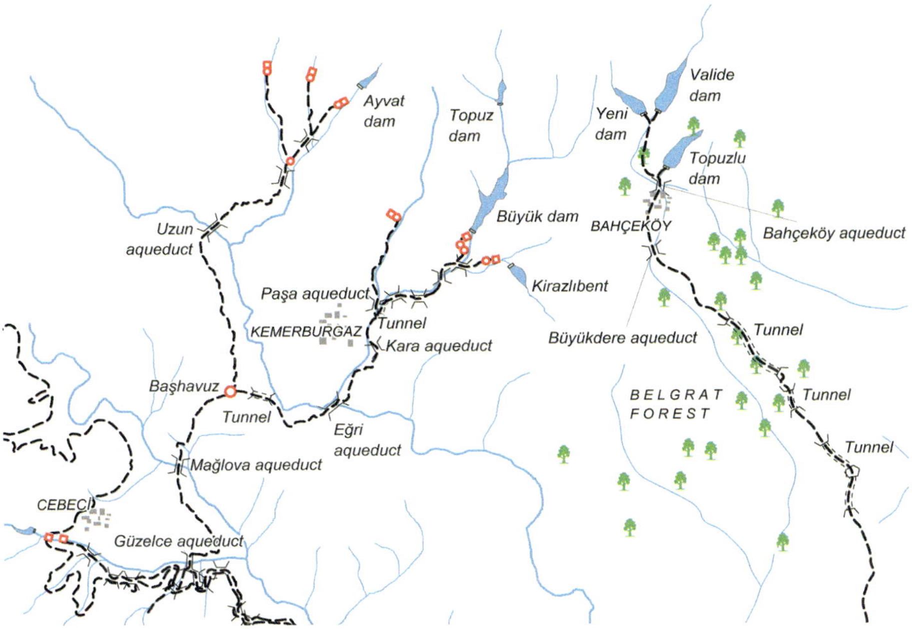

Fig. 2.30 Location of the Historical Istanbul Dams with the upper parts of the Kirkcesme and Taksim Water Conveyances. (U.Ozis and Y.Arisoy)

Fig. 2.31 Yeni Dam supplying the Taksim system. (Photo by Y.Ozdemir)

The majority of dams constructed until these centuries were embankment dams, so that the masonry dams of Istanbul deserve a special place with regard to the historical development of dams.

2.6.4.2 Other Ottoman Dams

There are also some newer dams of the nineteenth century like Samlar and Elmali I dams around Istanbul (Cecen 1987; Ozis 1999), and the 23 m high Maden dam near Karasu. The Semali embankment dam near Amasya is considered as an Ottoman dam. It is mentioned in certain sources that the imperviousness of the dam was achieved by covering the upstream face with ox-skins (Ozis 1999).

2.6.5 River Diversion

2.6.5.1 Sakarya – Sapanca Diversion

The engineer and architect Sinan planned in 1583 the diversion of the Sakarya river to the Marmara sea over Lake Sapanca for flood control, water power (mills), and river navigation purposes. This idea had roots in the sixth century but could not yet be realized until present times (Cecen 1981).

2.6.5.2 Gediz Diversion

The Gediz river was diverted to the outer bay in the late nineteenth century in order to prevent the closure of the Izmir Bay (Ozis 1994a; Buyukyildirim 2017).

2.6.6 Irrigation

Various irrigation systems under actual operation have their roots in Ottoman times, like the Surgu irrigation near Malatya from nineteenth century, and the Beysehir-Cumra irrigation south of Konya from early twentieth century (Bildirici 1994; Ozis 1994a; Ozis et al. 2009; Buyukyildirim 2017).

2.6.7 Water Power

The first electricity was generated in Turkey in 1902 in the Tarsus hydroelectric scheme (Ozis 1994a; Ozis et al. 2009, 2014a). The scheme used the elevation difference of the Berdan River at Tarsus falls.

2.7 Conclusion

Ancient hydraulic works, dating back to various civilizations of her last 4000 year long history, make Turkey one of the world's foremost open-air museums in this respect. Some of them are, with very few repairs or modifications, still in operation after several centuries or even millenia. Based on this tradition, Turkey has continued to harness her water resources during the Republican period.

Acknowledgment The authors gratefully acknowledge the support of the Ege and Dokuz Eylul Universities in their office work; they also appreciate the help of various institutions in their field investigations, such as DSI (State Hydraulic Works Authority), ISKI (Water and Sewerage Authority of Istanbul), the Municipality of Edirne, Turkish Armed Forces, Rural Services Authority, Mediterranean Irrigation Union, T.G.T. Power Stations Company, SU-YAPI Engineering and Consulting Company, ATA Construction Company, Austrian Archeological Institute, Ephesus Museum, Antakya Museum, Museum of Turkish and Islamic Art, Topkapi Palace Museum, and Suleymaniye Library.

For the encouragement of his work, the first author wishes to extend his gratitude to the late Ord. Prof.Dr. Ekrem Akurgal and to several personalities in the fields of archeology and history, as well as to the late Prof.Dr. Kazim Cecen and to many experts in the field of civil engineering, from Turkey and abroad.

The first author thanks the Frontinus-Gesellschaft e.V. (Berlin) for awarding him the Frontinus-Medal in 2014, with regard to his scientific work on the historical hydraulic structures in Turkey.

This text summarizes the studies and field investigations of the first author for more than 40 years, often in cooperation with several colleagues and students. The authors wish to thank their close colleague civil engineers for their contributions, many of them being also former students of the first author, whose names often appearing as co-authors of publications cited in the lists of references, especially to the late Prof. Dr. Orhan Baykan. The present text is an updated version of several previous publications of the authors, given in the list of references.

The authors extend their thanks to the Springer Institution and to the editors Prof. Dr. Nilgun Harmancioglu and Prof.Dr. Dogan Altinbilek for inviting them to prepare this chapter for the book "Water Resources of Turkey".

References

Acar S (2010) Bentler ve Sinan'in suyolu. Biryil, Istanbul

Akmandor N (1968) Koca Sinan'in planciligi, eserleri, muhendisligi. In: Bektas C (ed) Koca Sinan, Sinan Komitesi, Istanbul, p 49–62 & Turkiye Muhendislik Haberleri 157:1–6

Alkan A (2015) Tarihi Milas suyollari. In: 4. Su yapilari sempozyumu bildirileri, Antalya 2015, Insaat Muhendisleri Odasi, Antalya, p 540–549

Alkan A, Ozis U (1991a) Su muhendisligi tarihi acisindan Cevlik kanal ve tunelleri. Insaat Muhendisleri Odasi – Teknik Dergi 3:353–366

Alkan A, Ozis U (1991b) Cevlik canal and tunnels from the point of view of hydraulics engineering history. In: Chamber of Civil Engineers – Digest 91, Ankara, p 92–95

Alkan A, Ozis U (2013) Ancient tunnel and diversion channel in Seleuceia Pieria – a flood diversion system near Antiochia in Turkey. J Civil Eng Urban 1:1–5

Alkan A, Tanriover A, Ozis U (1999) Ege bolgesinde tarihi su yapilari ve Priene. In: Izmir Su Kongresi, 1999 Izmir, Muhendislik ve Diger Meslek Odalari Izmir Subeleri, Izmir, p 29–43

Alkan A, Ozdemir Y, Baykan O (2014) Inverted siphons in Western Anatolia. Frontinus-Gesellschaft Hefter 28:55–70

Alzinger W (1987) Beispiele antiker Wasserversorgungsanlagen: Ephesos. In: Die Wasserversorgung antiker Stadte, vol 2, Zabern, Mainz, p 180–184

Arisoy Y, Ozis U, Kaya B (1987) Lamas havzasi tarihi su getirme sistemleri. In: IX. Teknik Kongre, 1987 Ankara, vol 2, Insaat Muhendisleri Odasi, Ankara, p 363–376

Aylward W, Bieg G, Kayan I (2002) The aqueduct of Roman Ilion and the bridge across the Kemerdere valley in the Troad. Studia Troica 12:397–427

Baykan O (1999) Curuksu (Lycus) vadisindeki uc antik kent ve onlarin tarihsel su yapilari. In: Denizli'de sanayilesme ve kentlesme sempozyumu, 1997 Denizli, Makina Muhendisleri Odasi, Denizli, p 117–127

Baykan O, Baykan N (2015) Binyillar boyunca Anadolu'daki su muhendisligi. In: 4. Su yapilari sempozyumu bildirileri, Antalya 2015, Insaat Muhendisleri Odasi, Antalya, p 144–156

Baykan O, Cantilav T (1997) Cibyratis birligi tarihsel su iletimleri (Cibyra, Oenoanda, Bubon, Balbura). In: Turkiye Insaat Muhendisligi 14. Teknik Kongresi, 1997 Izmir, Insaat Muhendisleri Odasi, Izmir, p 1099–1114

Baykan O, Dag A (1994) Perge tarihsel su getirme sistemleri. In: DSI'nin 40. kurulus yili su ve toprak kaynaklarinin gelistirilmesi konferansi, vol 1, 1994 Ankara, Devlet Su Isleri, Ankara, p 63–72

Baykan O, Iskan H (2011) Patara eskil kenti su iletim ve dagitim sistemi. In: 2. Su Yapilari Sempozyumu, 2011 Diyarbakir, Insaat Muhendisleri Odasi, Diyarbakir, p 67–78

Baykan O, Kocakaya I, Alkaya D (1997) Patara tarihsel su iletimi. In: Turkiye Insaat Muhendisligi 14.Teknik Kongresi, 1997 Izmir, Insaat Muhendisleri Odasi, Izmir, p 1067–1082

Baykan O, Tanriover E, Ciftci I (2001a) Historical water supply system of Magnesia on the Meander. In: International symposium on water resources and environmental impact assessment, 2001 Istanbul, Istanbul Water and Sewerage Authority, Istanbul, p 139–148

Baykan O, Baykan N, Tanriover E (2001b) Denizli ili arkeolojik belgeleme calismalari (Eskil-Antik Suyollari). Turkiye Bilimler Akademisi – Kultur Envanteri Dergisi 1:33–44

Baykan O, Alkan A, Bacanli U, Baykan N, Ozis U (2011) Testing flood estimation methods on ancient closed conduits. In: International Balkans conference on challenges of civil engineering, 2011 Tirana, EPOKA University, Tirana

Baykan O, Baykan N, Bacanli U (2013) Bezirgan (Kas-Kalkan-Antalya) kapali havzasi bosaltim tunelinin hidrolojik ve hidrolik cozumlemesi. In: Taskin ve heyelan sempozyumu, 2013 Trabzon, TMMOB, Ankara, p 465–473

Belli O (1996) Dogu Anadolu Bolgesinde kesfedilen Urartu barajlarina toplu bir bakis. Belleten 229:638–751

Belli O (1997) Dogu Anadolu'da Urartu sulama kanallari – Urartian irrigation canals in eastern Anatolia. Arkeoloji ve Sanat, Istanbul

Bildirici M (1994) (second edn. 2007, third edn. 2009) Tarihi Su Yapilari: Konya, Karaman, Nigde, Aksaray, Yalvac, Side, Mut, Silifke. Devlet Su Isleri, Ankara

Bildirici M (2002) The water supply system of ancient Galata in Istanbul. In: Eighteenth International Congress, History Seminar, 2002 Montreal, International Commission on Irrigation and Drainage, Montreal, p 159–170

Bildirici M (2004) (second edn. 2009) Tarihi sulama, su depolama, taskin koruma tesisleri. Devlet Su Isleri, Ankara

Bildirici M (2008) Kilyos water supply system with suterazis. In: Ohlig C (ed) Cura Aquarum in Jordanien, 2007 Petra, Deutsche Wasserhistorische Gesellschaft, Siegburg, p 341–343

Bildirici M (2014) Karaman Ermenek Yukari Caglar – Water tunnels and water supply system of ancient Sbede. In: Ohlig C, Tsuk T (eds) Cura Aquarum in Israel II, 2012 Israel. Deutsche Wasserhistorische Gesellschaft, Siegburg, pp 275–284

Bildirici M, Bildirici O (1996) Historical irrigation in the plain of Eregli of Konya. In: Sixteenth International Congress, History Seminar, 1996 Cairo, International Commission on Irrigation and Drainage, Cairo, p 179–194

Burdy J, Lebouteiller P (1998) L'acqueduc romain de Xanthos. In: Anatolia Antique – Eski Anadolu, vol 6, Institut Francais d'Études Anatoliennes Georges Dumézil, Istanbul, p 227–248

Burdy J, Taslialan M (1997) L'acqueduc d'Antioche de Pisidie. In: Anatolia Antique – Eski Anadolu, vol 5, Institut Francais d'Études Anatoliennes Georges Dumézil, Istanbul, p 133–166
Burney C (1972) Urartian irrigation works. Anatol Stud 22:179–186
Buyukyildirim G (1994a) Antalya bolgesi tarihi su yapilari. Devlet Su Isleri, Ankara
Buyukyildirim G (1994b) Selge antik kenti su yapilari. In: DSI'nin 40.kurulus yili su ve toprak kaynaklarinin gelistirilmesi konferansi, 1994 Ankara, vol 1, Devlet Su Isleri, Ankara, p 23–36
Buyukyildirim G (1997) Perge kenti tarihsel su yapilari. In: Turkiye Insaat Muhendisligi 14. Teknik Kongresi, 1997 Izmir, Insaat Muhendisleri Odasi, Izmir, p 1051–1066
Buyukyildirim G (2017) 20. yuzyilda su isleri ve Antalya. Devlet Su Isleri, Ankara
Cangiri A, Akpinar M (1994) Icel-Silifke-Kizilgecit tarihi su yapilari. In: DSI'nin 40.kurulus yili su ve toprak kaynaklarinin gelistirilmesi konferansi, 1994 Ankara, vol 1, Devlet Su Isleri, Ankara, p 47–61
Cecen K (1968) Die turkischen Talsperrenbauten. Wasser und Boden 2:34–37
Cecen K (1979a) A great hydraulic engineer at the XII. Century: Al-Jazari. In: XVIII. IAHR Congress, 1979 Cagliari, vol 6, International Association for Hydraulic Research, Cagliari, p 690–693
Cecen K (1979b) Osmanische Wasserbauten in Istanbul. In: Leichtweiss-Institut fur Wasserbau, Mitteilungen, vol 64, Braunschweig, Technische Universitat, p 1–21
Cecen K (1981) Osmanlilar devrinde Karadeniz-Sakarya-Izmit baglantisi. In: I.Uluslararasi Turk-Islam Bilim ve Teknoloji Tarihi Kongresi, 1981 Istanbul, vol 5, Istanbul Teknik Universitesi, Istanbul, p 225–248
Cecen K (1984) Istanbul'da Osmanli devrinde su tesisleri, Istanbul Teknik Universitesi Bilim ve Teknoloji Tarihi Arastirma Merkezi, Istanbul
Cecen K (1986a) Suleymaniye su yollari, Istanbul Teknik Universitesi Bilim ve Teknoloji Tarihi Arastirma Merkezi, Istanbul
Cecen K (1986b) 16.yuzyilda Istanbul su tesisleri, In: II. International Congress on the History of Turkish-Islamic Science and Technology, 1986 Istanbul, vol 3, Istanbul Technical University, Istanbul, p 105–122
Cecen K (1987) Seldschukische und osmanische Talsperren. In: Garbrecht G (ed) Historische Talsperren. Wittwer, Stuttgart, pp 275–295
Cecen K (1988) Mimar Sinan ve Kirkcesme Tesisleri. Istanbul Su ve Kanalizasyon Idaresi. Istanbul
Cecen K (1990) Sinan's water supply system in Istanbul. Istanbul Su ve Kanalizasyon Idaresi, Istanbul
Cecen K (1991a) Halkali Sulari. Istanbul Su ve Kanalizasyon Idaresi, Istanbul
Cecen K (1991b) Uskudar Sulari. Istanbul Su ve Kanalizasyon Idaresi, Istanbul
Cecen K (1992) Taksim ve Hamidiye sulari. Istanbul Su ve Kanalizasyon Idaresi, Istanbul
Cecen K (1994) Ulkemizdeki tarihi su tesisleri ve bunlarin teknik yonden degerlendirilmesi, In: DSI'nin 40.kurulus yili su ve toprak kaynaklarinin gelistirilmesi konferansi, 1994 Ankara, vol 1, Devlet Su Isleri, Ankara, p 11–22
Cecen K (1996a) Roma suyollarinin en uzunu. Turkiye Sinai Kalkinma Bankasi, Istanbul
Cecen K (1996b) The longest roman water supply line. Turkiye Sinai Kalkinma Bankasi, Istanbul
Cecen K (ed) (1999) Osmanli Imparatorlugu'nun dorugu – 16.Yuzyil teknolojisi. Istanbul Su ve Kanalizasyon Idaresi, Istanbul
Cecen K (2000) (Kolay C, ed) Istanbul'un Osmanli donemi suyollari. Istanbul Su ve Kanalizasyon Idaresi, Istanbul
Cezeri E (1196) Kitab-i hiyel. Topkapi Sarayi Arsivi, Istanbul & Facsimile 1990: Ismâ'il ibn al-Razzâz Jazarî, Olaganustu mekanik araclarin bilgisi hakkinda kitap. Kultur Bakanligi, Ankara
Cinaroglu A (2007) Alaca Hoyuk Hitit baraji onarimi. In: Tarihi eserlerin guclendirilmesi ve gelecege guvenle devredilmesi sempozyumu-1, 2007 Ankara, Insaat Muhendisleri Odasi, Ankara, p 105–107
Commito A, Rojas F (2010) The aqueducts of Aphrodisias. In: Ratté C, Staebler PD (eds) The Aphrodisias regional survey, vol 5, Zabern, Mainz, p 289–307

Coulton JJ (1987) Roman aqueducts in Asia Minor. In: Macready S, Thompson FH (eds) Roman architecture in the Greek world. The Society of Antiquaries, London, pp 72–84
Crouch DP (1993) Water mangement in ancient greek cities. New-York, Oxford
Crouch DP (1996) Priene's streets and water supply. In: De Haan N, Jansen G (eds) Cura Aquarum in Campagna, 1994 Pompeii. Peeters, Leiden, pp 137–143
Crouch DP (2004) Geology and settlement: Greco-Roman patterns. New-York, Oxford
Crouch DP, Gungor T, Alkan A (1997) Priene and Pompeii, cities in geolgically active settings. In: Marinos PG, Koukis GC, Tsiambasos GC, Stoujaras GC (eds) Engineering geology and the environment. Balkema, Rotterdam, pp 3107–3111
Dalman KO (1933) Der Valens-Aquadukt in Konstantinopel. IstForsch, Bamberg
Dinckal N (2001) Die 'Compagnie des eaux de Scutari et Kadi-keui' und die Versorgung der asiatischen Seite Istanbuls durch das Wasserwerk Elmali (1888–1904). In: Dinckal N, Mohajeri S (eds) Blickwechsel 2000. Technische Universitat, Berlin, pp 117–137
Dinckal N (2005) Istanbul und das Wasser – Zur Geschichte der Wasserversorgung und Abwasserentsorgung von der Mitte des 19.Jahrhunderts bis 1966. Oldenbourg, Munchen
Doring M (2012) Die antiken Wasserbauten von Antiochia, Turkei. WasserWirtschaft 1+2:10–16
DSI (1994) Samsun Bolge Mudurlugu sinirlari icindeki tarihi su yapilari, Devlet Su Isleri, Ankara
DSI (ed) (2008) Tarihi su yapilari konferansi. Devlet Su Isleri II.Bolge Mudurlugu, Izmir
Emre K (1993) The Hittite dam of Karakuyu. In: Prince Masaka HMT (ed) Essays on anatolian archeology, Wiesbaden, p 1–42
Eyice S (1979) Byzantinische Wasserversorgungsanlagen in Istanbul. In: Leichtweiss-Institut fur Wasserbau, Mitteilungen, vol 64, Braunschweig, Technische Universitat, p 1–31
Fahlbusch H (1982) Vergleich antiker griechischer und romischer Wasserversorgungsanlagen, MInstWasser 73, Braunschweig
Fahlbusch H (1987a) Elemente griechischer und romischer Wasserversorgunsanlagen. In: Die Wasserversorgung antiker Stadte, vol 2, Zabern, Mainz, p 133–163
Fahlbusch H (1987b) Beispiele antiker Wasserversorgungsanlagen: Aspendos. In: Die Wasserversorgung antiker Stadte, vol 2, Zabern, Mainz, p 172–175
Fahlbusch H (1987c) Beispiele antiker Wasserversorgungsanlagen: Perge. In: Die Wasserversorgung antiker Stadte, vol 2, Zabern, Mainz, p 193–195
Fahlbusch H (1987d) Beispiele antiker Wasserversorgungsanlagen: Side. In: Die Wasserversorgung antiker Stadte, vol 2, Zabern, Mainz, p 218–221
Fahlbusch H (1987e) Die Wasserversorgung des hellenistischen Pergamon. In: Leichtweiss-Institut fur Wasserbau, Mitteilungen, vol 97, Braunschweig, Technische Universitat, p 65–98
Firatli N (1951) Ankara'nin ilkcag su tesisati. Belleten 59:249–259
Forchheimer P (1890) Die altere Wasserversorgung von Konstantinopel. Zeitschrift des Vereins Deutscher Ingenieure 35:868–878
Forchheimer P (1923) Wasserleitungen. FiE III, Wien, p 224–255
Forchheimer P, Strzygowski J (1893) Die byzantinischen Wasserbehalter von Konstantinopel. Beitrage zur Geschichte der byzantinischen Baukunst und zur Topographie von Konstantinopel, Wien
Garbrecht G (1975) Bewasserungswirtschaft in der Geschichte: Urartu, 900-600 v. Chr Zeitschrift fur Bewasserungswirtschaft 1:63–74
Garbrecht G (1976) Wasserwirtschaftliche Anlagen des antiken Pergamon. Wasser und Boden 3:45–50
Garbrecht J (1977) Hydraulic towers in roman pressure conduits. In: XVII. IAHR Congress Proceedings, 1977 Baden-Baden, vol 6, International Association for Hydraulic Research, Baden-Baden, p 778–780
Garbrecht G (1985) Wasser – Vorrat, Bedarf und Nutzung in Geschichte und Gegenwart. Hamburg, Rohwolt
Garbrecht G (ed) (1987a) Historische Talsperren 1. Wittwer, Stuttgart
Garbrecht G (1987b) Die Wasserversorgung des antiken Pergamon. In: Die Wasserversorgung antiker Stadte, vol 2, Zabern, Mainz, p 11–48

Garbrecht G (ed) (1991) Historische Talsperren 2. Wittwer, Stuttgart
Garbrecht G (1995) Meisterwerke antiker Hydrotechnik. Teubner, Stuttgart
Garbrecht G, Fahlbusch H (1975) Wasserwirtschaftliche Anlagen des antiken Pergamon: Kaikos Leitung, MInst-Wasser 44, Braunschweig
Garbrecht G, Fahlbusch H (1978) Wasserwirtschaftliche Anlagen des antiken Pergamon: Umbau und Neubau der Kaikos Leitung, MInstWasser 60, Braunschweig
Garbrecht G, Holtorff G (1973) Wasserwirtschaftliche Anlagen des antiken Pergamon: Die Madradagleitung, MInstWasser 37, Braunschweig
Genc S, Baykan O, Baykan N, Bacanli U (2010) Bezirgan (eskil Pirha kenti) kapali havzasi eskil tuneli akaclama olayi. In: Karahan H, Baykan O (eds) VI. Ulusal Hidroloji Kongresi, 2010 Denizli, Pamukkale Universitesi, Denizli, p 77–87
Gerger R, Kurkcuoglu K (1997) Sanliurfa'daki tarihi su yapilari. In: Turkiye Insaat Muhendisligi 14.Teknik Kongresi, 1997 Izmir, Insaat Muhendisleri Odasi, Izmir, p 1129–1144
Grewe K (1994) Die romische Wasserleitung nach Side (Turkei). Antike Welt 2:192–203
Grewe K (1998) Licht am Ende des Tunnels: Planung und Trassierung im antiken Tunnelbau. Zabern, Mainz
Grewe K (2010a) Meisterwerke antiker Technik. Zabern, Mainz
Grewe K (2010b) Zwei fruhe Wasserbauten in Anatolien. Kolner Jahrbuch 43:281–287
Grewe K (2014) Aquadukte – Wasser fur Roms Stadte. Regionalia, Rheinbach
Grewe K, Kessener P (2007) A stone relief of a water-powered stone saw at Hierapolis. In: Énergie hydraulique et machines élévatrices d'eau dans l'antiquité, 2007 Napoli, p 227–234
Grewe K, Ozis U, Baykan O, Atalay A (1994) Die antiken Flussuberbauungen von Pergamon und Nysa (Turkei). Antike Welt 4:348–352
Grewe K, Ozis U, Alkan A (2010) Umleitung mitten durch den Fels (Der Titus-Tunnel bei Cevlik). Antike Welt 6:56–62
Hecht K (1975, 1976, 1978, 1979, 1983) Wasserwirtschaftliche Anlagen des antiken Pergamon – Aquadukte der Kaikos-Leitung, MInst-Wasser 45, 54, 61, 64, 78, Braunschweig
Hepbostanci G, Alkan A, Ozdemir Y (2015) Van yoresi tarihi su yapilari, In: 5.Tarihi eserlerin guclendirilmesi, gelecege guvenle devredilmesi sempozyumu, 2015 Erzurum, vol.I, Insaat Muhendisleri Odasi, Erzurum, p 251–260
Hill R (1974) The book of knowledge of ingenious mechanical devices. Reidel, Dordrecht
Hodge T (1992) Roman aqueducts and water supply. Duckworth, London
Inal I (2009) Hittite dam from 1250 BC towards today – MO 1250'den gunumuze Hitit baraji. Devlet Su Isleri, Ankara
Iskan H, Baykan O (2013) Neue Ergebnisse zur Wasserleitung von Patara/Turkei. In: Wiplinger G (ed) Historische Wasserleitungen, 2011 Wien, Frontinus-Gesellschaft, Berlin, p 93–103
ISKI (1983) Tarih boyunca Istanbul sulari ve Istanbul su ve kanalizasyon sorunu. Istanbul Su ve Kanalizasyon Idaresi, Istanbul
IWA (Koyuncu I, Sen Z, Ozturk S, Altintas M, Ozturk I, eds) (2012) Third IWA specialized conference on water and wastewater technologies in ancient civilizations. International Water Association, Istanbul
Izmirligil U (1979) Die Wasserversorgunsanlagen von Side. In: Leichtweiss-Institut fur Wasserbau, Mitteilungen, vol 64, Braunschweig, Technische Universitat, p 1–25
Izmirligil U (1983) Samsat (Samosata) su yolu arastirmasi, 1981. In: IV.Kazi Sonuclari Toplantisi, 1982 Ankara, Kultur ve Turizm Bakanligi, Ankara, p 345–356
Kayan I (2000) The water supply of Troia. Studia Troica 10:135–144
Kessener P (2001) The Aspendos Aqueduct, an outstanding example of Roman hydraulic technology. Frontinus-Gesellschaft Hefter 24:95–108
Kessener P, Piras S (1997) The pressure line of the Aspendos aqueduct. Adalya II:159–187
Kreiner R (2013) Die Muhlen des Degirmendere Aquaduktes von Ephesos und des Aquaduktes von Anaia/Kadikalesi (Turkei). In: Wiplinger G (ed) Historische Wasserleitungen, 2011 Wien, Frontinus-Gesellschaft, Berlin, p 131–138

Kurkcu M (2014) L'urbanisme et les aménagements hydrauliques de Termessos, Dissertation, Université de Paris-Sorbonne
Kurkcu M (2015a) Ariassos ve Kremna'da gozlenen su sistemleri ve sarniclar hakkinda bir degerlendirme, Mediterranean Journal of Humanities, I:301–311
Kurkcu M (2015b) Phaselis su sistemleri ve hidrografisi uzerine genel bir degerlendirme, Phaselis, I:69–79
Kurkcuoglu AC (1992) Sanliurfa su mimarisi. Kultur Bakanligi, Ankara
Lassus J (1983) Das fließende Wasser von Antiochia. In: Boucher JP (ed) Journées d'études sur les aqueducs romains – Tagung uber romische Wasserversorgungsanlagen, 1977 Lyon, Belle Lettres, Paris, p 207–229
Mays LW (ed) (2010) Ancient water technologies. Springer, Dordrecht
Murphy D (2013) The aqueduct of Elaiussa Sebaste in rough Cilicia. In: Wiplinger G (ed) Historische Wasserleitungen, 2011 Wien, Frontinus-Gesellschaft, Berlin, p 71–84
Neve P (1969/70) Eine hethitische Quellgrotte in Bogazkoy. In: Istanbuler Mitteilungen 19/20, Tubingen, p 97–107
Nikolic M (2008) Cross disciplinary investigation of ancient long-distance water pipelines. Dissertation, University of Victoria
Ogun B (1970) Van'da Urartu sulama tesisleri ve Samram (Semiramis) kanali. Anadolu, Ankara
Ortloff CR, Crouch DP (1998) Hydraulic analysis of a self-cleaning drainage outlet at the hellenistic city of Priene. J Archeol Sci 25:1211–1220
Ortloff CR, Crouch DP (2001) The urban water supply and distribution system of the ionian city Ephesos in the roman imperial period. J Archeol Sci 28:843–860
Ortloff CR, Kassinos A (2003) Computational fluid dynamics investigation of the hydraulic behavior of the Roman inverted siphon system in Aspendos, Turkey. J Archaeol Sci 30:417–428
Ozand E (1968) Koca Sinan'in en buyuk muhendislik eserlerinden biri olan Istanbul'un Kirkcesme tesisleri. In: Bektas C (ed) Koca Sinan, Sinan Komitesi, Istanbul, p 34–47. & Turkiye Muhendislik Haberleri 157:8–10
Ozbay F (1998) Olba/Diokaisareia su sistemi. Olba 1:121–129
Ozbay F (2001) Elaiussa Sebaste ve Korykos su sistemi. Olba 4:145–161
Ozis U (1977) The ancient dams of Istanbul. Int Water Power Dam Constr 7:49–51, 8:44–47
Ozis U (1981a) The ancient Istanbul dams within the perspective of dam technology history. In: International Congress on the History of Turkish-Islamic Science and Technology, 1981 Istanbul, vol 3, Istanbul Teknik Universitesi, Istanbul, p 19–30
Ozis U (1981b) Anadolu'da su kaynaklarinin gelistirilmesinin dunu, bugunu, yarini. In: Su ve toprak kaynaklarinin gelistirilmesi konferansi 1, 1981 Ankara, Devlet Su Isleri, Ankara, p 8–33
Ozis U (1982) An outlook on ancient cisterns in Anatolia. In: International conference on rain water cistern systems, 1982 Honolulu, University of Honolulu, Hawai, p 9–15
Ozis U (1984a) (second ed.: 1987) Su muhendisligi acisindan Anadolu'daki eski su yapilari. Dokuz Eylul Universitesi, Izmir
Ozis U (1984b) Historische Staumauern und Fernwasserleitungen von Istanbul. WasserWirtschaft 4:211–215
Ozis U (1987) Historical parallels in the water supply development of Roma and Istanbul. In: Wunderlich WO, Prins JE (eds) Water resources developments in perspective. Balkema, Rotterdam, pp 35–44
Ozis U (1991) Alabanda und seine antike Wasserversorgung. Antike Welt 2:106–113
Ozis U (1994a) Su muhendisligi tarihi acisindan Turkiyedeki eski su yapilari. Devlet Su Isleri, Ankara
Ozis U (1994b) Anadolu'daki tarihi su yapilarina genel bir bakis. In: DSI'nin 40.kurulus yili su ve toprak kaynaklarinin gelistirilmesi konferansi, 1994 Ankara, vol 1, Devlet Su Isleri, Ankara, p 1–10
Ozis U (1994c) Historische Wasserbauten in Anatolien: Ein Fundstellen- und Literaturverzeichnis. Frontinus-Gesellschaft Hefter 18:89–108
Ozis U (1995) Caglar boyunca Anadolu'da su muhendisligi. Insaat Muhendisleri Odasi, Istanbul

Ozis U (1996) Historical water schemes in Turkey. Water Resour Dev 3:347–383
Ozis U (1998) Wasserbauten im Laufe von 4000 Jahren in der Turkei. In: Terzioglu A, Bayazit M (eds) In Memoriam – Prof.Dr. Kazim Cecen – Anma Kitabi – Gedenkschrift, Renk Ajans, Istanbul, p 137–150
Ozis U (1999) Historical dams in Turkey. Turkish national committee of international commission on Large Dams, Ankara
Ozis U (2001) Ein Einblick in die historischen Wasserversorgungsanlagen Istanbuls. In: Dinckal N, Mohajeri S (eds) Blickwechsel 2000. Technische Universitat, Berlin, pp 73–86
Ozis U (2015a) Water works through four millenia in Turkey. In: EWRA ninth world congress proceedings, 2015 Istanbul, European Water Resources Association, Istanbul, Keynote paper 4:1–13
Ozis U (2015b) Water works through four millenia in Turkey. Environ Process 3:559–573
Ozis U (2017) Historical water works in Turkey, In: International Water Association, Second IWA regional symposium – the past, present and future of world's water resources, Proceedings CD, 2017 Cesme, IWA & Izmir Institute of High Technology, Izmir, p 7–12
Ozis U, Arisoy Y (1986) Edirne water conveyance system constructed by the great Mimar Sinan. In: II.International congress on the history of Turkish-Islamic science and technology, 1986 Istanbul, vol 2, Istanbul Technical University, Istanbul, p 135–144
Ozis U, Arisoy Y (1987) Mimar Sinan'in suyollari. Dokuz Eylul Universitesi, Izmir
Ozis U, Arisoy Y (1996) Water conveyance systems of the great Architect Sinan. In: Aktas-Yasa A (ed) Uluslararasi Mimar Sinan Sempozyumu, 1988 Ankara, Ataturk Kultur, Dil ve Tarih Yuksek Kurumu, Ankara, p 241–256
Ozis U, Arisoy Y (2000) Sinan's aqueducts. In: International symposium: Bridging Large Spans (BLS) – from antiquity to present, 2000 Istanbul, International Association of Shell and Spatial Structures & Mimar Sinan University, Istanbul, p 11–20
Ozis U, Arisoy Y (2003) Muhendis ve Mimar Sinan'in suyollari. In: Dolen E, Kacar M (eds), Ihsanoglu E (pres), Turk Teknoloji Tarihi, 2001 Istanbul, Turk Bilim Tarihi Kurumu & ISKI, Istanbul, p 259–277
Ozis U, Atalay A (1999) Fernwasserleitungen von Ephesos. In: Friesinger H, Krinzinger F (eds); Brandt B, Krierer KR (red) 100 Jahre Osterreichische Forschungen in Ephesos, 1995 Wien, Osterreichische Akademie der Wissenschaften, Wien, p 405–411
Ozis U, Benzeden E (1977) Historical hydraulic works in Anatolia. In: XVII. IAHR congress proceedings, 1977 Baden-Baden, vol 6, International Association for Hydraulic Research, Baden-Baden, p 744–747
Ozis U, Harmancioglu N (1980) Some ancient water works in Anatolia. In: Gunay G (ed) International seminar on karst hydrogeology, Oymapinar 1979, Devlet Su Isleri, Ankara, p 380–395
Ozis U, Atalay A, Hasal, M, Atalay (Utku) V (1979a) Antike Fernwasserleitungen von Alabanda und Gerga. In: Leichtweis-Institut fur Wasserbau, Mitteilungen, vol 64, Technische Universitat, Braunschweig, p 1–8
Ozis U, Harmancioglu N, Baykan O, Unal I, Tolkun, M. (1979b) Flood flows and capacities of the historical Pergamon and Nysa tunnels in Anatolia. In: XVIII. IAHR Congress, 1979 Cagliari, vol 6, International Association for Hydraulic Research, Cagliari, p 695–698
Ozis U, Atalay A, Becerik M, Ozdikmen K (1998) Historical water conveyance systems to Ephesus. In: From Past to Present, Selcuk, First international symposium, 1997 Selcuk, Selcuk Municipality & Ege University, Izmir, p 125–136
Ozis U, Ozdemir Y, Kosova A, Corduk A (1999) Izmir'in tarihi su getirme sistemleri. In: Izmir su kongresi, 1999 Izmir, Muhendislik ve Diger Meslek Odalari Izmir Subeleri, Izmir, p 45–56
Ozis U, Atalay A, Becerik M, Ozdikmen K (2005a) Tunnelstrecken in Qanatbauweise: der Kenchrios-(Degirmendere)-Fernwassereitung nach Ephesus. Frontinus-Gesellschaft Hefter 26:293–300
Ozis U, Atalay A, Becerik M, Ozdikmen K (2005b) Aqua Iulia: Die Kenchrios-(Degirmendere)-Fernwasserleitung von Ephesos. In: Brandt B, Gassner V, Ladstatter S (eds) Synergia –

Festschrift fur Friedrich Krinzinger, vol.I, Osterreichisches Archaologisches Institut, Wien, p 213–219
Ozis U, Baykan O, Atalay A, Arisoy Y, Alkan A, Ozdemir Y (2006) Su yapilarinin tarihi gelisimi. In: Ozis U, Turkman F, Baran T (eds) Su yapilari C.I – Su kaynaklarinin gelistirilmesi, Dokuz Eylul Universitesi, Izmir, p 1.1–1.36
Ozis U, Arisoy Y, Alkan A, Ozdemir Y (2007) Brucken und Wasserbauten aus seldschukischer und osmanischer Zeit in der Turkei. Spur, Zurich
Ozis U, Baykan O, Atalay A, Arisoy Y, Alkan A, Ozdemir Y (2009) Water bridging civilizations through four millenia in Turkey. State Hydraulic Works – Ankara & Fifth World Water Forum – Istanbul
Ozis U, Harmancioglu N, Alkan A (2010) Tarihi su tunelleri ve benzeri ortme yapilarinin hidrolik kapasiteleri ve karsi gelen taskinlarin donegelim sureleri. In: Karahan H, Baykan O (eds) VI. Ulusal Hidroloji Kongresi, 2010 Denizli, Pamukkale Universitesi, Denizli, p x–xxi
Ozis U, Baykan O, Atalay A, Arisoy Y, Alkan A, Ozdemir Y (2014a) Historische Wasserbauten in der Turkei. WasserWirtschaft 7+8:83–86
Ozis U, Atalay A, Ozdemir Y (2014b) Hydraulic capacity of ancient water conveyance systems to Ephesus. Water Sci Technol Water Supply 6:1010–1017
Ozis U, Arisoy Y, Ozdemir Y, Alkan A (2016) Fernwasserleitungen des Osmanischen Hauptarchitekts Sinan. WasserWirtschaft 1:34–39
Ozis U, Alkan A, Ozdemir Y (2018a) Bedeutende romische Fernwasserleitungen in der Turkei. In: Wiplinger G, Letzner W (eds) Wasserwesen zur Zeit des Frontinus, Bauwerke – Technik – Kultur, Trier 2016, Frontinus Gesellschaft & Peeters, Berlin, p 141–158
Ozis U, Alkan A, Ozdemir Y (2018b) Muhendis ve Mimar Sinan. Dokuz Eylul Universitesi – Muhendislik Fakultesi, Izmir
Pamir H, Yamac I (2012) Antiokheia ad Orontes su yollari. Adalya 15:33–68
Pinar I (2011) Izmir'in su yollari (with transl. of Weber 1899), Buyuksehir Belediyesi, Izmir
Piras S (2001) Aspendos – Historical and archeological aspects. Frontinus-Schriftenreihe Hefter 24:109–113
Sahin F (2016) Die Wasserversorgung der Stadt. In: Iskan H, Schuler C, Aktas S, Reizenstein D, Schmolder-Veit A, Kocak M (eds) Patara. Lykiens Tor zur romischen Welt. Zabern, Darmstadt, p 59–63
Sarikaya HZ, Eroglu V, Altay A (2001) The history of sewerage services in Istanbul. In: Dinckal N, Mohajeri S (eds) Blickwechsel 2000. Technische Universitat, Berlin, pp 139–150
Schnitter N (1979) Antike Talsperren in Anatolien. Leichtweiss-Institut fur Wasserbau, Mitteilungen 64:1–8
Schnitter N (1990) Der Architekt Sinan als Wasserbauer. Wasser, Energie, Luft – Eau, Enérgie, Air 10:300–302
Schnitter N (1994) A history of dams. Balkema, Rotterdam
Stenton EC, Coulton JJ (1986) Oinoanda, the water supply and aqueduct. Anatol Stud 36:15–59
Strobel K (2013) Das Wassermanagement hethitischer und hellenistisch-romischer Zeit in Zentralasien. In: Wiplinger G (ed) Historische Wasserleitungen, 2011 Wien, Frontinus-Gesellschaft, Berlin, p 55–70
Tolle-Kastenbein R (1990) Antike Wasserkultur. Beck, Munchen
Tomasello F (1991) L'acquedotto romano e la necropoli presso l'istmo. Bretschneider, Roma
Turk S, Baykan O, Tanriover E, Kaya B (2010) Likya bolgesi eskil su muhendisligi sistemleri. In: Karahan H, Baykan, O (eds) VI. Ulusal Hidroloji Kongresi, 2010 Denizli, Pamukkale Universitesi, Denizli, p 110–125
Tuttahs G (1998) Milet und das Wasser, ein Beispiel fur die Wasserwirtschaft einer antiken Stadt. Universitat GH Essen, Essen
Tuttahs G (2001) Hinweise auf zentrale Wasserversorgungsanlagen der antiken Stadt Milet vor der romischen Kaiserzeit (1.-3. Jh.n.Chr.), Frontinuss-Gesellschaft Hefter 24:23–49
Tuttahs G (2007) Milet und das Wasser. Deutsche Wasserhistorische Gesellschaft, Siegburg

Tutuncuoglu F, Benzeden E (1979) Tarihi Topuzlu Bendi hazne hidrolojisi. In: I. Ulusal hidroloji kongresi, Istanbul Teknik Universitesi, Istanbul, p 420–428
Unal M (2001) Izmit antik su sistemleri ve Pasasuyu. Kocaeli Dokumantasyon Merkezi, Izmit
Unutmaz I (2013) Anadolu'da antik donemden gunumuze su muhendisligi harikalari. Wilo, Istanbul
Uytterhoeven I (2013) Aqueducts as suppliers of private water facilities in late roman Asia Minor. In: Wiplinger G (ed) Historische Wasserleitungen, 2011 Wien. Frontinus-Gesellschaft, Berlin, pp 139–162
Viollet PL (2000) L'hydraulique dans les civilisations anciennes. Ponts et Chaussées, Paris
Ward-Perkins JB (1955) The aqueduct of Aspendos, British School in Rome 23:115–123
Weber G (1898) Die Hochdruckwasserleitung von Laodikea. JdI 13:1–13
Weber G (1899) Die Wasserleitungen von Smyrna. JdI 14:4–25
Weber G (1904) Wasserleitungen in kleinasiatischen Stadten. JdI 19:86–101
Weber G (1905) Wasserleitungen in kleinasiatischen Stadten II. JdI 20:202–210
Wilberg W (1923) Der Aquadukt des C. Sextilius Pollio. In: FiE III, Wien, p 256–262
(Wiplinger 2006a, DSI 2008, IWA 2012, Wiplinger 2016 include several papers by different authors on various ancient water works in Turkey)
Wiplinger G (ed) (2006a) Cura Aquarum in Ephesus, 2004. Osterreichisches Archaologisches Institut, Wien & Leuven, Peeters
Wiplinger G (2006b) Wasserlabyrinth unter Ephesos. Antike Welt 6:69–77
Wiplinger G (2008) The Degirmendere Aqueduct to Ephesos. Anodos 8:393–400
Wiplinger G (2010) Die Wasserversorgung von Ephesos in byzantinischer Zeit. In: Daim F, Drauschke J (eds) Byzanz, das Romerreich im Mittelalter, Roman-German Centralmuseum, Mainz, p 593–613
Wiplinger G (ed) (2013a) Historische Wasserleitungen 2011 – Gestern – Heute – Morgen. Frontinus-Gesellschaft, Berlin
Wiplinger G (2013b) Der Degirmendere Aquadukt von Ephesos und seine Zukunft. In: Wiplinger G (ed) Historische Wasserleitungen, 2011 Wien, Frontinus-Gesellschaft, Berlin, p 105–129
Wiplinger G (2013c) Bypasse fur Reparaturarbeiten an Aquadukten, In: Die Wasserversorgung im antiken Rom, Frontinus-Gesellschaft, Berlin, p 197–207
Wiplinger G (ed) (2014) International Congress on the history of water management and hydraulic engineering in the Mediterranean region – Programme and Abstract, Frontinus-Gesellschaft & Osterreichisches Archaologisches Institut, Antalya
Wiplinger G (ed) (2016) Liciae Pamphyliae Pisidiae, Antalya 2014, Frontinus Gesellschaft, Berlin & Peeters, Leuven
Wiplinger G, Letzner W (eds) (2018) Wasserwesen zur Zeit des Frontinus, Bauwerke – Technik – Kultur, Trier 2016, Frontinus Gesellschaft & Peeters, Berlin
Yenigun K, Kurkcuoglu AC, Yazgan MS, Gerger R, Ulgen U (2013) From ancient to present: development of drinking water resources system in Sanliurfa. Water Sci Technol Water Supply 3:646–655
Yungul N (1957) Taksim suyu tesisleri, Istanbul Belediyesi Sular Idaresi, Istanbul

Chapter 3
Water Resources Development

Dogan Altinbilek and Murat Ali Hatipoglu

Abstract The annual technical and economical exploitable water potential of Turkey is calculated as 112 billion m^3 and 18 billion m^3 of this amount belongs to groundwater. Annual freshwater consumption is about 54 billion m^3, of which 74% is used for agriculture, 13% for domestic uses, and 13% for industrial uses. This sum corresponds to the development of only 48.2% of the available exploitable potential. Turkey has made great efforts to develop water resources for irrigation, power generation, flood control, and other purposes during the Republic era. The creation of dams and reservoirs has enabled Turkey to save the water from its brief seasons of rainfall to be used throughout the year. Construction of major projects are ongoing to utilize the available water potential of Turkey. On the other hand, taking into consideration Turkey's population of 80 million, the quantity of water per capita is 1400 m^3, which is only about one fifth of that of the countries in North America and Western Europe. According to water scarcity studies, Turkey will be facing a scarcity threat in near future. Turkey, like many countries, faces challenges in efficiently developing and managing its limited water resources while maintaining water quality and protecting the environment.

Keywords Water potential · Republic era · Exemplary projects · Water scarcity

D. Altinbilek
Past Director General, State Hydraulics Work (DSI), Ankara, Turkey

Past Vice President, World Water Council, Marseille, France

Civil Engineering Department, Middle East Technical University, Ankara, Turkey
e-mail: hda@metu.edu.tr

M. A. Hatipoglu (✉)
Investigation, Planning and Allocation Department, General Directorate of State Hydraulic Works (DSI), Ankara, Turkey
e-mail: mhatipoglu@dsi.gov.tr

N. B. Harmancioglu, D. Altinbilek (eds.), *Water Resources of Turkey*, World Water Resources 2, https://doi.org/10.1007/978-3-030-11729-0_3

3.1 General Characteristics

Turkey, being at the crossroads of Europe and Asia, is located between 26°–45° eastern longitudes and 36°–42° northern latitudes. The country, extending for around 1.650 km from west to the east and 650 km from north to the south, has a total surface area of 78 million ha. Turkey has borders with Iran, Georgia, Armenia, Syria, Iraq, Bulgaria and Greece. It is surrounded by the Black Sea, Mediterranean and Aegean Seas.

Turkey has a semi-arid climate with extremes in temperatures. In central parts of the country, winters are cold with late springs, while the coastal regions have a warm Mediterranean climate. Precipitation figures exhibit great variance throughout the country. The average annual rainfall is 574 mm, ranging from 250 mm in the southeast to more than 3000 mm in the northeast Black Sea area (Fig. 3.1).

Turkey's territory is divided into 25 drainage basins. Geographically, there is a large variation in the average annual precipitation, evaporation, and surface runoff parameters in the country. The rivers often have irregular regimes. Approximately 70% of the total precipitation falls between October and April, and summer months prevail less rainfall. The average annual precipitation figure for Turkey corresponds to an average of 450 billion m^3 of water per year.

Turkey's average annual runoff is approximately 181 billion m^3. However, under current technical and economic constraints, the annual exploitable potential is calculated as 112 billion m^3, including 18 billion m^3 of groundwater (DSI 2016).

Taking into consideration Turkey's population of 80 million, the quantity of water per capita is 1400 m^3. Countries regarded as being rich in water resources have 8–10 thousand m^3 water per capita per year. The available water per capita in Turkey is about one-fifth of that in water-rich countries.

Fig. 3.1 Annual total precipitation in Turkey (1981–2010). (MGM 2018)

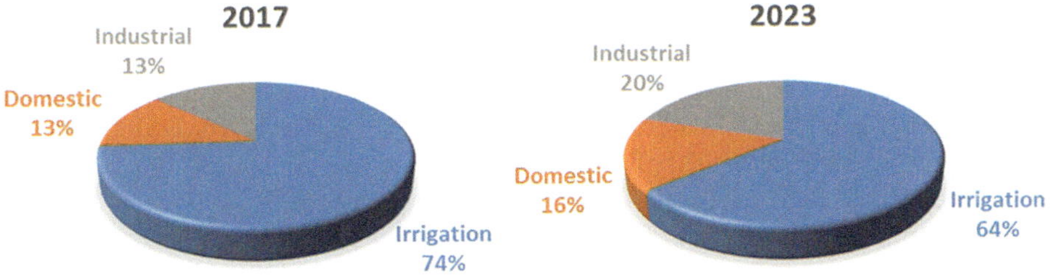

Fig. 3.2 Water consumption by sectors in Turkey

Of the 450 billion m^3 of annual precipitation, 274 billion m^3 evaporates from the surface and transpires through plants. 69 billion m^3 of precipitation directly recharges aquifers. There is a continuous interaction between surface runoff and groundwater, but it is estimated that a net of 28 billion m^3 of groundwater feeds the rivers. The amount of evapotranspiration is relatively larger with a ratio of 55%, considering the other water budget elements.

As previously mentioned, the annual average potential of economically exploitable water resources is 112 billion m^3. Annual freshwater consumption is about 54 billion m^3, of which 74% is used for agriculture, 13% for domestic uses, and 13% for industrial uses. This sum corresponds to the development of only 48.2% of the available exploitable potential. 38.5 billion m^3 (71.4%) of the consumed water is provided from surface waters and 15.5 billion m^3 (28.6%) from groundwater. It is projected that the share of water use for irrigation will reduce to 64% by 2023 (Fig. 3.2) (OSIB 2017b).

Groundwater studies have been carried out since 1956. According to investigations performed until 2017, the mean annual total safe yield is estimated at 18 billion m^3/year. Currently, on a yearly basis, 10 billion m^3 of the groundwater reserve is allocated for irrigation and 5.5 billion m^3 for domestic and industrial purposes.

The total land resources of Turkey is 78 million ha and almost one third of this, 28 million ha, can be classified as arable land. Recent studies indicate that 8.5 million ha is economically irrigable with the available technology. Presently, about 6.5 million ha of irrigation infrastructure has been developed. 4.21 million ha of this amount has been equipped with irrigation infrastructure by the General Directorate of the State Hydraulic Works (DSI), while the remaining 1.3 million ha have been developed by the General Directorate of Rural Services (KHGM) and Special Provincial Administrations. Approximately 1 million ha have been developed by small- scale privately owned irrigation schemes. It is targeted that the remaining 2 million ha of irrigation area will be irrigated by 2023 through DSI's efforts (DSI 2017).

The majority of irrigation is realized as gravity irrigation, which leads to low water efficiency. The use of water saving irrigation techniques (sprinkler and drip irrigation systems) has gained momentum in the last decade, but these still occur in limited ratios (about 23%) with respect to the overall irrigation practices. The proportion of the closed system in irrigation networks has reached 89% in projects under construction. The Turkish government has provided subsidies and interest-free loans to farmers in order to promote water saving irrigation technologies.

Domestic water is the second most important water use, with a total of 7 billion m^3 water abstracted from natural resources. By of the end of 2016, DSI completed 193 water supply projects supplying an approximate annual total of 3.87 billion m^3 of domestic and industrial water to 42 millions of people (DSI 2017).

Hydropower generation is another important water using sector. According to DSI, Turkey's gross theoretical hydropower potential is 433 billion kWh/year. As of 2017, the technically exploitable hydropower potential is reported as 216 billion kWh/year, while the economically feasible potential stands around 180 billion kWh/year with the development of new projects. There are currently 620 hydropower plants with the total installed capacity of 27,311 MW, and the average annual electricity production is about 95.3 billion kWh, providing approximately 52.9% of the total economically feasible potential (DSI 2017).

3.2 Development of Water Resources During the Republic Era

Anatolian settlements, known as the "cradle of civilizations", have always been founded on the banks of rivers and close to water sources since ancient times. Throughout history, there have been communities that had the opportunity to benefit from rivers. Anatolia is located at the crossroads of many civilizations, with many water facilities remaining from various periods of the last 4000 years, and some of them are still operating in good condition. During the Ottoman period, the construction of water structures was carried out generally by the foundations, and the first modern irrigation and drainage project, the Cumra Project, was realized between 1908 and 1914. Organized and continuous studies of water works were initiated with the establishment of the General Directorate of Public Works (Umur-u Nafia Muduriyet-i Umumiyesi) in 1914. Irrigation, reclamation, flood control, navigation, water storage, and distribution were among the duties of this General Directorate (Demir 2001).

After the foundation of the Turkish Republic in 1923, there have been great efforts to develop the country through the utilization of natural resources. Waters Directorate, and Bursa, Adana, Ankara, Edirne and Izmir Water Regional Directorates, were established under the General Directorate of Public Works in 1925 (Fig. 3.3). The map shown in Fig. 3.3 is a very old map remaining from the early years the Republic era but which still remains in the archives of DSI. Unfortunately, water resources projects couldn't gain adequate acceleration due to insufficient monetary allocations and lack of available water observations. The occasion of a severe drought that prevailed in 1929 led to the establishment of Waters General Directorate (Sular Umum Mudurlugu).

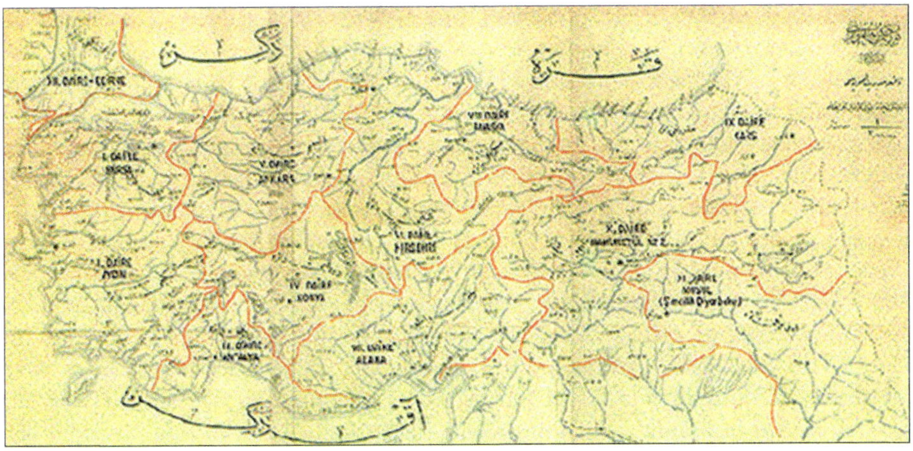

Fig. 3.3 Regional distribution of Water Directorates in the early years of republic era. (DSI 2018a)

Fig. 3.4 Cubuk-1 dam. (DSI 2018b)

The Water Works General Directorate (Su Isleri Reisligi) was established in 1939. After this date, the importance of water issues was understood well, and prefeasibility, planning, gauging, and water level recording studies of water resources were achieved.

The first dam built after the establishment of the Turkish Republic in 1923 was Cubuk-1 Dam, which was constructed from 1930 to 1936 to meet the water demand of Ankara (Fig. 3.4). Following the completion of Cubuk Dam, construction of

Golbasi Dam (1938) in Bursa, Gebere Dam (1941) in Nigde, Sihke (1948) in Van, and Porsuk1 (1949) in Eskisehir were started. Some projects on lakes were also carried out in this period, such as Golcuk in Isparta, Kesis, Doni, and Ermenis in Van, Isikli in Denizli, Marmara in Manisa, and Eymir lakes in Ankara.

As in many countries, the structural framework of water resources management in Turkey has been gradually formed in line with the goals of overall development. Studies for the development of water resources in Turkey during 1930s were initiated especially for the realization of small-scale irrigation projects. In the early years of the Republic, technology and productivity were low in agriculture, and infrastructure and public services were inadequate. Although the size and scope of these studies were expanded in a relatively short period of time, it was not until the late 1940s that basin wide hydrologic assessment and master plan studies were started. Activities in this respect gained a momentum and accelerated with the establishment of a number of institutions. These studies served as a blueprint and constituted a solid foundation at each stage of the basin-wide development efforts throughout Turkey (Eroglu 2007). Master plan and individual studies were followed by intensive design and construction works, which have so far contributed to the realization of many large and small dams.

Comprehensive water planning activities were accelerated in a systematic manner in the 1950s with the establishment of the General Directorate of State Hydraulic Works (DSI). DSI is the primary executive state agency responsible for planning, design, construction, and operation of hydraulic structures in order to develop the nation's overall water resources in a sustainable manner.

DSI was established by Law No. 6200 on 18th December 1953. As a public agency, it is responsible for four major tasks;

- supply water to settlements for domestic and industrial use;
- take necessary measures to prevent flood hazards from life and property losses;
- equip all economically irrigable land with modern irrigation facilities; and
- develop the technically and economically viable hydroelectric energy potential.

The major systematic aspect of water related activities in Turkey is central planning. At the national level, Five Year Development Plans (FYDP) are aimed at ensuring the optimum distribution of all kinds of resources among various sectors of the economy. With the establishment of the State Planning Organization (SPO) in 1960, comprehensive planning activities have been carried out in the country, which included the construction of physical structures to meet energy and food needs for the increasing population, as well as realizing socioeconomic development goals expected to provide welfare for citizens. From that moment, Turkey has made considerable progress in increasing water supply (Kibaroglu et al. 2011).

In the first plan period (1963–1967), about 400,000 ha area was irrigated, and this amount reached 1 million ha in 1980 by the projects developed by DSI. During that period, more focused attention was given to socio-economic development, based on water and land resources. Accordingly, 4.85 million ha of irrigation networks were constructed and opened to operation in Turkey till 2003. 2.7 million ha of this amount were developed by the General Directorate of DSI and the remaining by

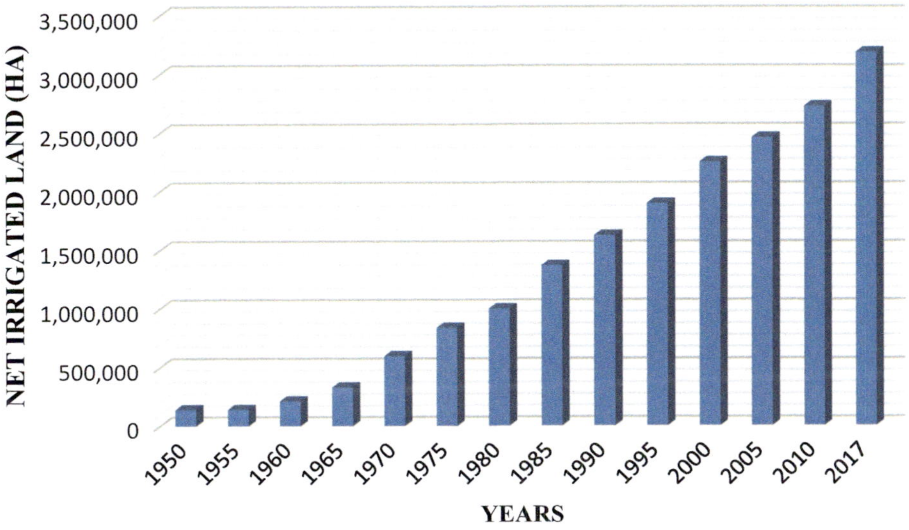

Fig. 3.5 Development of irrigation projects by DSI

the General Directorate of Rural Services (GDRS), and other institutions. However, the share of agriculture in total capital investments tended to decline over time, and relatively lower investment funds were allocated in agriculture compared to other sectors. In 1963, the share of agriculture in the total fixed capital investments was 12%, while it dropped to 5.3% in 1998 (DSI 2004). Today, of the 28 million ha of arable land, almost 6.5 million ha is irrigated by DSI (Fig. 3.5) and other institutions, and the objective is to increase the irrigated areas to 8.5 million ha by 2023 (OSIB 2017b).

Following the publication of Law No. 1053 in 1968, the protocols were signed between DSI with the two largest cities of Turkey, Istanbul and Ankara Municipality in 1969 and 1970, respectively. Then, the General Directorate of State Hydraulic Works started planning, project and construction studies on domestic water and industrial water to be supplied to cities exceeding 100,000 inhabitants. At that time, the number of cities with a population of over one hundred thousand was 14. Between 1968 and 2003, with the decisions of the Council of Ministers at various dates, in accordance with Article 10 of Law No. 1053, DSI was authorized to provide domestic water and industrial water to 43 cities. In 2003, 17 out of 43 domestic water projects were in operation, 6 of which were under construction, and the remaining 20 were in the final project and planning phase. As of the beginning of 2017, 193 facilities were built to provide domestic water for 42 million people.

The first power plant that was established and operated before the Republic era was the hydroelectric power plant with an installed capacity of 60 kW, built by Tarsus Municipality in 1902 (DSI 2004). After the declaration of the Republic, priority was given to national resources to meet the energy needs. As a result of these policies, the total installed capacity reached 78 MW, and the production potential

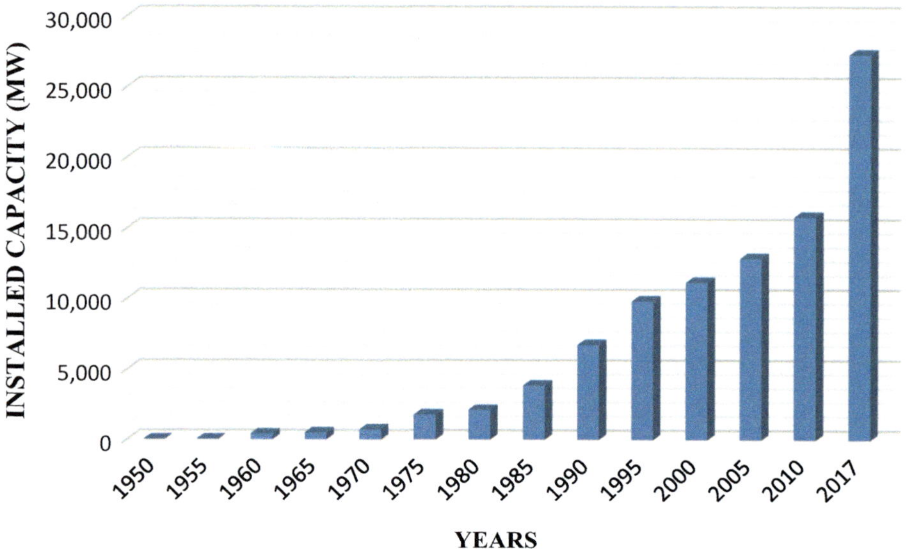

Fig. 3.6 Development of hydroelectric installed capacity

increased to 106 million kWh in 1930. However, the installed capacity of hydroelectric power plants was only 3.2 MW in these years.

In 1935, the Electricity Survey Administration (EIE) was established in order to determine the electricity needs of the country and to conduct surveys to meet the needs of hydroelectric and other energy sources. Between 1935 and 1953, the planning studies related to determination of the hydroelectric energy potential were performed, and the feasibility studies of Seyhan, Sariyer, Hirfanli, Kesikkopru, Demirkopru, Kemer dam, and HEPP's (hydroelectric power plants) were carried out by EIE. In 1940, according to the report prepared by EIE, there were 202 power plants with a total installed capacity of 242 MW. Only 28 of these power plants were hydroelectric power plants, and the share of HEPPs in the total energy production was 3.2% (DSI 2004). In 2003, energy generation was 26 billion kWh, and this amount reached 95.3 billion kWh/year with an installed capacity of 27,311 MW by 2017 (Fig. 3.6).

3.3 Exemplary Projects Completed Recently

In recent decades, Turkey has made great progress in water resource development for irrigation, power generation, flood control, and other purposes. The creation of dams and reservoirs has enabled Turkey to save the water from its brief seasons of rainfall to be used throughout the year for irrigation, energy, drinking, and sanitation purposes. Multipurpose water infrastructures have also enabled Turkey to regulate the flow of its rivers and to release sufficient amounts of water to downstream countries even during the dry seasons. Construction of major projects are ongoing to utilize the water potential of the country. Some recently completed or ongoing projects are summarized below.

3.3.1 The Turkish Republic of Northern Cyprus (TRNC) Water Supply Project

As an island, Cyprus has very limited water resources. Almost the whole water need of the island is met through groundwater resources. The existing groundwater potential of Cyprus diminishes day by day due to excessive withdrawal of groundwater resources, resulting in sea water intrusions into aquifers.

"The Turkish Republic of Northern Cyprus (TRNC) Water Supply Project" was launched in 2015 to satisfy the long-term water demand of the TRNC. This project will supply municipal and irrigation water from the south of Turkey to Northern Cyprus via a suspended pipeline across the Mediterranean Sea.

The project is technologically unique in that such a long-distance transmission line, with approximately 80 km distance of sea transition, is a first in the world. Sea transition was implemented by a very special 1.6 m diameter polyethylene pipe, hanging in 250 meters depth from the sea level by tying the pipe with anchoring cables to the sea bottom (Fig. 3.7).

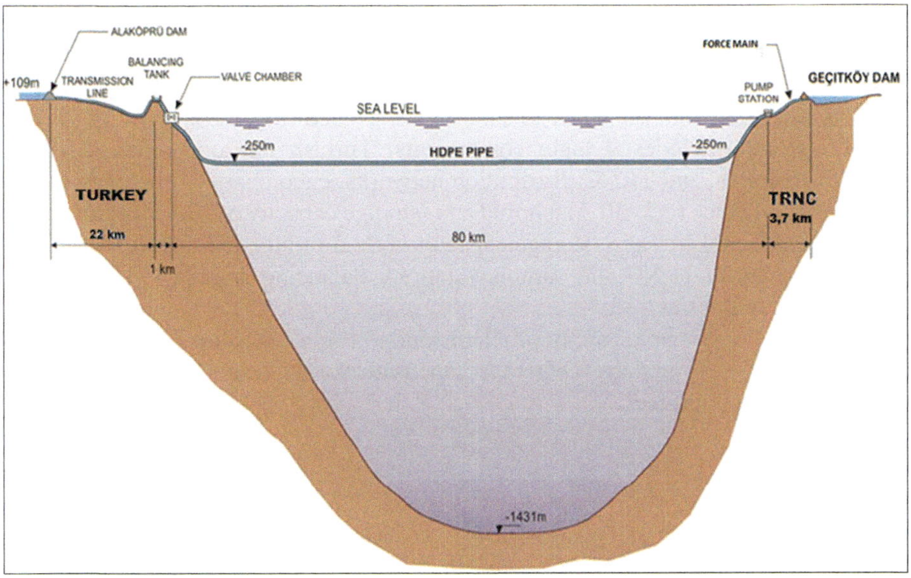

Fig. 3.7 Schematic presentation of the Turkey-Northern Cyprus (TRNC) water supply project

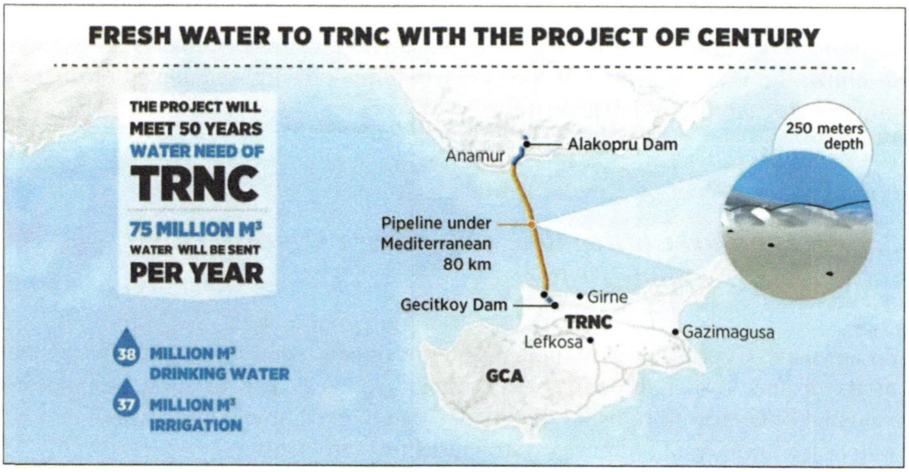

Fig. 3.8 TRNC water supply project. (Hidropolitik Akademi 2018)

Annually 75 million m^3 of water to be taken from Alakopru Dam, constructed on the Anamur Dragon Stream in Turkey, will be transmitted to the Gecitkoy Dam in the TRNC through an 80 km transmission pipeline. With the accomplishment of the project, completed by the State Hydraulic Works (DSI), TRNC is enabled to meet the water demand for the next 50 years. The transmitted water will be used for drinking, industrial and irrigation purposes and will be a significant contribution to the economic development of the region (Fig. 3.8).

Out of 75 million m^3 water to be transmitted from Alakopru Dam to the TRNC, 38 million m^3 water will be used as drinking water, and 37 million m^3 water will be used for drip irrigation of 7000 ha of agricultural land.

The pipes to be installed will be connected on the surface of the sea by means of vessels and filled with sea water to sink on the sea bed (Fig. 3.9).

The project includes 4 main components: Turkish territories, Sea Crossing, TRNC territories, and TRNC distribution networks, encompassing the construction of the concrete face rock-fill Alakopru Dam (storage capacity of 130.5 million m^3) in Turkey and Gecitkoy Dam (storage capacity of 26.5 million m^3) in TRNC, a water treatment plant in TRNC and pumping stations, balancing tanks, and transmission lines in both countries.

TRNC, having water scarcity problems due to limited resources of groundwater and surface water, will be able to compensate for its water need through the realization of this project.

Fig. 3.9 Installation of pipes. (DSI 2018c)

3.3.2 Istanbul Melen Water Supply Project

Istanbul, the largest industrial, trading, tourism and cultural center of Turkey, is a megacity with a population of 15 million. Istanbul is located on both the European and the Asian continents, divided by the Bosphorus Strait. The coastal location of Istanbul has its advantages and disadvantages in terms of water supply and wastewater disposal.

The current population of the Istanbul metropolitan area is estimated at approximately 15 million. Population growth in the city is almost twice the overall rate for the whole of Turkey because of large in-migration. The population of the city has experienced an average annual growth rate of 4.5% over the last half a century.

Istanbul experienced severe water shortages in the early 1990s. A population of around 400,000 annually migrates to Istanbul. It can be said that Istanbul's growth is equal to that of a medium sized city in Europe. Thus, Istanbul's problems such as potable water, sewerage, and treatment of wastewater remained unsolved for many years (Altinbilek 2006).

The annual water need of Istanbul in 2017 was 1082.04 million m^3/year. The water demand in 2040 is estimated to be 1790.32 million m^3/year. The available water resources developed on the European side are 400 million m^3/year and 566 million m^3/year on the Asian side. At present, 65% of the population lives on the European side and 35% lives on the Asian side. Contrary to the population distribution, potentially more water is available on the Asian side. To overcome the geographical misdistribution, water is transported under the sea from the Asian side to the European side.

The 1971 Master Plan for Istanbul water supply suggested that 11 water sources could be developed for Istanbul. Six of these have already been put into service. The General Directorate of State Hydraulic Works (DSI) has begun two large projects in

Table 3.1 Istanbul Melen water supply project (DSI 2012)

Stage	Supplied Water hm³/year	Supplied Water m³/s	Water Intake/ Storage Facility	Transmission Main Capacity (m³/sec)	Pumping Capacity (m³/sec)	Water Treatment Plant Capacity (m³/day)
I	268	8.5	Regulator intake	8.5	8.5	720,000
II	307	10	Dam	15	10	800,000
III	307	10		15	10	800,000
IV	307	10		–	10	800,000
Total	1180	38.5		38.5	38.5	3,120,000

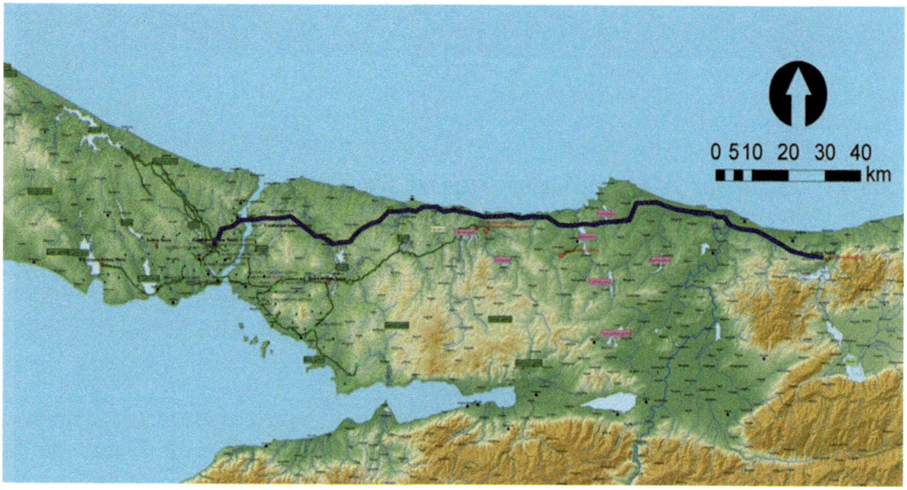

Fig. 3.10 The Melen system

order to meet the water needs of Istanbul in the medium and long term, namely the Yesilcay project and the Melen project.

The Melen System is developed for covering the medium term and long term water demand of Istanbul. The system, which is divided into four stages (Table 3.1) to supply 1.18 billion m^3 of water per year, is expected to secure Istanbul's water needs until 2040. Its source is the Melen River, which is located 170 km to the east of Istanbul. Raw water will be pumped to a treatment plant via a transmission line, and treated water will be conveyed to a service reservoir via a tunnel and a pipeline under the Bosporus (Fig. 3.10).

The Melen system will supply 268 million m^3 of water per year to Istanbul in its first stage. At the end of the fourth stage, the system will provide 1.18 billion m^3 water per year to Istanbul.

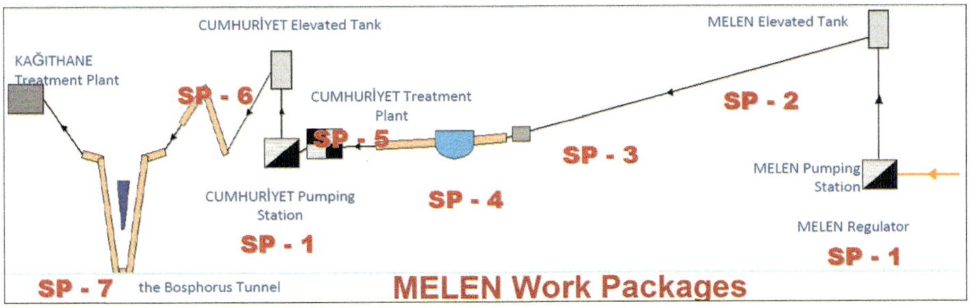

Fig. 3.11 Istanbul water supply project. (DSI 2012)

Fig. 3.12 Melen dam. (DSI 2018d)

The first stage of the Melen project involves 189 km pipelines, 33 km tunnels and a water treatment capacity of 720,000 m^3/day. The domestic water demands of an additional 2.75 million people will be supplied (Fig. 3.11).

At the second stage of the project, Melen Dam (Fig. 3.12) will be constructed. Units will be added to pump stations and water treatment works, and the second pipeline will be laid down.

3.3.3 GOLSU Project

The GOLSU project, initiated with the slogan of "1000 Small dams in 1000 Days", which aims to increase the water storage capacity of Turkey and to realize irrigation projects in a short time in rural areas other than large irrigation projects, was started on 04/04/2012 and concluded by the end of 2014.

The first small dam in the Republican era was Van-Merkez Sihke dam, which was built in 1958. During the 58 years between 1954 and 2012, 461 small dams were built, while 1000 small dams and irrigation projects were realized with the Golsu Project over a 3-year period. Approximately 170,000 ha of agricultural lands were irrigated, 587 million m^3 of water was stored, 27 million m^3 of drinking water was supplied, and 17.1 thousand ha of land were protected by the project.

The GOLSU Project has social, economic and environmental dimensions and is regarded as a nationwide socio-economic development project. Benefits of the GOLSU Project can be summarized as follows;

- provide employment opportunities for 450,000 people,
- contribute to the development of agriculture and animal husbandry by preventing the migration from the countryside,
- increase in income in the amount of 1.7 Billion TL per year,
- prevention of flood damages and soil erosion,
- supply of animal drinking water,
- keeping the groundwater potential at a safe reserve level,
- dissemination of nationwide aquaculture production,
- reducing costs of production and providing energy savings by converting pumped irrigations to gravitational irrigations,
- supply of water for firefighting,
- creation of recreational areas,
- dissemination of afforestation,
- reduction of the adverse effects of global climate change by increasing the current storage capacity of the country.

After the successful implementation of the GOLSU project, the second stage, "1071 Small Dams Project", started on 01/01/2016 and will be completed in 2019. Through this project, it is planned to irrigate 320,000 ha of agricultural land and store approximately 1.8 billion m^3 of water in the small dams.

3.3.4 The Southeastern Anatolia Project (GAP)

Southeastern Anatolia Region of Turkey was economically the least developed region and remained far behind the rest of the country for years. The region covering Adiyaman, Batman, Diyarbakir, Gaziantep, Mardin, Siirt, Sanliurfa, Sirnak and Kilis provinces in the Southeastern Anatolia Region is defined as the "GAP Region"

Fig. 3.13 Provinces included in Southeastern Anatolia Project (GAP)

(Fig. 3.13). Surrounded by Syria in the south and by Iraq in the southeast, GAP has a total area of 75,358 square kilometers, which constitute 9.7% of Turkey's total area. GAP lies at the lower reaches of the Euphrates (Firat) and Tigris (Dicle) rivers within Turkey. The Euphrates and Tigris, as the two main branches of the Shatt-al-Arab basin, are the major river systems running through Southeastern Turkey. The GAP area is rich in water and soil resources; the Euphrates and Tigris rivers represent over 28% of Turkey's surface water. The mean annual flow of the Euphrates River is estimated to be 33.6 billion m^3, and that of the Tigris River is estimated to be 50.9 billion m^3 (excluding the flow of Karun River from Iran). Some 98% of the Euphrates River runoff originates in the highlands of Turkey, while the rest of its catchment in lower arid regions makes little contribution to the river (Altinbilek 2004).

The development potential of both the Euphrates and Tigris Rivers was recognized in the 1960s, and the idea of harnessing their waters for irrigation and hydropower generation emerged. Towards the end of the 1970s, the General Directorate of State Hydraulic Works (DSI) planned the 'Southeastern Anatolia Project'—a series of land and water resources development projects on the two rivers. Through a Master Plan in 1989 and a significant revision in 2002, the Southeastern Anatolia Project was transformed from a land and water resources development project into a large-scale, multi-sectoral regional development project to be implemented in nine of Turkey's provinces, which came to be known as the Southeastern Anatolia Region.

The Southeastern Anatolia Project is the most comprehensive regional development project ever implemented in Turkey. Beyond the dams, hydroelectric power plants, and irrigation schemes on the Euphrates and the Tigris rivers, GAP, as an integrated project, envisages the development of communication, housing, industry, education, health, tourism and other services. On the basis of the relative position of the region in the nation's socio-economy, the development potentials and problems, and the national development aspirations, the objectives for the development of the GAP region are set as follows.

- To develop all the land and water resources in the region, in order to achieve accelerated economic and social development,
- To alleviate disparity between the region and other regions by increasing production and welfare levels in the region,
- To increase the productivity and employment capacity in the region,
- To meet the increased need for infrastructure resulting from population explosion and urbanization,
- To organize economic and physical infrastructure in rural areas in such a way as to utilize the resources in the most useful ways and to direct urban growth to desired directions,
- To contribute to the national objectives of sustained economic growth and export promotion by efficient utilization of the region's resources.

As a large-scale development project related to water, GAP is a combination of 13 projects for agriculture, energy, and domestic water supply. At full development of these projects, of which 6 are located in Tigris basin and 7 in Euphrates basin, 27 billion kWh of hydroelectric energy will be generated annually with an installed capacity of 7500 MW. 1.7 million ha of land will be irrigated, and 22 dams and 19 HEPPs will be constructed on the main and tributary branches of Tigris and Euphrates rivers.

Ataturk dam (Fig. 3.14) is the most significant structure within the GAP Project. Ataturk Dam, functional since 1992, is widely considered to be not only the largest dam in Turkey, but also one of the largest in the world. It is a rock-fill dam with a clay core. Its height from the foundation is 169 m, and the volume of the reservoir is 48,700 hm^3. The dam generates 8900 GWh of electric power per year and supplies irrigation water to Sanliurfa, Harran, Mardin, Ceylanpinar, and Siverek-Hilvan plains.

Ataturk Dam povided many benefits for people living in the project area as well as for the country as a whole. Its direct benefits have been in terms of energy generation and irrigated agriculture, with indirect benefits through the promotion of urban, industrial, agricultural and commercial activities, mostly in Sanliurfa Province (Altinbilek and Tortajada 2012).

The social and economic impacts of the construction of the Ataturk Dam and its reservoir have been substantial through a variety of pathways. Both the dam and the reservoir have acted as an engine for economic growth and development in a historically underdeveloped area that has flourished since the construction of the dam.

Fig. 3.14 Ataturk dam. (DSI 2018e)

Fig. 3.15 Provinces included in Konya plain project (KOP)

3.3.5 Konya Plain Project (KOP)

Konya Plain Project (KOP) region includes small and large size closed plains of the Central Anatolia. The areas of Konya, Karaman, Aksaray, and Nigde provinces are included within the Konya plains (Fig. 3.15).

Konya plain has typical climatic features of the Central Anatolia. The average annual precipitation of the basin is 398 mm which is less than Turkey's average. The basin is the most distressed region of Turkey with respect to water sufficiency and precipitation. The shortage in precipitation and the resulting drought affect nearly every sector and cause a slowdown in regional development, a decrease in the income of the farmers, problems in the provision of basic foods, serious losses in the industries in relation with agricultural production, and unemployment due to the decrease in production.

Konya Plain Project (KOP) is a sustainable and integrated development project for Central Anatolia. KOP, which is vital for the region, aims to meet the regional irrigation, domestic and industrial water needs, prevent excess groundwater extraction, ensure balance in the groundwater table, raise agricultural yields, introduce modern irrigation systems, promote stock breeding, protect the environment, and generate hydroelectric energy. Further, the project aims to minimize the differences in levels of development in the region and between regions. The project will serve to improve the competitive power of the region through increased economic and social capacity and will enhance the welfare of the citizens.

The stock of 3 million ha agricultural land in the region corresponds to 12.4% of that of the total country. The program, aimed to be completed prior to 2020, includes 14 large irrigation network systems, 3 drinking supply projects, and 1 hydropower plant. With the completion of the ongoing and planned irrigation projects, the irrigated area in the region will reach 1.1 million ha, creating new employment opportunities for nearly 100,000 people. In addition 164 million m^3 water will be supplied to 1.5 million residents.

3.3.6 Water for Africa

The continent of Africa is home to 54 countries where around 1/7 of the Earth's population lives. However, due to geographical conditions, long periods of drought, lack of infrastructure and difficult social and political situations, people are faced with problems in accessing clean and healthy drinking water. Ensuring water and water hygiene is an important issue for the development of Africa. Despite the world's advanced technology, the problem in this area remains a serious problem. Today's global challenges focus on issues that require immediate action for humanitarian needs worldwide.

Turkey assumes a humanitarian responsibility and provides financial and technical assistance in the water sector with a specific focus in Africa. The main target of the Turkish water aid is to ensure sustainable safe drinking water and sanitation for vulnerable people living mainly in the crisis areas without access to clean drinking water and improved sanitation. Turkey shares its own expertise by tailoring them to fit the specific needs and development priorities of the partner countries in Africa.

The Turkish Cooperation and Coordination Agency Directorate (TIKA) is in charge of coordination of all aid activities. TIKA's mission is to contribute to poverty eradication and sustainable development in partner countries.

The work in the field of water and sanitation in Africa can mostly be gathered under two main headings

- Well digging projects being conducted in rural areas to provide healthy drinking water.
- The improvement of water distribution networks towards providing healthy access to drinking water and maintaining water sanitation.

Through TIKA's 8 Program Coordination offices in Africa, a variety of projects have been implemented on a regional and countrywide basis in line with the needs of the community, with drinking water having the top priority (TIKA 2018).

Being the biggest organization in the field of water sector in Turkey, DSI, with the coordination of TIKA, has launched various works in African Countries to find solutions to the existing water shortage problem, to develop limited water sources, and to develop projects for the future. Thus, through cooperation of DSI with TIKA and together with the support of NGO's, 487 wells have been constructed to supply drinking water for 1,7 million people in Africa (namely in Ethiopia, Sudan, Djibouti, Niger, Mali, Burkina Faso, Mauritania and Somalia).

DSI has undertaken the development of large-scale hydraulic works as well and initiated the construction of a dam on the Ambouli River. The purpose of the dam is to provide domestic water to the capital city of Djibouti and to protect the city from potential floods. The rock-fill type dam project, with a talweg height of 38 m and a total reservoir volume of 14,37 hm^3, is due to be completed at the end of 2018. The total project cost of €11 million is fully financed by the Turkish Government (MoFWA 2017).

DSI has also provided technical assistance (e.g. advisory, investigation, detailed design) and granted technical equipment to many countries in Africa for the sustainable development and use of water resources.

Training programs in the field of water and sanitation to African countries have been another important activity coordinated by TIKA. DSI and Turkish Water Institute (SUEN) have been organizing training programs to share the Turkish knowledge and experience through lectures in planning in the water sector, integrated water resources management, drinking water, and wastewater treatment technologies.

3.4 Evaluation of Water Sufficiency

Water is the most vital and, more importantly, the finite and irreplaceable resource, which is fundamental to human life. However, due to population growth and water resources degradation, there is a need to preserve and manage the limited water resources.

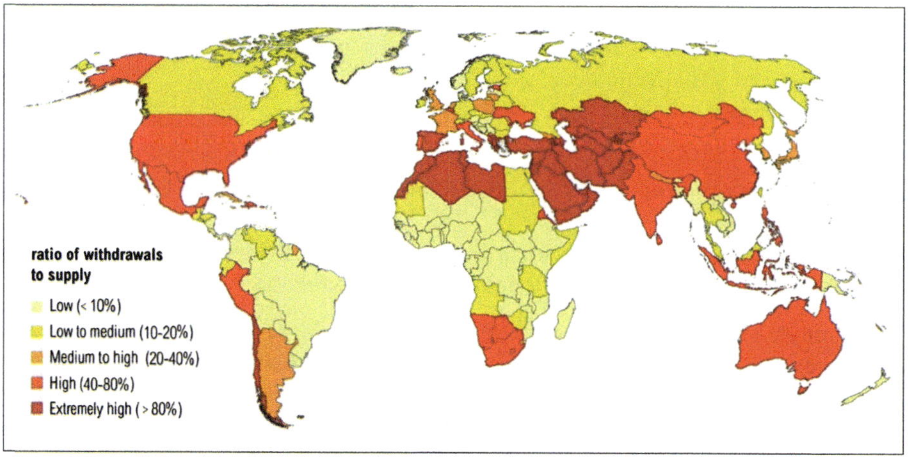

Fig. 3.16 Water stress by country. (WRI 2015)

The population of the world was 6.1 billion at the beginning of twenty-first century, and it is predicted to exceed 9 billion by 2050. While the population is increasing steadily, water resources in the world remain the same. This leads to an increase in water demand and a lack of water to meet the ever-increasing demand.

Water scarcity, which is generally accepted as the lack of access to adequate quantities of water for human and environmental uses, is increasingly being recognized in many countries as a serious and growing concern (IWMI 2012). Water scarcity is becoming an important problem in different parts of the world, and it is predicted to expand and affect a greater part of the world population and environment. In that context, many researchers and scientists have developed indices to forecast and measure the current water scarcity in different parts of the world.

The World Resources Institute (WRI) scored and ranked future water stress in 167 countries by 2020, 2030 and 2040 using climate models and socioeconomic scenarios. WRI predicted that 33 countries around the world will experience extreme water stress by 2040 (Fig. 3.16). It is possible to conclude from Fig. 3.16 that more than half of world population will be living under medium to extremely high water stress.

One of the frequently used measures of water scarcity is the 'Falkenmark indicator' or 'water stress index'. This method specifies water scarcity in terms of the total available water resources to the population of a region, measuring scarcity as the amount of renewable freshwater that can be used per capita per year. If the amount of renewable water in a country is less than 1700 m^3 per person per year, that country is said to be experiencing water stress; below 1000 m^3, it is said to be experiencing water scarcity, and below 500 m^3, absolute water scarcity (IWMI 2012).

The method developed by International Water Management Institute (IWMI) considers the portion of the renewable water resource available for human

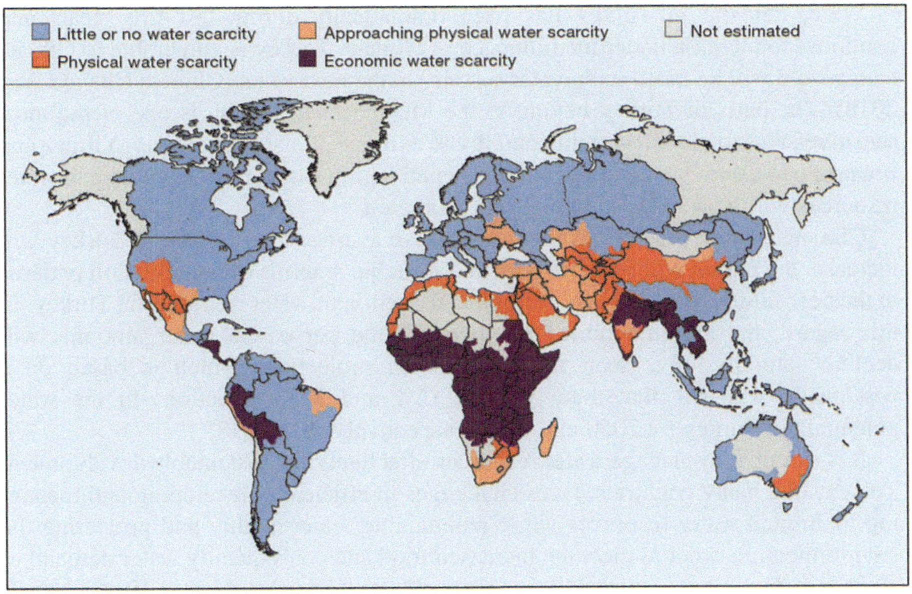

Fig. 3.17 Areas of physical and economic water scarcity. (IWMI 2007)

requirements with respect to the primary water supply provided. Instead of total water intake, it takes into account the water withdrawn for consumption and evaluates the country's adaptability to consumption, such as the capacity for infrastructure enhancement (Nepomilueva 2017). IWMI mapped countries based on water scarcity categories, and the results are presented in Fig. 3.17.

According to all water scarcity studies described above, Turkey will be facing scarcity in near future. Turkey has one of the highest levels of water security threat among the European countries, having its per capita water resources expected to decrease by nearly one third by the midcentury. The availability of water per capita in Turkey is only about one fifth of that of the countries of the North America and Western Europe. Turkey has 25 important hydrological basins, some of which are densely populated and face high or very high levels of water stress.

3.5 Problems and Outlook for the Future

Turkey has a semi-arid climate, and the temperature, precipitation, evaporation and surface runoff parameters are geographically variable. Water is not always at the right place at the right time to meet present and anticipated needs, so the country experiences a shortage of water (Turkey Country Report 2003).

Water demand in Turkey has risen dramatically during last fifty years and continues to increase under the influence of climate. Turkey is vulnerable to climate change and will be faced with water scarcity in the next years (Climate Change Post 2018). The parts of Turkey belong to the Mediterranean basin, is one of the most sensitive areas to global warming and future extreme climate conditions (Yilmaz and Imteaz 2014), and it is assumed that the negative impacts of climate change to water resources will be considerably high in this region.

Climate change projections indicate that the average temperature of Turkey will increase, the region will be more arid, and unstable in terms of precipitation patterns in the near future. This will also result in a reduction of water resources in Turkey. In this regard, the aforementioned numbers for the per capita water amounts will decline naturally. It is seen from the model projections which is based on a pessimistic scenario, that there will be 16% and 27% reductions in the water potentials in Turkey by 2050 and 2075, respectively (Sen 2013).

It is essential to manage water resources effectively for sustainable development. Turkey, like many countries, faces challenges in efficiently developing and managing its limited water resources while maintaining water quality and protecting the environment. In order to meet the increased food and consequently water demand of all sectors in parallel with rapidly growing and urbanizing population, Turkey needs to continue to develop its water resources (Baris and Karadag 2007).

Agricultural sector is the greatest user of water in Turkey. The inefficient use of water in agriculture results in over-abstraction of water from both surface and groundwater in several river basins. Currently, water delivery systems in irrigation schemes are comprised of open systems (77%) and piped systems (23%), and water losses due to leakage and evaporation in open canals is high, compared to that in closed systems. Additionally, surface irrigation methods (flooding, furrow, border, etc.) are widely applied in many regions (almost 70% of total irrigated areas). Irrigation efficiency, measured at scheme scale, is about 42% in irrigated areas operated by DSI (OSIB 2017b).

Turkey's agricultural sector needs to become more water efficient. Turkey has various national strategies, plans, and programs dealing with water resources management. The 10th Development Plan of Turkey is one of the main plans to set clear objectives and targets for the sustainable use and effective management of water resources. In order to achieve 2023 targets and the objectives of the 10th Development Plan, "Enhancing Efficiency of Water Use In Agriculture Programme" is enacted by the Ministry of Forestry and Water Affairs. The targets of this program include:

- The percentage of irrigated lands where water saving modern irrigation techniques are used with respect to the total irrigated lands that are developed via DSI investments is targeted to increase from 20% to 25% during the Development Plan period.
- Increasing the percentage of irrigated lands by DSI from 62% to 68% and irrigation efficiency from 42% to 50% during the Plan period

- Increasing the number of water saving modern irrigation systems in use by 10% per annum
- Decreasing the use of groundwater by 5% during the Plan period.

The current hydrometric network is not sufficient in terms of quality and quantity, and reliable data are inadequate for hydrological forecasts (OSIB 2017a). This prevents monitoring of flood or drought situations and constrains taking preventive measures to reduce damages.

Water resources management in Turkey has a very fragmented character in terms of administrative and legal structures. There are too many regulations related to surface waters, and there are several organizations responsible for water management issues in Turkey. It is often possible to encounter duplicate work for the same purpose, as many institutions and organizations operate in the same field. Because cooperation among institutions is problematic, it is observed that different management plans are developed for the same area, and simultaneous applications of these plans have negative consequences (OSIB 2017a).

There are some studies regarding the restructuring of water legislation in Turkey in recent years. But, a single wide-scoped water law has not come into force yet. This brings several challenges and problems for the management of water resources. Current laws and regulations should be gathered under a single framework and a "water law" should be issued.

References

Altinbilek D (2004) Development and management of the Euphrates and Tigris Basin. Int J Water Resour Dev 20(1):15–33

Altinbilek D (2006) Water management in Istanbul. International Journal of Water Resources Development 22(2):241–253

Altinbilek D, Tortajada C (2012) The Ataturk dam in the context of the Southeastern Anatolia (GAP) project, impacts of large dams: a global assessment, water resources development and management, pp 171–179

Baris ME, Karadag A (2007) Water resources management issues and recommendations. J Appl Sci 7(24):3900–3908

Climate Change Post (2018) Fresh water resources Turkey, https://www.climatechangepost.com/turkey/fresh-water-resources/. Accessed 12 July 2018

Demir A (2001) Su ve DSI Tarihi, Devlet Su Isler Vakfi, Ankara

DSI (2004) Dunden Bugune DSI, Ankara

DSI (2012) Water and DSI, Ankara

DSI (2016) Turkiye'nin Yuzeysuyu Potansiyeli, Ankara

DSI (2017) DSI Faaliyet Raporu

DSI (2018a) http://www.dsi.gov.tr/kurumsal-yapi/hakkimizda. Accessed 12 July 2018

DSI (2018b) http://www.dsi.gov.tr/projeler/cubuk-baraji. Accessed 12 July 2018

DSI (2018c) http://www.dsi.gov.tr/haberler/2015/07/24/. Accessed 12 July 2018

DSI (2018d) http://www.dsi.gov.tr. Accessed 12 July 2018

DSI (2018e) http://www.dsi.gov.tr/projeler/ataturk-baraji. Accessed 12 July 2018

Eroglu V (2007) Water resources Management in Turkey, international congress on River Basin management, Antalya, Vol 1, p 321–332, March 2007

Hidropolitik Akademi (2018) http://www.hidropolitikakademi.org/en/turkey-launches-project-to-supply-fresh-water-to-cyprus.html. Accessed 12 July 2018
IWMI (2007) Water for food, water for life: a comprehensive assessment of water Management in Agriculture. Earthscan/Colombo, London
IWMI (2012) http://www.iwmi.cgiar.org/News_Room/pdf/Understanding_water _scarcity.pdf. Accessed 12 July 2018
Kibaroglu A, Scheumann W, Kramer A (2011) Turkey's water policy. National framework and International Cooperation. Springer, Berlin
MGM (2018) https://www.mgm.gov.tr/veridegerlendirme/aylik-normal-yagis-dagilimi.aspx. Accessed 12 July 2018
MoFWA (2017) Turkey's water aid perspective, republic of Turkey The Ministry of Forestry and Water Affairs, November 2017
Nepomilueva D (2017) Water scarcity indexes, Bachelor's Degree Thesis, Helsinki Metropolia University of Applied Sciences
OSIB (2017a) Su Kaynaklarinin Gelistirilmesi ve Hidroloji Calisma Grubu Raporu, 2. Ormancilik ve Su Surasi, 5–7 Mayis 2017
OSIB (2017b) Sulama Calisma Grubu Belgesi, 2.Ormancilik ve Su Surasi, 5–7 Mayis 2017
Sen OL (2013) A holistic view of climate change and its impacts in Turkey, December, 2013
TIKA (2018) http://www.tika.gov.tr/en/news/tikas_water_projects-8592. Accessed 30 July 2018
Turkey Country Report (2003) Ankara, March 2003
WRI (2015) http://www.wri.org/blog/2015/08/ranking-world%E2%80%99s-most-water-stressed-countries-2040. Accessed 12.07.2018
Yilmaz AG, Imteaz MA (2014) Climate change and water resources in Turkey: a review. Int J Water 8(3):299–313

Chapter 4
Climate and Drought in Turkey

Murat Turkes

Abstract The aim of this section of the book is to make a scientific synthesis of the climate of Turkey along with the aridity and drought conditions, in a contemporary physical geographical, hydro-climatological and meteorological approach. The Mediterranean macroclimate, including much of Turkish territory, mainly results from the seasonal alternation between mid-latitude frontal cyclones associated with polar air masses during the winter, the subtropical high pressure systems link to subsiding maritime tropical air mass over the mid-north Atlantic, and the continental tropical air masses over the Northern Africa during the summer. In addition to these pressure and circulation conditions, the north-western surface and/or circulation-based extensions of the Asiatic monsoonal low pressure system cause generally very dry and hot conditions over the Eastern Mediterranean and the Middle East regions, including the southern and central parts of Turkey. According to the Köppen-Geiger climate classification, the southern and western portions (Marmara, Aegean, Mediterranean and South-eastern Anatolia regions) of Turkey are characterized mainly with the dry summer sub-tropical Mediterranean climate (Csa). All of these regions are also influenced by mid-to-high degree drought probability and drought risk, while the Black Sea Region is characterized with a mid-latitude temperate climate with a low-level drought probability and risk. The probabilities of being extremely-dry evidently indicate maximum values on the coast of Mediterranean Sea and over the border zone between Turkey and Syria with a highest probability of about 0.27. Continental steppe and cold snowy climates are dominated over most of the Central and Eastern Anatolia regions.

Keywords Mediterranean and Turkey climates · North Atlantic oscillation · Composite precipitation anomalies and 500 hpa anomalous circulations · Drylands · Drought probability · Vulnerability/risk of Turkey

---

M. Turkes (✉)
Center for Climate Change and Policy Studies, Bogazici University, Istanbul, Turkey
e-mail: murat.turkes@boun.edu.tr

N. B. Harmancioglu, D. Altinbilek (eds.), *Water Resources of Turkey*, World Water Resources 2, https://doi.org/10.1007/978-3-030-11729-0_4

4.1 Introduction

The present Mediterranean-type macroclimates of the Earth mainly result from the seasonal alternation between mid-latitude frontal cyclones associated with polar air masses during the winter, and subtropical high pressure systems related with subsiding maritime and continental tropical air masses during the summer. When considering the 'true' Mediterranean macro-climate, in addition to these pressure and circulation conditions, the north-western surface and/or circulation-based extensions of the Asiatic monsoonal low pressure system (Fig. 4.6) cause generally very dry and hot conditions over the Eastern Mediterranean and the Middle East regions, including the southern and central parts of Turkey, during the 5-month warm period of the year from May to September (Turkes 1998, 1999, 2003, 2010).

Nature, magnitude and variability of precipitation amounts and air temperatures are generally associated with the location, variation and activity of the atmospheric centers of action throughout the year (except summer) over most of the Mediterranean regions and Turkey (Iyigun et al. 2013; Kutiel et al. 2001; Kutiel and Turkes 2005; Turkes 1996, 1998, 2003; Turkes et al. 2002a; Turkes and Erlat 2003, 2005, 2008, 2009; Turkes and Tatli 2009; Turkes et al. 2009; Xoplaki 2002; Xoplaki et al. 2004; Erlat and Turkes 2012, 2013; etc.). In addition to the humid-temperate Black Sea climate and the continental inland climates, there are various Mediterranean and Mediterranean-like climate types in Turkey. Two major factors cause this evident climate diversity in Turkey. First, Turkey is situated in a transition zone that is under the influence of various atmospheric disturbances and weather types originating from polar and tropical regions. The second factor is the complexity of topographical features, rapid elevation changes within short distances, the land-sea interactions, and finally thermodynamic influences and modifications caused directly by the Mediterranean Sea at the south and the Black Sea at the north. The zonal soils in Turkey are distributed over the country in accordance with the spatial distribution patterns of the major climate types along with the lithology (parent material), geomorphology and vegetation formations.

The aim of this chaper is to provide a comprehensive picture of the climate types and aridity, and spatial and temporal variability in dryness and drought conditions of Turkey and its relatively nearby surrounding regions with different physical geographical conditions. These regions are characterized mainly with the mid-latitude and the Mediterranean macro climates primarily according to the Köppen-Geiger climate classification, the United Nations Environmental Program and the United Nations Convention to Combat Desertification Aridity Index (UNEP/UNCCD *AI*), and the Standardized Precipitation Index (*SPI*). Climatological and meteorological assessments have consisted of the dominant permanent and/or semi-permanent atmospheric pressure and circulation conditions. This chapter also includes some basic points of general soil characteristics of Turkey with respect to climate.

4.2 Data and Methodology

4.2.1 Data

For the station-based climate data of Turkey, long-term average monthly precipitation totals and monthly mean temperature data were calculated by using the long-term monthly precipitation totals and monthly mean temperature series, which were originally developed by Turkes (1996, 1999, 2013) and Turkes et al. (2002b), respectively. Detailed information on meta-data and homogeneity analyses applied to long-term precipitation and temperature series of Turkey can be found in Turkes (1996, 1999, 2013) and Turkes et al. (2002b, 2009), respectively.

4.2.2 Methodology

4.2.2.1 The Köppen-Geiger Climate Classification System

The Köppen-Geiger Climate Classification System (Köppen 1936; Köppen and Geiger 1954) is the most widely used tool of climatic classifications for physical geographical and environmental (climatological, biogeographical, ecological, etc.) purposes. We accepted in this section the criteria that follow Köppen's last publication about his classification system in the Köppen-Geiger Handbook (Köppen 1936), with the exception of the boundary between the temperate (***C***) and cold (***D***) climates, because we used the Köppen-Geiger climate data computed by Peel et al. (2007). Peel et al. (2007) followed Russell (1931) and used the temperature of the coldest month > 0 °C, rather than > −3 °C as used by Köppen in defining the temperate – cold climate boundary (see Wilcock 1968; Essenwanger 2001 for a history of this modification).

The Köppen-Geiger climate system consists of a shorthand code of letters designating major climate groups, subgroups within the major groups, and also subdivisions to differentiate particular seasonal characteristics of air temperature and precipitation. In the system, five major climate groups are determined by capital letters as follows, where the groups ***A***, ***C*** and ***D*** have sufficient heat (energy) and precipitation amounts for growth of forests and wood-land vegetation formations:

- ***A* Humid Tropical Climates**: Mean air temperature of the coldest month is above 18 °C ($T_{\text{cold}} \geq 18$ °C). Tropical ***A*** climates have no winter season. Annual rainfall amount is large enough to exceed annual potential evapotranspiration (*PET*) amount.
- ***B* Arid (Dry) Climates**: *PET* exceeds precipitation on the average conditions throughout the year. No water surplus occurs in the annual water balance; consequently, no permanent streams develop in the Arid ***B*** climate regions, with some regional exceptions such as in the Central Anatolia region of Turkey characterized mainly by a ***BS*** climate.

- ***C*** **Mid-latitude Warm Temperate (Mesothermal) Climates:** The coldest month has a mean air temperature below 18 °C, but above 0 °C (0 °C < T_{cold} < 18 °C); at least 1 month has a mean temperature above 10 °C (T_{hot} > 10 °C). Temperate ***C*** climates hence have both an evident summer and an evident winter season.
- ***D*** **Cold (Snow) (Microthermal) Climates**: The coldest month mean air temperature is under 0 °C ($T_{cold} \leq 0$ °C), and the mean air temperature of the warmest month is above 10 °C (T_{hot} > 10 °C). The 10 °C isotherm also coincides approximately with pole-ward limit of forest growth.
- ***E*** **Polar (Ice) Climates**: Mean air temperature of the coldest month is under 0 °C (T_{cold} < 0 °C), and the mean temperature of the warmest month is below 10 °C (T_{hot} < 10 °C). The polar ***E*** climates have no true summer.

In the Köppen-Geiger climate system, subgroups of the five major groups are indicated by a second letter according to the following codes:

- ***S*** **Steppe Climates**: Semi-arid climates with long-term mean annual precipitation amount varying from 350 mm to 700 mm at the low latitudes. Exact precipitation boundary is determined by a formula considering temperature.
- ***W*** **Desert Climates**: An arid climate mostly having mean annual rainfall amount less than 250 mm. The exact boundary of steppe climate is also determined by a formula.
- The letters ***S*** and ***W*** are applied only to the arid (dry) ***B*** climates yielding two important combinations as ***BS*** and ***BW***.
- ***f*** **Moist Climate Without Dry Season**: Adequate precipitation amount in all months. This modifier letter is applied to ***A***, ***C*** and ***D*** groups of major climates.
- ***m*** **Monsoon Climate with a Short Dry Season**: Tropical rainforest climate in spite of short and dry winter in the monsoonal precipitation regime cycle, particularly in the monsoon Asia. It applies only to ***A*** climates.
- ***w*** **Dry Winter Climate**: Dry season happens in winter of the respective hemisphere (low-sun season).
- ***s*** **Dry Summer Climate**: Dry season is in summer of the respective hemisphere (high-sun season), mostly associated with the subtropical high pressure systems (e.g. Azores high for the Mediterranean Basin and its surrounding regions). It applies to ***C*** and ***D*** major climates.

From combinations of the two letter groups, 13 distinct climates of the Köppen-Geiger classification appear as in the Table 4.1. Finally, Köppen added a third letter to the code group of his classification (not given here).

4.2.2.2 The UNEP/UNCCD Aridity Index

Arid, semi-arid and dry sub-humid climates were defined as “areas, other than polar and sub-polar regions, in which the ratio of annual precipitation to potential evapotranspiration falls within the range from 0.05 to 0.65” for the purposes of the United

Table 4.1 Two-letter Group Climates of the Köppen-Geiger Classification with a Short Description[a]

- ***Af***: **Tropical rainforest climate**
- ***Am***: **Tropical monsoon climate** (a variant of ***Af*** with a short dry season)
- ***Aw***: **Tropical savanna climate**
- ***BW***: **Desert climate**
- ***BS***: **Steppe climate**
- ***Cs***: **Temperate rainy climate with dry summer** (humid mesothermal) or **Mediterranean climate** (dry summer subtropical)
- ***Cw***: **Temperate rainy climate with dry winter** (humid mesothermal)
- ***Cf***: **Temperate rainy climate without dry season** (humid mesothermal) or **humid temperature west coast**
- ***Ds***: **Cold snowy forest climate with dry summer** (humid microthermal).
- ***Dw***: **Cold snowy forest climate with dry winter** (humid microthermal)
- ***Df:*** **Cold snowy forest climate humid in all seasons** (humid microthermal)
- ***ET*** **Polar tundra climate**
- ***EF*** **Polar forest climate**

[a]*The groups with in the Köppen-Geiger climate system, framed with red boxes, show climate types that are found in Turkey*

Nations Convention to Combat Desertification (UNCCD 1995). In the present study, *UNEP/UNCCD Aridity Index (AI)* is used as one of the base methods for determining dry-land types in Turkey and assessing their vulnerability to the desertification processes.

Following UNEP (1993), *AI* is written as:

$$AI = \left(\frac{P}{PET}\right) \tag{4.1}$$

where *P* and *PET* are annual precipitation and potential evapotranspiration totals (mm), respectively. The *PET* values were estimated by taking the approach used in the WATBUG program into consideration, which was developed by Willmott (1977) for climatic water budget, and Thornthwaite's climate classification (1948). The *AI* values below 1.0 show an annual moisture deficit in average climate conditions. The criteria in Table 4.2 were used to characterize the dry-lands of Turkey (Turkes 1999).

4.2.2.3 Drought Probability Based on Standardized Precipitation Index

In the classical approach for obtaining the Standardized Precipitation Index (*SPI*), the cumulative distribution function (*CDF*) of precipitation totals are formed from the fitted frequency distribution according to McKee et al. (1993, 1995), and then the probabilities from the fitted *CDF* are transformed to standard normal distribution by using inverse standard normal distribution. Consequently, this is a method consisting of a transformation of one probability distribution to another. In this common methodology, the gamma distribution is widely used (e.g. Thom 1966; McKee et al. 1993, 1995; Wilks 1995; Guttman 1998, 1999; Turkes and Tatli 2009, etc.).

Table 4.2 Dry land (arid climate) types in Turkey according to the Aridity Index (AI) and their vulnerability to desertification. (Turkes 1999)

Aridity criteria	Dry land type	Assessment
$0.20 \leq AI < 0.50$	Semi-arid areas	Vulnerable to desertification
$0.50 \leq AI < 0.65$	Dry sub-humid areas	Vulnerable to desertification
$0.65 \leq AI < 0.80$	Sub-humid areas	Vulnerable to desertification

On the other hand, Turkes and Tatli (2009) proposed a new *SPI* approach for obtaining the local-time means from maxima and minima envelopes of the monthly total precipitation series for calculation of the *SPI* series. This proposed new *SPI* technic is an analogical methodology of the empirical mode decomposition (*EMD*). It was initially proposed for the study of ocean waves by Huang et al. (1998) and Wu et al. (2007). This approach was made use of by many researchers in the atmosphere and climate sciences (e.g., Zhu et al. 1997, 2001; Coughlin and Tung 2001, 2004; Turkes and Tatli 2009). The basic (climatological) drought probability used in this study is based on the monthly series of the proposed new *SPI*, which had been calculated before by Turkes and Tatli (2008, 2009) for the 96 stations of Turkey.

4.3 Results and Synthesis

4.3.1 Physical Geographical Settings

Turkey is located between 36–42° northern latitude and 26–45° eastern longitude in the Northern Hemisphere (Fig. 4.1). Accordingly, there is a 76-minute local time difference between the west and the east of the country. According to the Turkish Statistics Institute, Turkey's total surface area is 785,347 km^2 and the total area of Turkey without natural lakes and is 769,604 km^2 (TurkStat 2016). Turkey is composed of seven geographical regions. According to the ranks of their surface area, these are of the Eastern Anatolia Region, Central Anatolia Region, Black Sea Region, Mediterranean Region, Aegean Region, Marmara Region and the Southeastern Anatolia Region (Fig. 4.1). Because it is located in the southern part of the mid-latitude geographical belt towards the sub-tropical belt, four seasons are experienced distinctively during the year. The territory of Turkey is located between Europe and Asia. Turkey's land borders are 2949 km long in total. The coastal border of the peninsula, which is surrounded by three sides with the Black Sea at the north, the Aegean Sea at the west and the Mediterranean at the south, is 8592 km except the islands.

Turkey is a country of highlands with an average altitude of 1141 m at an average slope of 17%, varying generally from 3% to 30% (Elibuyuk and Erkan 2010). This figure of the average altitude of the country is higher than the average altitude of each continent on the Earth. Turkey's most flat region is the South-eastern Anatolia; the lowest region is the Marmara; the greatest sloppy (the most gradient) region is the Black Sea; the highest one is the East Anatolia Region (Fig. 4.1).

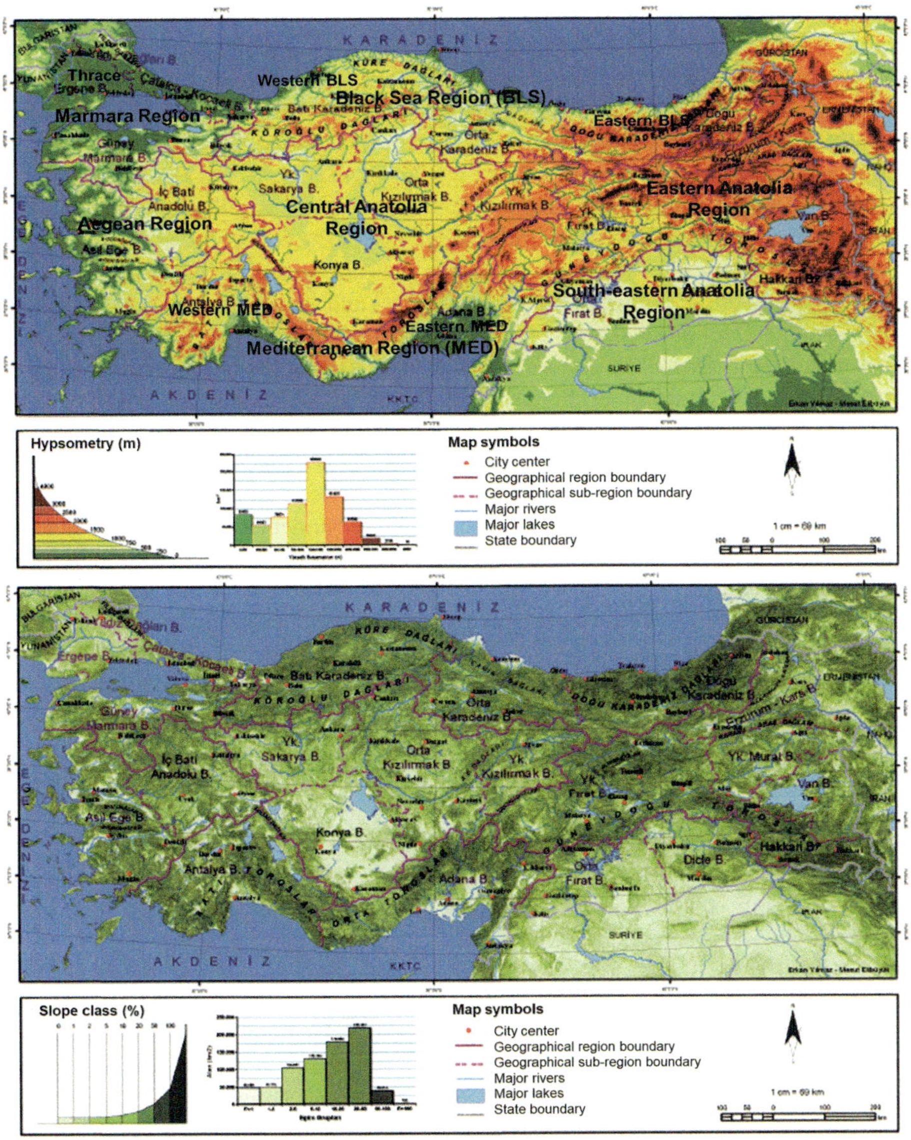

Fig. 4.1 (**a**) Altitude steps (hypsometry in meter) and (**b**) Slope classes (in percent, %) the boundary lines of the geographical regions and sub-regions (shortly B on the maps) of Turkey. (Elibuyuk and Erkan 2010)

The geological evolution of Turkey, which is situated on the Alpine-Himalayan mountain ranges or orogenetic belt geologically, had begun towards the late Palaeozoic period and continues its completion today, during the post Quaternary period. Mountains cover a great portion of the country's land. It is clearly seen that the high mountain chains, which are namely the North Anatolian Mountains at the north, and the Mediterranean Taurus Mountains and the South-eastern Tauruses at the south, stretch along the northern and southern coastal zones from east to west making wide arches. Those mountain ranges at the Mediterranean and Black Sea regions are generally perpendicular to the coastal zones. The middle part of the country consists of the Anatolian Plateau (Fig. 4.1). This plateau separates the high mountain ranges at the north and south from each other, although they traverse them in the east. For this reason, the eastern part of the country is higher and more mountainous than the remaining parts. On the other hand, there are many plains and low basins in the inner parts of the country, and coastal plains and delta-flood plains along the coastal zone. Deltas formed by the Kizilirmak and Yesilirmak rivers and the Adana plain can be given as examples of the large coastal plains in the country. According to the orogenetic belt classification studies made for Turkey, Turkey as a whole is located in the Mediterranean section of the Alpine orogenetic belt (Ilhan 1976; Ketin 1983, 1994). This orogenetic belt was situated between the Eurasian plate at the north and the African and Arabian platforms at the south. Plate tectonics theory provides a better understanding of the geologic and morphotectonic evolutions of Turkey and its nearby surrounding. The plate tectonics is realized by the approaching and collision of two plates in the subduction zones, causing sediment deposits and magmatic rocks having a thickness of thousands of meters (Ketin 1994). The complex Alpine–Himalayan orogenetic belt, on which Turkey also is located, had formed as a result of the faster movement and approaching of large pieces of Gondwanaland plate in the south (i.e., African, Arabian, Indian, etc.) than that of the Eurasian plate (including the platforms of Russia, Siberia, etc.) and by folding, faulting and uplifting of the Alpine–Himalayan geosyncline's deposits by the orogenic process (Erinc 1982). The Arabian–Eurasian convergence is an important process for better understanding of the geologic/geomorphologic evolution and neo-tectonic fundamentals of Turkey. Final collision of the Arabian–Eurasian (here, Anatolian sub-plate) had taken place in the Miocene epoch (Sengor 1980).

Ketin (1966), in his study to determine the tectonic units of the Anatolian Peninsula, divided Turkey into four tectonic units, which are namely the Pontid (the northern Anatolia and Marmara regions), the Anatolid (the Central Anatolia region), the Torid (Tauruses and other Mediterranean mountains), and the Border Folds (the South-eastern Tauruses and South-eastern Anatolia region). Ketin (1966) found out that tectonics and orogenic development in the Anatolian Peninsula had slowly progressed from north to south. The first severe orogenic movements had started in the northern Anatolia and Marmara regions and, in the following period, had spread towards the Central Anatolia, Tauruses and at last towards the South-eastern Tauruses, which is the youngest orogenic mountain belt of Turkey geologically. The South-eastern Tauruses tectonic unit had completed its main formation in the late Miocene epoch, and even the Pliocene layers joined into the folding movement. It is evident that the old Central Anatolia land region had played a role of inner massif after the Eocene epoch and made a significant contribution to the orography of the Taurus Mountains (Ketin 1966, 1983; Erinc 1982).

4.3.2 General Meteorology and Climate Dynamics

In this sub-section, general meteorological and climate dynamic aspects and the occurrence and classification of the macro Mediterranean and Turkish climates are explained. Mediterranean climates are mostly considered under the climates of **Group** *C* Mid-latitude warm temperate (mesothermal) climates according to the Köppen's climate classification (Table 4.1). The mid-latitude climates are controlled mainly by both tropical and polar air masses and weather systems according to the Strahler's (1973) genetic climate classification (Group II). This classification was mainly developed, based on the global and regional atmospheric circulation conditions, air-masses, and mid-latitude cyclones and anticyclones, etc. (Figs. 4.2 and 4.3) (Turkes 2010). The mid-latitude temperate (mild) climates are located in parts of the latitude range between 30° and 60° in either hemisphere, where the term mild (temperate) refers to the winter temperatures and not necessarily to those of the summer.

These warm temperate climates are those of the middle latitudes, dominating the planetary polar front zone during at least half of the year, particularly from the late autumn to middle-spring, over which both tropical and polar air masses and weather patterns extend and dominate. The geographical zone in which the mid-latitude climates prevail is mainly subject to frontal low-pressure systems (mid-latitude cyclones). Most of the precipitation, in particular in winter, in these climate regions occurs associated with the frontal precipitation events through the mid-latitude cyclones. Average annual total precipitation amounts of the warm temperate climates are modest, ranging from about 350 mm on the equator-ward margin and in some inland areas (in far eastern inland areas) to about 650 mm at the pole-ward margin. However, it is over 1000 mm over the high-mountainous regions of the northern Mediterranean basin, for example on the Mediterranean Sea coast of Turkey and southern wind-ward side of the Alps, due to the topographically induced precipitation in addition to the frontal precipitation events (Turkes 2010).

4.3.2.1 Occurrence and Spatial Patterns of the Mid-Latitude Temperate Climates

The mid-latitude temperate climates have the greatest temporal variability, varying from a few days to several weeks. Seasonal variability (i.e. seasonality) is also evident in these climates, whereas the seasonality lacks both in the tropics with the almost constant heat gain and the almost-continuous cold of the polar (i.e. Arctic and Antarctic) regions. The mid-latitude regions are mainly characterized with frequently alternating incursions of tropical (T) and polar (P) air masses throughout the year (Fig. 4.1). This apparent regional circulation dynamic produces more convergence movements than anywhere else with the exception of the equator. Seasonal alternation of the tropical and polar air-masses controls changes and variability in frequency and intensity of the Rossby waves and associated weather

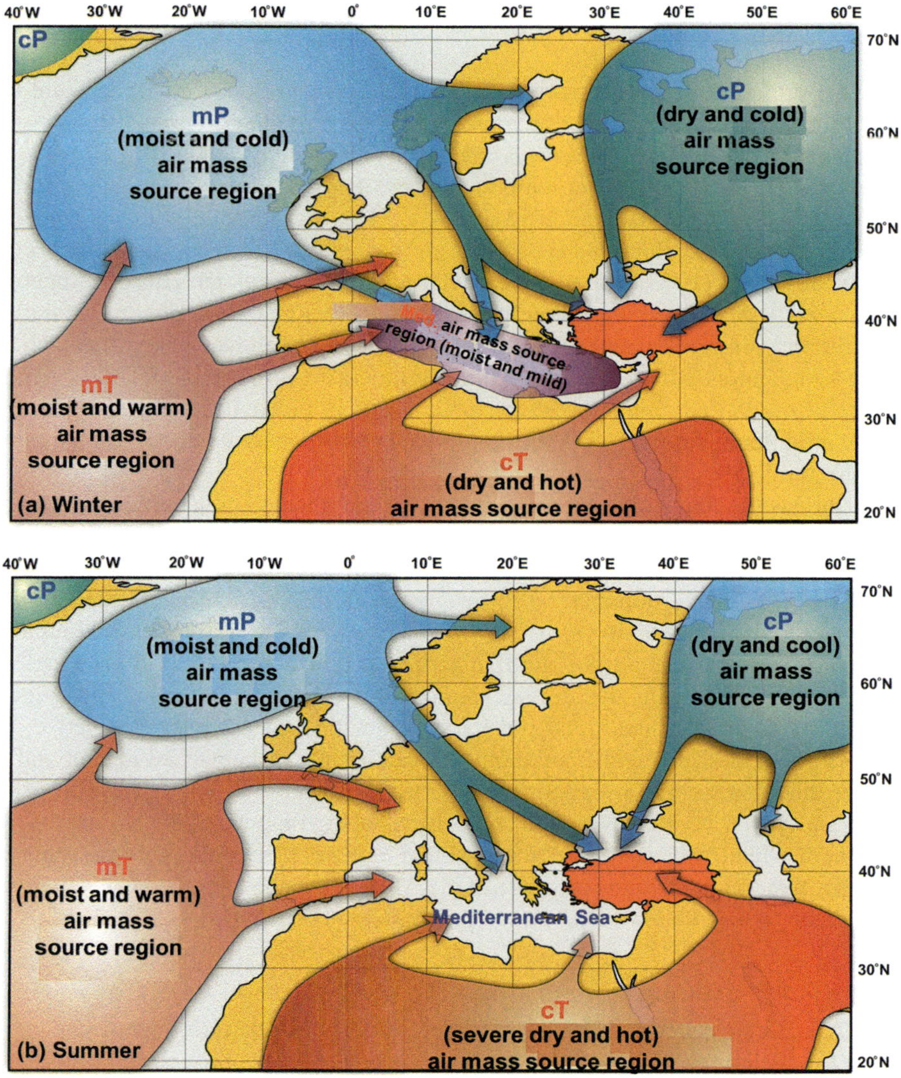

Fig. 4.2 Seasonal changes in climatology of air masses over the Mediterranean Basin and surrounding regions due to changes in planetary and synoptic scale surface and tropospheric pressure and wind systems driven by the apparent movements of Sun and associated migration of the Intertropical Convergence Zone (ITCZ) between (**a**) Winter and (**b**) Summer. (rearranged in English from Turkes 2010)

systems of mostly dynamic originated cyclonic and anticyclonic nature (Erinc 1969; Barry and Chorley 2003; Turkes 1998, 2010; Sahin et al. 2015; Lolis and Turkes 2016). The seasonal rhythm of the surface air temperature regime in the Mediterranean climate is usually more prominent than that of the precipitation regime.

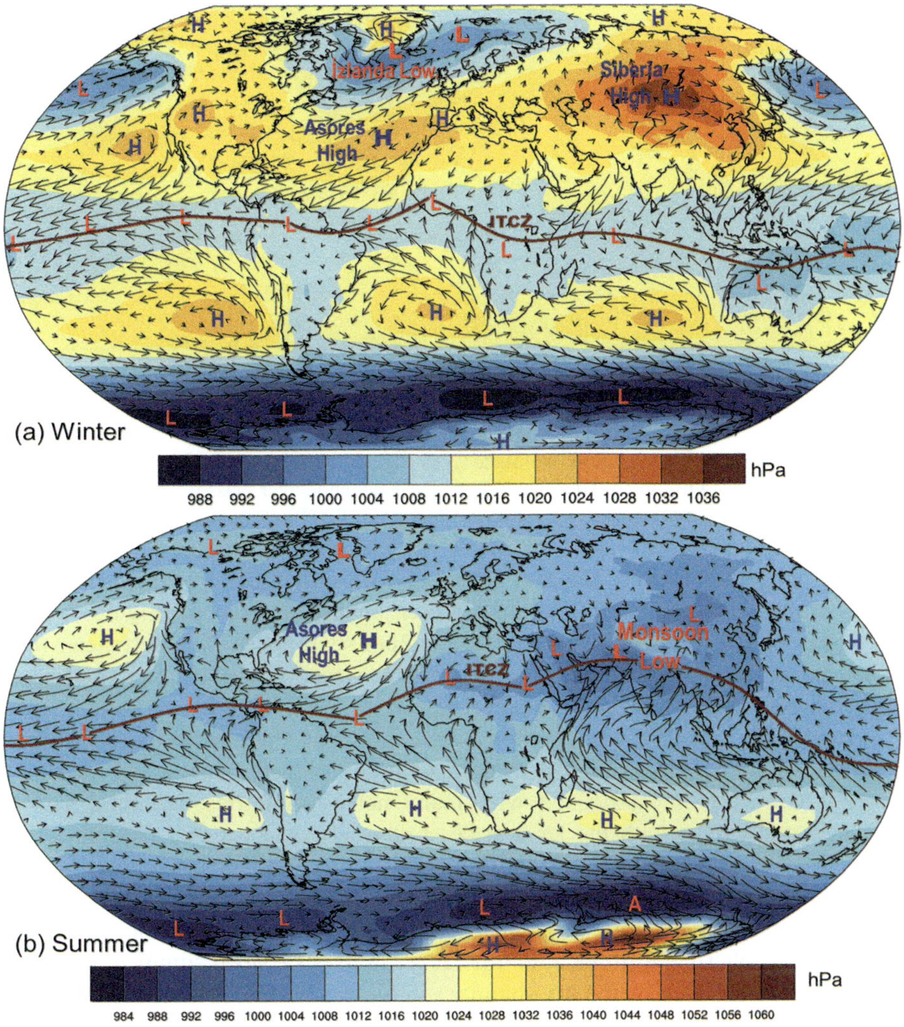

Fig. 4.3 Seasonal changes in climatology of the global surface air pressure and wind circulation systems associated with the apparent migration of the Intertropical Convergence Zone (ITCZ) between (**a**) Winter and (**b**) Summer. (rearranged in English from Turkes 2017a)

The mid-latitude climates occupy the equator-ward margin of the mid-latitudes, occasionally extending into the subtropics and being elongated pole-ward in some western coastal areas due to the high topographic barriers. They clearly constitute a transition between warmer tropical climates of the south and colder mid-latitude climates of the farther north. Summers in these climates are long and warm or hot, while winters are usually short and relatively mild. High variability in both precipitation total amounts and seasonal distribution over the ***C climates*** is characteristics.

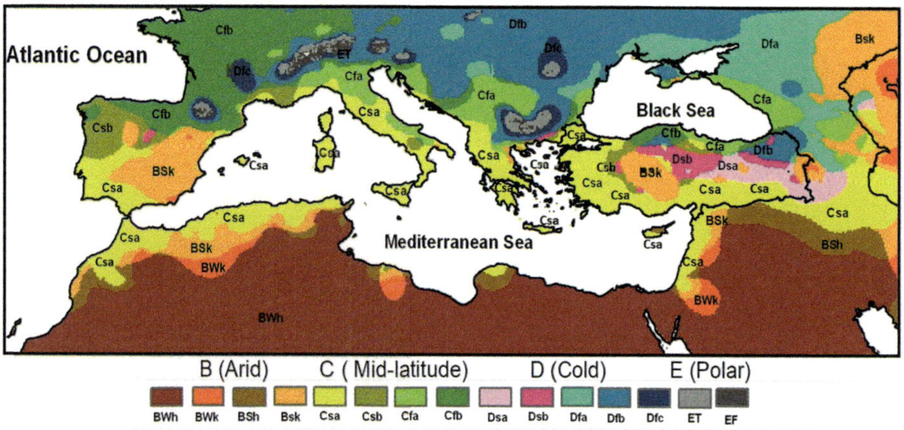

Fig. 4.4 Geographical distribution of climate types over the Mediterranean Basin and surrounding regions according to the three-letter symbols of the Köppen-Geiger classification. [From Turkes (2017a) re-drawn by making use of the data by Peel et al. (2007)]

The temperate warm ***C climates*** are subdivided into three types primarily on the basis of precipitation amounts (limits) and seasonality, and secondarily on the basis of some air temperature limits and temperature seasonality: Mediterranean (***Csa***, ***Csb***), humid subtropical (Cfa, Cwa), and marine west coast (Cfb, Cfc), all of which are found in the Mediterranean Basin (Fig. 4.4).

The Mediterranean basin is significantly important for understanding the climate, climate variability, hydrological cycle, and future climate change and variability over a large region, associated with many nearby countries including Turkey, because of its links with the large-scale atmospheric and oceanic teleconnections and circulations (Sahin et al. 2015). The genesis of the Mediterranean macroclimate (shortly the Mediterranean climate) is mainly associated with the seasonal migrations of the mostly dynamic-originated pressure systems (i.e. centers of action) (Fig. 4.3) and the alternating tropical and polar air masses between summer and winter (Fig. 4.2). Consequently, the dry summer subtropical Mediterranean climates can be mostly found along the west coasts between about 25° and 40° latitudes as a natural part of the mid-latitude temperate climates (Turkes 2017a).

The largest Mediterranean climate region of the Earth is found in the Mediterranean Basin, which can thus be called as the 'true' Mediterranean macroclimate mainly due to the topographic features of the subtropical westward of the continents (Turkes 2017a). In the Atlantic-ward of the southern Europe and the Mediterranean Sea Basin, there are no north to south oriented mountain chains, and the Mediterranean climates penetrate a great distance from the west basin of the Mediterranean Sea, with some parts of the Iberia Peninsula, to the East Basin of the Mediterranean Sea and the Mediterranean coastal and inland regions of the Middle East region (Fig. 4.4). Although the term Mediterranean is used to describe the climate synonymous with the region surrounding the Mediterranean Sea, it is not exclusive to this region. The Mediterranean climates are also found, for example, in the northern Iran,

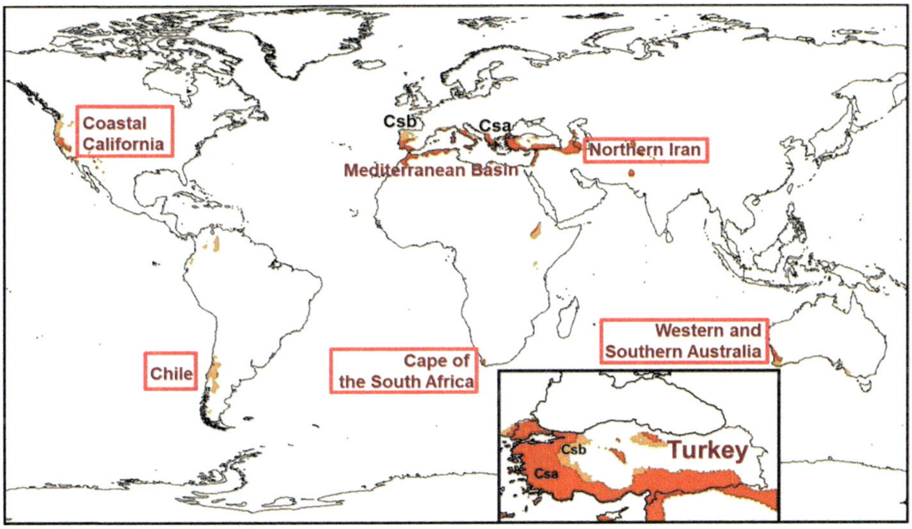

Fig. 4.5 Geographical distribution of the Mediterranean and the Mediterranean-like climate regions over the world. (Turkes 2010)

California, Chile, Australia, New Zealand and South Africa, in addition to the 'true' Mediterranean regions of Portugal, Spain, Coastal North-west Africa (Morocco, Tunisia, Algeria), France, Italy, Greece, Turkey, Lebanon and Israel (Fig. 4.5) (Turkes 2010; Turkes et al. 2011). These regions are mostly dominated in summer by dry, stable and subsiding air from the eastern portions of dynamic originated subtropical highs, which is also the sinking branch of the Hadley cell circulation. In winter, the wind and pressure systems shift equator-ward as a result of the equator-ward migration of the Rossby waves and associated polar jet streams and upper air westerlies with the polar front. The Mediterranean climate regions are influenced by the westerlies with their trailing mid-latitude frontal cyclones (Turkes 2010).

Seasonality is the most dominant and distinctive character and factor of the Mediterranean climates (Turuncoglu et al. 2018). The Mediterranean climate tends to alternate wet and dry seasons, because it is located over the transitional zone between the dry west coast tropical desert, mainly related with the subtropical high pressure systems, and the descending segment of the tropical Hadley circulation cell, and the wet west coast climate mainly related with the polar front and associated mid-latitude cyclones.

As we shortly discussed the mid-latitude climates above, there are many atmospheric features that can be considered for dominating the Mediterranean climates. For instance, the Rossby waves, which are formed by the upper air troughs and lows, and the upper air ridges and highs, controlled penetration of polar air masses [continental polar (cP), maritime polar (mP), and very rarely continental Arctic (cA)] towards equator at certain months of the year, and tropical air masses [continental tropical (cT), and maritime tropical (mT)] towards poles at other certain months of the year (Fig. 4.2). In this respect, the seasonal movements arising from mainly the Sun's seasonal migration (i.e., Sun's apparent movement) and thus the amount and intensity of the Sun's radiation resulting in an energy exchange between the poles and the equatorial belt are the main mechanisms (Turkes 2010). On the other hand, due to the seasonal movements of the inter-tropical convergence zone (ITCZ) associated closely with the movements of the Sun (Fig. 4.3), the Rossby waves would be closer to the equatorial belt in winter than that in summer. Thus, polar originated weather systems can more strongly fluctuate to further equator-ward in winter than in summer. This fluctuation is also responsible in enhancing the seasonal energy contrasts globally, particularly over the continents.

In the mid-latitudes, while the planetary-scale Rossby waves, the upper air jet streams governing also westerlies and the frontal cyclones are the most powerful features of these climates, these atmospheric controls are also themselves controlled by topography, continentality, land-sea distribution and interactions, air masses, and their thermo-dynamics and mechanical modifications arising from the physical geographical features of the Earth surfaces (Turkes 2010). Consequently, there is no simple explanation of the mid-latitude climates including the Mediterranean climate as a whole.

Almost all precipitation comes from the frontal cyclones, except for the late spring and early summer convective instability showers and thunderstorms in inland Mediterranean climates of the Anatolia Peninsula and the Middle-East region. On the other hand, the Alp Mountains of the South Europe and the North Anatolia and Taurus mountains of Turkey create stronger influence on the westerly and northerly air-flows and associated mid-latitude frontal cyclones by means of the orographic lifting of the moist air masses. This mechanism leads adiabatic cooling and condensation of water vapor of air masses, and occurrence of the precipitation over the mountains, particularly by the time of that faced to southerly and westerly moist air masses (Turkes 1996, 1998, 2010).

4.3.2.2 Dry-Summer Subtropical Mediterranean Climates

The Mediterranean climate, ***Cs*** of the main Köppen classification ***Group C*** and the climate number 8 of the Strahler's classification ***Group I*** (Strahler 1973; Turkes 2010) are mainly characterized with warm to hot dry summers and cool to cold wet winters.

Most Mediterranean climates of the Earth are classified as ***Csa***, which means that summers are hot with mid-summer monthly averages between 24 °C and 29 °C and high maximums above 38 °C. Average cold-month temperatures are about 10 °C with occasional minimums below freezing temperatures.

The coastal Mediterranean climate regions have much milder summers than the inland Mediterranean climate regions. This is mostly due to sea breezes in some coastal areas as a result of being adjacent to cool currents associated with the permanent anticyclonic circulation of the subtropical high pressure systems located over the subtropical oceans of the Earth. These coastal Mediterranean climates are classified as ***Csb***. In the coastal areas, the hot-month mean air temperatures lie between 16 °C and 21 °C. ***Csb*** winters are slightly milder than ***Csa*** winters, the former having cold-month mean air temperatures of about 13 °C. ***Csb*** regions also have higher humidity, frequent advection and evaporation fogs, and occasional low *stratus*, *altostratus* and *nimbostratus* overcast.

Marine west coast ***Cf climates*** located at the pole-ward of the Mediterranean climate region (Fig. 4.4) occur as a result of persistent influences by the mid-latitude frontal cyclones during the year. Consequently, marine-west coast climate regions are mostly characterized with windy, cloudy and rainy conditions almost around the year. On the average conditions, these climates are cooler than their equator-ward neighbor climates.

Based on all the explanations above, we may summarize the Mediterranean climates with following five distinctive characteristics (Turkes 2010; Turuncoglu et al. 2018):

(1) About half of the modest annual precipitation amount falls in winter, whereas summers are mostly virtually rainless.
(2) Winter temperatures are unusually mild for the mid-latitudes except some eastern and inland regions; summer air temperatures vary from hot to warm.
(3) Cloudless skies and intensive sunshine (shortwave solar radiation) are typical particularly in summer months.
(4) It has major influences from sea and land distribution and the interactions between sea and lands, in addition to the ocean-atmosphere interaction, during the year particularly in the 'true' or 'actual' Mediterranean macro climate region, etc.
(5) The major characteristics of the Mediterranean climate is the high temporal variability at seasonal and inter-annual to centennial scales due to following factors (Turkes 2010):

    - It extends in a transition region between temperate and cold mid-latitudes and tropics (i.e. subtropical zone);
    - It faces significant circulation (associated pressure and wind systems characterizing mid-latitude and tropical/monsoonal (Fig. 4.6) weather and climate, respectively) changes between winter and summer;
    - It is closely associated with several atmospheric oscillation and/or teleconnection patterns during the year, where they vary depending on seasons, such as North Atlantic Oscillation (NAO), Arctic Oscillation (AO), Mediterranean Oscillation (MO), El Niño-Southern Oscillation (ENSO), and North Sea – Caspian Pattern (NCP), etc. (e.g., Kutiel and Turkes 2005; Turkes 1998; Turkes and Erlat 2003, 2005, 2006, 2008, 2009, 2018a, b; Trigo et al. 2006; Erlat and Turkes 2012, 2016; Sahin et al. 2015, etc.).

Regarding the last bullet above, recent studies indicated that the NAO is one of the major atmospheric sources for the spatial and temporal variability of the

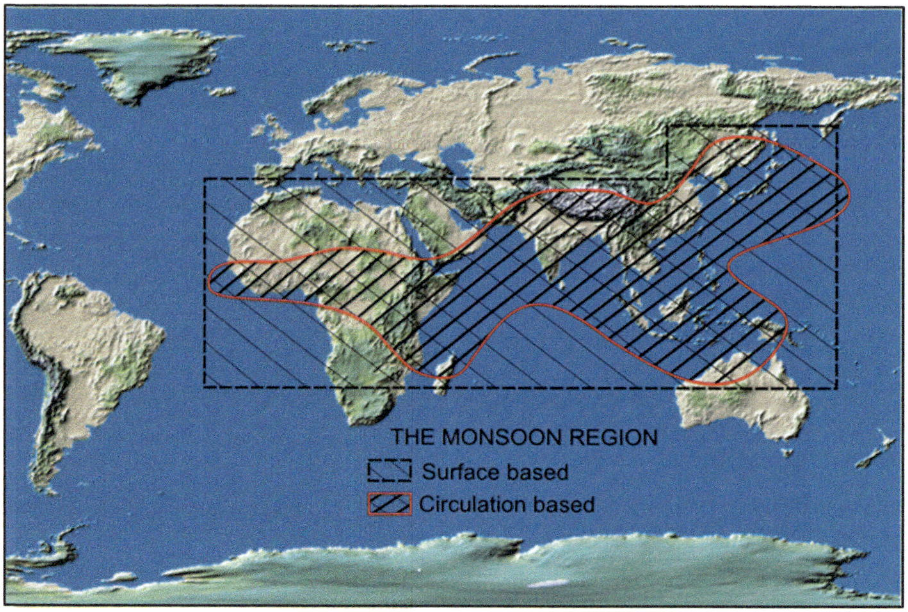

Fig. 4.6 Schematic illustration for the geographical extends of the direct circulation-based and indirect surface-based influences of the major monsoon systems (Monsoonal Lows) during the year, and the area of their observed influences on the Mediterranean Basin and Turkey. [Henderson-Sellers and Robinson (1986): re-arranged and plotted on a new base map by making some modifications]

precipitation conditions in Turkey, including significant wet periods and meteorological droughts (Turkes and Erlat 2003, 2005, 2006; Sahin et al. 2015; Turkes 2014a, b, etc.).

The NAO, which is associated with the surface pressure see-saw or oscillation among two large-scale dynamic centers of action, so called the Azores High and the Icelandic Low, is a well-known atmospheric teleconnection pattern (Turkes and Erlat 2003). The NAOIs are developed in order to evaluate behavior of the NAO and regional climate anomalies linked to the extreme NAO episodes. A NAO index (NAOI) is generally arranged based on the difference of normalized sea level pressure (SLP) anomalies between a station in the area of the Azores and another one in Iceland. Another NAOI is also calculated by using the difference of the normalized SLP between a station located in the Iberian Peninsula and another in Iceland (Turkes and Erlat 2005). The Ponta Delgada-Reykjavk (PD-R) NAOI was used in this study, because Turkes and Erlat (2005, 2006) found that the PD-R NAOI is the best NAOI among the three NAOI compared [i.e. PD-R, L-S(R) and Gibraltar-Reykjavk (G-R)], with respect to the ability of controlling variability in precipitation series of Turkey, particularly for winter precipitation.

When the NAO is in its extreme phases, both the Icelandic Low and the Azores High are well developed during cool/cold period of the year, especially in winter

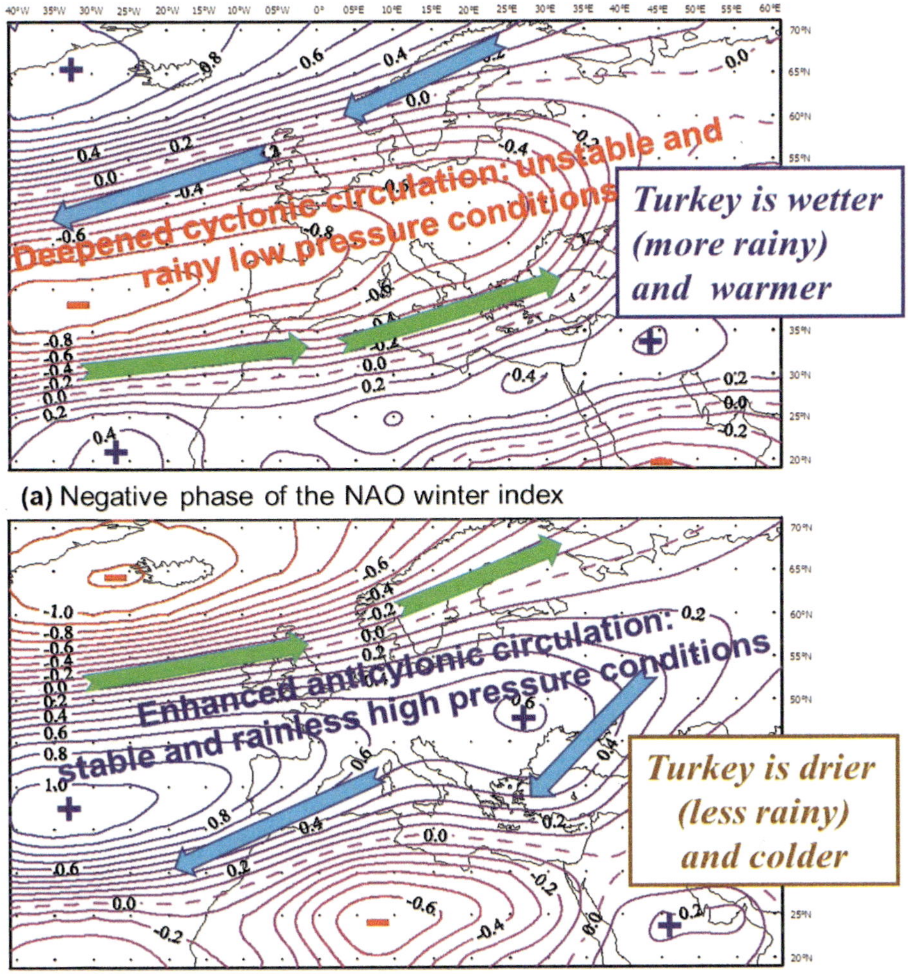

(a) Negative phase of the NAO winter index

(b) Positive phase of the NAO winter index

Fig. 4.7 Geographical distributions of composite winter 500-hPa geopotential height anomalies during (**a**) the negative phase and (**b**) the positive phase of the Ponta Delgada–Reykjavik (PD-R) NAO winter index. [re-drawn and re-arranged from Turkes and Erlat (2009)]

(Fig. 4.7). The negative phase of the NAO indicates stronger-than-average westerly and south-westerly circulation over the subtropical north-east Atlantic, North Africa and the Mediterranean basin towards Turkey, and the stronger-than-average north-easterly circulation across Scandinavia and the mid-latitude and sub-Arctic north-east Atlantic, particularly in winter, spring and annually (Fig. 4.7a). On the contrary, the positive phase of the NAO shows the increased westerlies over the mid-latitudes and Scandinavia and increased easterly and north-easterly circulation over a large conveyer zone from Turkey to the subtropical Atlantic via the Mediterranean basin and North Africa (Figs. 4.7b and 4.8).

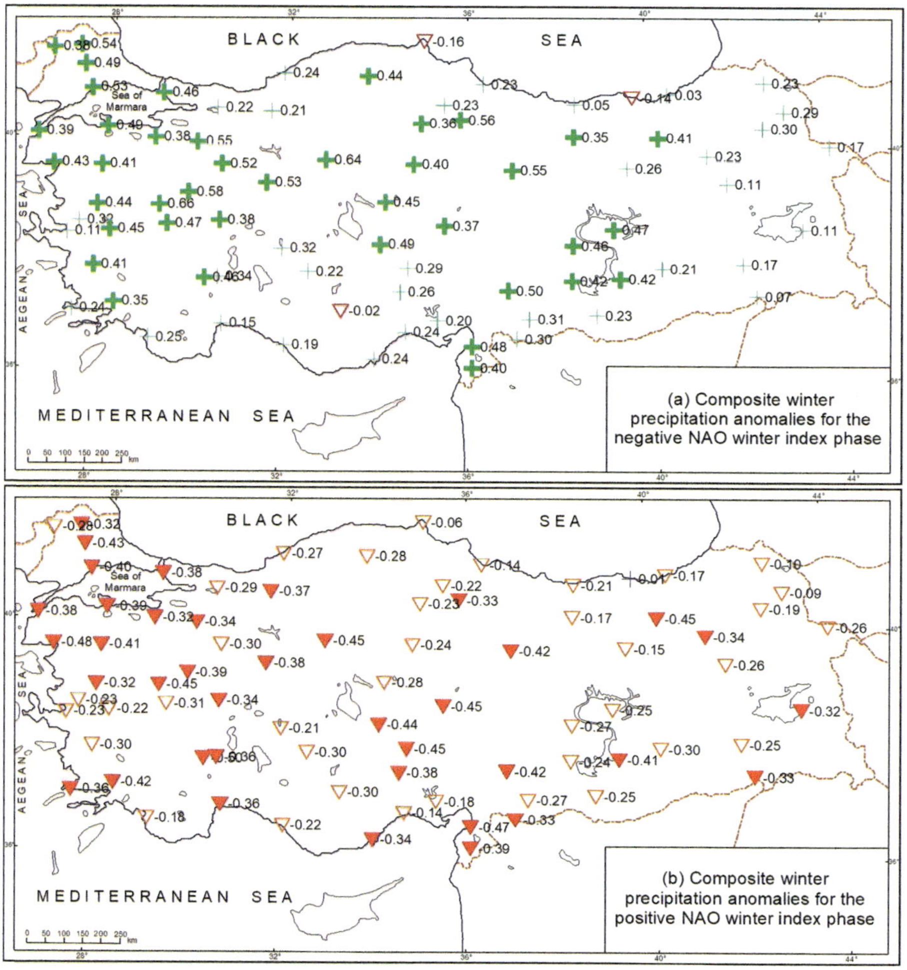

Fig. 4.8 Geographical distribution patterns of composite precipitation anomalies, (**a**) during the weak and (**b**) the strong phases of the NAO winter index. (Turkes and Erlat 2003)
Bold plus symbols (filled inverse triangles) show significant wetter (drier) than long-term average precipitation conditions at the 0.05 level of significance, according to Cramer's test

Individual widespread and significant wet conditions and meteorological drought events in the cold months of the year in Turkey, particularly in winter, are mostly controlled by the extreme NAO conditions (Figs. 4.7, 4.9 and 4.10). According to the studies performed by Turkes and Erlat (2003, 2005, 2006) and Turkes (2014a, b, 2016), the 500-hPa circulation corresponding to the negative NAOI phase brings above long-term average precipitation to Turkey in winter (Fig. 4.8a), spring, autumn, and annually. This circulation is associated with the NAO pattern in which the 500-hPa geopotential level is anomalously high in the area of the Icelandic Low and anomalously low across the regions of the Azores High and Europe in

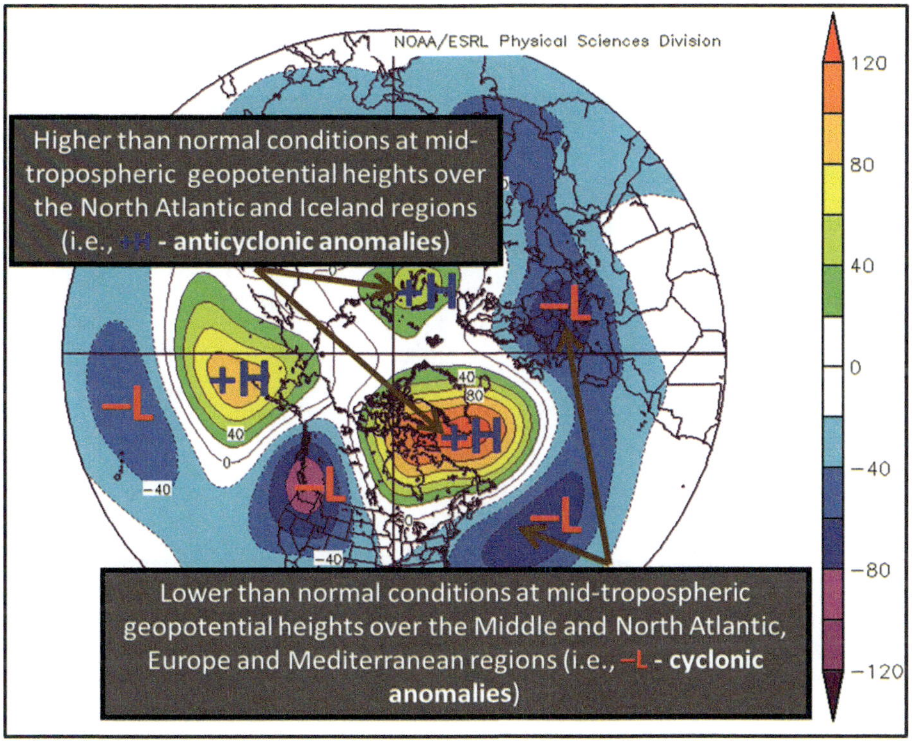

Fig. 4.9 Anomalies pattern of 500 hPa geopotential height level during the 1969 winter season (December 1968 to February 1969) with respect to the 1981–2010 climatology over the Regions of the Europe, Mediterranean and Turkey, which pointed out an example of the anomalies mid-tropospheric circulation anomalies associated with the weak (negative) phase of the year-to-year variability in the NAO. In Turkey, this pattern leaded occurrence of the geographically coherent large-scale and significant *positive precipitation anomaly* in 1969 (**strong wet year**). [See: Fig. 4.11(a)]

general (Fig. 4.7a). On the contrary, the anomalous circulation pattern corresponding to the positive NAOI phase over the North Atlantic and Europe is responsible for the drier than long-term average precipitation conditions in Turkey, particularly in winter (Fig. 4.8b), when the 500-hPa geopotential level is anomalously low over the area of the Icelandic Low and anomalously high across the subtropical and mid-latitude north-east Atlantic and the European regions (Fig. 4.7b). In other words, when the NAO is in its extreme low-index (negative) value, winters in Turkey tend to be generally wet (rainy) (Fig. 4.8a), whereas winters in Turkey tend to be experienced generally with drought (rainless) conditions when the temporal pattern of the NAO is in its extreme high-index (positive) value (Fig. 4.8b).

Considering the extreme NAO events for the period 1940–2015, it is very likely that widespread strong wet conditions in some or most regions of Turkey during the winters of 1940–1942, 1956, 1963, 1966, 1969, 1970, and 2010–2011 were detected

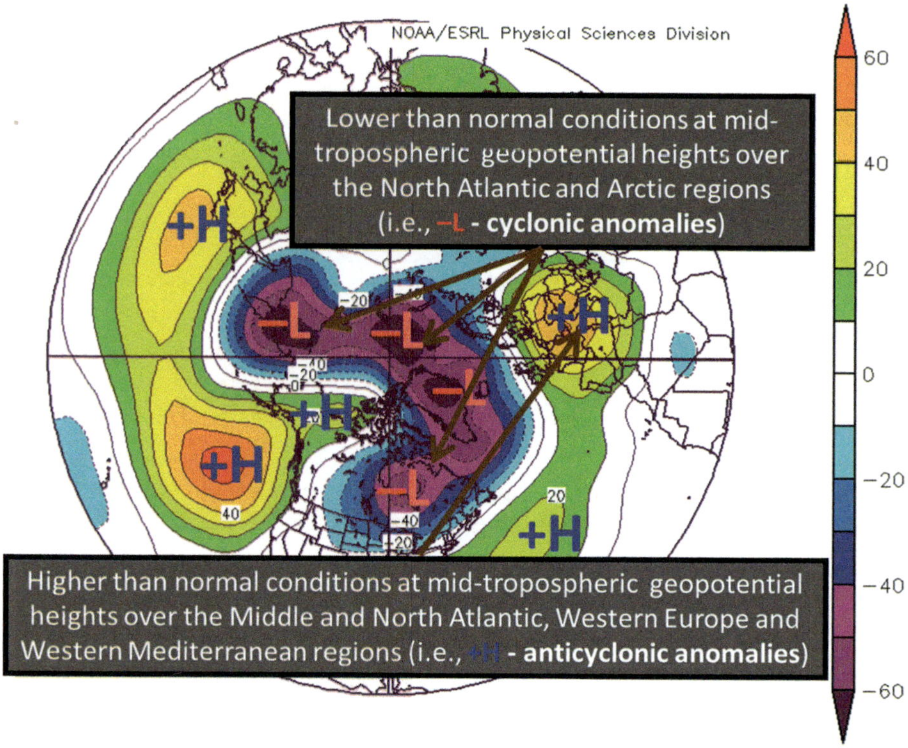

Fig. 4.10 Anomalies pattern of 500 hPa geopotential height level during the 1989 winter season (December 1988 to February 1989) with respect to the 1981–2010 climatology over the Regions of the Europe, Mediterranean and Turkey, which pointed out an example of the anomalies mid-tropospheric circulation anomalies associated with the strong (positive) phase of the year-to-year variability in the NAO. In Turkey, this pattern leaded occurrence of the geographically coherent large-scale and significant *negative precipitation anomaly* in 1989 (**strong drought or dry year**). [See: Fig. 4.11b]

to match the extreme low-index values in the same winter seasons of at least two NAOIs. On the other hand, it is possible to explain that geographically coherent severe droughts in some or most regions of Turkey during the winters of 1943, 1957, 1973, 1974, 1983, 1989, 1990, 1992, 1993, 1994, 2007–2008, and 2013 were closely related with the extreme high-index values of the NAO indices. In this respect, according to Turkes and Erlat (2005) for instance, during the weakest winter index of the three NAOIs in 1969 (Fig. 4.9), positive precipitation anomalies (wet conditions) exhibited a coherent distribution pattern over much of Turkey, indicating a mixed distribution of very much above and extremely above long-term average precipitation conditions (Fig. 4.11a). On the other hand, it is very likely that geographically the largest severe drought events occurred in the winter of 1973 (not shown here) and in the successive winters of 1989 and 1990. The most severe and coherent widespread drought event (Fig. 4.11b) appeared during the strongest NAO event in 1989 (Fig. 4.10).

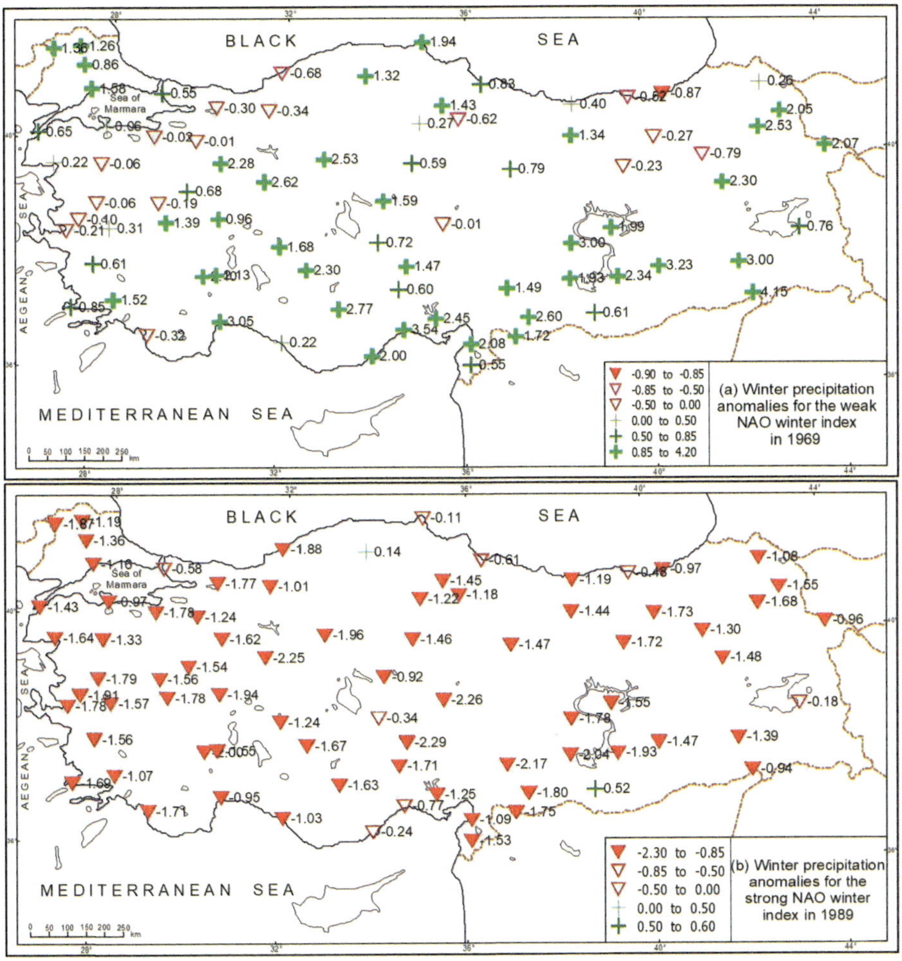

Fig. 4.11 Geographical distribution patterns of normalized winter precipitation anomalies in Turkey, during the weak and strong NAO winter indices in the years of (**a**) 1969 and (**b**) 1989. (Turkes and Erlat 2005). *Bold plus symbols (filled inverse triangles) show considerable wetter (drier) than long-term average precipitation conditions*

4.3.2.3 Turkish Climate

As we already discussed above, the major climate of the Mediterranean Basin including Turkey arises mainly from an unequivocal seasonal alteration from winter to summer between mid-latitudes and sub-tropical/tropical pressure and wind systems (circulation patterns). Thus, there is a highly seasonal precipitation regime in the western and southern regions of Turkey, characterized mainly by dry and hot summers (Turkes 2010). During the summer, the weather and climate in most regions of Turkey, particularly in the Mediterranean and the continental steppe

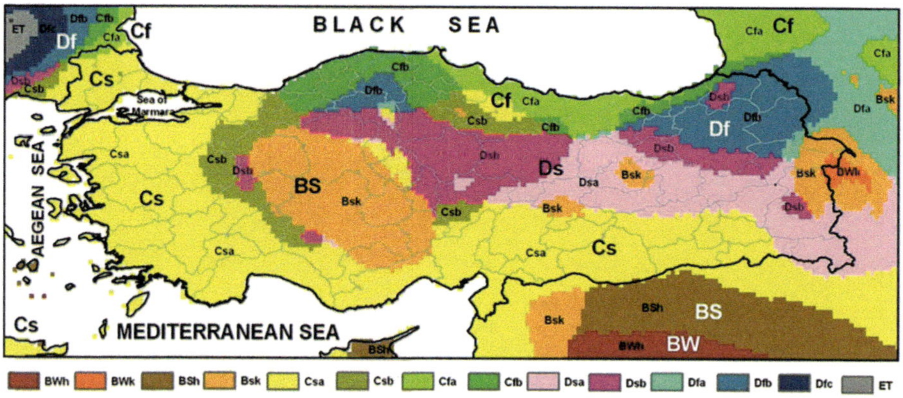

Fig. 4.12 Geographical distributions of climate types in Turkey based on the first, second and third hand letters classification of the Köppen-Geiger climate system. (Turkes 2010)

climate types over the western, central and southern Turkey, are influenced by the pole-ward descending branch of the tropical Hadley cell producing clear, stable, warm and dry cloudless conditions associated with the subsiding and adiabatically warm air masses. Cold fronts or troughs of the trailing mid-latitude cyclones and the convective instability thunder-storms, which are associated with the dynamically originated deep and cold-nuclei upper-air lows and troughs and related with relatively cold air masses, cause occurrence of the summer convectional rainfall over the continental Central and Eastern Anatolia regions along with the inland Mediterranean climate regions. Types of precipitation events (i.e. hydrometeors) are mostly strong rainfall, rain-showers, hail-storms and thunder-storms with heavy rain and hail events.

In winter, however, trailing mid-latitude cyclones are responsible for most of the precipitation events and amounts over the regions. Topographic influences over the mountain chains facing the prevailing weather systems, such as Alps and Atlas Mountains in general, and the North Anatolian Mountains, the Mediterranean Taurus Mountains, and the South-east Tauruses in the north and south of Turkey, significantly contribute to increase the amount, frequency and intensity of the precipitation over these regions with the temperate rainy west coast (Black Sea in Turkey) and the Mediterranean climates, respectively. Cold spells are usually linked to both mP and cP air masses. Winter mP air masses are considerably less cold than cP air, and it may give cool, cloudy and wet conditions, while the cP air may produce cold, stable and dry conditions (Turkes 2010).

The climate of Turkey according to the Köppen-Geiger climate system is considerably diverse (Fig. 4.12), as in many other climate classifications of Turkey. When only the first and the second-hand letters in classification of the Köppen-Geiger climate system are considered, the following major climate types can be identified (Turkes 2010):

(1) **Sub-tropical steppe climate *BS*** (mostly ***BSk***) in Turkey is found in the mid-part of the continental Central Anatolia region and the Van-Igdir district over most-eastern part of the continental Eastern Anatolia region (Fig. 4.12).
(2) **Temperate rainy or humid temperature west coast climate without dry season *Cf*** (mostly ***Cfa*** and ***Cfb***) (humid mesothermal) dominates the Black Sea coastal region of Turkey with the exception of the western sub-region (Fig. 4.12).
(3) The Marmara, Aegean, Mediterranean and the South-eastern Anatolia regions and the western and southern portions (i.e. some parts of the Sakarya and Konya sub-regions) of the continental Central Anatolia Region belong to the **dry summer subtropical Mediterranean climate or temperate rainy climate with dry summers** (humid mesothermal) ***Cs*** (mostly ***Csa***) (Fig. 4.12). Yearly soil-moisture deficiency is not characteristic in **Group *C* climates** in general, whereas seasonal soil-moisture deficiency, particularly in summer months, is evident in the Mediterranean ***Csa*** and ***Csb*** climates due to changes in the hemispheric and regional circulation, air mass and pressure systems producing dry conditions in the summer months.
(4) On the other hand, a **cold snowy forest climate with dry summers** (humid microthermal) ***Ds*** (mostly ***Dsa*** and ***Dsb***) takes place over a relatively large zone on the mid-northern portions of the continental Central and Eastern Anatolia regions of Turkey; whereas a **cold snowy forest climate humid in all seasons** (humid microthermal) ***Df*** (mostly ***Dfb***) exists over relatively small areas seen in the northern portions of the continental Central Anatolia region and the north-eastern Anatolia sub-region (mostly Erzurum-Kars sub-region) of Turkey (Fig. 4.12).

In theory, and according to the UNEP/UNCCD Aridity Index (*AI*), *AI* values below 1.0 generally show an annual moisture (soil water) deficit in average climatic conditions, whereas *AI* values above 1.0 generally show an annual moisture surplus (Fig. 4.13).

The maritime influences on the weather and climate conditions of Turkey are characterized mainly with the mP and mT air masses (Fig. 4.2). In winter, Mediterranean air masses carried by the westerly air flows (W, SW and NWerly) tend to decrease towards the continental inland regions, including the Central, Eastern and South-eastern Anatolia regions of Turkey. Major changes of the inland Mediterranean regions reflect a decrease of precipitation amounts in winter and the number of rainy days, and an increase of precipitation amounts in spring months, particularly with continentality, such as that in the mid-west and south parts of the Central Anatolia Region and the South-eastern Anatolia Region of Turkey. The Northern Anatolia Mountains, on the other hand, separate the Mediterranean, steppe and cold snowy forest climates from the west coast temperate rainy (Black Sea in Turkey) climate, while the Tauruses and the South-eastern Taurus Mountains separate the Mediterranean climate (***Csa***) from the more continental Mediterranean climate (***Csb***), steppe and cold snowy forest climates to the north and the east of the Anatolian Peninsula (Fig. 4.12).

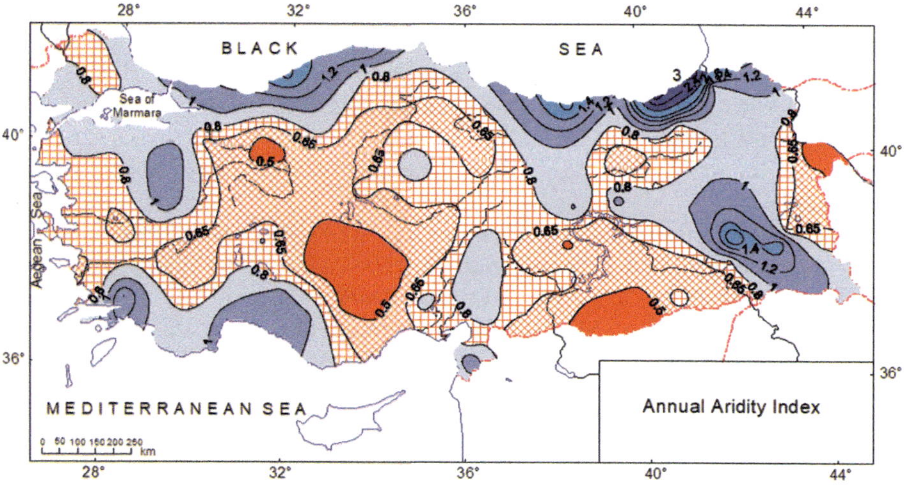

Fig. 4.13 Geographical distribution of the UNEP/UNCCD aridity indices of the 151 climatological and meteorological stations in Turkey, in which arid, semi-arid, dry sub-humid and sub-humid areas of the country are highlighted by hatching with the red color. [Re-arranged in English from Turkes (2013)]

In Turkey, dry sub-humid climatic conditions extend throughout most of the Continental Central Anatolia and South-eastern Anatolia regions, some part of the eastern Mediterranean, and eastern and western parts of the Continental Eastern Anatolia region; while the semi-arid climatic conditions are found only in the Konya Plain and the Igdir district of the Eastern Anatolia region (Fig. 4.13). The areas having values $0.65 < AI < 0.80$, where an annual soil moisture deficit exists, are also concentrated around the semi-arid and dry sub-humid areas of Turkey. The arid-lands in Turkey having *AI* values between 0.20 and 0.80 are likely to have been influenced by the desertification processes (Table 4.2), by considering the existing hydroclimatological conditions, human-induced land degradation, observed and projected climate change and variability, etc. (e.g., Ozturk et al. 2012, 2015; Topcu et al. 2010; Sen et al. 2012; Turkes 1999, 2010, 2013, 2017b, c; Dengiz 2018; Turkes and Akgunduz 2011).

4.3.3 Climatological Drought Probabilities in Turkey

In this sub-section, the long-term climatological *SPI* probabilities in Turkey, calculated by using the monthly *SPI* series, are discussed through considering the SPI classes of below-normal, normal, above-normal, generally dry and wet, extremely-wet and extremely-dry, as given in Table 4.3 (Turkes and Tatli 2008, 2009). Drought is described without dividing it into being meteorological, agricultural and hydrological drought as "lack of water and/or water deficiency occurred when the level of

Table 4.3 Classification of the *SPI* values for calculating the climatological *SPI* probabilities

SPI values	Classification
≥ 2.00	Extremely wet
≥ 1.00	Above normal
> 0.0	Wet
$-0.99 - 0.99$	Normal
< 0.0	Dry
≤ -1.00	Below normal
≤ -2.00	Extremely dry

natural water availability used by several systems on the Earth surface is below long-term average or the normal level at the regional scale and for a particular time period. In general, drought events are considered as three dimensional natural events that are characterized with components of magnitude/severity, length/frequency and geographical distribution pattern" (Turkes 2010, 2013, 2017c).

Drought occurs in conjunction with climate change and variability in various time periods and may last consequently a few years or more than a few years. Drought has no one mean. Drought events are divided into four main groups; meteorological, agricultural, hydrological and socio-economical droughts. Description of a drought in general is the meteorological drought above mentioned. Agricultural drought occurs in conjunction with loss of soil moisture around plant root level, and it causes loss of plant and agricultural yield. Hydrological drought occurs in conjunction with the decreased amount or level of water, which is lower than amount demanded, in hydrological systems or in reservoirs in a region or district. Meteorological, agricultural and hydrological drought may turn into a socio-economical drought, when the magnitude of a drought event and the length of time period with drought increase. This state of a drought event is also related with low degree of coping ability of the system (agriculture, energy, ecosystem, urban and socio-economical, etc.), and lacks of drought risk management and drought management plans or weak implementation of these management plans (Turkes 2014b, 2017c).

Drought event is a serious threat effecting human life and health, socio-economic and ecological systems directly or indirectly at several degrees. Long-term drought events negatively affect agriculture, forestry, livestock, ecosystems, and hydroelectric energy, etc., by decreasing timely, qualified and sufficient water supplies. As a summary, the well-known adverse effects of geographically larger and significant drought events longer than 2 years can be listed as follows:

(i) Decreasing amount of ground water;
(ii) Decreasing water amount and level at stream-flows, reservoirs and water structures, etc.;
(iii) Increasing water pollution; salinization of water and soil;
(iv) Increasing soil pollution due to agricultural chemicals;
(v) Increasing soil erosion, land degradation and desertification;
(vi) Increasing the risk of wild bush and forest fires;

(vii) Damages to forest and rural wetland ecosystems and weakening and loss of biodiversity;
(viii) Increasing vulnerability level of natural vegetation and agricultural ecosystems due to adverse effects of disease, fires, wind deflation, and other related factors.

Spatial distributions of climatological probabilities for *below-normal* (the *SPI* values ≤ -1) monthly *SPI* values are illustrated in Fig. 4.14a. Maximum probabilities obtained for the below-normal conditions are nearly 0.35 on the Mediterranean cost of Turkey, and 0.37 along the Turkey-Syria border. On the other hand, throughout the Black Sea coastal region, the probabilities of being below-normal are generally 0.15 and a little below.

The probability of *being normal* (the *SPI* values within the interval of ± 1) reveals a spatial pattern where the displayed probabilities increase from the western and southern regions of Turkey (Fig. 4.14b), in which the Mediterranean rainfall regime with a high inter-seasonal and inter-annual variability is dominant, to the Black-Sea rainfall region, where the annual precipitation totals are greater and seasonality and year-to-year variability are lower than almost all of other regions of Turkey (Turkes

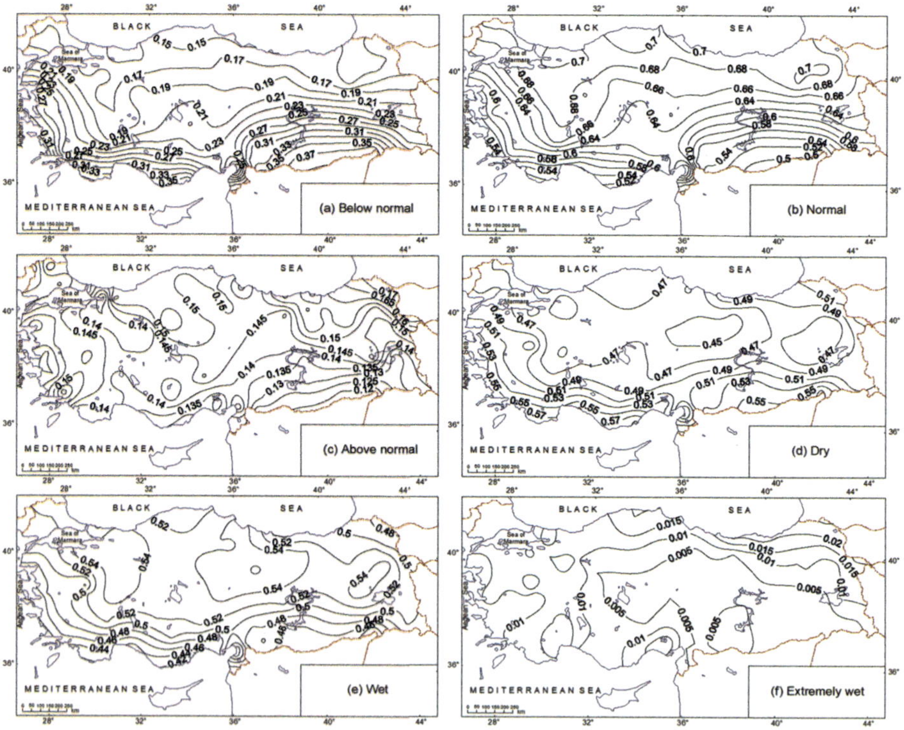

Fig. 4.14 Geographical distributions of the probabilities obtained for various *SPI* classes of monthly total precipitation series from 96 stations in Turkey. [Re-arranged in English from Turkes and Tatli (2008)]

1999, 2003, 2013). Probabilities of being normal over the central and the northern regions of the country have ratios varying from 0.65 to 0.70. Minimum probabilities of being normal are 0.52 on the Mediterranean coasts, and they are approximately 0.50 along the Turkey-Syria border.

The probabilities of being *above-normal* (the *SPI* values ≥ 1) show minima over the western and southern regions of Turkey (Fig. 4.14c), where the seasonality and inter-annual variability in total precipitation series are high in all seasons. The probabilities over these areas are within the interval of 0.12–0.14, and they decrease from the southern regions towards the coastal belt of the Black Sea rainfall region and the north-eastern Anatolia. These patterns are consistent with the previous studies of Turkes (1996, 1998, 2003, 2013, etc.) in terms of the main characteristics of the precipitation climatology in Turkey.

When the probabilities of monthly *SPI* values *being dry* (the *SPI* values <0) in general are considered, the probabilities of being dry are detected as 0.57 on the Mediterranean coastal belt and 0.55 over the Turkey-Syria border area. Probabilities tend to decrease from the Mediterranean rainfall region towards the north and reach a minimum value of 0.45 in the continental Central Anatolia Region (Fig. 4.14d).

The minimum probabilities of monthly *SPI* values *being wet* (the *SPI* values >0) in general are found as 0.44 over the Mediterranean coasts and 0.46 over the continental Mediterranean (South-eastern Anatolia) Region, whilst the probabilities of being wet reach a maximum level of about 0.54 over the eastern Marmara, western Black Sea, continental Central, and Eastern Anatolia rainfall regions (Fig. 4.14e).

The probabilities of *being extremely-wet* (the *SPI* values ≥ 2) are significantly low over Turkey, as is expected for most of the Mediterranean macro climate regions and more generally in the subtropical climate belt (Fig. 4.14f). The probabilities of being extremely-wet are at the level of 0.01 and below over the majority of Turkey. The maximum probabilities of being extremely-wet are about 0.02 in the middle and eastern Black Sea sub-regions and in the northern part of the north-eastern Anatolia sub-region, respectively.

One of the most significant and remarkable results for the *SPI* probabilities in Turkey is detected in the probabilities of *being extremely-dry* (the *SPI* values ≤ -2) (Fig. 4.15). The probabilities of being extremely-dry evidently depict maximum values on the Turkish shores of the Mediterranean Sea and over the border zone between Turkey and Syria with the highest probability of about 0.27 in the Harran and Akcakale plains, which are climatologically one of the most arid and desert-like environments of Turkey (Turkes 2003, 2013).

The probabilities of being extremely-dry reach their maximum values of about 0.21–0.23 on the Mediterranean coast of Turkey and 0.27 over the Turkey-Syria border. The estimated probabilities tend to decrease clearly from the southern regions towards the Black Sea shores. The lowest probabilities of being extremely-dry are detected as about 0.03 in the western and eastern Black Sea sub-regions and the north-eastern Anatolia sub-region of Turkey, respectively (Fig. 4.15).

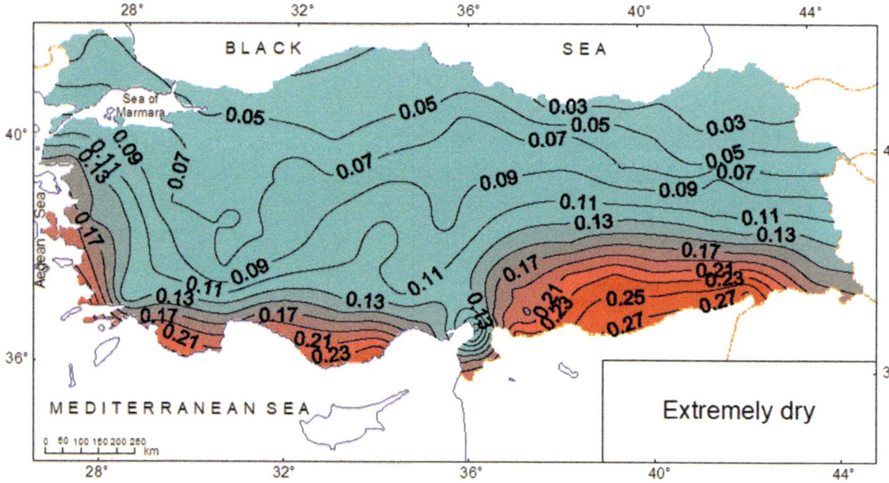

Fig. 4.15 Geographical distributions of the drought probabilities from the *Extremely-dry SPI* class of monthly total precipitation series from 96 stations in Turkey. [Re-arranged in English from Turkes and Tatli (2008)]

4.3.4 Drought Vulnerability and Risk

This sub-section aims at performing a vulnerability and risk assessment of Turkey based on a recent paper of Turkes (2017c), by considering the recent trends and approaches on the subject for disaster risk management of the weather and climate extremes and disasters towards climate change adaptation and reduction of its impacts.

Economic and social losses caused by weather and climate-related extreme events and disasters have been increasing with a great spatial and inter-annual variability in many regions of the world and in Turkey (Turkes and Erlat 2018b). The character, severity and impact powers of the extreme weather and climate events and disasters closely depend, not only on exposure and vulnerability levels that are varying across temporal and spatial scales, but also on economic, social, geographic, demographic, cultural, institutional, governance, and environmental and ecological factors (Turkes 2017c). A large part of Turkey is located over the dry-summer subtropical macro climate belt characterized with many climate and climate-related problems arising from its nature. In this respect, Turkey is a mid to high climate risk-prone country, with respect to not only the present climate and climate variability, but also to future climate change and variability. Turkey is also one of the South-eastern Europe and Eastern Mediterranean countries that are considerably vulnerable to desertification processes varying mid to high degrees (CEM 2017; Turkes 2017b, c).

In a study performed by Turkes (2017b), a Social Vulnerability Index (*SVI*) was used for determining "Turkey's 'Potential' Drought Hazard/Disaster Risk." The *SVI* is a tool that was originally developed and applied in the United States to help emergency response planners, public health experts, and officials geographically identify and map the communities that will most likely need support before, during, and after a hazardous event.

The estimation for Turkey's Vulnerability Index in terms of the impacts of drought events was based on five vulnerability factor classes including the following:

(i) Education status with one indicator (indicators not given here);
(ii) Demography with four indicators;
(iii) Economic activity with six indicators mostly related to agricultural activities and production;
(iv) Public health and health protection status and public coping capacities with five indicators;
(v) Natural environmental characteristics with three indicators.

In this frame, based on a well-known drought risk formula and the climatological drought occurrence probabilities estimated from the *SPI* series of Turkey, "Turkey's Drought Hazard/Disaster 'Potential' Risk Modelling and Valuation" was performed and mapped on the main administrative city provinces by using the spatial analysis tools of a geographical information system (GIS) (Fig. 4.16b).

When the socioeconomic and environmental conditions of the drought vulnerability in Turkey is considered, relatively the lowest and the highest vulnerability scores can be ranked as follows in terms of geographical regions, sub-regions, and the city provinces of Turkey (Fig. 4.16a):

The city provinces with relatively the lowest drought vulnerability levels include Edirne and Kirklareli provinces in the Western Thrace (European part of the Marmara Region); Bilecik, Bolu, Karabuk, Zonguldak and Bartin provinces in the Western Black Sea sub-region along with the city provinces in the Eastern Black Sea sub-region (Giresun, Trabzon, Rize and Artvin); and the Erzincan, Bayburt, Tunceli and Elazig provinces, and Erzurum and Ardahan provinces in the mid-west and north-east portions of the Eastern Anatolia Region, respectively (Fig. 4.16a). The city provinces with relatively the highest drought vulnerability scores consist of Istanbul and the surrounding cities, and Bursa in the Marmara Region in general. Others include Afyonkarahisar, Izmir and the surrounding city provinces except the provinces Mugla, Usak and Kutahya; Antalya, Adana, Hatay and Kahramanmaras provinces in the Mediterranean Region characterized with a true dry and hot summer Mediterranean climate, with a Köppen-Geiger climate (***Csa***), along with almost all city provinces in the South-eastern Anatolia Region with a continental Mediterranean rainfall regime (Gaziantep, Adiyaman, Sanliurfa, Diyarbakir, Mardin and Batman). This vulnerability level is seen also in almost all eastern and south-eastern portions of the Eastern Anatolia Region, including the cities of Sirnak, Mus, Van, Agri and Hakkari provinces (Fig. 4.16a).

Based on a special classification of the monthly *SPI* anomalies in order to estimate the drought risk of Turkey, Turkes (2017c) analyzed and evaluated the Drought Hazard-Disaster Risk (*DHDR*) under the conditions of Below-normal "Exact Drought Probability" for the administrative city provinces of Turkey (Fig. 4.16b). According to this special classification, relatively the lowest and the highest risk evaluations derived from the *DHDR* under the conditions of Below-normal "Exact Drought Probability" can be summarized as follows:

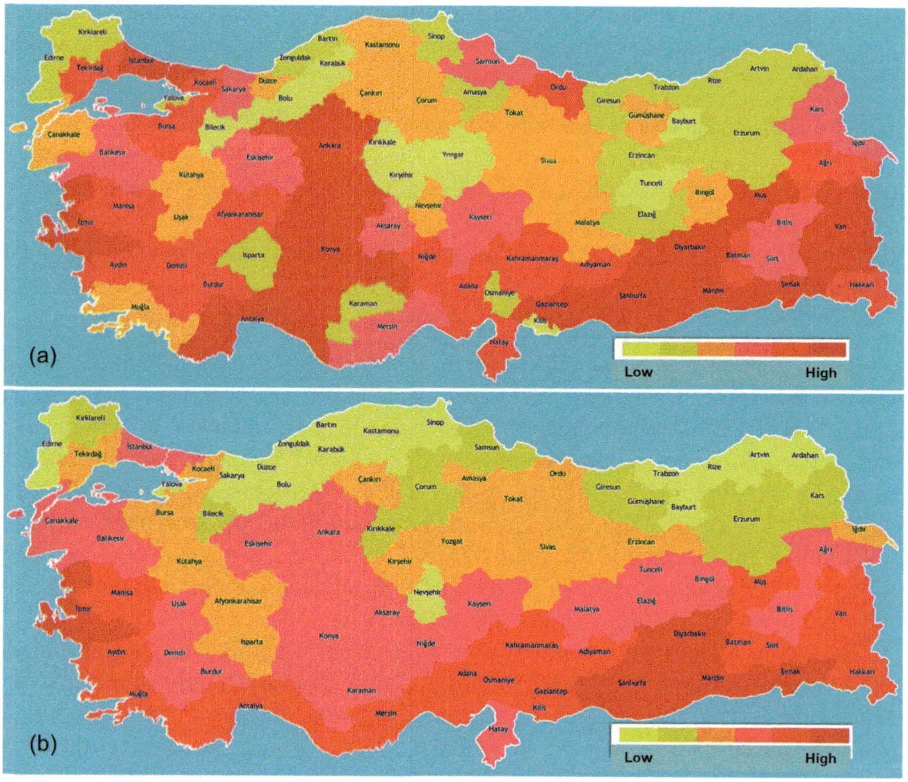

Fig. 4.16 Geographical distributions of (**a**) drought hazard-disaster *SVI* and of (**b**) evaluations for the Drought Hazard-Disaster Risk under the conditions of Below-normal "**Exact Drought Probability**" based on the *SPI*, both of which were estimated for Turkey's administrative city provinces. [Re-arranged in English from Turkes (2017b)]

The relatively highest *DHDR* scores of the geographical regions/sub-regions and the city provinces in Turkey are: Izmir (the highest), Manisa, Aydin and Mugla city provinces in the Aegean Region; Antalya, Mersin, Adana, Osmaniye and Kahramanmaras in the Mediterranean Region; Sanliurfa, Diyarbakir, Mardin (the highest ones) and Kilis, Gaziantep, Adiyaman, Siirt and Batman in the South-eastern Anatolia Region; Sirnak (the highest), Siirt, Hakkari, Van and Mus in the Eastern Anatolia Region; Ankara, Eskisehir, Konya, Aksaray, Karaman, Nigde and Kayseri in the Central Anatolia Region; and Istanbul, Canakkale and Balikesir in the Marmara Region, respectively (Fig. 4.16b). On the other hand, relatively the lowest *DHDR* levels of the geographical regions/sub-regions and the city provinces in Turkey under the conditions of Below-normal "Exact Drought Probability" consist of Edirne (the lowest) and Kirklareli in the Western Thrace of the Marmara Region along with Bilecik province of the Eastern Marmara sub-region; Corum, Samsun, Giresun and Gumushane and the all the city provinces from Sakarya to Rize and Artvin (the lowest ones) except the city provinces of Amasya, Tokat and Ordu in the

Black Sea Region; Nevsehir (the lowest), Kirikkale and Corum in the Central Anatolia Region; and Ardahan (the lowest), Kars and Erzurum in the Eastern Anatolia Region, respectively (Fig. 4.16b).

4.4 General Soil Characteristics of Turkey with Respect to Climate

4.4.1 Climate and Soil

Climate is usually considered the dominant factor on soil formation. However, the soil parent material still claims an impressive number of adherents. The natural topography and geomorphologic units, which are particular relief groups developed under the specific morphoclimatic or morphogenetic regions of the Earth, are also important for soil formation. In recent years, the role of vegetation in soil formation has aroused further interest. Yet, because systematic quantitative studies of the relationships between soil properties and all soil-forming factors have not been adequate yet to explain all these relationships, all the remaining factors can be kept constant in order to detect the role played by each soil-forming factor.

However, if we would consider the soil as a physical system as proposed by Jenny (1994), we may add substances to or remove from it, because the soil system is an open system. Every soil system is characterized by properties that we may designate by symbols, such as *s1, s2, s3, s4, s5*, etc. For example, *s1* may indicate nitrogen content, *s2* acidity, *s3* apparent density, *s4* amount of calcium (Ca), *s5* pressure of carbon dioxide (CO_2), etc. The simplest method of expressing the assumptions of interrelationships is given in form of the Eq. (4.2) as follows (Jenny 1994):

$$F\,(s1, s2, s3, s4, s5, \bullet\bullet\bullet) = 0 \tag{4.2}$$

On the other hand, when the soil and environment relationship is considered, it is generally realized that the soil system is only a part of a much larger system that is composed of the most upper part of the lithosphere, the lowest part of the atmosphere (i.e. lower troposphere), and a considerable part of the biosphere. Consequently, properties of the soil system that are universally recognized as soil characteristics should be considered. They include the following: soil climate (soil temperature and soil moisture, etc.), kind and number of soil organisms, and topography, or the shape of the surface of the soil system. These properties can be indicated by special symbols (cl' = climate, o' = organisms, t' = topography) (Jenny 1994), which are included in the Eq. (4.3)) as:

$$F\left(cl', o', t', s1, s2, s3, \bullet\bullet\bullet\right) = 0 \tag{4.3}$$

The following set of individual equations of soil-forming factors can also be written:

$$s = fcl\ (climate)\ o, t, p, ti, \ldots \tag{4.4}$$

$$s = fo\ (organisms)\ cl, t, p, ti, \ldots \tag{4.5}$$

$$s = ft\ (topography)\ cl, o, p, ti, \ldots \tag{4.6}$$

$$s = fp\ (parent\ material)\ cl, o, t, ti, \ldots \tag{4.7}$$

$$s = fti\ (time)\ cl, o, t, p, \ldots \tag{4.8}$$

The subscripts indicate that the remaining soil-forming factors do not vary. For instance, in order to study accurately the soil-climate relationships, it is necessary that comparisons should be restricted to soils of identical origin and the time of soil formation, etc. On the other hand, the climate factor is so complex that no single numerical value can be assigned to a given climate. It becomes necessary to work with individual climatic variables, the most important of which are air humidity (H), air temperature (T), and soil moisture (m). Treating these three important climate elements as independent variables, an approximate equation may be written as follows:

$$s = f\ (H, T, m)\ o, t, p, ti, \ldots \tag{4.9}$$

4.4.2 *Climate-Related Land Degradation*

Soils in the semiarid and the dry-sub humid climate regions (drylands) of Turkey are vulnerable to land degradation, either because they have poor resistance to erosion, or because of their chemical and physical properties. A preliminary study on the soil moisture conditions of the soils in Turkey was published first by Caglar (1937). He investigated the soil-water relationships under the conditions of the continental Central Anatolia Region. The hydrology, climate, soil conditions, microorganisms and soil–water relations, the quality of water used for irrigation and characteristics of the waters in Turkey had also been discussed in this publication. In fact, the majority of the publications on soil water conservation studies in Turkey were concentrated on soil erosion in the past (e.g. Yamanlar 1956; Celebi 1970, 1971; Mermut et al. 1981, etc.). In this respect, for instance, sand dune rehabilitation studies were undertaken for the sand dune areas in Adana and Mersin-Tarsus districts of the Eastern Mediterranean sub-region of Turkey in order to enhance carbon sequestration and soil formation conditions at the stone pine canopies/root-zones of arenosols (Akca et al. 2010; Polat and Kapur 2010).

The most important and well-known soil degradation and desertification district in Turkey is Karapinar town (Konya). The Karapinar town is located over the plateaus and plains between in Konya and Eregli districts of Turkey. It is characterized with a semi-arid steppe climate of the continental Central Anatolia Region, and annual average precipitation amount is about 250–300 mm over the district (Turkes 1999, 2003, 2013). Because of the semi-arid climate, no forests naturally grew up in the district, and steppe vegetation dominated over the land. Consequently, the flat terrain had been used as a pasture for long years and especially sheep rising was practiced here. In areas where the vegetation was destroyed as a result of intensive grazing and drought events, the thin layer of the soil (mainly clay and silt) was removed (Groneman 1968). The remaining sand was removed by the wind deflation in short distances and turned the land into a large continental dune. So the ancient dune-like landscape was probably built up, mainly due to the fact that the native steppe vegetation was damaged by over grazing, trampling and by other human activities (Groneman 1968). The wind eroded parts of this ancient dune-like landscape altered it in many places and formed many blowouts. The shifting sand accumulated downwind and built up new dunes during the 1960s in many places (Fig. 4.17). In fact, this increasing activity in the continental dune field began to affect the Karapinar settlement as of 1956. Consequently, the Karapinar town and its some nearby villages had faced the danger of abandonment due to immigration in the early 1960s. Approximately 4300 hectare (ha) areas were covered with an inland dune in the south-southwest of the district, developed due to severe wind erosion and deflation (Groneman 1968; Carkaci 1999; Yildirim 1999; Akca et al. 2000; Yildirim and Akay 2010; Kantarci et al. 2011, etc.).

Fig. 4.17 The degraded land by strong wind erosion and deflation, and the Barkan-like Sand Dune formations in Karapinar district during the late 1950s and the 1960s. (Carkaci 1999)

Fig. 4.18 A general view of a nearby village of Karapinar Town in early 1960s. (Carkaci 1999)

Finally, the soils lost their yield capacity and sand dunes occurred as the result of erosion. It was observed that clouds of sand and dust storms rose, and vehicles on the Konya-Adana Highway were dragged, and the paint of the cars was totally or partially damaged. Children could not go to school because of sand storms, machines did not work, and the incidence of ear-nose-throat diseases increased among the people (Anonymous 2007). Winds that cause erosion in the district are mainly south-south-westerly. Finally, these significantly adverse conditions forced the local community to leave their places as they were threatened by this disaster (Fig. 4.18). Grazing and picking woody plants is still forbidden in the protected and reclamation areas. In the past, all the farmer villages had been vacated, except the village Kindam, which is situated about three kilometers southwest from Karapinar (Groneman 1968). Formerly, all the deserted villages had been summer camps (yayla in Turkish). In winter, the summer camps were deserted, and then the farmers started living in the towns.

Following this destruction, the necessary technical sand-dune preventing and/or controlling works and actions began in 1962/63 by the former General Directorate of Soil-Water, VI. Regional Directorate of Konya Soil-Water, Chief Engineering of Wind Erosion Plan and Application Group (Anonymous 2007, 2010; Carkaci 1999).

Reasons for the increased wind erosion and deflation in the Karapinar district can be summarized as follows (Groneman 1968; Carkaci 1999; Akca et al. 2000; Yildirim and Akay 2010; Kantarci et al. 2011):

- The sandy material originated from old lake bed mobilized, following the drying up of the lake;

- The hot and semi-arid continental climate conditions along with the high precipitation seasonality and year-to-year variability;
- Impetuous destruction of pastures because of overgrazing and the use of pasture plants for fuel,
- Excessive tillage, particularly by the Shock-Disc Plough that degrades the soil structure and buries productive surface horizon of the soil, and
- Karapinar is located on a strong wind course.

The first step taken against erosion in the district was the establishment of the "Association for Saving Karapinar from Erosion" in 1959. Afterwards, studies were started by the former Directorate of Konya Soil-Water Research Institute in 1962. A team was formed of technical personnel, and an area of 16,000 ha was taken under control by enclosing it with a wire fence. Then, 3000 ha of this area was assigned to the Turkish Armed Forces to be used for military purposes. The remaining 13,000 ha area was divided into four sections, based on the problems observed. Soil improvement practices started in this area considering the degree of the problem. The former Directorate of Konya Soil-Water Research Institute maintained its studies continuously for 10 years, and when the improvement studies were completed, the area was assigned to Konya Institute of Soil and Water Research Directorate in 1973 to be used for protection control, research and production studies. Today, 4300 ha of this land is given back to farmers, and studies are continued in the 8700 ha under the control of the government (Fig. 4.19) (Yildirim 1999). The research station in Karapinar town is now entitled as "Research Institute of the Directorate of Konya Soil and Water Resources".

Fig. 4.19 A present view of the Karapinar wind erosion area that has been opened to sustainable and productive agricultural practices for many years. (Photograph: Murat Turkes 17.06.2013)

4.4.3 Soil Geography

Generalized soil map of Turkey (Cullu et al. 2018), which is based on the World Reference Base for Soil Resources (WRB), is given in Fig. 4.20. These soil data compilations were subsequently analyzed in a GIS environment, and the soils were classified in accordance with the WRB (IUSS Working Group WRB 2015) procedures by Cullu et al. (2018), which are primarily intended to facilitate the correlation of national and local systems with international soil classification schemes.

According to this classification, the most widely distributed soils of Turkey (i.e. mapped WRB soil groups) are the Cambisols/Cambisols-Leptosols (63.55%), Fluvisols (9.50%), Calcisols/Calcisols-Leptosols (9.39%), Vertisols (5.15%), Alisols-Acrisols-Podsols (3.25%), Kastanozems (2.85%), and Luvisols (2.05%). The group of Cambisols/Cambisols-Leptosols is detected almost in all the geographical regions of Turkey, with the exception of the coastal zone of the Black Sea Region, south-west Anatolia, the mid-south part of the Central Anatolia Region, and the southern part of the South-eastern Anatolia Region, the last two regions of which are mainly characterized with the Calcisols/Calcisols-Leptosols soil groups. Cullu et al. (2018) pointed out that the Cambisols, Calcisols, Vertisols, and Fluvisols are the most appropriate and extensively managed soil ecosystems, particularly for field crops and cereals. Furthermore, the Calcisols, Luvisols, Acrisols-Alisols-Podsols, and Arenosols are significantly suitable for the production of some crops, such as the

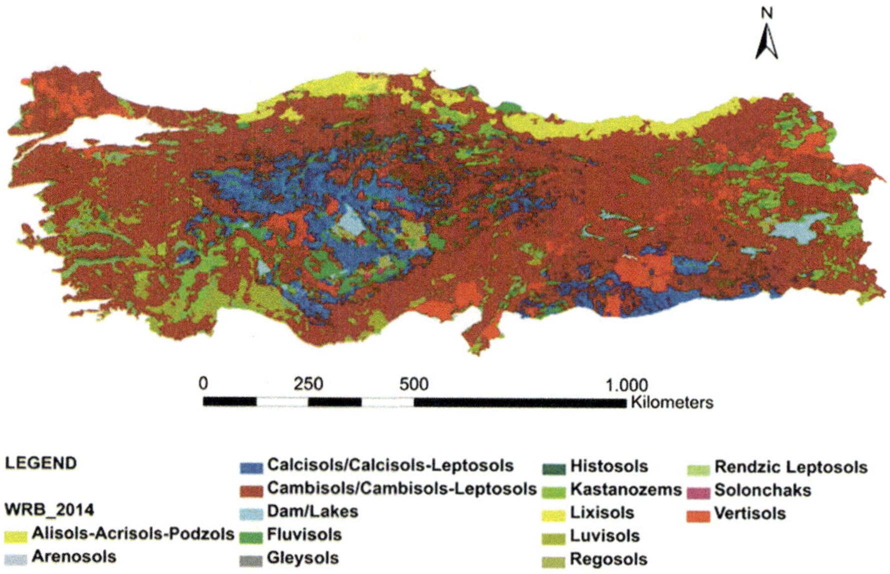

Fig. 4.20 Geographical distributions of the generalized WRB soil groups of Turkey. (Cullu et al. 2018)

olives, carobs, apricots, hazelnuts, and stone pine. On the other hand, because the Solonetz, Solonchaks, Gleysols, and Regosols are the most problematic soil ecosystems for agricultural production, they need to be treated by supplementary activities for sustainable use.

Acknowledgements The authors are grateful the NOAA/OAR/ESRL/PSD (Boulder, Colorado) for providing "the 20th Century Reanalysis V2 data" at "http://www.esrl.noaa.gov/psd/data/20thC_Rean/ and http://www.esrl.noaa.gov/psd/data/gridded/data.20thC_ReanV2.html", and the Turkish Meteorological Service for providing the climatological and meteorological data.

References

Akca E, Kapur S, Serin M, Cevik B, Eswaran H (2000) Karapinar: a case study of rehabilitation of a wind eroded area in Turkey. 13–17 June, 2000, Desertification meeting, Konya, pp 202–210

Akca E, Kapur S, Tanaka Y, Kaya Z, Bedestenci HC, Yakti S (2010) Afforestation effect on soil quality of sand dunes. Polish J Environ Stud 19(6):1109–1116

Anonymous (2007) Yeseren Col Karapinar. Republic of Turkey, Ministry of Agriculture and Rural Affairs, General Directorate of Agricultural Research, Soil and Water Resources Research Institute of Konya (in Turkish)

Anonymous (2010) Karapinar is greener now. Booklet. Konya Directorate of Research Station of Soil, Water and Combating Desertification, Karapinar, Konya

Barry RG, Chorley RJ (2003) Atmosphere, weather and climate, 8th edn. Routledge Taylor and Francis Group, London, 421 p

Çaglar KO (1937) Anadolu Sartlarina gore Su ve Toprak Munasebetlerinin Arastirilmasi. Ankara Yuksek Ziraat Enstitusu, Ankara

Carkaci A (1999) Examples of good practices of Turkey for combating wind erosion (Examples of Karapinar). Directorate of Research Institute of Soil Water and Combating Desertification, Karapinar, Konya

Celebi H (1970) The soil losses induced by wind erosion and some physical and chemical properties of soils of Sivas Ulas State Farm. Guven Publisher, Ankara. 46 p (in Turkish)

Celebi H (1971) Study on the resistance to soil erosion in the Kars meadow and Sarikamis Forest soils. Agric Eng 53:18–22 (in Turkish)

CEM (2017) Turkiye Collesme Modeli, Teknik Ozet. Collesme ve Erozyonla Mucadele (CEM) Genel Mudurlugu, Ankara

Coughlin KT, Tung KK (2001) QBO signal found at the extratropical surface through northern annular modes. Geophys Res Lett 28:4563–4566

Coughlin KT, Tung KK (2004) Eleven-year solar cycle signal throughout the lower atmosphere. J Geophys Res Atmos 109:D21105.1–D21105.7

Cullu MA, Gunal H, Akca E, Kapur S (2018) Soil geography. In: Kapur S, Akca E, Gunal H (eds) The soils of Turkey. Springer World Soils Book Series. Springer, pp 105–110. ISBN: 978-3-319-64390-8. https://doi.org/10.1007/978-3-319-64392-2

Dengiz O (2018) Potential impact of land use change on land productivity dynamics with focus on land degradation in a sub-humid terrestrial ecosystem. Theor Appl Climatol 133(1–2):73–88. https://doi.org/10.1007/s00704-017-2162-1

Elibuyuk M, Erkan Y (2010) Altitude steps and slope groups of Turkey in comparison with geographical regions and sub-regions. Cografi Bilimler Dergisi (J Geogr Sci) 8:27–55

Erinc S (1969) Klimatoloji ve Metodlari. Istanbul Universitesi Cografya Enstitusu, Yayin No. 35, Istanbul

Erinc S (1982) Jeomorfoloji, Cilt: I. Genisletilmis ucuncu baski. Istanbul Universitesi Edeb. Fak. Yay. No: 2931. Edebiyat Fakultesi Matbaasi, Istanbul

Erlat E, Turkes M (2012) Analysis of observed variability and trends in numbers of frost days in Turkey for the period 1950–2010. Int J Climatol 32(12):1889–1898. https://doi.org/10.1002/joc.2403

Erlat E, Turkes M (2013) Observed changes and trends in numbers of summer and tropical days, and the 2010 hot summer in Turkey. Int J Climatol 33(8):1898–1908. https://doi.org/10.1002/joc.3556

Erlat E, Turkes M (2016) Dates of frost onset, frost end and the frost-free season in Turkey: trends, variability and links to the North Atlantic and Arctic oscillation indices, 1950−2013. Clim Res 69:155–176. https://doi.org/10.3354/cr01397

Essenwanger OM (2001) Classification of climates. World survey of climatology volume 1C (General climatology 1C), Epilogue by Landsberg, H.E. Elsevier, Amsterdam

Groneman AF (1968) The soils of the wind erosion control camp area Karapinar, Turkey. A part of Great Konya Basin soil research and trainings project report, Vageningen

Guttman NB (1998) Comparing the palmer drought index and the standardized precipitationindex. J Am Water Resour Assoc 34:113–121

Guttman NB (1999) Accepting the standardized precipitation index: a calculation algorithm. J Am Water Resour Assoc 35:311–322

Henderson-Sellers A, Robinson PJ (1986) Contemporary climatology, 1st edn. Longman Scientific & Technical, Essex/New York, 439p

Huang NE, Shen Z, Long SR, Wu MC, Shih HH, Zheng Q, Yen NC, Tung CC, Liu HH (1998) The empirical mode decomposition and the Hilbert spectrum for nonlinear and nonstationary time series analysis. Proc R Soc London Ser A Math Phys Eng Sci 454:903–995

Ilhan E (1976) Geology of Turkey, ODTU Muhendislik Fakultesi Yayin No. 51, Ankara, 239 pp (in Turkish)

IUSS Working Group WRB (2015) World reference base for soil resources 2014, update 2015 international soil classification system for naming soils and creating legends for soil maps. World soil resources reports No. 106. FAO, Rome, 203p

Iyigun C, Turkes M, Batmaz I, Yozgatligil C, Gazi VP, Koc EK, Ozturk MZ (2013) Clustering current climate regions of Turkey by using a multivariate statistical method. Theor Appl Climatol 114:95–106. https://doi.org/10.1007/s00704-012-0823-7

Jenny H (1994) Factors of soil formation: a system of quantitative pedology. Foreword by Ronald Amundson. Dover Publications, New York

Kantarci MD, Ozel HB, Ertek M, Kirdar E (2011) An assessment on the adaptation of 6 tree species to steppe habitat during Konya-Karapinar sand-dune afforestations. Bartin Orman Fakultesi Dergisi (J Bartin For Fac) 13(19):107–127

Ketin I (1966) Anadolu'nun Tektonik Birlikleri. MTA Dergisi 66:20–34

Ketin I (1983) Turkiye Jeolojisine Genel Bir Bakis. Istanbul Teknik Universitesi (ITU) Kutuphanesi Sayi: 1259. ITU Matbaasi, Gumussuyu, Istanbul

Ketin I (1994) Genel Jeoloji, Yerbilimlerine Giris. Dorduncu Baski, Istanbul Teknik Universitesi (ITU) Vakfi Yayinlari No. 22, Istanbul

Köppen W (1936) Das geographisca System der Klimate. In: Kööppen W, Geiger G (eds) Handbuch der Klimatologie. 1. C. Gebr, Borntraeger, pp 1–44

Köppen W, Geiger G (1954) Klima der Erde (map). Justus Perthes, Darmstadt, Germany. American distributor, A. J. Nystrom, Chicago

Kutiel H, Turkes M (2005) New evidence about the role of the North Sea – Caspian pattern (*NCP*) on the temperature and precipitation regimes in continental Central Turkey. Geogr Ann Ser B 87:501–513

Kutiel H, Hirsch-Eshkol TR, Turkes M (2001) Sea level pressure patterns associated with dry or wet monthly rainfall conditions in Turkey. Theor Appl Climatol 69:39–67

Lolis CJ, Turkes M (2016) Atmospheric circulation characteristics favoring extreme precipitation in Turkey. Clim Res 71:139–153. https://doi.org/10.3354/cr01433

McKee TB, Doesken NJ, Kleist J (1993) The relationship of drought frequency and duration of time scales. Presented at the eighth conference on applied climatology. American Meteorological Society, Anaheim CA, pp 179–186

McKee TB, Doesken NJ, Kleist J (1995) Drought monitoring with multiple time scales. Presented at the ninth conference on applied climatology. American Meteorological Society, Dallas TX, pp 233–236

Mermut AR, Tanju O, Hatipoglu F, Baskaya HS, Aktas M (1981) Annotated bibliography on Turkish soil science 1928–1976. Documentation of Turkish Research and Technical Centre, 1007 p (in Turkish)

Ozturk T, Altinsoy H, Turkes M, Kurnaz ML (2012) Simulation of temperature and precipitation climatology for Central Asia CORDEX domain by using RegCM 4.0. Clim Res 52:63–76. https://doi.org/10.3354/cr01082

Ozturk T, Ceber ZP, Turkes M, Kurnaz ML (2015) Projections of climate change in the Mediterranean Basin by using downscaled global climate model outputs. Int J Climatol 35:4276–4292. https://doi.org/10.1002/joc.4285

Peel MC, Finlayson BL, McMahon T (2007) Updated world map of the Köppen-Geiger climate classification. Hydro Earth Syst Sci 11:1633–1644

Polat O, Kapur S (2010) Soil quality parameters in Arenosols under Stone Pine (Pinus pinea L.) plantations in the Turan Emeksiz sand dune Area. Cukurova University, Institute of Science. J Sci Eng 3(2):167–174. Adana (in Turkish)

Russell RJ (1931) Dry climates of the United States: I climatic map. Univ Calif Publ Geogr 5:1–41

Sahin S, Turkes M, Wang S-H, Hannah D, Eastwood W (2015) Large scale moisture flux characteristics of the Mediterranean basin and their relationships with drier and wetter climate conditions. Clim Dyn 45(11):3381–3401. https://doi.org/10.1007/s00382-015-2545-x

Sen B, Topcu S, Turkes M, Sen B, Warner JF (2012) Projecting climate change, drought conditions and crop productivity in Turkey. Clim Res 52:175–191. https://doi.org/10.3354/cr01074

Sengor AMC (1980) Turkiye'nin Neotektoniginin Esaslari. Turkiye Jeoloji Kurumu Yayini. DSI Basim ve Foto-Film Isletme Mudurlugu Matbaasi, Ankara

Strahler AN (1973) Introduction to physical geography, 3rd edn. Wiley, New York

Thom HCS (1966) Some methods of climatological analysis. Technical note no. 81. WMO, Geneva

Thornthwaite CW (1948) An approach toward a rational classification of climate. Geogr Rev 38:55–94

Topcu S, Sen B, Turkes M, Sen B (2010) Observed and projected changes in drought conditions of Turkey. In: Options Méditerranéennes, Series A, Mediterranean Seminars 2010, No. 95 – Economies of drought and drought preparedness in a climate change context. CIHEAM, Paris, pp 123–127

Trigo R, Xoplaki E, Zorita E, Luterbacher J, Krichak S, Alpert P, Jacobeit J, Saenz J, Fernandez J, Gonzalez-Rouco F, Garcia-Herrera R, Rodo X, Brunetti M, Nanni T, Maugeri M, Turkes M, Gimeno L, Ribera P, Brunet M, Trigo I, Crepon M, Mariotti A (2006) Chapter 3: Relations between variability in the Mediterranean region and mid-latitude variability. In: Lionello P, Malanotte-Rizzoli P, Boscolo R (eds) Mediterranean climate variability. Elsevier Developments in Earth & Environmental Sciences 4, Amsterdam, pp 179–226

Turkes M (1996) Spatial and temporal analysis of annual rainfall variations in Turkey. Int J Climatol 16:1057–1076

Turkes M (1998) Influence of geopotential heights, cyclone frequency and southern oscillation on rainfall variations in Turkey. Int J Climatol 18:649–680

Turkes M (1999) Vulnerability of Turkey to desertification with respect to precipitation and aridity conditions. Turk J Eng Environ Sci 23:363–380

Turkes M (2003) Spatial and temporal variations in precipitation and aridity index series of Turkey. In: Bolle HJ (ed) Mediterranean climate – variability and trends, Regional climate studies. Springer, Heidelberg, pp 181–213

Turkes M (2010) Climatology and meteorology, 1st edn. Kriter Publisher – Publication No. 63, Physical Geography Series No. 1. ISBN: 978-605-5863-39-6, 650 + XXII pp., Istanbul (in Turkish)

Turkes M (2013) Iklim Verileri Kullanilarak Turkiye'nin Collesme Haritasi Dokumani Hazirlanmasi Raporu. Birinci Baski, T.C. Orman ve Su Isleri Bakanligi, Collesme ve Erozyonla Mucadele Genel Mudurlugu Yayini, ISBN: 978-6054610-51-8, 57p, Ankara

Turkes M (2014a) Kuraklik Olaylarinin Iklim Degisikligi ve Collesme Acisindan Onemi ve Turkiye'deki 2013–2014(?) Kurakliginin Sinoptik Klimatolojik/Meteorolojik ve Atmosferik Baglantilari. 3 Mart 2014. Hidropolitik Akademi Iklim Degisikligi ve Kuraklik Calismalari, Ankara. http://www.hidropolitikakademi.org/wp-content/uploads/2014/03/Climate_Change_2014_and_Dara_C%C3%B6llesme.pdf

Turkes M (2014b) Turkiye'deki 2013-2014 kurakliginin ve klimatolojik/meteorolojik nedenlerinin cozumlenmesi (Analysis of 2013-2014 drought in Turkey and its climatological and meteorological reasons). Konya Toprak Su Dergisi 2:20–34 (in Turkish and English)

Turkes M (2016) Kuresel Iklim Degisiklikleri ve Baslica Nedenleri ile Dunya'da ve Turkiye'de Gozlenen ve Ongorulen Iklim Degisiklikleri ve Degiskenligi. Icinde: "Kuresel Iklim Degisikligi ve Etkileri" Engin Ural Anisina (Ed: M. Somuncu). Turkiye Cevre Vakfi Yayini, Ankara, pp 71–115

Turkes M (2017a). General climatology: fundamentals of atmosphere, weather and climate, revised 2nd edn. Kriter Publisher Physical Geography Series No. 4. ISBN: 978-605-9336-28-4, xxiv + 520 p, Kriter Publisher, Berdan Matbaasi, Istanbul (in Turkish)

Turkes M (2017b) Birlesmis Milletler Collesme ile Savasim Sozlesmesi ve Turkiye'nin Durumu-Uyumu. Icinde: Uluslararasi Cevre Rejimleri (Der. G. Orhan, S. C. Mazlum, Y. Kaya), 1. Baski, pp 169–197. Dora Basim-Yayin Dagitim Ltd. Sti., Bursa. ISBN: 9789752447530

Turkes M (2017c) Drought vulnerability and risk analysis of Turkey with respect to climatic variability and socio-ecological indicators. Ege Cografya Dergisi 26(2):47–70 (in Turkish with an English abstract, figure and table captions)

Turkes M, Akgunduz AS (2011) Assessment of the desertification vulnerability of the Cappadocian district (Central Anatolia, Turkey) based on aridity and climate-process system. Int J Hum Sci 8:1234–1268

Turkes M, Erlat E (2003) Precipitation changes and variability in Turkey linked to the North Atlantic oscillation during the period 1930-2000. Int J Climatol 23:1771–1796

Turkes M, Erlat E (2005) Climatological responses of winter precipitation in Turkey to variability of the North Atlantic oscillation during the period 1930-2001. Theor Appl Climatol 81:45–69

Turkes M, Erlat E (2006) Influences of the North Atlantic oscillation on precipitation variability and changes in Turkey. Nuovo Cimento Della Societa Italiana Di Fisica C-Geophysics and Space Physics 29:117–135. https://doi.org/10.1393/ncc/i2005-10228-8

Turkes M, Erlat E (2008) Influence of the Arctic oscillation on variability of winter mean temperatures in Turkey. Theor Appl Climatol 92:75–85. https://doi.org/10.1007/s00704-007-0310-8

Turkes M, Erlat E (2009) Winter mean temperature variability in Turkey associated with the North Atlantic oscillation. Meteorog Atmos Phys 105:211–225. https://doi.org/10.1007/s00703-009-0046-3

Turkes M, Erlat E (2018a) Observed change and trends in record and absolute maximum and minimum air temperatures of Turkey during 1950-2015, and associated atmospheric conditions favouring these variations. Int J Climatol Accepted, in print

Turkes M, Erlat E (2018b) A scientific assessment of observed changes and trends in extreme weather and climate events in the world and Turkey. In: Ucal M (ed) Climate change and green perspective: green economy, green growth. Heinrich Böll Stiftung Foundation, Istanbul, pp 5–38

Turkes M, Tatli H (2008) Turkiye'de kuraklik olasiliklarinin standartlastirilmis yagis indisi (SPI) kullanilarak saptanmasi ve iklimsel degiskenlik acisindan degerlendirilmesi. Kuresel Iklim Degisimi ve Su Sorunlarinin Cozumunde Ormanlar Sempozyumu Bildiriler Kitabi (Ed., Ünal Akkemik), 13–14 Aralik 2007. Istanbul Universitesi Orman Fakultesi, Istanbul, pp 55–62

Turkes M, Tatli H (2009) Use of the standardized precipitation index (SPI) and modified SPI for shaping the drought probabilities over Turkey. Int J Climatol 29:2270–2282

Turkes M, Sumer UM, Kilic G (2002a) Persistence and periodicity in the precipitation series of Turkey and associations with 500 hPa geopotential heights. Clim Res 21:59–81
Turkes M, Sumer UM, Demir I (2002b) Re-evaluation of trends and changes in mean, maximum and minimum temperatures of Turkey for the period 1929-1999. Int J Climatol 22:947–977
Turkes M, Koc T, Saris F (2009) Spatiotemporal variability of precipitation total series over Turkey. Int J Climatol 29:1056–1074
Turkes M, Kurnaz ML, Ozturk T, Altinsoy H (2011) Climate changes versus 'security and peace' in the Mediterranean macroclimate region: are they correlated? In: Proceedings of International human security conference on human security: new challenges, new perspectives, 27–28 October 2011, Istanbul, pp 625–639
TurkStat (2016) Turkey in statistics, 2015. Turkish Statistical Institute, Publication Number 4431, Printing Division, Ankara
Turuncoglu UU, Turkes M, Bozkurt D, Onol B, Sen OL, Dalfes HN (2018) Climate of Turkey. In: Kapur S, Akca E, Gunal H (eds) The soils of Turkey. Springer World Soils Book Series. Springer, pp 25–44. ISBN: 978–3–319-64390-8. https://doi.org/10.1007/978-3-319-64392-2
UNCCD (1995) The United Nations convention to combat desertification in those countries experiencing serious drought and/or desertification, particularly in Africa. Text with annexes. United Nations Environment Programme (UNEP), Geneva
UNEP (1993) World atlas of desertification. United Nations Environment Programme (UNEP), London
Wilcock AA (1968) Köppen after fifty years. Ann Assoc Am Geogr 58:12–28
Wilks DS (1995) Statistical methods in the atmospheric sciences: an introduction. Academic Press, San Diego
Willmott CJ (1977) WATBUG: a FORTRAN IV algorithm for calculating the climatic water budget. University of Delaware, Delaware
Wu ZH, Huang NE, Long SR, Peng CK (2007) On the trend, detrending, and variability of nonlinear and nonstationary time series. Proc Natl Acad Sci USA 104:14889–14894
Xoplaki E (2002) Climate variability over the Mediterranean. PhD dissertation. University of Bern, Switzerland
Xoplaki E, Gonz'alez-Rouco JF, Luterbacher J, Wanner H (2004) Wet season Mediterranean precipitation variability: influence of large scale dynamics. Clim Dyn 23:63–78
Yamanlar O (1956) Land classification on the basis of soil conservation in Marmara Region, Turkey with special reference to Yalova district. Istanbul Univ Pub. No. 697. Fac Forest No. 42, 98 p (in Turkish)
Yildirim AI (1999) The greening desert Karapinar. In: Cereal symposium, June 8–11, 1999, Konya, pp 440–448 (in Turkish)
Yildirim AI, Akay A (2010) The greening desert of Karapinar: an example from Turkey. In: Second International symposium on sustainable development, International Burch University-IBU Campus, Sarajevo, pp 546–551
Zhu X, Shen Z, Eckermann SD, Bittner M, Hirota I, Yee JH (1997) Gravity wave characteristics in the middle atmosphere derived from the empirical mode decomposition method. Geophys Res Lett 102:16545–16561
Zhu X, Yee JH, Talaat ER (2001) Diagnosis of dynamics and energy balance in the mesosphere and lower thermosphere. J Atmos Sci 58:2441–2454

Internet References

http://www.esrl.noaa.gov/psd/data/20thC_Rean/
http://www.esrl.noaa.gov/psd/data/gridded/data.20thC_ReanV2.html

Chapter 5
Surface Water

Hafzullah Aksoy

Abstract Turkey is a country with considerable water resources in a region where water is a strategic element due to its scarcity. With the increase in the population and demand, water scarcity has accelerated, making water a more vital element. Due to the considerably large size of the country and to the spatial and temporal uneven distribution of water resources, water demand in some regions exceeds the available water for a period of time that creates a water shortage. Surface water hydrology in Turkey is a great challenging problem as it has the greatest share (80%) in the water potential of the country. The total annual surface flow is 186.6 billion m^{3}, almost all born from the river basins within the territory of Turkey and harvested with the rainfall-runoff coefficient of 0.37 from the 574.0 mm annual rainfall. Surface water resources of Turkey are divided between 25 river basins; most stay within the country. There are also transboundary river basins where Turkey is either the upstream country or the downstream country. Among the river basins, the Euphrates-Tigris has the highest contribution, which is almost 1/3 of the surface water potential of Turkey. The northeastern part of Turkey receives the highest precipitation while the least amount of rainfall falls on Central Anatolia. A quite high number of hydrometric gauges have been established to record the quantity and quality of water; however, more are emerging, considering the topographical and geographical diversity in Turkey, to observe such hydrometeorological variables as streamflow, precipitation, evaporation, snow depth, etc. Due to the great spatial variability in the hydrometeorological conditions, some regions in Turkey are flood-prone while others can be affected by extreme droughts. Trend analysis has shown that extremes in the streamflow become more pronounced with larger maxima and lower minima than before. Also, lakes are important fresh water bodies for domestic use as well as irrigation, farming, industry, fisheries, etc. Water level in many lakes however decreases so that the availability of water in the lakes are under risk. Integrated water resources development strategies are needed at country-scale to balance the demand with the available water, considering also the decreasing trend in water availability and the increasing trend in the demand.

H. Aksoy (✉)
Department of Civil Engineering, Istanbul Technical University, Istanbul, Turkey
e-mail: haksoy@itu.edu.tr

N. B. Harmancioglu, D. Altinbilek (eds.), *Water Resources of Turkey*, World Water Resources 2, https://doi.org/10.1007/978-3-030-11729-0_5

Keywords Flood envelope curve · Flow duration curve · Euphrates-Tigris · Flood · Low flow · River basin · Surface water · Trend analysis · Turkey

5.1 Introduction

Turkey is a country which has a land roughly of 780,000 km^2, with a topography varying from the mean sea level up to 5165 m. The climatological diversity shared by seven regions over the country makes its hydrology complex. Precipitation, evaporation and surface runoff vary greatly among the regions. Annual precipitation calculated by the kriging method is 574.0 mm in Turkey (DSI 2017a). It changes from 325 mm (in Konya, Central Anatolia) to 2312 mm (in Rize, Eastern Black sea region) at point scale (DSI 2016a). The annual surface flow is 186.6 billion m^3 that result from a rainfall-runoff coefficient of 0.37. Almost all is born from the river basins within the territory of Turkey; only 5.8 billion m^3 of surface water comes from neighboring countries. Surface water contributes to 80% of the total water potential of Turkey.

Temporal variation in meteorological conditions is quite significant as well. The least annual precipitation record is 63.3 mm observed in 1933 in Himmetdede (Kayseri, Central Anatolia), while the peak was recorded as 4043.3 mm in 1931 in Rize (Eastern Black Sea) (DSI 2016a). Due to the great spatial and temporal variability in precipitation, water demand of different sectors is also variable from region-to-region and from year-to-year in the same region. Water demand is naturally higher in regions with less water. Therefore, it is not an exaggerated statement to say that surface water hydrology in Turkey is a challenging problem for researchers, practitioners, and decision-makers as well as law-makers.

In this chapter, surface water resources in Turkey are studied. First, the river basins and the hydrometric network of Turkey are introduced. Basin- or country-scale high and low flows are then analyzed, and characteristics and frequency analysis of floods and low flows are updated. After the trend analysis on low, high and mean flows, information about lakes as a source of surface water of Turkey is presented. The chapter ends with a brief discussion and provides future ideas on surface waters of the country.

5.2 River Basins

Turkey has a considerable number of rivers and lakes as surface water resources. It is divided into 25 river basins after the upper Euphrates and Tigris river basins are combined to be evaluated as one basin (Fig. 5.1 and Table 5.1). Most of the river basins stay fully within the territory of Turkey, while some are shared with the neighbor countries. The river basins have drainage areas ranging from 6306.2 km^2

Fig. 5.1 River basins in Turkey

for Lake Burdur closed basin to 176142.7 km^2 for the combined Euphrates-Tigris river basins, with shares of 0.81% and 22.55% in the total drainage area of Turkey, respectively. The long-term average of precipitation in the river basins changes from 390.1 mm in Konya closed basin in Central Anatolia to 1000.1 mm in the Eastern Black Sea basin at annual scale. Due to the topography, morphometry, climatology, meteorology and land use-land cover characteristics of the basins and many other factors, the runoff coefficient assumes the minimum value of 0.0852 in Lake Burdur closed basin and the maximum value of 0.8372 in Antalya basin, which are two neighboring basins. At the country level, the runoff coefficient is 0.3749 calculated as the area weighted average. Based on the calculated runoff coefficients, the contribution of the basins remains between 40.57 mm (the least in the Lake Burdur closed basin) and 720.58 mm (the most in the Eastern Black Sea basin). When contribution to the total water potential of Turkey is concerned in terms of runoff volume, Lake Burdur with 0.14% is the least productive basin, and the combined Euphrates-Tigris river basin with 30.65% contribution is the most productive. Finally, when the yield figures are compared, it is seen that the range changes from 1.29 L/s-km^2 (in Lake Burdur and Akarcay closed basins) to 22.85 L/s-km^2 (in the Eastern Black Sea basin). In the following paragraphs, river basins in Turkey are briefly introduced one by one, based on DSI (2016b, 2017a).

The Meric-Ergene river basin (Number 1 in Table 5.1) is located in the European part of Turkey. It discharges its surface waters into the Aegean Sea. The basin has a dendritic river network with one outflow into the sea. The surface water network is mainly composed of small rivers and creeks most of which are in intermittent character when Meric is excluded. The main river Meric (Evros in Greek and Maritza in Bulgarian languages) is a transboundary river with its upstream in Bulgaria and downstream in Greece and Turkey, which is, at the same time, a boundary river between Greece and Turkey. Tributaries named Arda and Tunca, both coming from Bulgaria through Greece, confluence with Meric, which then

Table 5.1 Characteristics of river basins of Turkey (DSI 2017a)

		Drainage area		Precipitation		Runoff			Yield
Basin number	Basin	km^2	%	mm	Runoff coeff	mm	hm^3	%	L/s-km^2
1	Meric-Ergene	14444.1	1.85	591.7	0.2156	127.55	1842.3	1.02	4.04
2	Marmara	23107.2	2.96	693.9	0.4703	326.32	7540.3	4.17	10.35
3	Susurluk	24332.0	3.12	649.8	0.2673	173.68	4226.0	2.34	5.51
4	Northern Aegean	9973.6	1.28	606.9	0.2479	150.46	1500.6	0.83	4.77
5	Gediz	17034.0	2.18	578.5	0.1568	90.69	1544.8	0.85	2.88
6	Kucuk Menderes	7059.7	0.90	611.1	0.1223	74.74	527.6	0.29	2.37
7	Buyuk Menderes	26133.2	3.35	598.7	0.1897	113.60	2968.7	1.64	3.60
8	Western Mediterranean	21223.9	2.72	739.9	0.4438	328.36	6969.1	3.85	10.41
9	Antalya	20330.8	2.60	768.6	0.8372	643.50	13082.9	7.24	20.41
10	Lake Burdur	6306.2	0.81	476.0	0.0852	40.57	255.8	0.14	1.29
11	Akarcay	7982.6	1.02	476.3	0.0856	40.78	325.5	0.18	1.29
12	Sakarya	63357.8	8.11	463.8	0.1755	81.41	5158.0	2.85	2.58
13	Western Black Sea	28929.8	3.70	761.1	0.4503	342.69	9914.0	5.48	10.87
14	Yesilirmak	39628.0	5.07	538.7	0.3083	166.10	6582.2	3.64	5.27
15	Kizilirmak	82197.3	10.53	451.3	0.1650	74.46	6120.4	3.39	2.36
16	Konya	50037.8	6.41	390.1	0.1356	52.91	2647.5	1.46	1.68
17	Eastern Mediterranean	21807.0	2.79	582.0	0.6493	377.87	8240.2	4.56	11.98
18	Seyhan	22241.6	2.85	576.2	0.5295	305.10	6785.9	3.75	9.67
19	Asi	7912.4	1.01	829.5	0.2763	229.17	1813.3	1.00	7.27
20	Ceyhan	21598.5	2.77	649.1	0.5258	341.30	7371.6	4.08	10.82
21	Euphrates-Tigris	176142.7	22.55	565.3	0.5566	314.63	55419.8	30.65	9.98
22	Eastern Black Sea	22844.6	2.93	1000.1	0.7205	720.58	16461.4	9.11	22.85
23	Coruh	20248.7	2.59	705.5	0.4933	348.03	7047.2	3.90	11.04
24	Aras	28114.6	3.60	483.5	0.3076	148.74	4181.8	2.31	4.72
25	Lake Van	17977.0	2.30	518.7	0.2427	125.90	2263.3	1.25	3.99
	Total	**780965.1**	**100.00**					**100.00**	
	Average			**574.0**	**0.3749**	**231.49**	**180790.2**		**7.34**

passes by meandering along the border between Greece and Turkey until it discharges into the sea. Another tributary of Meric is the Ergene, which collects all surface water resources in the eastern part of the basin fully within the territory of Turkey. There are several lakes in the watershed, some are saline and none is significant in terms of surface water potential. Lands in the watershed are used mainly for agriculture. Therefore, surface water in the watershed is heavily consumed for agriculture as well as domestic, industrial and energy uses. In the observation period of 1981–2010, the river basin had been the driest in 1994 and the wettest in 2010. It has the highest flow in March and the lowest in August.

The Marmara basin (Number 2 in Table 5.1) is a non-dendritic river network that covers all surface water courses other than Susurluk, which discharge into the Marmara Sea. Surface water courses linked with the Black Sea in the European part of the country, as well as the Asian side of the peninsula in the eastern side of Istanbul, all the way to the Sakarya river basin are also included within the Marmara watershed. The watershed is composed of three parts; northern Marmara, southern Marmara and eastern Marmara. One of the main rivers in the basin is the 100 km-long Biga Cayi (Kocabas Cayi) in the western part. This particular river has excessive flow in summer for irrigational use in the Biga plain. Another river in the basin is Gonen Cayi (105 km long approximately) that flows into the Erdek Bay in the Marmara Sea after it is joined by its tributaries. Other rivers are Karasu, Haramidere, Sazlidere, Alibey and Kagithane (in the northern Marmara), Kiraz, Kocadere, Kozanli, Serindere, Yatak Deresi and Safran (in the eastern Marmara). Lakes in the basin are Kucukcekmece, Buyukcekmece and Terkos in the northern Marmara part. Kucukcekmece Lake is a lagoon connected with the sea while Buyukcekmece and Terkos are important water supply resources for the Greater Municipalitan area of Istanbul. In the observation period of 1981–2010, the basin had been the driest in 1985 and the wettest in 1998. It has the highest flow in February and the lowest in September.

Susurluk (Number 3 in Table 5.1) is a dendritic river basin that collects water from rivers named Kocacay, Simav Cayi, Mustafa Kemalpasa Cayi and Nilufer, and discharges into the Marmara Sea northerly. Among the rivers in the basin, Nilufer is 172 km long and comes from the southeastern part of the basin. Floods are frequently observed due to high water level in the river in winter and spring, and they cause agricultural losses. Orhaneli is another river that receives important groundwater springs and joins Emet River. In rainy seasons, Simav Cayi is a flood-prone river along its 320 km-length, while it has very low flows in summer months. A quite high number of lakes exists in the basin, which are Manyas, Uluabat, Dalyan, Arap Ciftligi, Karagol, Kilimli and Aynali. The river basin has important plains with dry and irrigated agriculture. Although the non-irrigated lands are in majority, the tendency towards irrigation becomes more frequent. Losses due to evaporation from the irrigated agricultural lands and lakes are quite important in the basin. In the observation period of 1981–2010, the basin had been the driest in 2001 and the wettest in 1981. It has the highest flow in March and the lowest in September.

Northern Aegean (Number 4 in Table 5.1) is another non-dendritic river basin in the Western part of Turkey, with main rivers Bakircay, Havran Cayi, Tuzla Deresi and Karamenderes. Bakircay is 150 km long and has a flow with a subcritical hydraulic regime. Other rivers in the basin are Madra, Havran, Tuzla and Akcin, which quickly rise after snowmelt and mostly get dry in summer. No major lakes exist in the basin in terms of water availability. The basin has plains used as agricultural lands that require water for irrigation. In the observation period of 1981–2010, the basin had been the driest in 1990 and the wettest in 1982. It has the highest flow in February and the lowest in September.

Another river basin in the western part of Turkey is Gediz (Number 5 in Table 5.1), named after the dendiritic network of the main river Gediz with its 350 km length. Passing through the Gediz plain, the river has a high number of tributaries and also receives groundwater springs, which are not effective on the summer flow regime of the river but could support floods, particularly when they are combined with snowmelt. Gediz River has an irregular flow regime with a quick response to precipitation. Therefore, plains could be flooded although the yield of the watershed is very low in summer. Lakes in the river basin are Marmara, Karagol, Golcuk and Sazli. The basin has a great importance in the agricultural production of Turkey. There are quite a high number of irrigation projects receiving water from irrigation-purpose dams such as Demirkopru, Avsar and Buldan. In the observation period of 1981–2010, the basin had been the driest in 1992 and wettest in 1981. It has the highest flow in February and the lowest in October.

Kucuk Menderes River basin (Number 6 in Table 5.1) is located in the south of Gediz River basin. The main river in the basin is Kucuk Menderes, a 140 km-meandering river westernly flowing in a hydraulically subcritical regime. It receives water from its right and left banks from the tributaries Camli, Egri, Uzun, Gelinbay, Aktas, Kurkdere, Kocamahmut, Hamidiye and Keles Cayi. The river has a narrow basin with a decreasing slope as it reaches the plain downstream from the mountainous upstream. This causes floods in the winter season. Lakes in the basin are Cakal and Gebeli. The basin is a home to important agricultural activites in plains with irrigation facilities. In the observation period of 1981–2010, the basin had been the driest in 1992 and the wettest in 1981. It has the highest flow in February and the lowest in August.

Buyuk Menderes River basin (Number 7 in Table 5.1) in the Aegean region is the largest of Western Anatolia. It discharges into the Aegean Sea through the network of Buyuk Menderes River, which is the most important and the longest (529 km) river of the region. Buyuk Menderes River collects water from its tributaries such as Banaz, Curuksu, Akcay and Cine. Important natural lakes in the basin are Bafa, Koca, Isikli, Capali, Alpaslan and Sercin. In plains along the Buyuk Menderes valley, irrigation is practiced heavily for agriculture. In the observation period of 1981–2010, the basin had been the driest in 1992 and the wettest in 1984. It has the highest flow in February and the lowest in July.

Western Mediterranean hydrologic basin (Number 8 in Table 5.1) is in the southwestern corner of Turkey. Due to its nondendritic structure, the basin flows into the Mediterranean Sea with the individual rivers Dalaman, Esen, Sarnic, Basgoz

(Akcay) and Alakir. Important natural lakes in the watershed are Koycegiz, Sulungur, Kargin, Avlan, Golhisar, Karagol, Sogut, Calti, Kovak, Kocagol, Tuzla, Yazir, Cigli, Ikizgoller and Yesil. Some of the lakes are open for touristic purposes. Agriculture is practiced in the plains of the basin for which irrigation is needed upto a certain level. In the observation period of 1981–2010, the basin had been the driest in 2001 and the wettest in 1981. It has the highest flow in January and the lowest in September.

The Antalya basin (Number 9 in Table 5.1) is in the southern part of Turkey. It has the highest runoff coefficient among all basins by converting 83.72% of the precipitation into runoff. It discharges its water into the Mediterranean Sea by individual rivers Aksu, Manavgat, Karpuz, Alara, Kargi, Dim, Karaman, Duden, Kopru and Alara. Lake Egirdir in the northern part of the basin is an important fresh water provider. Rivers Pupa, Bahtiyar and Sucullu and karstic water resources flow into the lake. Water excess in Lake Egirdir flows into Lake Kovada, which itself flows into Antalya Bay through Aksu River or by means of the karstic connection to the Mediterranean Sea. Two karstic lakes in the basin are Ilvat and Dipsiz. Dry and irrigated agriculture is commonly practiced in the plains. In the observation period of 1981–2010, the basin had been the driest in 1991 and the wettest in 1981. It has the highest flow in January and the lowest in September.

Lake Burdur (Number 10 in Table 5.1) is a closed basin in southwestern Anatolia, which covers three lakes, namely Burdur, Aci and Salda. The basin is the smallest of the 25 basins in terms of the size of its drainage area. It composes less than 1% of the total surface area of Turkey. The least contribution (0.14%) to the surface water budget of the country belongs to this basin, not only because of its size but also due to its lowest runoff coefficient (8.52%). The basin shares the minimum yield of 1.29 L/s-km^2 with the Akarcay closed basin. Rivers recharging Lake Burdur get dry in summer as they are not long enough and as they are also used for irrigation before reaching the lake. Although used in irrigation like other rivers, Bozcay collects water from springs and flows northerly, with peaks in winter, and becomes negligible without getting dry in summer. It floods hazardously from time to time. Bayindir and Keciborlu are two other creeks flowing into Lake Burdur, which is diminishing gradually. Kestel is another lake in the watershed, which has a karstic structure. Located in the west of Lake Burdur, the densely saline Lake Aci is used neither for fisheries nor irrigation. A tectonic lake Salda has saline water; however, due to important spring discharges into the lake, its water could convert into fresh water locally or for some periods in the year. Golcuk is a small crater lake recharged by precipitation and used for irrigation and domestic water needs of Isparta city. Karatas, Ulupinar and Yapisli are other lakes in the basin. Plains in the basin are used for agriculture that requires water for irrigation. In the observation period of 1981–2010, the basin had been the driest in 1997 and the wettest in 1984. It has the highest flow in March and the lowest in October.

Akarcay (Number 11 in Table 5.1) is another closed basin in the western part of Central Anatolia. The basin shares the minimum yield of 1.29 L/s-km^2 with the Lake Burdur closed basin and follows the Lake Burdur basin as the second smallest in terms of contribution to the surface runoff potential of the country. Lakes Aksehir and Eber in the eastern part of the basin are the receiving bodies; Akarcay flowing easterly is the main river. Despite the size of its drainage area and length, Akarcay is not that important in terms of water potential as it is in a dry region and has no considerable water input from sources and mountains. Kali River is one of the main tributaries of Akarcay. In summer, it has no flow at its confluence with Akarcay, which itself gets dry at the downstream section by the end of the summer season. Other rivers in the basin are Degirmendere, Agik, Yalvacbeli and Engili. Lake Aksehir with its drainage area of 7340 km^2 is the receiving body of the closed basin. Eber is a shallow lake, which is 4 m-deep at maximum, has an inlet from Akarcay and an outlet to Lake Aksehir. Karamuk and Emre are two more lakes in the basin. In the observation period of 1981–2010, the basin had been the driest in 2001 and the wettest in 2002. It has the highest flow in April and the lowest in October.

Sakarya River basin (Number 12 in Table 5.1), named after its main river, flows northerly into the Black Sea. Sakarya River confluences with many rivers (to list at the direction of flow); Seyitsuyu, Pinarbasi, Porsuk, Kanlikopru, Ankara Cayi, Kirmir, Catak, Karasu, Bozuyuk, Goksu and Mudurnu. The Central Anatolian characteristics of the upstream watershed are preserved in the downstream, which has general characteristics of the Black Sea region. At lower sections in the downstream, the river is navigable. Lake Sapanca in the North has an outlet to Sakarya River. Lakes Mogan and Eymur are in Ankara, used mainly for recreational purposes. Lakes Gokce, Akgol, Acarlar, Karasu, Caticak, Beylikahir, Asaği and Kucuk Akgol are other lakes in the river basin. In the observation period of 1981–2010, the basin had been the driest in 1994 and the wettest in 1998. It has the highest flow in March and the lowest in October.

The Western Black Sea basin (Number 13 in Table 5.1), as its name depicts, is at the western part of the Black Sea region in Turkey. It is composed of several rivers and creeks, the most important being Filyos River in the western part of the basin, which collects water from rivers Arac, Melen, and Devrek. Filyos is a flood-prone river in flood seasons. Other rivers in the basin are Devrekani, Aydos, Karasu, Kanli, Kocairmak, Aydinlar and Buyuk Melen. Lakes in the basin are Efteni (Melen), Abant, Yenicaga, Yedi Goller, Aksaz, Sarikum, Sunnet, Karagol, Karamurat, and Cubuk (in Bolu). In the observation period of 1981–2010, the basin had been the driest in 1994 and the wettest in 1998. It has the highest flow in April and the lowest in September.

Named after the main river, Yesilirmak River basin (Number 14 in Table 5.1) has a dendritic flow network flowing into the Black Sea in Samsun. Yesilirmak River has two main tributaries, which are Cekerek with a confluence on the left bank, and Kelkit with a further downstream confluence on the right. Natural lakes in the basin are Ladik, Semenlik, Dumanli, Kargali, Kaz, Orta Ova, Borabay and Bora. In the observation period of 1981–2010, the basin had been the driest in 2001 and the wettest in 1988. It has the highest flow in April and the lowest in September.

Kizilirmak River basin (Number 15 in Table 5.1), named after the main river, Kizilirmak, has a network of water courses from the eastern and northern parts of Central Anatolia region to flow into the Black Sea as a delta. At its headwater in the eastern part of the Central Anatolia, it flows westerly and collects quite a high number of surface water sources, such as Aciirmak, Tecer, Kalin, Cat, Sarmisakli, Akkosanozu, Coruhozu, Balaban and Terme (listed in the flow direction), before its confluence with Delice Cayi that collects the right-hand side water courses in the upper part of the basin. Further downstream, it confluences with Devres and Gokirmak Rivers. Natural lakes within the basin are Seyfe, Aci, Tuzla, Balik, Duden, Todurge, Gocekler, Lota, Buyuk Gol, Niger, Corak, Caglar and Hanif. In the observation period of 1981–2010, the basin had been the driest in 2005 and the wettest in 1984. It has the highest flow in April and the lowest in September.

Konya (Number 16 in Table 5.1) is a closed basin in the Central Anatolian region. This is the driest basin in Turkey with the least annual precipitation. There are different bodies that receive water. The most important lakes in the basin are Lakes Tuz and Beysehir, the second and the third largest lakes in Turkey, respectively, in terms of surface area. Rivers that recharge Lake Tuz are Pecenek, Melendiz and Pinarbasi. Other rivers in the basin are Carşamba, Ivriz, Zanapa, Ibrala and Ayranci. As an important water body for water needs of the surrounding agricultural lands, Lake Beysehir has an outlet into Lake Sugla, which is another lake in the Konya basin. Lake Cavuscu receives water from the Cepesli River. Lakes Ambar and Tuzla are recharged by groundwater sources. Other lakes in the basin are Aci, Ilgin, Obruk, Besgoz Golu, Hotamis, Suleymanli and Timras. In the observation period of 1981–2010, the basin had been the driest in 2001 and the wettest in 2002. It has the highest flow in March and the lowest in August.

The Eastern Mediterranean basin (Number 17 in Table 5.1) is a nondentritic basin in the southern part of Turkey. The main river is Goksu, one of the largest rivers in southern Anatolia. Pirinc is a tributary of Goksu before it confluences with Ermenek River. Discharge of the Göksu River is high with rainfall in winter and with snowmelt in spring and summer. The river has a deep valley that prevents its use for irrigation. Another river in the basin is Tarsus in the eastern part of the basin, and another one is Anamur. Rivers in the karstic western part of the basin are short in length and small in drainage area size. Lamas and Alata are two more rivers in the eastern part of the basin. There are two lakes in the basin, which are Akgol and Paradeniz, both of which connect to the sea; these lakes are essentially lagoons. In the observation period of 1981–2010, the basin had been the driest in 2007 and the wettest in 1981. It has the highest flow in April and the lowest in September.

Seyhan River Basin (Number 18 in Table 5.1) in southern Anatolia is composed of the river network of Seyhan River, which is one of the most important rivers in the region. The river has two main branches, Zamanti and Goksu. Being the headwater of the Seyhan River, Zamanti collects tributaries Karagoz, Tarcin, Zindan and Alagoz among many others before its confluence with Goksu River, which receives water from Aga Deresi and Yanik Cay. After the confluence of Zamanti and Goksu Rivers, almost all contribution to the Seyhan River is from the rivers of the west bank, which are Dogan, Eglence, Korkun and Cakit. In terms of natural lakes, the basin has almost none other than two glacier lakes Yedi Goller and Dipsiz, and two smaller size lakes Cigli and Karagol, all at the upstream part of the basin. The basin has the most important agricultural lands in Turkey. There still exists a great water demand for agriculture although many irrigation projects have already been realized.

In the observation period of 1981–2010, the basin had been the driest in 2007 and the wettest in 1981. It has the highest flow in April and the lowest in September.

Asi River basin (Number 19 in Table 5.1) is the most southern basin in Turkey, which flows into the Mediterranean Sea at Iskenderun Bay. It has mainly the dendritic network of the Asi River (Oronthes with its ancient name) which has its headwaters in Syria. Within the basin, there are also rivers short in length and small in drainage size, flowing individually into the Mediterranean Sea. The Asi river collects water from Karasu and Afrin Cayi, the former having its water course fully within the territory of Turkey, the latter being a transboundary river flowing first from Turkey to Syria and then from Syria to Turkey. The Asi River has a seasonal flow regime that is high in wet seasons; it even floods in winter. Lakes in the basin are Balik (also called Gölbaşi) and Yenişehir (also called Güneyde). More importantly, Lake Amik, with about 200 km^2 surface area, has been drained for agricultural purposes and converted into a lake with a much smaller surface area of 62 km^2. It is a receiving body for the Karasu and Afrin rivers to overflow into the Asi River. However, after the drainage, the lake has been so shallow that it even gets dry in summer months. In the observation period of 1981–2010, the basin had been the driest in 2007 and the wettest in 1987. It has the highest flow in February and the lowest in August.

Another river basin in southern Anatolia is Ceyhan River basin (Number 20 in Table 5.1) that flows southerly into the Mediterranean Sea at the western coast of Iskenderun Bay. Starting with the Goksun River, that collects water from Hurman, Sarsap and Nargile Rivers as the headwaters in the north, it receives also Ergenez and Surgulu rivers among many others in the eastern and western parts of the basin. From this point onward, the river is contributed mainly by rivers in the west (Andirin, Kesis, Savran, Sambas, Ceperca) and two others in the east (Hacigulu and Karacay). At the lowerlands of the basin in Cukurova which is a very important irrigational plain in Turkey, contribution into the river becomes weak compared to the upstream rivers, and losses are high at that part of the river. Yet, it still has high enough discharge as the river basin has three fourths of its drainage area at water-reach high regions, and more than half is in the mountainous northern upper part under the effect of snow in winter. The river has a high flow-regime with rainfall in lowlands of the basin. Flow in the river is high in the spring and summer months because of snowmelt in the upperlands of the basin. Natural lakes in the basin are Golbasi, Azapli, Inekli, Humasir, Akyatan, Agyatan, Esemen and Tuz. There are also lagoons in the Ceyhan River delta. In the observation period of 1981–2010, the basin had been the driest in 2001 and the wettest in 1988. It has the highest flow in April and the lowest in October.

The Euphrates-Tigris river basin (Number 21 in Table 5.1) is composed of the Euphrates and Tigris river transboundary basins combined. This is the largest basin in terms of the size of its drainage area (22.55%) and its contribution (30.65%) to the surface water potential of Turkey. The combined Ephrates-Tigris river basin has the highest flow in April and the lowest in September.

The Euphrates River basin, with its headwater in Eastern Anatolia, flows through Southeastern Anatolia before crossing the border between Turkey and Syria. The

Euphrates river basin itself is again the largest basin in terms of both the size of drainage area and its contribution to the surface water potential of Turkey. Flowing all the way in Syria, it crosses the border of Syria with Iraq and confluences with Tigris River in northern Baghdad, finally discharging into the Persian Gulf. The river basin in the territory of Turkey before it crosses the border with Syria, has three parts; upper Euphrates (the headwater part), middle Euphrates, and lower Euphrates. It has two main branches, Karasu in the West and Murat Suyu in the East. To name a few tributaries before the confluence of Murat with the Euphrates, they are Serceme, Pulk, Tuzla, Girdim, Meydanli, Karabucak, Calti, Lik, Arapkir, Munzur, Pulumur and Peri Suyu. Murat Suyu is contributed by many tributaries until it reaches the reservoir of Keban Dam on the Euphrates River. At the lower parts of the Euphrates River basin, there are tributaries short in length and small in drainage size, coming from the West (the right bank) and reaching the Euphrates before it crosses the border. Tributaries in the Eastern part of the basin (Culap Suyu and Habur Cayi, for example) cross the border before reaching the Euphrates. Natural lakes in the Euphrates basin are Hacli, Gokpinar, Yedi Goller, Kesis, Agir, Sazlica, Kazan, Hazar and Kaz. In the observation period of 1981–2010, the basin had been the driest in 2001 and the wettest in 1988.

The Tigris River basin is composed of the network of Tigris River that starts with its headwater in Eastern Anatolia and flows through Southeastern Anatolia. The Tigris River has the second largest drainage area and surface runoff contribution after the Euphrates. It is a transboundary river as it crosses to Iraq and, at the same time, a boundary river between Turkey and Syria before it reaches the territory of Iraq. Finally, it reaches the Persian Gulf after it joins Euphrates within Iraq. River Zap, with its headwater within the territory of Turkey, contributes to the Tigris through its confluence in Iraq. The Tigris River is contributed by the left-hand side tributaries, the major of which are Batman, Botan and Hezil. The tectonic type Lake Hazar is the only lake in the basin except for upto 30 glacier-type small lakes at Cilo Mountains in the Eastern part of the basin. In the observation period of 1981–2010, the basin had been the driest in 2008 and the wettest in 1988. It has the highest flow in April and the lowest in October.

The Eastern Black Sea basin (Number 22 in Table 5.1) is the collection of all rivers flowing into the Black Sea between the Yesilirmak delta and the boundary of Turkey with Georgia. Rivers in the intervening zone between the Kizilirmak and Yesilirmak deltas are also counted within this basin. This is the wettest basin in Turkey with 1000.1 mm annual precipitation and highest yield of 22.85 L/s-km^2. Rivers in the basin flow in the north direction into the Black Sea individually, among which Harsit River could be an exemption with a considerable drainage area. The rivers in the watershed are short in length and small in size of drainage area. However, they always have high discharges as the region is wet throughout the year. Other rivers in the basin are Iyidere, Melet, Aksu, Terme, Pazar, Gelevera, Değirmendere, Karadere and Firtina. Lakes in the basin are Sera, Karagol-Arhavi, Karagol-Rize, Cermes, Uzungol, Karagol-Giresun, Gaga, Cakirgol, Karanlik Goller, Gernek, Balik, Uzungol-Samsun, Liman, Geri, Sirmenlik and Karabogaz. In the observation period of 1981–2010, the basin had been the driest in 2001 and the wettest in 1993. It has the highest flow in May and the lowest in September.

The Coruh River basin (Number 23 in Table 5.1) has a transboundary river with a dendritic network in the Northeast. By crossing the border between Turkey and Georgia at Muratli, it flows into the Black Sea in Georgia through a large delta. It is one of the important basins in terms of hydroelectric energy potential. There are many tributaries contributing the main river, among which Fencul, Tortum, Oltu and Berta can be listed in order from upstream to downstream. The river is frozen in winter at headwaters in Oltu and Berta. At the headwaters of the main river, snow is dominant as the typical character of Eastern Anatolian basins, while in the lower parts, the basin changes into a Black Sea-character basin with rain and snow. Therefore, the river has high discharges in spring and autumn and carries high amount of sediment downstream. The yield throughout the year is due to high precipitation over the basin, springs and lakes in the basin, and crater lakes at headwater part of the river. Lakes in the river basin are Tortum, Karagöl-Ardanuc, Karagol-Savsat, Karagol-Borcka and Karagol-Yusufeli. In the observation period of 1981–2010, the basin had been the driest in 2001 and the wettest in 1993. It has the highest flow in May and the lowest in September.

Aras River Basin (Number 24 in Table 5.1) is a transboundary basin at the border of Turkey with Armenia, Nakhchivan Autonomous Republic of Azerbaijan and Iran. The main watercourse in the basin is the easterly flowing Aras, which has the southerly flowing Arpacay as a branch. Arpacay and, after the confluence, Aras flow as boundary rivers between the above countries. To the north of the Aras River basin, Kura River flows into Armenia by crossing the border. Rivers in the basin are under the effect of a snow regime. They are frequently frozen in winter. Lakes in the basin are Cildir, Aktas, and Balik. Smaller size lakes, such as Aygir, Kuyucak, Gulyuzu, Mizko and Deniz, can also be listed. In the observation period of 1981–2010, the basin had been the driest in 2001 and the wettest in 2004. It has the highest flow in April and the lowest in September.

Lake Van Closed Basin (Number 25 in Table 5.1) is in Eastern Anatolia at the border with Iran. Within the basin, Lake Van and Lake Ercek exist. Lake Van is the largest lake in Turkey. It is a saline lake. The lake is recharged mainly by rivers in the northern and eastern parts of the basin. Major rivers recharging the lake are Bendimahi, Hosap Suyu, Karasu, Zilan and Delicay. Immediately to the east of Lake Van is Lake Ercek, another closed basin within the Lake Van basin and the receiving body of Memedik (Ozalp) River. Lake Van is an important and particular lake with a surface area of 3443.5 km^2, drainage area of 16.096.4 km^2, water surface elevation of 1650 m, and the maximum depth of about 460 m (DSI 2016b). Due to the climate in the basin, evaporation loss from the lake is less than the inflow into the lake. When sedimentation is also considered, any rise in the water level could be understandable. Other lakes within the basin are Nazik, Ercek, Nemrut, Akgol, Sodali, Hidirmentes and Aygir. In the observation period of 1981–2010, the basin had been the driest in 2001 and the wettest in 1993. It has the highest flow in May and the lowest in September.

5.3 Hydrometric Network

Due to the general fact that hydrometeorological data are of crucial importance in river basin management and hydraulic design, a hydrometric network has been established all over Turkey since 1936. First time in 1925, regional offices for water resources development were established under the Ministry of Public Works (Bayazit and Avci 1997), and preliminary studies on river basin planning started in 1932. The first state agency responsible for streamflow measurement, hydropower planning and design was EIE (General Directorate of Electrical Power Resources Survey and Development Administration), which was founded in 1935. Tasks on hydrometric measurements on main rivers were carried out by EIE. The agency was abolished in 2011, and its departments were transferred into different state organizations mostly connected to State Hydraulic Works (DSI). DSI is the major state agency being responsible for water projects since 1954 when it was founded and operates a project-based hydrometric work. The unification of EIE with DSI almost doubled the hydrometric network operated by DSI.

The existing hydrometric network is composed of streamflow gauging stations, suspended sediment gauges, water quality sampling, lake water level gauges, meteorological stations and snow gauges (Fig. 5.2). Some of the gauges take daily measurements while, in some, monthly observations are made. The number of hydrometric stations in the 25 river basins of Turkey is 3500 (after EIE has joined DSI in 2011), as detailed in Table 5.2. The number of streamflow gauging stations increased from about 300 in 1960 to 1331 in 2014. Two steep increases in the number of installed streamgauges were observed in the 1960s and 1980s. In addition to the hydrometric network of DSI, General Directorate of Meteorology (MGM) operates an observation network composed of 1674 meteorological stations (DSI 2017a). The distribution of streamflow gauges taking daily measurements over river basins is given in Table 5.3.

5.4 Flood Characteristics

In Turkey, floods are considered as the second most disastrous natural hazard after earthquakes. Inventory of rainfall and floods has shown that hazardous floods are mostly observed in the Black Sea, Mediterraean and Western Anatolian regions during the flood period which extends from March to July. Floods in Turkey are directly linked to the topography, land use, land cover, urbanization level, and rainfall regime of the river basin. As Turkey is mainly under the effect of northwestern and southern climatological conditions, and as the mountains are parallel to

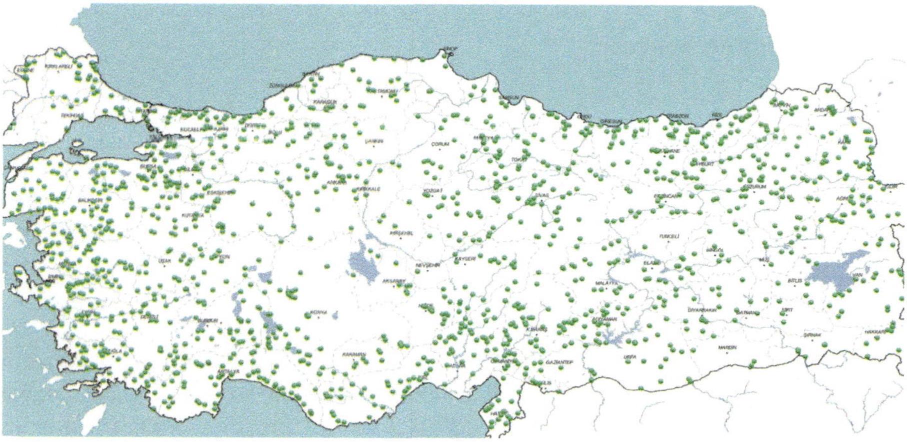

Fig. 5.2 Hydrometric network of Turkey

Table 5.2 Hydrometric network of Turkey (DSI 2017a)

Hydrometric gauge	Measurement frequency	Number of gauges	Online gauges
Streamflow gauge	Water level measurement (daily)	1331	733
	Discharge measurement (once a month)		
Streamflow gauge	Discharge measurement (once a month)	1648	
Lake water level gauge	Daily measurement of water level and other parameters	98	48
Meteorology (DSI)	Precipitation and Evaporation	187	
Meteorology (MGM)	Precipitation, evaporation, temperature, wind speed, air humidity, air pressure, etc.	1674	
Snow	Snow depth, snow water equivalance	252	21
Sediment	Sampling from the river	155	
Small hydro-power gauge		536	

the Black Sea in the north and Mediterranean in the south, wet fronts are prevented from moving into the inner regions of the country. This increases rainfall intensity at the coast and makes the northern and southern coastal regions flood-prone. Another factor, the snowmelt in flood-prone regions, increases flooding in terms of its magnitude and frequency.

Floods are not related to meteorological or climatological conditions alone. In particular, such countries under economic development as Turkey are faced with continuous and intensive industrial development and urbanization as well as diverse human activities in various parts of river basins. This disrupts the hydrological

Table 5.3 Distribution of streamflow gauges with daily measurements over river basins (DSI 2017a)

Basin number	Number of gauges	Area per gauge (km^2)	Basin number	Number of gauges	Area per gauge (km^2)	Basin number	Number of gauges	Area per gauge (km^2)
1	27	535	10	7	901	19	26	304
2	83	278	11	8	998	20	40	540
3	68	358	12	86	737	21	185	952
4	33	302	13	66	438	22	77	297
5	36	473	14	70	566	23	30	675
6	14	504	15	103	798	24	56	502
7	58	451	16	54	927	25	22	817
8	59	360	17	41	532			
9	49	415	18	33	674	Overall	1331	587

balance in the entire basin and results in flood disasters that cause high numbers of losses of life and property. With the growth of settlement in river basins, new roads are opened and new facilities are established so that the land structure is changed. Furthermore, due to unsuitable agricultural practices, the land is used more intensively, the forests and meadows are being destroyed, and the flood disasters become more frequent and larger in magnitude. As a result, meteorological and topographic characteristics are the main natural factors that cause floods; nevertheless, the geology that triggers landslides has an equivalent importance (Eris and Aksoy 2014).

When the flood inventory of Turkey is analyzed, historical floods in Table 5.4 have been recorded as the most destructive in terms of the number of life losses. Nevertheless, according to DSI (2017b), Turkey faces less destruction compared to many other countries in terms of the flooded area and life and property losses although it is considered among the countries with a high number of floods. Floods are the most frequent in the Eastern Black Sea, Mediterranean and Aegean regions at the country-scale. However, Edirne in the Thrace Region at the border with Bulgaria and Greece is the most flood-prone province due mostly to the effect of the uncontrolled water release from the upstream dams.

It is more likely that floods in Turkey are experienced in the rivers without flood control measures. Unauthorized settlement around the riverbeds increases the flood risk, hence loss of life and damage in property. Downstream measures become inadequate to prevent floods in the absence of upstream measures and of basin-scale projects. In many cases, it is necessary to take actions even in areas where flood protection measures are not required.

Since 1954 when it was founded, DSI has built 7838 flood protection facilities, including 68 flood-purpose dams. With these facilities, around 1.8 million ha of lands in Turkey have been protected against floods. However, the number of floods opposedly increased. In 41 years from 1975 to 2015, 1209 floods were recorded, that caused a loss of 720 lives and damaged 894,474 ha of lands (Fig. 5.3). As a general

Table 5.4 Most destructive floods in Turkey at a chronological order (DSI 2017b)

Date	Place	Causalties/ deaths
23 August 1956	Sincan village (Adiyaman, Southeastern Anatolia)	79
11 September 1958	Hatip Creek (Ankara, Central Anatolia)	169
20 November 1974	Kirfabesbegi (Silopi, Mardin, Southeastern Anatolia)	38
30 August 1979	Amasya, Tokat and Corum provinces (northern part in Central Anatolia)	59

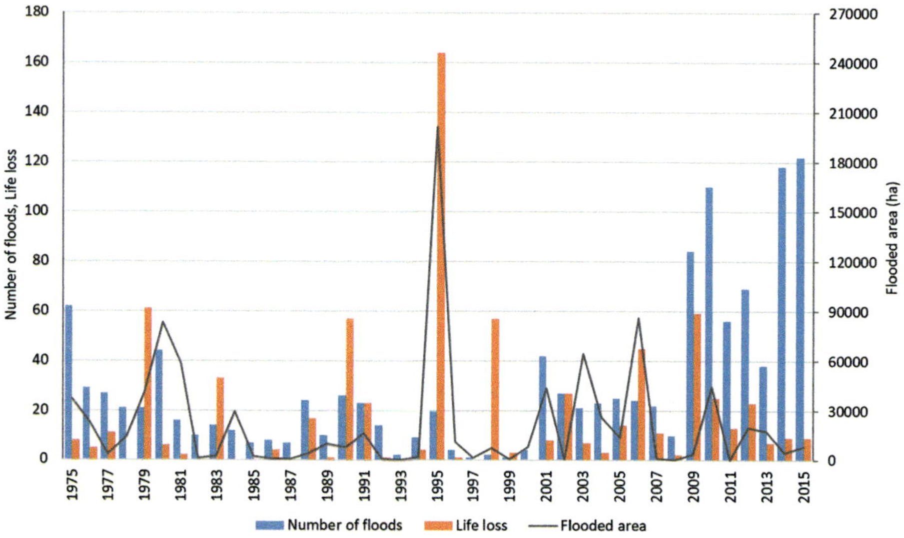

Fig. 5.3 Number of floods and life losses, and flooded area between 1975 and 2015

statement, it is seen that, despite the increasing number of floods in recent years, a significant reduction in the number of losses of life and in the affected lands has been achieved by practicing structural and non-structural flood measures (DSI 2017b).

Recently, a system called TAMBIS (a Turkish acronym for Taskin, Ariza ve Mudahale Mekansal Bilgi Sistemi to be translated as Flood, Damage and Intervention Spatial Information System) has been established by DSI. The system allows early warning of floods and notifications about any breakdown in or intervention to the system. DSI has built thousands of water structures throughout the country to serve citizens in order to combat floods by following the latest possibilities of technology. TAMBIS, as a most updated technological project, is open to the use of citizens. It imports data from citizens before, during or after the flood event and aims to make a warning at the right moment, at the right point, and to the right person such that the hazard is minimized. With the system, data on the time, place and magnitude of the flood disaster is delivered to the citizens through an SMS, email or a photograph via mobiles or a smart phone application.

5.5 Flood Frequency Analysis

Frequency analysis of maximum floods is performed and published as a frequency atlas by DSI (1994). In this atlas, flood discharge per unit area, and 2, 5, 10, 25, 50 and 100-year return period floods estimated by the best-fit distribution are provided for 1366 streamflow gauging stations with minimum 5 years of observation. Probability distribution functions considered are two- and three-parameter lognormal (LN2, LN3), two-parameter gamma (G2), log Pearson type III (LP3), and Gumbel (G). The distributions are tested with Kolomogorov-Smirnov and Chi-square statistics. Number of gauging stations best-represented by each probability distribution function are given by DSI (1994) as in Table 5.5. The LP3 is observed to be dominant in representing the maximum floods in river basins of Turkey.

5.6 Envelope Curves for Maximum Floods

Flood frequency analysis and probable maximum flood are two methods generally applied in flood estimation. However, for both methods, hydrological and meteorological information and data are needed. Therefore, the methods are not applicable in basins with inadequate data or no data at all. Alternatively, the envelope curve of maximum floods is used. It is a graph that plots the discharge against the basin area and proposes a unique relationship between the maximum flood discharge and the area of the hydroclimatologically homogeneous basin. The envelope curve is the upper limit for floods which are expected to be well lower than the envelope at the site. The envelope curve can be used in estimating the maximum flood in a basin of a certain size in the region interested (Bayazit and Onoz 2004).

DSI has published envelope curves for all river basins in Turkey (DSI 1994). Bayazit and Onoz (2004) studied the flood envelope curves of hydroclimatologically homogeneous regions in Turkey, using data until year 2000, except for Akarcay, Konya (closed), Asi, Eastern Black Sea and Van (closed) basins for which no meaningful curves were obtained. The envelope curves of the remaining 20 basins were grouped together to form eight envelopes (Table 5.6, Figs. 5.4 and 5.5).

Envelope curves of four regions (Thrace-Marmara, Aegean, Western Mediterranean and Central Anatolia) in Fig. 5.4 are close to each other, whereas those in other regions in Fig. 5.5 are rather different. Flood discharge per unit area is the smallest in Coruh and Aras river basins, the highest in the Western Mediterranean regions for basins in size less than 1000 km^2 and in Tigris region for larger basins. A steep increase in the flood flow exists in Euphrates where the basin has a size above

Table 5.5 Number of streamflow gauging stations with the chosen probability distribution functions (PDFs) of the maximum floods (DSI 1994)

Basin number	Basin	Probability distribution function					
		G	G2	LN2	LN3	LP3	Total
1	Meric-Ergene	7	6	2	7	14	36
2	Marmara	14	10	6	3	28	61
3	Susurluk	10	11	4	6	23	54
4	Northern Aegean	9	5	0	2	11	27
5	Gediz	7	5	3	3	11	29
6	Kucuk Menderes	1	3	1	4	4	13
7	Buyuk Menderes	10	7	1	5	29	52
8	Western Mediterranean	15	9	4	5	14	47
9	Central Mediterranean	11	9	4	2	15	41
10	Lake Burdur (Closed)	1	2	2	0	8	13
11	Akarcay	2	2	2	1	8	15
12	Sakarya	30	16	7	8	44	105
13	Western Black Sea	15	10	4	4	30	63
14	Yesilirmak	23	13	6	11	23	76
15	Kizilirmak	31	18	6	8	62	125
16	Konya (Closed)	18	16	5	11	39	89
17	Eastern Mediterranean	6	5	2	3	16	32
18	Seyhan	8	6	2	3	10	29
19	Hatay	3	3	1	2	9	18
20	Ceyhan	14	11	2	6	17	50
21	Euphrates-Tigris (Euphrates)	45	22	14	13	60	154
	Euphrates-Tigris (Tigris)	19	14	5	2	19	59
22	Eastern Black Sea	19	9	3	7	36	74
23	Coruh	14	8	2	2	12	38
24	Aras	20	6	4	2	20	52
25	Lake Van (Closed)	3	3	1	1	6	14
	Total	**355**	**229**	**93**	**121**	**568**	**1366**

10,000 km^2. The flood envelope curve for all basins in Turkey is also plotted in Figs. 5.4 and 5.5. It is represented by the curves of the Thrace-Marmara and Eastern Mediterranean regions for river basins smaller than 1000 km^2 (Fig. 5.4), and by the curves of Tigris and Euphrates regions for larger basins (Fig. 5.5). In Turkey,

$$Q = 1.81A^{1.22} \quad \text{for } A \leq 300\,\text{km}^2 \tag{5.1}$$

$$Q = 79A^{0.5} \quad \text{for } 300\,\text{km}^2 \leq A \leq 10000\,\text{km}^2 \tag{5.2}$$

$$Q = 7900 \quad \text{for } 10000\,\text{km}^2 \leq A \tag{5.3}$$

Table 5.6 Grouping flood envelope curves for river basins in Turkey (Bayazit and Onoz 2004)

Region	Basins
Thrace-Marmara	Meric-Ergene, Marmara, Susurluk
Aegean	Northern Aegean, Gediz, K. Menderes, B. Menderes
Western Mediterranean	Western Mediterranean, Antalya, Burdur
Central Anatolia	Sakarya, Western Black Sea, Yesilirmak, Kizilirmak
Eastern Mediterranean	Eastern Mediterranean, Seyhan, Ceyhan
Coruh-Aras	Coruh, Aras
Euphrates	Euphrates
Tigris	Tigris

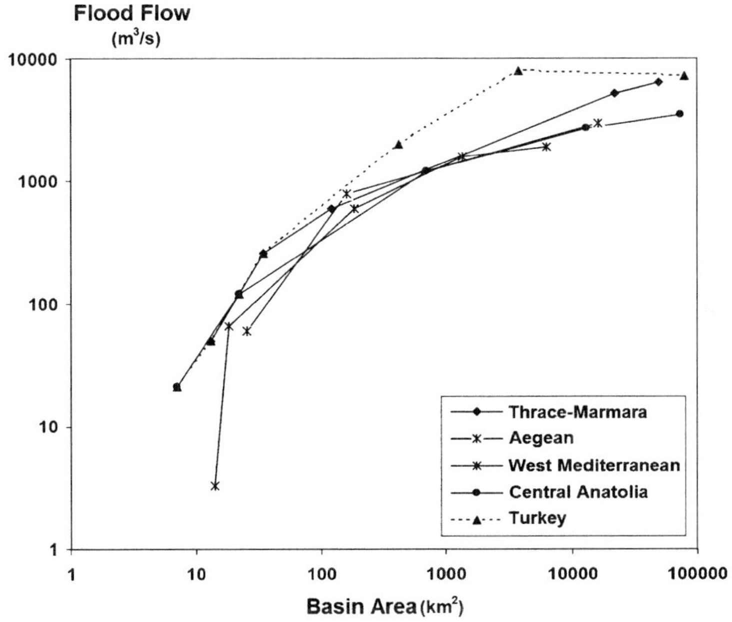

Fig. 5.4 Flood envelope curves of the Thrace-Marmara, Aegean, Western Mediterranean and Central Anatolian regions. (Bayazit and Onoz 2004, 2008)

are proposed in which Q is the flood discharge in m^3/s and A the basin area in km^2 (Bayazit and Onoz 2004).

5.7 Low Flow Characteristics

Low flow frequency analysis is a useful practice for estimating the probability of water availability in streams during critical dry periods. It is employed, at the same time, in water supply planning to determine allowable water transfers and withdrawals. The use of low flows is extended to the determination of minimum

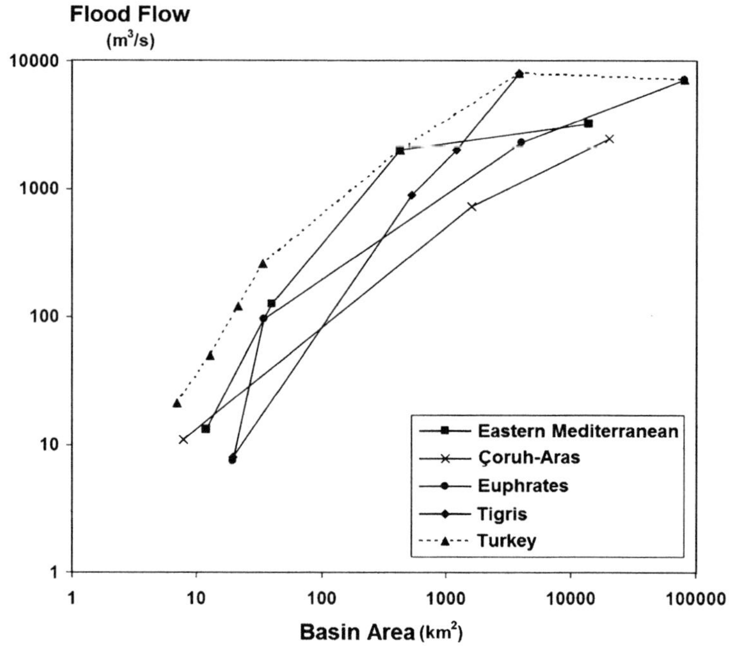

Fig. 5.5 Flood envelope curves of the Eastern Mediterranean, Coruh-Aras, Euphrates and Tigris regions. (Bayazit and Onoz 2004, 2008)

downstream release from hydropower plants, water supply and cooling systems etc. A certain duration low flow discharge with a certain return period is an important information for the engineering design practice. As an example, Q_{7-10}, the 7-day low flow with 10 years of return period, is one of the reference low flow indices used for many design practices, such as the protection or regulation of water quality from waste water discharges or waste load allocations (Riggs et al. 1980), the regulation of water withdrawals and discharges into streams (Carter and Putnam 1974), and the comparison of the impacts of climate change and irrigation on low surface streamflows (Eheart and Tornil 1999; Eheart et al. 1999). It is also used as a local extinction flow (Ontario Ministry of Natural Resources 1994) and is considered as the worst case-scenario in water quality modelling (Mohamed et al. 2002).

For characterizing and evaluating low flows, flow duration curves, recession curve analysis, low flow indices and frequency analysis have been frequently used (Bayazit and Onoz 2008). A quite high number of probability distribution functions are used for the purpose of low flow frequency analysis (Hewa et al. 2007; Liu et al. 2015). In Turkey, low flows have been investigated in 1990s at river basin-scale through research studies of Bulu et al. (1997), Bulu and Aksoy (1998) for Meric-Ergene river basin in the Thrace Region, the European part of Turkey; Sertbas (1996), Bulu and Onoz (1997) and Saris (2016) for the Sakarya and Meric river basins in the north-western part of the country; Duran (2000) for the Aegean Region in the west, Saracoglu (2002) for the Mediterranean region in the south; Yurekli et al.

(2005) for the Yesilirmak river basin in the north; and Koken (2009) for the Tigris river basin in the southeast. The 2-parameter Weibull and lognormal (W2 and LN2) were taken as candidate probability distribution functions together with the 3-parameter lognormal (LN3). Selected probability distribution functions were checked for their suitability to fit each sequence of D-day low flow. A D-day low flow is defined as the minimum value of the average of daily flows taken over D sequential days. One D-day low flow is calculated per a year. Among the tested probability distribution functions, W2 fitted most of the river basins.

A low flow frequency analysis case study of Aksoy et al. (2018) and Eris et al. (2019), using the up-to-date data of intermittent and non-intermittent rivers in four river basins from different regions in Turkey, is summarized here. In the case study, four basins are considered; Meric-Ergene in the Thrace region, the northwestern part of the country (Number 1 in Table 5.1), Gediz in the Aegean region, Western Anatolia (Number 5 in Table 5.1), and Seyhan and Ceyhan in the Eastern Mediterranean region, Southern Anatolia (Numbers 18 and 20 in Table 5.1). Frequency analysis of D = 1, 7, 14, 30, 90 and 273-day low flows of each stream gauge is performed by checking the suitability of probability distribution functions; Weibull (W2), Gamma (G2), Generalized Extreme Value (GEV), Log-Normal (LN2) from the 2-parameter probability distribution function family; and Weibull (W3), Gamma (G3), Log-Normal (LN3) from the 3-parameter probability distribution function family. The GEV probability distribution function conforms to low flows by being the best-fit among the selected probability distribution functions in the four river basins (Table 5.7). Table 5.7 shows, for example, that 7-day low flows in Meric-Ergene basin are best-fit by W2 in one gauging station, by GEV in 7 and by LP3 in 4 stations. The G2 and LN2 probability distribution functions have never been the best in fitting the 7-day low flow sequences of streamflow gauging stations in the Meric-Ergene river basin. Gauging stations with 10 non-zero D-day low flow data at minimum are considered. Due to the intermittent character of the region, the number of gauging stations considered in the frequency analysis increases when D goes from 1 to 273. With the use of the best-fit probability distribution function, the low flow-duration-frequency curves are determined for any D-day low flow discharge of any given return period. An example low flow-duration-frequency curve is seen in Fig. 5.6.

5.8 Rainfall-Runoff Coefficient

In Turkey, when the hydrological year (from October 1st of the previous calendar year to September 30th of current year) is considered, the annual total precipitation is 581.1 mm, which is equal to 454117.7 hm^3 when converted into volume of water (Table 5.8). Due to hydrometeorological seasonality, the highest precipitation falls in December while the runoff peak is observed in April when the runoff coefficient is at its maximum (0.668) during the year. The lowest runoff coefficient is 0.159 calculated for October.

Table 5.7 Number of streamflow gauging stations with the chosen probability distribution functions (PDFs) of the D-day low flows (Aksoy et al. 2018)

	Meric-Ergene						Gediz						Seyhan						Ceyhan					
PDF	1	7	14	30	90	273	1	7	14	30	90	273	1	7	14	30	90	273	1	7	14	30	90	273
G2	0	0	1	1	0	1	2	0	1	1	0	5	1	0	2	1	1	1	1	2	1	2	3	4
LN2	0	0	0	0	0	1	0	1	1	1	0	1	0	0	0	0	0	1	0	0	0	1	0	1
W2	0	1	0	0	0	0	1	0	0	1	0	0	0	0	0	0	0	0	1	2	2	2	2	0
GEV	5	7	6	5	13	11	6	7	8	6	7	11	13	14	12	11	12	10	21	16	17	20	18	13
LP3	5	4	5	10	6	13	5	6	4	5	10	6	2	2	2	4	3	4	9	12	12	8	10	15
Total	10	12	12	16	19	26	14	14	14	14	17	23	16	16	16	16	16	16	32	32	32	33	33	33

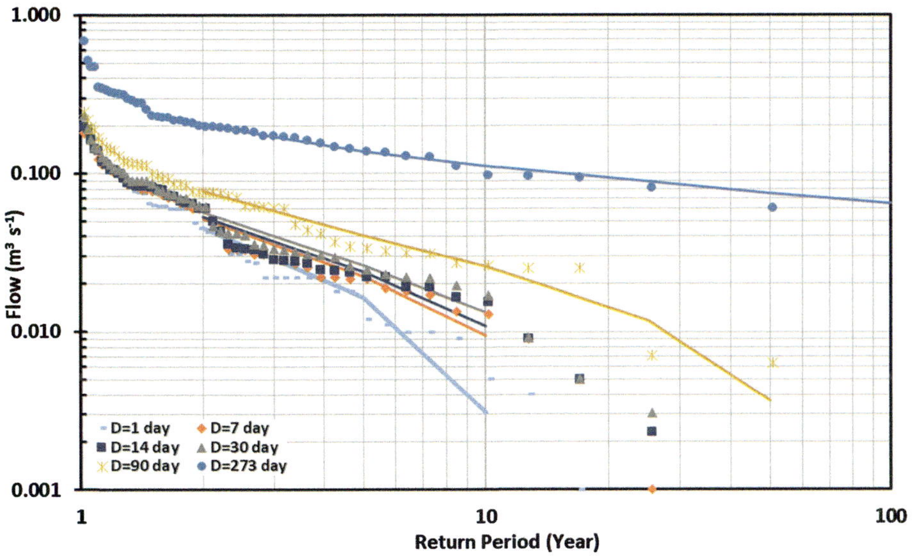

Fig. 5.6 Low flow-duration-frequency curve (Gauging station D01A031 in Meric-Ergene basin)

5.9 Flow Duration Curve

Another important tool to determine the discharge required in river basin management practices is called the flow duration curve. The shape of flow duration curves is influenced by various factors. A flow duration curve with a steep slope throughout indicates a highly variable stream whose flow comes mainly from quick runoff, while a flat slope shows the dominance of the groundwater discharge. Based on the time series of the natural annual and monthly streamflow data at the country-scale for the period of 1981–2010, flow duration curves are plotted as in Fig. 5.7. The variability of the annual streamflow is observed from the slope of the flow duration curve. At monthly scale, however, the variability in high flows produced by quick runoff is remarkable. At the lower section of the monthly flow duration curve, the variability is reduced because of the low flow born from the slow motion processes such as groundwater. The monthly flow duration curves are under the seasonal effects of the year, wet and dry climatic conditions.

5.10 Trend Analysis

Trend analysis of hydrologic variables, and of streamflow in particular, is greatly important in hydrological practice. Streamflow is under the effect of not only the natural changes but also antropogenic activities. It is therefore important to investigate whether streamflow records exhibit any evidence of trends. Similarly, not only

Table 5.8 Monthly and annual precipitation, runoff volume and runoff coefficient in Turkey (DSI 2016b)

	O	N	D	J	F	M	A	M	J	J	A	S	Annual
Prec. (mm)	49.2	72.0	78.1	67.3	62.9	60.4	59.0	49.1	30.8	16.7	13.9	21.7	581.1
Prec. (hm^3)	38,449	56,267	61,034	52,594	49,155	47,201	46,107	38,371	24,070	13,051	10,863	16,958	454,118
Runoff (hm^3)	6105	9107	13,760	14,557	15,948	23,725	30,798	25,008	13,532	7856	5738	4857	170,991
Runoff Coef.	0.159	0.162	0.225	0.277	0.324	0.503	0.668	0.652	0.562	0.602	0.528	0.286	0.377

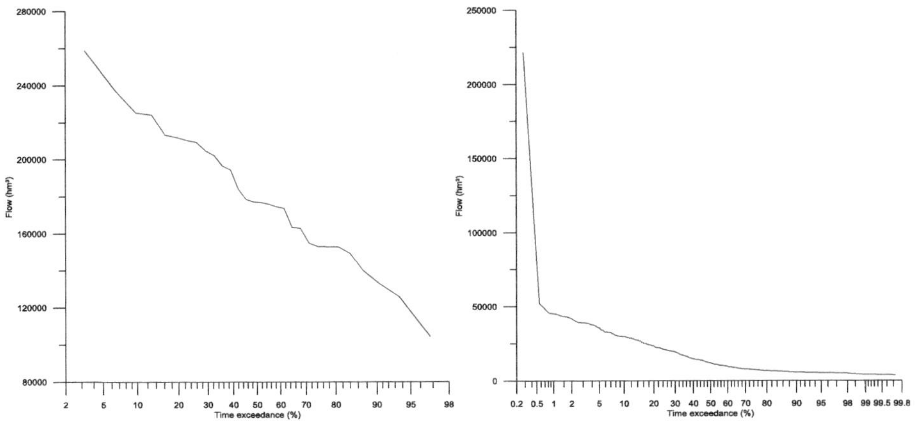

Fig. 5.7 Flow duration curve of annual (left panel) and monthly (right panel) flows of Turkey

the over-the-year changes but also within-the-year changes are common. For instance; in the northwest, the Thrace region that covers the European part of Turkey, there is a significant variation in streamflow throughout the year with relatively long and dry summers. Superimposed on this varied hydrometeorological basis is a wide variety of water use and management practices. Therefore, regions such as the Thrace are vulnerable to hydrological extremes, droughts and floods, which finally return back as environmental issues (Aksoy et al. 2005).

Trends in maximum, mean and low flows calculated from the daily streamflow of nearly 100 gauging stations all over Turkey were investigated by Cigizoglu et al. (2005). Trend analysis was carried out, using parametric and nonparametric tests, and were applied on maximum, mean, 1-day, and 7-day low flows calculated from the daily data annually. Trend existence was detected in majority of the rivers in the western and southern Turkey and partly in the central and eastern regions. Trends in the mean and low flows were more common compared to maximum flows. When a few stations are exempted, flows showed a decreasing trend. Statistically significant decrease in the mean and low flows (for some cases in the maximum flows) become obvious in the western, central, and southern river basins. More importantly, results are found in agreement with the trends in precipitation.

As far as the annual mean flow is concerned, the parametric test displayed trend existence at 27 gauging stations. All of the detected trends are negative (decreasing mean flow). Gauging stations with a trend are mostly located in the Western Anatolian basins, some in the Mediterranean region, and only one in the southeastern part of the country. No trend exists in the remaining part of the country, based on the results of the parametric test. Similarly, the non-parametric test has shown negative trends in all gauging stations where a trend exists. In general, both tests show that the annual mean flow in the western and southern basins are under the effect of a decrease in time. The regions with trend are the northwestern part (Marmara region), western part (significant part of the Aegean region), southern part (significant part of

Mediterranean region), and western part of Central Anatolia (Sakarya basin in particular). Both tests provided results quite close to each other.

For the annual maximum flow, the parametric test detected trends at 14 out of 96 gauging stations: 3 positive and 11 negative trends. No trend was detected in the maximum flows of 12 basins (almost half of the total number of hydrologic basins in Turkey). The nonparametric test found trends at 15 gauging stations. Only four stations had a positive trend, whereas the remaining trends were of a decreasing type. Gauging stations where both tests display trend existence in maximum flows are situated in the northwestern and northern parts of Turkey.

For low flows, the 1-day and 7-day averages are basically analyzed. As defined previously, the 1-day low flow is the lowest daily mean flow of a water year, while the 7-day low flow is the lowest of the consecutive overlapping 7-day mean values within the water year. The parametric test detected trends in the 1-day low flows of 44 gauging stations out of 96 streamflow gauging stations. Only four stations experienced an increasing trend, while the remaining 40 stations displayed a decreasing trend. For the 7-day low flows, trend was found at 43 stations among which only three showed an increasing trend. The number of gauging stations with trends in the 1-day and 7-day low flows was found the same, 39 out of 96. Trend existence in low flows was detected intensively in the western and southern basins. The central and eastern regions displayed trend existence only partially. Results of the nonparametric test pointed out 42 stations with trend existence both in the 1- and 7-day low flows. Although this is not a full agreement between the two tests, the general conclusion was that trends were widely spread in the western and southern river basins of Turkey, whereas only some parts of the central and eastern regions showed trend existence.

Trend analysis on streamflow data of river basins in Turkey has indicated significant trends which are found to be negative in general although positive trends have also been notified. Negative trends in the streamflow may be attributed to the observed decrease in rainfall and to the increase in temperature. As a particular example, negative trends associated with the streamflow in the Meric-Ergene basin (Number 1 in Table 5.1) could be linked to the foreseen decrease in rainfall and increase in temperature (Aksoy et al. 2005). When the trend analysis is taken as a whole, it is seen that trend existence is more common in the mean and low flows compared to the maximum flows. Trends are generally of a decreasing type when a few increasing trends are ignored. Statistically significant decrease is detected mainly in the annual mean and low flows (also in maximum flows of some gauging stations) in the western and southern basins rather than in the central and eastern parts of the country.

A most updated piece of work by DSI (2016b) and Ozdemir and Erkus (2017) gives the general picture in terms of the existence or no-existence of trends in the annual streamflow in each river basin (Fig. 5.8). Based on the non-parametric Mann-Kendall trend test applied at the 10% significance level, streamflow data in Turkey have either no trend (13 basins) or have a decreasing trend (11 basins). A positive trend is applicable only in one (Aras River in Northeast) out of 25 basins in Turkey. Most of the basins with no trend have indeed insignificant decreases in flows

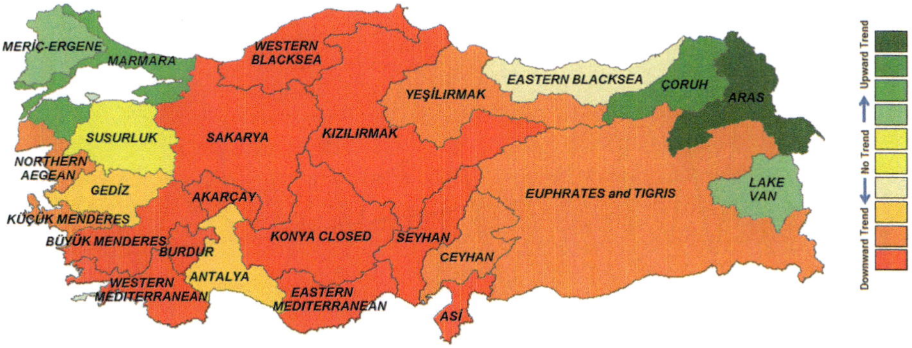

Fig. 5.8 Trend Analysis of Natural Annual Flows at Basin-scale. (DSI 2016b)

(Northern Aegean, Kucuk Menderes, Yesilirmak, Ceyhan, Euphrates-Tigris, Gediz, Antalya and Eastern Black Sea). Similarly, there are insignificant increasing trends in Meric-Ergene, Marmara, Coruh and Lake Van. No trend was detected in the Susurluk river basin. The basin-scale trend analysis shows that the country is under the effect of a decreasing trend in the annual flow when the northeastern and northwestern basins are excluded. Trend analysis repeated at the country-scale shows that a significant negative trend has started in 1988, the wettest year within the period of 1981–2010 (Fig. 5.9). The wet and dry years (1988 and 2001) are clearly observed.

5.11 Lakes

In Turkey, natural lakes extend over an area of about 10,000 km^2. Lakes in Turkey are concentrated in four regions: Lakes District in southwestern Anatolia (Egirdir, Burdur, Beysehir ve Acigol), Southern Marmara (Sapanca, Iznik, Ulubat, Kus), Lake Van and Lake Tuz regions (Fig. 5.10) (SYGM 2017). Table 5.9 summarizes natural lakes with their characteristics and basins. Some of the lakes are deep while others are only a few meters deep and shallow. Lake Van is the deepest with a maximum depth reaching 451 m; Lake Tuz is among the shallow lakes with only 2 m of maximum depth. In terms of water chemistry, some lakes have fresh water used in irrigation or consumed for domestic use, while the most contain saline water at different concentrations. Lakes drained for different purposes such as Lake Amik in the Asi river basin are also listed.

Other than natural lakes, there exist 706 dam reservoirs in Turkey, among which Ataturk Dam (817 km^2), Keban Dam (675 km^2), Karakaya Dam (268 km^2) on the Euphrates River; Hirfanli Dam (263 km^2) and Altinkaya Dam (118 km^2) on the Kizilirmak River should be mentioned first due to their size. Among others are Kesikkopru (on Kizilirmak River); Almus, Hasan Ugurlu and Suat Ugurlu (on Yesilirmak River), Hasan Polatkan and Gokcekaya (on Sakarya River); Demirkopru

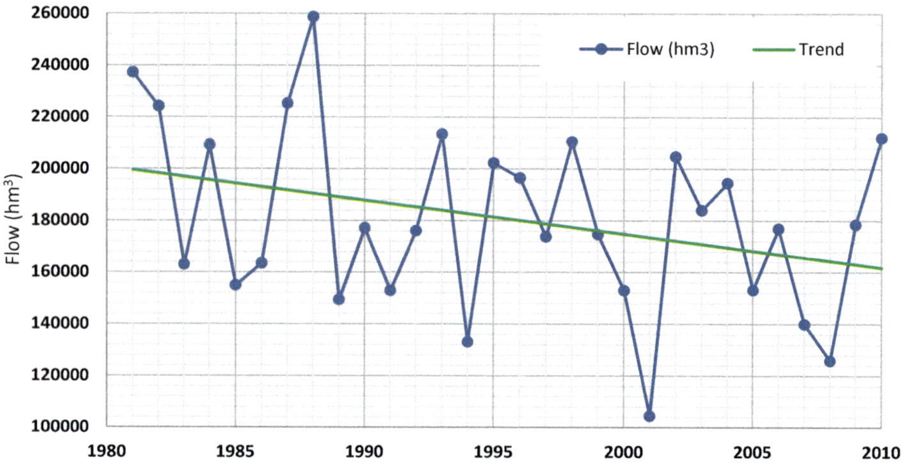

Fig. 5.9 Trend analysis of natural annual flows at country-scale. (DSI 2017a)

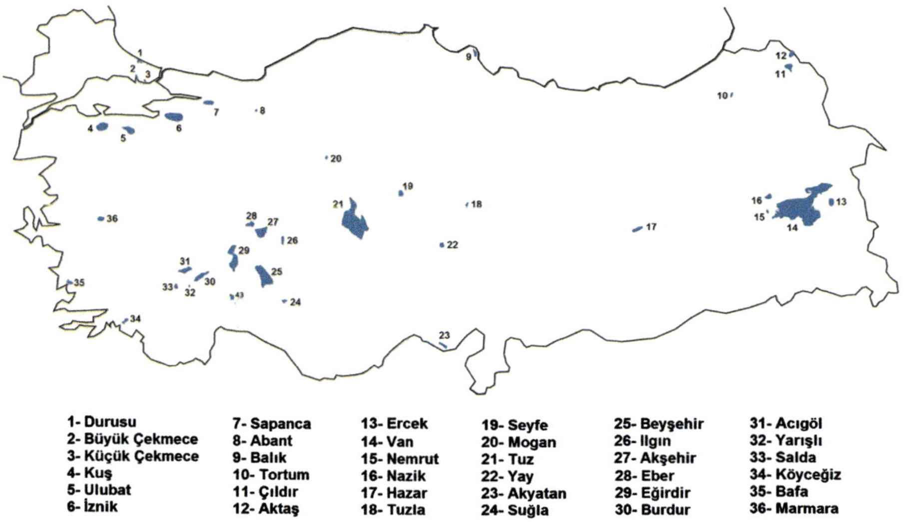

Fig. 5.10 Major lakes in Turkey

(on Gediz River); Kemer and Adiguzel (on Buyuk Menderes River); Seyhan (on Seyhan River); Aslantas and Berke (on Ceyhan River); and Oymapinar and Manavgat (on Manavgat River).

Table 5.9 Natural lakes in Turkey

Lake number as in Fig. 5.10	Lake	Basin number	Type	Surface area (km^2)	Elevation (m)	Max depth (m)	Water
1	Durusu (Terkos)	2	Coastal	25	50	11	Fresh
2	Buyukcekmece[a]	2	Coastal	11 /28.5	0	3.5 / 8.6	Saline/ Fresh
3	Kucukcekmece	2	Coastal	16	3	20	Saline
4	Kus (Manyas)	3	Tectonic	166	15	5	Fresh
5	Ulubat (Apolyont)	3	Tectonic	134	5	4	Fresh
6	Iznik	2	Tectonic	298	85	65	Fresh
7	Sapanca	12	Alluvial	47	4	61	Fresh
8	Abant	13	Landslide	1.28	1298	15	Fresh
9	Balik	15	Lav	34		100	Saline
10	Tortum	23	Landslide	8	1100	95	Fresh
11	Cildir	24	Lav	115	1959	130	Fresh
12	Hozapin (Aktas)	24	Tectonic	14	1794	Shallow	Saline
13	Ercek	25	Lav	98	1803	15	Saline
14	Van	25	Volcanic-Lav	3713	1648	451	Saline
15	Nemrut	25	Crater	12	2247	155	Fresh
16	Nazik	25	Lav	48	1816	50	Fresh
17	Hazar (Golcuk)	21	Tectonic	86	1248	80	Saline
18	Tuzla	15	Tectonic	23	1138	12	Saline
19	Seyfe	15	Tectonic	15	1110	5	Saline
20	Mogan	12	Alluvial	6	972	5	Saline
21	Tuz	16	Tectonic	1500	925	2	Saline
22	Yay	15	Tectonic	37	1071	2	Saline
23	Akyatan	18	Coastal	35	4	Shallow	Saline
24	Acigol	10	Tectonic	153	836	2	Saline
25	Sugla	16	Tectonic-karstic	125	1040	2	Fresh
26	Beysehir	16	Tectonic-karstic	656	1211	70	Fresh
27	Cavuscu (Ilgin)	12	Tectonic-karstic	51	1019	Shallow	Fresh
28	Aksehir	11	Tectonic	353	958	4	Saline
29	Eber	11	Tectonic	126	967	3	Fresh
30	Egirdir	9	Tectonic-karstic	468	916	13	Fresh
31	Burdur	10	Tectonic	200	85.4	110	Saline
32	Yarisli	10	Tectonic	16	950	Shallow	Saline

(continued)

Table 5.9 (continued)

Lake number as in Fig. 5.10	Lake	Basin number	Type	Surface area (km^2)	Elevation (m)	Max depth (m)	Water
33	Salda	10	Tectonic	45	1139	Shallow	Saline
34	Koycegiz	8	Alluvial	52	8	25	Saline
35	Bafa	7	Alluvial	60	2	45	Saline
36	Marmara	5	Alluvial	34	71	Shallow	Fresh
Drained lakes							
	Amik	19	Tectonic	60	81	4	Fresh
	Avlan	8	Karstic	8	1024	Shallow	Fresh
	Karagol	8	Karstic	23	1050	Shallow	Fresh
	Kestel	10	Karstic	25	779	4	Fresh

Adopted from Hosgoren (1994)
[a]Being a natural coastal saline lake Buyukcekmece has been dammed for the purpose of domestic use of fresh water for the Istanbul city. Figures in the table refer the natural and dammed lake, respectively

5.12 Concluding Remarks

At the first glance, numerous surface water resources are seen flowing into the sea through either large rivers, or intermittent creeks, or ephemeral water courses as well as lakes. Turkey has 25 river basins, each with its own hydrological character. Most of the river basins are within the country while there also exist transboundary rivers; Turkey is the upstream country in some while, in the others, it is the downstream country. Some rivers within the transboundary river basins flow as boundary rivers as well for a while. There are river basins which discharge into a closed lake. Most, on the other hand, flow into the sea either with a dendritic structure or as a combination of individual rivers.

Turkey has geographically heteregenous wet and dry regions, which affect the hydrology of the country. Not only floods but also droughts have been recorded in the past. They are expected to become more severe in the future with the existing trend. Low flows are expected to become lower in the whole country when a few river basins in the northwest and northeast of the country are excluded.

The hydrometric network in Turkey needs to be improved in terms of the number and the spatial distribution of monitoring stations, based on the topographical character of the country and the needs for the future water resources planning. In this sense, the ability of the hydrometric network should be extended to measure additional meteorological variables such as evaporation and snow, lake water levels, and reservoir operation data to be used in determination of the water budget of river basins. However, when no data exist, which is not an infrequent case in Turkey, empirical tools such as the envelope curves or the intensity-duration-frequency curves become helpful tools for water resources studies.

Surface water in Turkey is a quite important problem with its share of 80% within the total water potential of the country. It is a challenging issue at the same time due to the non-homogeneous structure of the country. The heterogeneity brings, at least, the necessity of different prevention measures to be taken against hydrological disasters such as floods and droughts. It is because of the same reason that the surface water resources are also faced with important management problems. Technical solutions could always work; nevertheless, central and regional administrative offices and well-established legislation would make the problem simpler to solve.

Acknowledgements Assistance and support given by Assoc. Prof. Dr. Ebru Eris Duman to me in preparing this chapter are greatly appreciated.

References

Aksoy H, Unal NE, Alexandrov V, Dakova S, Yoon J (2005) Hydroclimatological analysis of Northwestern Turkey with links to climate change. Int J Climatol 28(8):1047–1060

Aksoy H, Onoz B, Cetin M, Yuce MI, Eris E, Selek B, Aksu H, Burgan HI, Esit M, Orta S, Cavus Y (2018) Hidrolojik Havzalarda Dusuk Akimlar ve Kuraklik Analizi, National Union of Geodesy and Geophysics of Turkey, Project no. TUJJB-TUMEHAP-2015-01, Final Report

Bayazit M, Avci I (1997) Water resources of Turkey: potential, development and management. Water Resour Dev 13(4):443–452

Bayazit M, Onoz B (2004) Envelope curves for maximum floods in Turkey. Digest, Teknik Dergi 15:927–931

Bayazit M, Onoz B (2008) Taskin ve Kuraklik Hidrolojisi. Nobel Yayin Dagitim, Ankara

Bulu A, Aksoy H (1998) Low flow and drought studies in Turkey. In: UNESCO FRIEND-AMHY, Low Flows Expert Meeting, 10–12 June, Belgrade, Yugoslavia, p 133–141

Bulu A, Onoz B (1997) Frequency analysis of low flows by PPCC test in Turkey. IAHS Publ. no. 246, pp 133–140

Bulu A, Onoz B, Aksoy H, Cokgor S, Cigizoglu HK (1997) Trakya Bolgesi Dusuk Akimlarinin Hidrolojik ve Istatistik Analizi. ITU Arastirma Fonu, Istanbul

Carter RF, Putnam SA (1974) Low flow frequency of Georgia streams. Open File Report 77-127, USGS Atlanta, Georgia

Cigizoglu HK, Bayazit M, Onoz B (2005) Trends in the maximum, mean and low flows of Turkish rivers. J Hydrometeorol 6:280–290

DSI (1994) Frequency analysis of maximum flows of river basins of Turkey (Turkiye Akarsu Havzalari Maksimum Akimlar Frekans Analizi, MAFA), T.C. Bayindirlik ve Iskan Bakanligi, Devlet Su Isleri Genel Mudurlugu, Etut ve Plan Dairesi Baskanligi, Ankara

DSI (2016a) 2015 Haritali Istatistik Bulteni, Kisim: I (Haritali Bulten), T.C. Orman ve Su Isleri Bakanligi, Devlet Su Isleri Genel Mudurlugu, Strateji Gelistirme Dairesi Baskanligi, G. Yayin No: 991, Grup No: VIII, Ozel No: 177, DSI Destek Hizmetleri Dairesi Baskanligi Basim ve Foto-Film Sube Mudurlugu, Ankara

DSI (2016b) Turkiye'nin Yuzeysel Suyu Potansiyeli (Taslak Raporu), Orman ve Su Isleri Bakanligi, DSI Genel Mudurlugu, Etut Planlama ve Tahsisler Dairesi, April 2016, Ankara

DSI (2017a) Su Kaynaklarinin Gelistirilmesi ve Hidroloji Calisma Grubu Raporu, Orman ve Su Isleri Bakanligi, DSI Genel Mudurlugu, 2. Ormancilik ve Su Surasi, 5–7 May 2017, Afyonkarahisar

DSI (2017b) Taskin Yonetimi Calisma Grubu Raporu, Orman ve Su Isleri Bakanligi, DSI Genel Mudurlugu, 2. Ormancilik ve Su Surasi, 5–7 May 2017, Afyonkarahisar

Duran SD (2000) Low flow hydrology and application to Aegean region, Msc Thesis, Istanbul Technical University, 138 p (in Turkish)

Eheart JW, Tornil DW (1999) Low-flow frequency exacerbation by irrigation withdrawals in the agricultural midwest under various climate change scenarios. Water Resour Res 35:2237–2246
Eheart JW, Wildermuth AJ, Herricks EE (1999) The effects of climate change and irrigation on criterion low streamflows used for determining total maximum daily loads. J Am Water Resour Assoc 35:1365–1372
Eris E, Aksoy H (2014) Rainfall triggered landslide in Trabzon province, Turkey. In: Ferrari E, Versace P (eds) Proceedings of the Mediterranean Meeting on "Monitoring, modelling and early warning of extreme events triggered by heavy rainfall", University of Calabria, Cosenza, Italy, June 26–28, pp 253–259
Eris E, Aksoy H, Onoz B, Cetin M, Yuce MI, Selek B, Aksu H, Burgan HU, Esit M, Yildirim I, Unsal Karakus E (2019) Frequency analysis of low flows in intermittent and non-intermittent rivers from hydrological basins in Turkey. Water Sci Technol – Water Supply 19(1):30–39
Hewa GA, Wang QJ, McMahon TA, Nathan RJ, Peel MC (2007) Generalized extreme value distribution fitted by LH moments for low-flow frequency analysis. Water Resour Res 43: W06301
Hosgoren MY (1994) Turkiye'nin golleri (Lakes of Turkey), Turk Cografya Dergisi 29:19–51 (in Turkish with English abstract)
Koken E (2009) Regionalization of low flow characteristics of the Tigris Basin in Turkey. Msc Thesis, Dokuz Eylul University, Izmir, Turkey
Liu D, Guo S, Lian Y, Xiong L, Chen X (2015) Climate informed low-flow frequency analysis using nonstationary modeling. Hydrol Process 29:2112–2124
Mohamed M, Stednick JD, Smith FM (2002) Comparison of field measurements to predicted reaeration coefficients, k2, in the application of a water quality model, QUAL2E, to a tropical river. Water Sci Technol 46:47–54
Ontario Ministry of Natural Resources (1994) Natural channel systems, an approach to management and design. Queen's Printer for Ontario, Ontario Ministry of Natural Resources, Peterborough, Ontario, 103p
Ozdemir AD, Erkus MK (2017) Trend analysis of total annual flows of Turkey basins, Abstract Book of IX. National Hydrology Congress, 4–6 October 2017, Diyarbakir, p 89
Riggs HC, Caffey CE, Orsborn JF, Schaake JC, Singh KP, Wallace JR (1980) Characteristics of low flows, report of an ASCE Task Committee. J Hydraul Eng 106(5):717–731
Saracoglu O (2002) Low flow hydrology and application in Mediterranean region, Msc Thesis, Istanbul Technical University, 132 p (in Turkish)
Saris F (2016) Low flow analysis in Porsuk Creek basin. Istanbul Universitesi, Edebiyat Fakultesi Cografya Dergisi 33:73–81 (in Turkish)
Sertbas YD (1996) The investigation of the most suitable probability distribution for the low flows of Sakarya Basin River flows. Msc Thesis, Istanbul Technical University, Istanbul, Turkey
SYGM (2017) Goller ve Sulak Alanlar Eylem Plani, Orman ve Su Isleri Bakanligi. Su Yonetimi Genel Mudurlugu, Ankara
Yurekli K, Kurunc A, Gul S (2005) Frequency analysis of low flow series from Cekerek Stream Basin. J Agric Sci 11(1):72–77

Chapter 6
Groundwater

Hasan Yazicigil and Mehmet Ekmekci

Abstract About 18% of the total water resources potential of Turkey is made up of groundwater resources. Significant portion of the streamflow of major rivers is supplied by groundwater through springs and baseflow. In 1960s and early 1970s, the financial capacity of Turkey did not allow construction of large dams for irrigation. Development of groundwater resources in alluvial plain aquifers where the agriculture was concentrated has been a priority. In 1990s, the building of large dams has been boosted and irrigation by surface waters preferred due to the lower operational cost. From 1990s, not enough funds have been allocated to explore and develop groundwater resources. In spite of its strategic significance, much more has been invested to investigate and develop the "visible" resource. This unbalanced policy of water resources management has reflected also in the organizational and institutional structure of Turkey. Groundwater resources of Turkey mainly occur in alluvial and karstic aquifers. Large coastal plains and deltas, grabens and pull-apart basins constitute the major alluvial aquifers. The thick and extensive carbonate rocks along the Taurus mountain belt favor formation of productive karst aquifers. The fractured rock aquifers are either low yield or of local importance. Igneous rocks have no permeability and they have very limited outcrops. Groundwater occurs in younger volcanic rocks with limited extension. However, volcanic rock aquifers at foothills of volcanoes, such as Erciyes and Nemrut, may supply a great amount of groundwater where they are recharged by snowmelt. Metamorphic rocks are hydrogeological barriers, in general. They may bear very little amounts of groundwater that might support aquatic ecosystems. Turkey has faced some water mismanagement problems whose consequences are observable in terms of the decline of groundwater levels, reduced spring and streamflows, desiccation of lakes and wetlands and loss of ecosystems. These consequences resulting mainly

H. Yazicigil (✉)
Department of Geological Engineering, Middle East Technical University, Ankara, Turkey
e-mail: hyazici@metu.edu.tr

M. Ekmekci
Hydrogeological Engineering Program, and International Research Center for Karst Water Resources, Hacettepe University, Ankara, Turkey
e-mail: ekmekci@hacettepe.edu.tr

N. B. Harmancioglu, D. Altinbilek (eds.), *Water Resources of Turkey*, World Water Resources 2, https://doi.org/10.1007/978-3-030-11729-0_6

from managing surface waters and groundwater resources separately, ignoring that they are interacting subsystems of the same and single source, are becoming more frequent and severe. Implementation of the EU-Water Framework Directive has helped, to a certain extent, to maintain the "good status" and to "recover" the degraded water resources and the ecosystems. The "safe yield" approach that has been used in groundwater management needs to be changed to a "sustainable yield" approach which considers also the ecological water needs. This can only be achieved by competent persons who are educated in hydrogeological characterization, conceptualization and modelling of groundwater systems.

Keywords Turkish groundwater · Turkish aquifers · Safe yield · Sustainable yield · Groundwater management

6.1 General Features

Geographically Turkey is defined as a transcontinental country in Eurasia and regarded as a natural bridge connecting the two continents, Europe to Asia (Fig. 6.1). The main part is named as Anatolia, which is located in Asia, while the smaller part, called Thrace takes place in Europe. The total surface area of the country is 780,532 km^2. It is surrounded by seas on three sides. Turkey is divided into seven geographical regions based on climatic, physical and land-cover characteristics; namely, Marmara, Aegean, Black Sea, Central Anatolia, Eastern Anatolia, Southeastern Anatolia and the Mediterranean. The economic development and the lifestyle of inhabitants of different geographical regions are also different. The importance of water to energy and food security in Turkey is immeasurable and water is now a priority for economic development in almost all regions of Turkey. About one third of the area in the country is cultivated.

The water resources are under two major stresses in Turkey: the population and climate change. The current population is about 80 million and it is projected to increase to about 95 million by the year 2050 (TUIK 2018). The water potential per capita is accordingly decreasing from 1610 m^3 per year per capita in 2007 to less than 1300 m^3 per year per capita in 2030. The distribution of the population throughout the country is not uniform, neither are the available water resources.

6.1.1 *Topography and Morphology*

Turkey is surrounded by the Mediterranean Sea in the south, the Black Sea in the north, and the Aegean Sea in the west. Generally speaking, Turkey is a mountainous country where large plains and flat features also exist at lowlands. The main mountain ranges extend along the Black Sea and the Mediterranean coasts. The

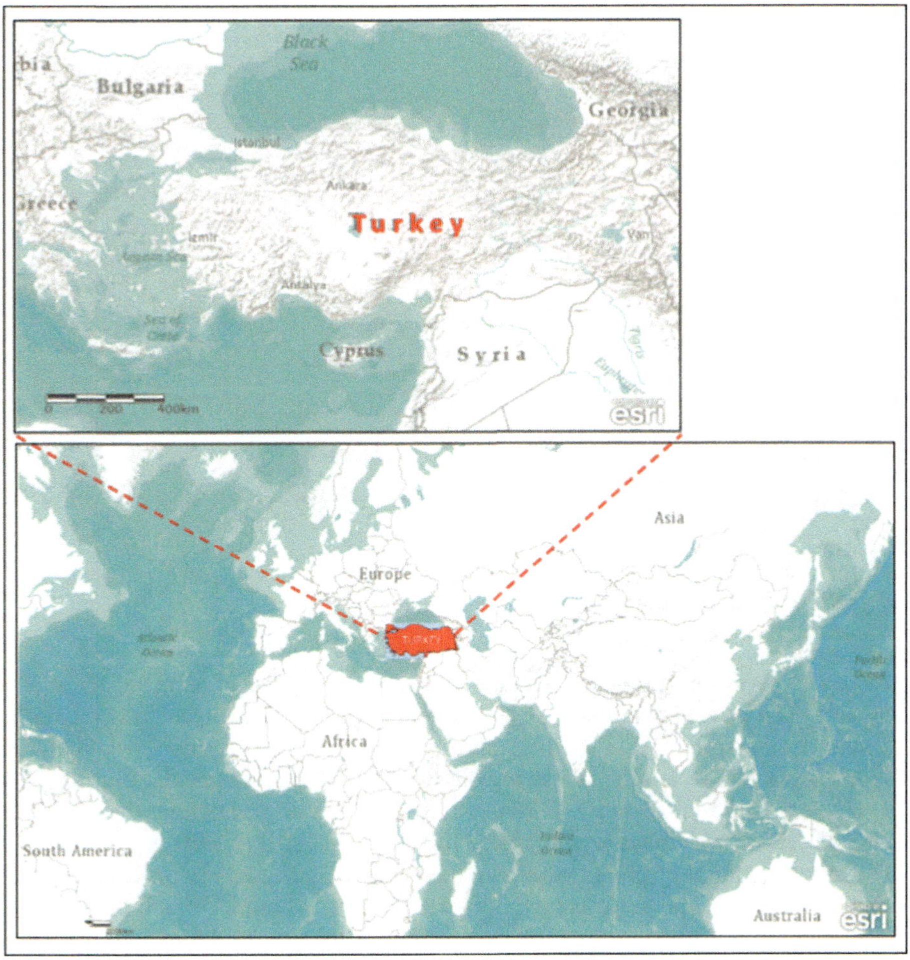

Fig. 6.1 Location of Turkey in Europe. (http://www.arcgis.com/home/webmap/viewer.html?useExisting = 1)

Anatolian plateaus extend between these mountain ranges. The mean altitude of Anatolia is about 1132 m. Only about one tenth of the territory is below 250 m which mainly forms the coastal area. The altitude of about half of the territory lies above 1000 m (Izbirak 2001). With this mean altitude, Turkey is considered as a highland country compared to the average altitudes of other countries. This topographical setting contains highland plateaus and highland plains, particularly in Eastern Anatolia, where the mean altitude (about 2000 m) is greater than the average of the country. The highest four mountains (>4000 m) are located in Eastern Anatolia. At the western part, low plains dominate, while Central Anatolia forms the closed basin type plateau with an average altitude of 1000 m above sea level (asl). A physical geographical map of Turkey is shown in Fig. 6.2.

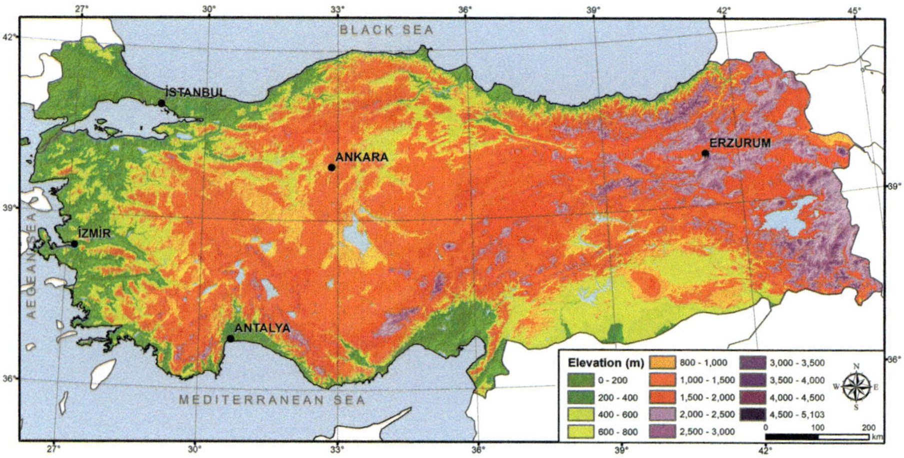

Fig. 6.2 Topographic map of Turkey. (Produced from Shuttle Radar Topography Mission (SRTM) data)

6.1.2 Climate and Hydrologic Basins

Turkey experiences four seasons but with varying types of climate across the country. The coastal geographical regions enjoy the maritime climate while continental climate prevails in Central and Eastern Anatolia. Central Anatolia and Southeast Anatolia are the driest regions with semi-arid continental climate. The Black Sea region in the north is humid and receives rain throughout the year. The average annual precipitation in Turkey for the period 1981–2010 is calculated as 574 mm. The highest rainfall occurs in the eastern Black Sea region with a rate of 2500 mm/year while it is below 300 mm/year in Central Anatolia. Owing to topographical variation, precipitation also significantly varies throughout the country (Fig. 6.3). Precipitation falls during winter and spring while summers and autumns are much drier. Topographical setting of the country also controls the variation of the average annual temperature. It is about 20 °C on the south coast and decreases significantly depending on the altitude and the distance from the sea. The average annual temperature falls to 4 °C in the interior regions. A certain portion of the precipitation occurs in the form of snowfall. Almost every part of Turkey, except the southern and western coastal areas, receives snowfall. The number of snowy days is higher in Eastern Anatolia. The snow coverage period is longer than 3 months; in some years, it may last for 4 months in this region. On the other hand, it is about 1 month in Central Anatolia.

Turkey is divided into 25 major hydrological (river) basins (Fig. 6.4), four of which are closed basins and five are transboundary. A countrywide water balance calculation has revealed that Turkey has a renewable water potential of 227 km^3, 186 km^3 of which is surface water and 41 km^3 is groundwater (DSI 2017a). The surface water potential of the major hydrologic basins and their watershed areas are

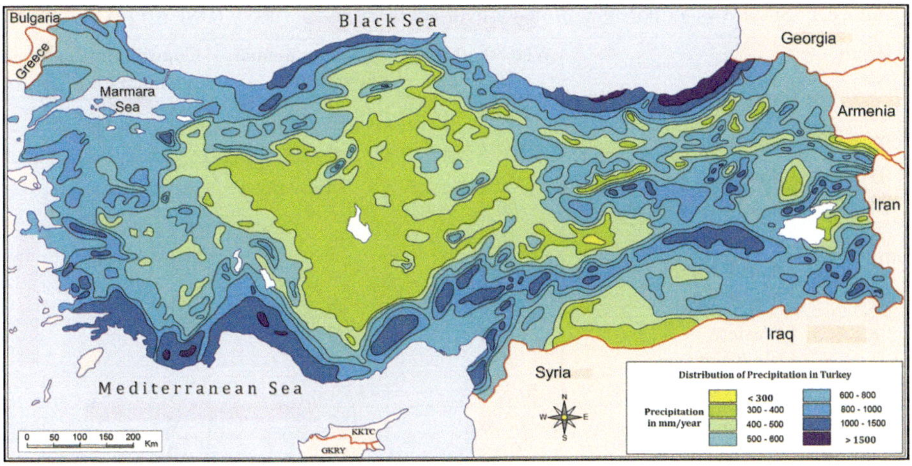

Fig. 6.3 Distribution of annual average precipitation in Turkey. (Modified from http://cografyaharita.com)

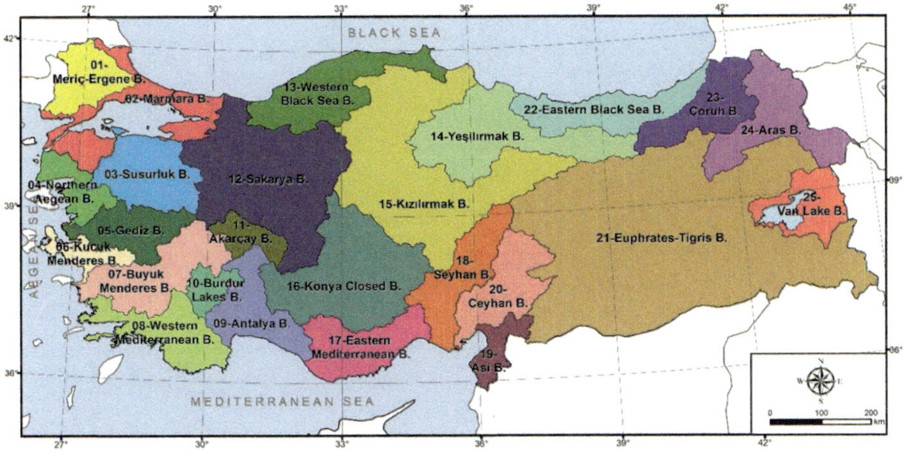

Fig. 6.4 River basins in Turkey. (DSI 2009)

tabulated in Table 6.1. The contribution of individual basins to the total surface water potential is depicted as a pie chart in Fig. 6.5. As seen from the chart, about 31% of the total surface water potential exists in the Euphrates-Tigris basin. Excluding the basin regions like the Antalya and Eastern Black Sea, the rest of the river basins has a share of less than 5% each. However, it is very difficult and probably incorrect to set a clear-cut distinction between surface water and ground water resources in Turkey due to the fact that a great majority of the surface waters are fed by ground water discharges.

Table 6.1 Surface water potential of major hydrologic basins in Turkey (DSI 2017b)

Basin No	Basin	Watershed area (km^2)	Average annual flow (km^3)	Contribution to total potential (%)
01	Meric – Ergene	14,444	1.8	1.0
02	Marmara	23,107	7.5	4.2
03	Susurluk	24,332	4.2	2.3
04	Northern Aegean	9974	1.5	0.8
05	Gediz	17,034	1.5	0.9
06	Kucuk Menderes	7060	0.5	0.3
07	Buyuk Menderes	26,133	3.0	1.6
08	Western Mediterranean	21,224	7.0	3.9
09	Antalya	20,331	13.1	7.2
10	Burdur Lake	6306	0.3	0.1
11	Akarcay	7983	0.3	0.2
12	Sakarya	63,358	5.2	2.9
13	Western Black Sea	28,930	9.9	5.5
14	Yesilirmak	39,628	6.6	3.6
15	Kizilirmak	82,197	6.1	3.4
16	Konya Closed	50,038	2.6	1.5
17	Eastern Mediterranean	21,807	8.2	4.6
18	Seyhan	22,242	6.8	3.8
19	Asi	7912	1.8	1.0
20	Ceyhan	21,599	7.4	4.1
21	Euphrates-Tigris	176,143	55.4	30.7
22	Eastern Black Sea	22,845	16.5	9.1
23	Coruh	20,249	7.0	3.9
24	Aras	28,115	4.2	2.3
25	Van Lake	17,977	2.3	1.3
	Total	**780,965**	**180.8**	**100**

Almost all rivers with large discharge rates, such as the Euphrates and Tigris in the Eastern-Southeastern Anatolia, Sakarya in Central Anatolia, Seyhan, Ceyhan, Manavgat, Koprucay, Aksu, Dalaman and Esencay in Southern Anatolia, and Gediz in Western Anatolia, originate from huge karstic springs that significantly contribute to their discharge rate (Yazicigil and Ekmekci 2003). This fact is also reflected in the graph depicting the relationship between the surface water potential and the size of the basin (Fig. 6.6). The Antalya Basin and the western Mediterranean Basin can be taken as an example. These basins are known for their well-developed karst systems. Lack of surface water is one of the typical properties of karst. In these basins, the

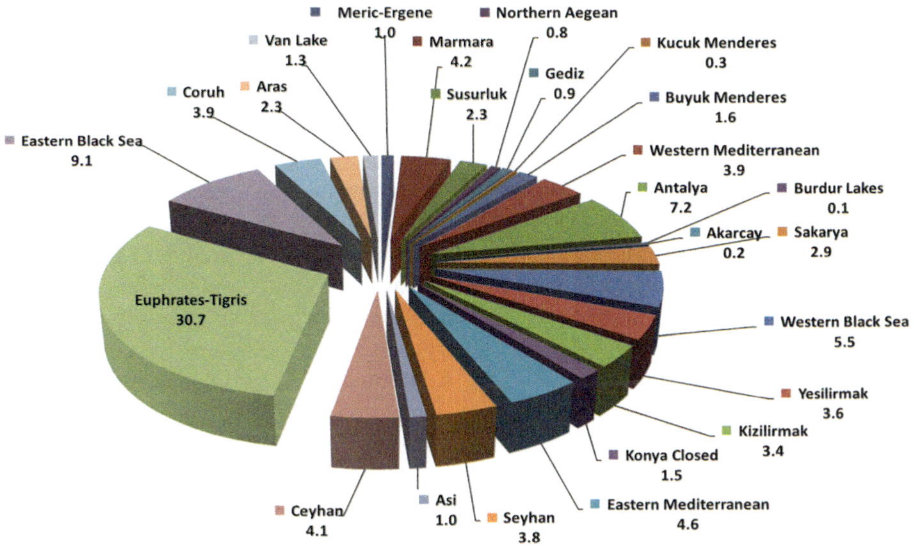

Fig. 6.5 Contribution of individual basins to the total surface water potential in percentage

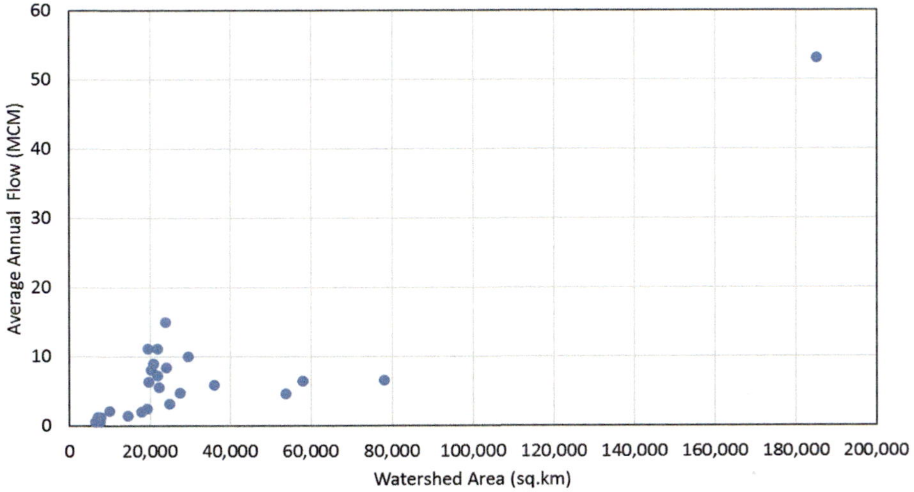

Fig. 6.6 Relationship between the surface water potential and the size of the basin

basin yield in terms of flow per unit area is relatively high. This is most probably because the surface flow in these basins is the outflow of karst groundwater. The flow measurement is made at the most downstream section. The groundwater outflow is measured at the surface and reported as stream flow. This misconception of the surface flow in basins where the base flow is significant is also reflected in water balance computations.

6.2 Geological Setting and Its Control on Hydrogeology

Turkey is characterized by a very complex geology, owing to its position between major tectonic plates and the relative motions of these plates. Plate margins, formed in Mesozoic to Cenozoic time, also cut Anatolia into fragments (Fig. 6.7). Okay and Tuysuz (1999) defines six major lithospheric fragments in Turkey: the Strandja, the Istanbul and the Sakarya Zones, the Anatolide-Tauride Block, the Kirsehir Massif and the Arabian Platform, each has its own geological characteristics. The Strandja, the Istanbul and the Sakarya Zones are known as the Pontides and separated from the southern Kirsehir Massif and the Anatolide-Tauride Block by the Izmir-Ankara-Erzincan suture (a former plate boundary). The Anatolide-Tauride Block is in contact with the Arabian Plate along the Assyrian-Zagros suture (see Fig. 6.7).

As a consequence, Turkey exhibits a wide variety of geological features with extremely complex geodynamics. The lithological units are very diverse not only in terms of age but also type. The age of lithological units varies from Precambrian to Holocene of all types of magmatic, sedimentary and metamorphic. In addition, as a result of the tectonic movements, overthrusts and allochthonous units are widespread, further complicating the geological structure of the country. As a distinct geological feature, the ophiolites and ophiolitic melanges cover large areas in Anatolia, along the suture zones of Triassic and younger oceans. These units are composed of lithological material of ophiolitic suit as well as of marine sedimentary

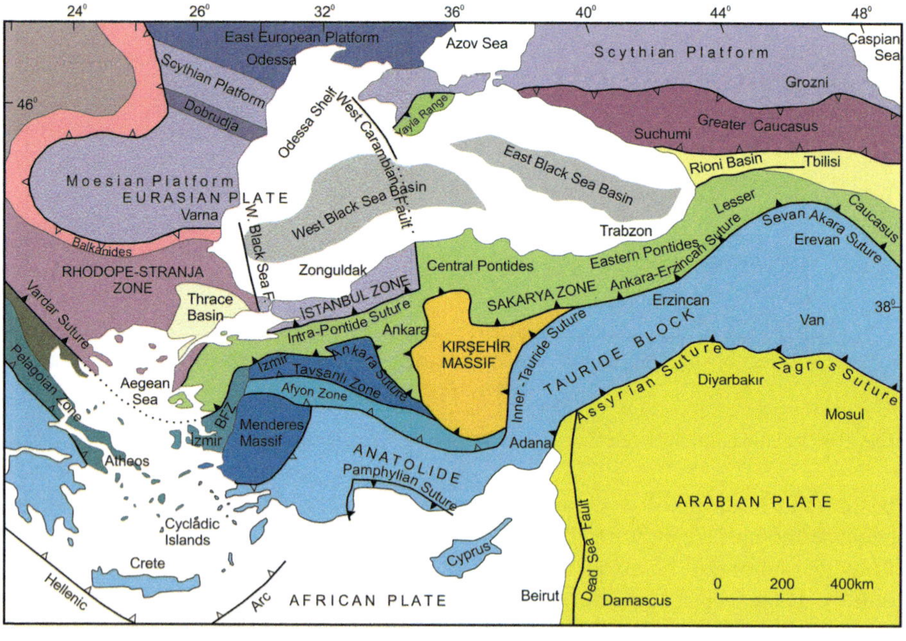

Fig. 6.7 Simplified tectonic map of Turkey and vicinity is showing the major plates and continental fragments. (Sahin et al. 2017; Okay and Tuysuz 1999)

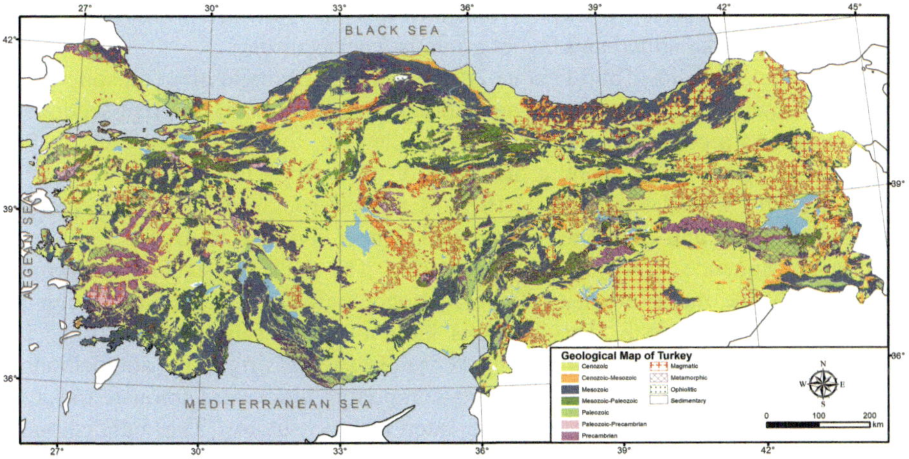

Fig. 6.8 Simplified geological map of Turkey. (MTA 1989)

units, such as flysch. Owing to the overthrusting, megablocks of carbonate rocks detached from carbonate platforms. The ophiolites and ophiolitic melanges do not favor aquifer formation, and they are hydrogeologically regarded as barriers to regional groundwater flow. However, the allocthonous megablocks of carbonate rocks are generally highly karstified and form productive aquifers. The geological map of Turkey given in Fig. 6.8 shows the diversity of the geological units by rock type and age.

The Istanbul Zone is a small continental fragment in the northwest of Anatolia. The major outcrops are of the Precambrian crystalline basement and the overlying sedimentary sequence. The basement is hydrogeologically impervious. Carbonates of shallow marine origin form the main aquifer in this zone. The conglomerates and sandstones are also the water bearing units and compose low yield, local aquifers. Marl, flysch, pyroclastites and shale are the impervious units.

The Strandja Zone is composed of a metamorphic basement overlain by Triassic sedimentary sequence. The basement is intruded by granitic rocks of Permian age. The Triassic sedimentary rocks are unconformably overlain by Late Cretaceous conglomerate and neritic limestone (Okay and Tuysuz 1999). Hydrogeologically, the Strandja Zone is covered by low permeability units. The marble in the metamorphic basement and the neritic limestone are the main lithological units where groundwater occurs.

The Sakarya Zone, which includes the Sakarya Continent and the Central and Eastern Pontides, is characterized by a complex geology owing to the widespread Triassic subduction-accretion complexes, also known as the Karakaya Complex in the western part (Tekeli 1981; Okay and Tuysuz 1999). This zone consists of a metamorphic basement overlain by a Jurassic-Eocene sequence. Apart from the limited existence of marbles, the Karakaya Complex does not contain lithological units that favor occurrence and movement of groundwater. The overlying Jurassic age limestone is karstified and favors groundwater occurrence.

The large Anatolide-Tauride zone is considered as a single block in spite of the variety of metamorphic, structural and stratigraphic features in different sections of this block. As a common stratigraphic characteristic, the block comprises "a Precambrian crystalline basement, a discontinuous Cambrian to Devonian succession dominated by siliciclastic rocks, a Permo-Carboniferous sequence of intercalated limestone, shale and quartzite, and a thick Upper Triassic to Upper Cretaceous carbonate sequence" (Okay and Tuysuz 1999). The thick and extensive carbonate sequence is the major element of the hydrogeological setting characterizing the Anatolide-Tauride Block. This block is known as the Taurus Karst Zone. The carbonate rocks that crop out in the block are of different origin: the authocthonous and the allocthonous. Allocthonous units comprising blocks of carbonate rocks are widespread along the Pamphylian and inter-Tauride sutures. The authocthonous units are mainly composed of thick carbonate rocks overlain by flysch-type sedimentary, impervious material.

The northernmost part of the Arabian Platform is within the southeastern territory of Turkey and is lithologically characterized by calcareous clastic units younger than Cretaceous, overlying a crystalline basement. Extensive emplacement of ophiolites and ophiolitic melanges over the platform took place during Late Cretaceous and Tertiary. The collision of the Arabian Platform with the Anatolides-Taurides during the Miocene has resulted in a secondary emplacement of allocthonous units over the Arabian Platform (Okay 2008). The authocthonous Eocene-Miocene carbonate rocks form the main aquifers, whereas the marls and the claystone bear groundwater in limited amounts and may form low yield, local aquifers.

There are two massifs, namely the Kirsehir Massif in central Anatolia and the Menderes Massif in western Anatolia, which should be mentioned in the framework of geology and hydrogeology of Turkey. The Kirsehir Massif consists of metamorphic and granitic rocks of Cretaceous age. The metamorphic rocks are composed of granulite, gneiss, micaschist, metaquarzite, marble and calcsilicate rocks. Late Cretaceous complex of basalt, radiolarian chert, pelagic limestone, sandstone and serpentinite tectonically overlay the metamorphic rocks (Seymen 1983; Okay and Tuysuz 1999). The marbles of the metamorphic rocks are the main water bearing formation. The Menderes Massif is another major metamorphic complex which is tectonically overlain in the south by the Lycian nappes, and in the northwest by a flysch zone, part of the İzmir-Ankara-Erzincan suture zone. The massif is divided into two main parts; the core and the cover. The core of the massif consists of Precambrian micaschists, and gneisses intruded by metagranites. The cover is composed of Paleozoic to Lower Tertiary metasedimentary rocks. From the oldest, the sequence starts with marble, quartzite and phyllite of Permo-Carboniferous age, continues with a thick sequence of Mesozoic marbles, and ends up with a thick platform limestone of Cretaceous age. This limestone is overlain by flysch sequence with serpentine blocks (Okay 2008). The Lycian nappes are emplaced tectonically on the flysch sequence. The major aquifers in the Menderes Massif occur mainly in the cover carbonate rocks and partly in the marbles of the core. The carbonate

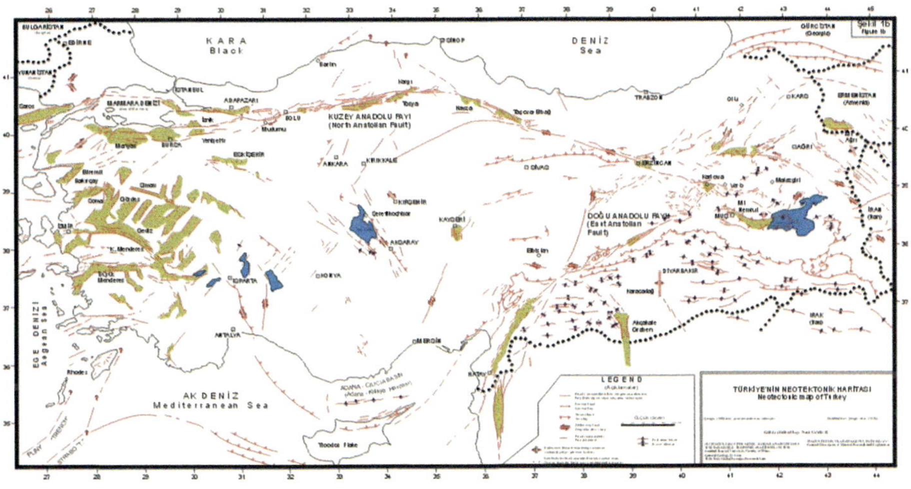

Fig. 6.9 Neotectonic map of Turkey. (Sengor et al. 1985)

megablocks of the Lycian nappes also form productive and extensive karst aquifers particularly in southwestern parts of the massif.

Tectonically, Turkey is located in a very active region of the earth. It is affected by different tectonic regimes which further complicated the geological setting of the country. Generally speaking, two major tectonic regimes are defined; the paleotectonic and the neotectonic regime. The neotectonic regime, starting from the Middle-Late Miocene has shaped the current geological setting of Turkey. The neotectonic features are shown in Fig. 6.9. The olive green areas on the map show the troughs filled with young alluvial deposits. It was also emphasized that karst evolution has been controlled to a great extent by the neotectonic development of the country (Ekmekci 2003).

The complex geology has also complicated the hydrogeological setting in Turkey. Considering the main objective of this chapter, the geological description is limited to the rock types exposed in the Turkish territory, because the hydrogeological setting of Turkey is outlined on the basis of the hydrogeological properties of the rock types.

The geology shown in Fig. 6.8 is further simplified for hydrogeological purposes as shown in Fig. 6.10. The geological units are grouped according to their rock types: sedimentary, magmatic and metamorphic. The sedimentary rocks are further subdivided into 4 groups; river alluvium, detrital formations, sandstone and carbonate rocks. However, the metamorphic carbonate rocks like marble are considered within the carbonate rocks. The magmatic rocks are also subdivided into crystalline and volcanic rocks.

Carbonate rocks constitute about one third of the country's surface area. They differ by age, but mostly they are of Mesozoic (Cretaceous and Jurassic) age. The Tertiary carbonate rocks are the second largest group of carbonate rocks. Distribution of carbonate rocks by age is shown on the map in Fig. 6.11.

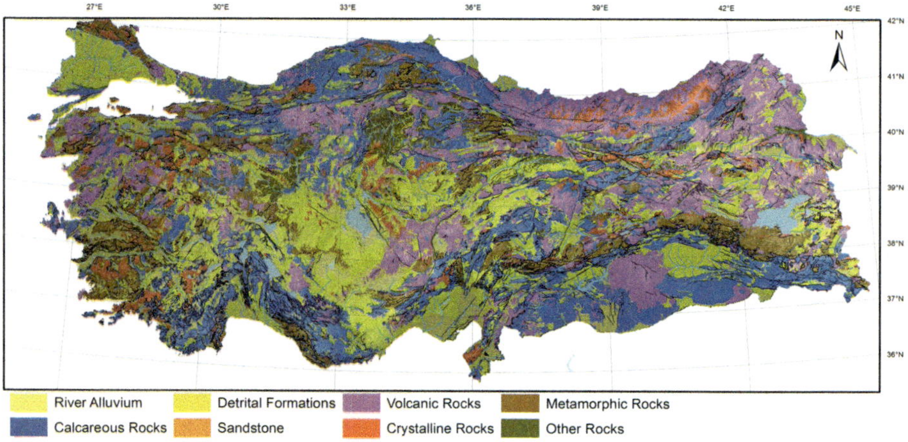

Fig. 6.10 Rock types exposed in Turkey

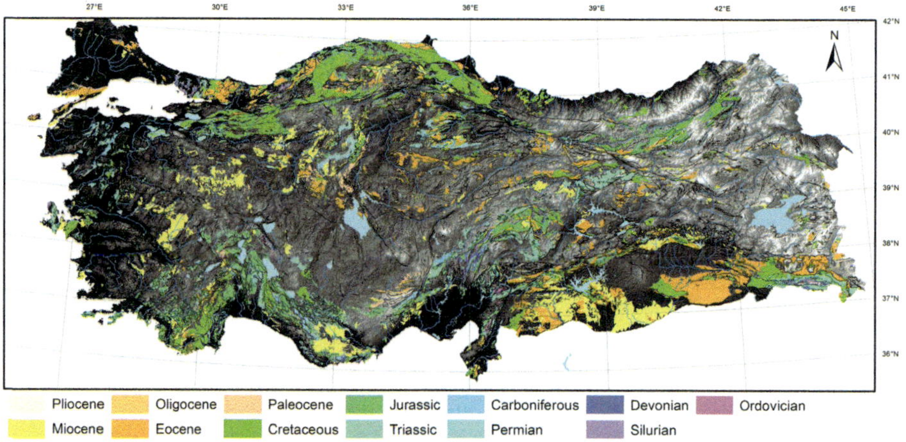

Fig. 6.11 Distribution of carbonate rocks by age

As shown in Fig. 6.10, the volcanic rocks are the second widespread rock type in Turkey, mainly exposed in Eastern Anatolia and Eastern Black Sea region. The volcanic rocks of the Eastern Black Sea region are older (Cretaceous-Paleocene) in general and are of intercalations of volcanosedimentary units. On the contrary, the volcanic rocks in Eastern and Southeastern Anatolia, and in some parts of the Aegean and Central Anatolia regions, are young (Plio-Quaternary) and of various character. Basaltic lava flows are common.

The yellow areas on the map in Fig. 6.8 indicate the detrital formations mainly of Neogene age. A great majority of the detrital formations, mostly those exposed in Central part of the country, are lacustrine deposits. They are composed of alternations of marl, claystone, sandstone and limestone layers. Thick young deposits fill the troughs and grabens in the Aegean region and pull-apart basins along the prominent fault zones (North and East Anatolian Fault zones), implying the active tectonic regime.

Metamorphic rocks are exposed around the massifs as they constitute the core of the massifs. As described above, the Menderes Massif in western Turkey and the Kirsehir Massif in Central Turkey are the two largest massifs in Turkey. Accordingly, the metamorphic rocks, mostly schists and gneisses, expose mainly around these massifs in the Aegean and Central Anatolia regions.

6.3 Water Bearing Formations and Groundwater Potential

The water bearing formations (aquifers) are grouped into four categories to give an understanding of the general hydrogeological conditions in Turkey: the alluvial deposits/gravel aquifers, karstic aquifers, volcanic rock aquifers and the fractured bedrock aquifers. In the late 1960s, hydrogeological mapping of the Turkish territory in 1/500000 scale has been conducted by the General Directorate of the State Hydraulic Works (DSI), and a map was published in 1970. Standard hydrogeological legend was employed in production of the 18 sheets of hydrogeological maps. Intergranular aquifers were colored in blue, the fractured and karstic aquifers in green and the non-aquifer formations in brown. The pale tones of each color indicate low yield and/or local potential. The dark tones represent high yield and/or extensive groundwater potential. A copy of the hydrogeological map produced through the above mentioned project is shown in Fig. 6.12. It is important to note that the colors in this map represent the aquifer formations and not the permeability of the units. Therefore, it is not surprising to see the same lithological formation in two different colors, brown and green. This means that the brown section of the formation may be permeable but with no saturated zone, while the green portion, on the other hand, suggests that a saturated zone has been developed.

Fig. 6.12 Hydrogeological map of Turkey in 1/500000 scale. (DSI 1970a)

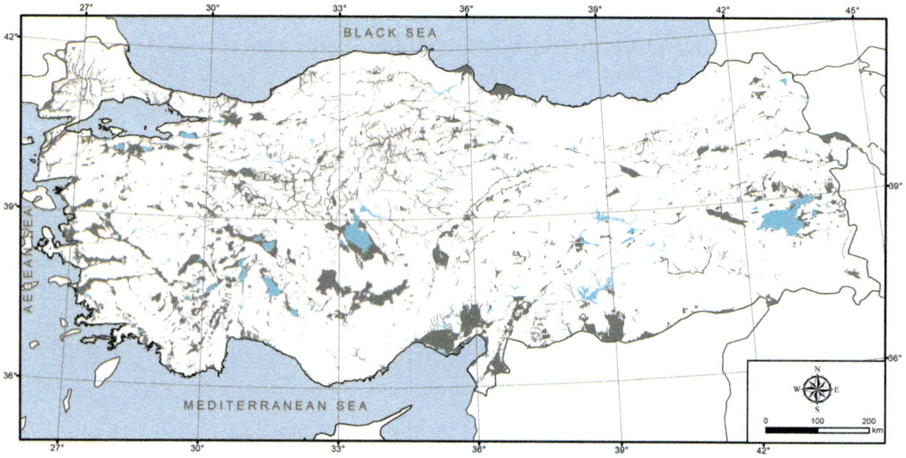

Fig. 6.13 Distribution of alluvial plain aquifers in Turkey

This map is still in use as it demonstrates the general hydrogeological setting of Turkey. The map exhibits a dark brown-dominated territory. This means that the majority of the rocks form low yield/local aquifers or impermeable formations. Blue color is prominent in delta areas of major rivers, indicating the coastal alluvial aquifers. Green areas represent the carbonate rock, mostly karstic aquifers.

6.3.1 Alluvial Aquifers

The alluvial deposits filling large river valleys, coastal plains, pull-apart basins, grabens, and inter-mountainous plains constitute the main intergranular aquifers in Turkey (Fig. 6.13). The importance of these plains stems from the fact that they are the main areas where irrigated agriculture in Turkey takes place. To date 195 plain aquifers were investigated and developed for groundwater use by the State Hydraulic Works (DSI), the main governmental body responsible for exploration and development of water resources. These aquifers are by far the most significant ground water reservoirs and have been used extensively for irrigational purposes over the past 40–50 years, ending up with severe declines in groundwater levels and reserves.

Being a peninsular country, Turkey is rich in coastal plains, which comprise significant alluvial aquifers. Rivers discharge into the Black Sea in the north, to the Aegean Sea in the west and to the Mediterranean Sea in the south, forming alluvial plains and large deltas on the coast. Although most of the plains are small, constituting aquifers of local importance, they are the main and the only water resource in their localities in most cases. Large coastal plains that make up the major coastal alluvial aquifers are those formed at the mouth of the major rivers. The Yesilirmak, Kizilirmak and Sakarya rivers on the Black Sea coast, Meric, Bakircay, Gediz,

Kucuk Menderes, Buyuk Menderes rivers on the Aegean Sea coast, Dalaman, Esencay, Manavgat, Koprucay, Aksu, Goksu, Berdan, Seyhan, Ceyhan and Asi rivers on the Mediterranean sea coast supply alluvium to construct coastal plains which potentially are extensive aquifers. The alluvial aquifers in coastal plains are usually of low yield and are threatened by salt water intrusion and agrochemical pollution.

The Aegean region is tectonically characterized by grabens dissecting the Menderes massif. The mountain ranges run perpendicular to the Aegean Sea. This favors the formation of large rivers with large basins. The Bakircay, Gediz, Kucuk Menderes and the Buyuk Menderes rivers are the major rivers draining the vast plains on the associated grabens. The alluvial plains therefore extend far inland to constitute large aquifers.

Owing to the active tectonic movements in the neotectonic period, several pull-apart basins with various sizes have developed along prominent strike-slip fault zones, such as the North Anatolian Fault Zone, the East Anatolian Fault Zone, and the Karasu rift valley, a transition between the Dead Sea Fault and the East Anatolian Fault zone, extending between Hatay and K.Maras provinces (Rojay et al. 2001). Lined up along the North Anatolian Fault Zone, 22 plains contain extensive or local alluvial aquifers. From west to east, Balikesir, Gonen, Karacabey, Bursa, Kemalpasa, Inegol, Gemlik, Iznik, Adapazari, Duzce, Bolu, Tosya, Vezirkopru, Suluova, Tasova, Erbaa, Niksar, Erzincan, Tercan, Erzurum and Pasinler plains are associated with the North Anatolian Fault Zone. The Caldiran, Malazgirt, Mus, Bingol, Elazig, Malatya, K. Maras, and Amik Plains are associated with the East Anatolian Fault Zone.

The groundwater in some graben and pull-apart basins having geothermal potential is threatened by geothermal fluids with high boron concentrations.

The alluvial aquifers in large river basins and pull-apart basins are generally high yield with spatially varying properties. The permeable layers of gravel and sand are usually not extensive and intercalated with silt and clay layers. The upper layers form unconfined aquifers when they are not covered by clayey layers. The lower layers are mostly semi-confined and rarely confined. Consequently, the storativity values vary from 0.001 to 0.2. The transmissivity values of the alluvial aquifers vary over wide ranges with values as low as 100 m^2/day to values higher than 1500 m^2/day, especially in gravel layers connected to large rivers. The specific capacity generally ranges between 0.1 l/s/m and 5 l/s/m. The groundwater in the plain aquifers is threatened by agrochemicals due to intensive agricultural activities.

In addition to coastal plains, the inland plains constitute significant intergranular aquifers. Some of the inland plains are associated with karst while several of them are of tectonic origin. A few of them are relics of paleolakes. Denizli, Burdur, Isparta, Simav, Afyon-Eber-Aksehir, Konya, Eregli, Aksaray, Develi, Kayseri, Turhal, Amasya, Cubuk, Eskisehir, Elbistan, Birecik, Suruc, Harran, Ceylanpinar, Yuksekova, Igdir, Kars, and Ardahan are among the inland plains in Turkey.

The Konya Plateau is also known as a plain and constitutes one of the largest aquifers in Central Anatolia. The watershed is more than 50,000 km^2, draining to the Salt Lake, which makes it a closed basin. It is a remnant of the Great Konya Lake of the Pleistocene period (Erol 1984). The Konya Plain aquifer is composed of karstic limestone of Neogene age. Located in the driest areas of Turkey, the basin has an annual water potential of about 4.3×10^9 m^3, about 57% of which is groundwater. A significant amount of the groundwater potential has originated from the melt of glaciers on the Taurus mountains, dating back to 8000 years (Sarikaya et al. 2011). On the other hand, the total water consumption is about 6.5×10^9 m^3/year, which exceeds the total input. The consequence is a drastic and persistent decline of groundwater level in the plain.

The Korkuteli-Bozova-Kestel plains and the Mugla Plain are also known to form a closed basin alluvial aquifer in the Mediterranean region. However, these plains are large karstic depressions known as poljes with alluvial deposits at their flat bottom. Elmali, Tefenni, Acipayam and Tavas plains are also associated with karst development.

6.3.2 *Karst Aquifers*

A great majority of the carbonate rocks of Turkey is karstified and constitutes the most productive aquifers. The following section giving details of karst hydrogeology of Turkey is extracted from the work by Ekmekci (2003).

Among a great variety of factors controlling the karst phenomenon tectonics, petrography, source of energy gradient and the type of the erosion base are the major ones. Although climate is among the controlling factors, in Turkey, the geodynamics seems to dominate the climatic effects and makes it less pronounced almost throughout the country. When all factors that control karst processes are considered together, two major types of karst can be defined in Turkey: a) Evolutionary karst which implies continuous karstification but at different stage of maturation, and b) Rejuvenated karst formed by reactivating a formerly developed and subsequently ceased karst structures either by an uplift and/or a drastic decline of the erosion base. Evolutionary karst implies that karst processes are continuous whereas the rejuvenated karst indicates an interruption in karstification. The Turkish territory can be divided into seven main karst provinces on the basis of evolution of karst. Each province has its own tectonic and paleogeographic history and therefore, the development of karst has been specific to that region.

The complexity in tectonic evolution is reflected in development of karst particularly in the Taurus mountain range. It is not possible to consider the karst in Taurus range under one single class. This is not because of the lithological variety but because of the differences in type and magnitude of tectonic effects. Both evolutionary and rejuvenated karst types exist in the Taurus range, where the western part is under the effect of regional submergence. The formerly developed non-mantled karst is being invaded by marine waters. As a consequence, the former vadose zone is becoming phreatic, giving rise to the occurrence of submarine karst features. Central Taurus region is more complex in terms of karst types. The eastern and western border sections of the Central Taurus region is characterized by rapid uplift, whereas the central part is under the effect of a slower uplift. Therefore, in central part of the region, juvenile, evolutionary karst is common while the border sections are characterized by paleokarst and rejuvenated karst.

Eastern Anatolia had been submerged by marine waters until Middle Miocene. The uplift of Eastern Anatolia caused the exposure of older lithologic units, some of which are karstified. The continuous uplift also caused erosion of the units covering the metamorphic basement, which also contains marbles. Therefore, apart from limestone karst, marble karst is also common in the Eastern Anatolian region. Westward escape of the Anatolian scholle, following the intersection of the Northern Anatolian Fault and the Eastern Anatolian Fault, lessen the rate of uplift and therefore the rate of erosion. Southeastern Anatolia is the border-folds province which has been under a compressional tectonic regime since the Early Miocene. However, the uplift is slower and karst is developed in limestone of Paleogene age. Capture of karst depressions by tributaries of the Euphrates and Tigris rivers, which incised their beds, suggests that the erosion base in this region is controlled by the main drainage systems.

The weakly active Central Anatolian province exhibits the best examples of rejuvenated karst. However, evolutionary karst also exists in areas that are not covered by Miocene deposits. In these areas, karst has almost completed its evolution. The impervious units are very shallow, and the carbonate rock masses are dissected by the drainage network. This is the reason for having relict karst in the northern and northwestern part of the Central Anatolian Province.

Similarly, the Black Sea region is tectonically weakly active but the karst is evolutionary and is controlled mainly by the Black Sea level changes. The karst aquifers classified by their development and the current hydrogeological conditions are shown in Fig. 6.14.

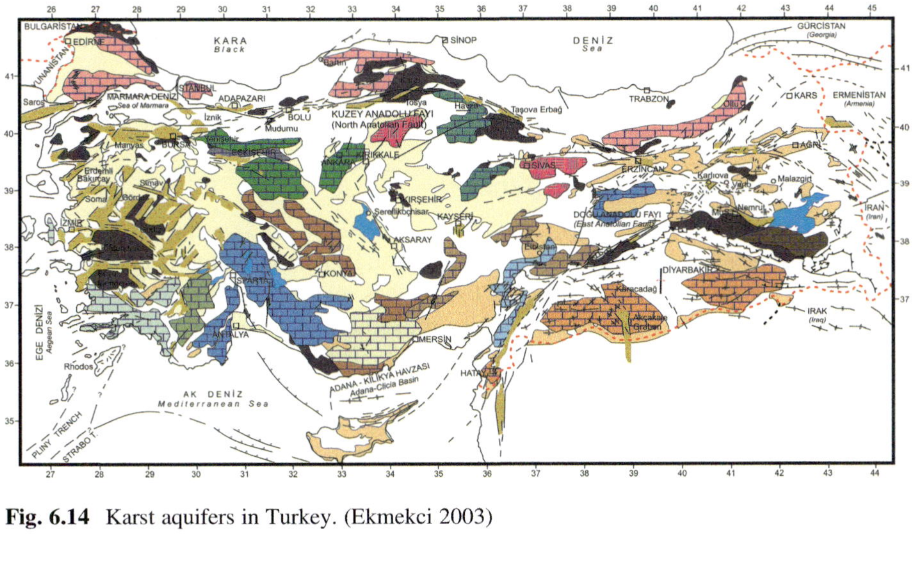

Fig. 6.14 Karst aquifers in Turkey. (Ekmekci 2003)

Although large volumes of previous work have been done on karst morphology and speleology (e.g. Bener 1965; Canik and Corekcioglu 1985; Nazik and Tuncer 2010; Nazik and Poyraz 2017), the hydrogeological characterization of aquifers has been limited to projects on development of karst water resources by hydrotechnical structures, mainly big dams. The main concern has been water leakage problems at dam sites and reservoir areas. Several groundwater tracing tests have been conducted mostly in the Taurus Mountain karst region. Groundwater velocities were calculated as high as 2500 m/day. The specific capacity of the wells penetrating the groundwater system is about 8 l/s/m. In some cases, pumping with a rate of 70 l/s does not create a drawdown in the well (Gunay 1985).

On the other hand, little is known on recharge areas, mode and rate of recharge, distribution of groundwater potential, groundwater flow velocities and direction of flow, extent and the boundary conditions of the karst aquifers and their interrelations with surface waters and other groundwater systems, and hydraulic characteristics of the aquifers. Pumping tests for hydraulic characterization are either very rare or not available. DSI has limited its investigations mostly to alluvial plain aquifers. This is mainly because of the facts that (a) the shallow alluvial aquifers were easier to explore and develop and (b) the groundwater potential of the shallow alluvial aquifers, together with the surface water resources, has been sufficient to meet the water demand. However, the increasing demand for more water has drawn the attention of the authorities to the productive karst aquifers. The lack of detailed hydrogeological investigations of karst aquifers has led to problems in effective management and protection of karst aquifers (Ford and Williams 2007). Being more vulnerable by their nature, the karst aquifers face problems of mismanagement and pollution, mainly because of the lack of clear understanding of the hydrological interconnections between karst groundwater and surface waters (Bakalowicz 2005). The groundwater potential of the karst aquifers is estimated to be high, because the rivers that make about 50% of the surface water potential are, to a great extent, fed by karstic groundwater. This portion of the surface water is in effect the discharge of the groundwater and should not be included in the surface runoff component of water balance calculations, but in the groundwater discharge component.

6.3.3 Volcanic Rock Aquifers

Although volcanic rocks are relatively widespread in Turkey, a small portion of them constitutes productive aquifers (Fig. 6.10). This is mainly because the andesitic and basaltic volcanic and volcanosedimentary rocks are composed of minerals susceptible to alteration by residues of clay minerals. These residues fill the pores and interstices, reducing the porosity and permeability of the rock. The older volcanic rocks have lower permeability due to argillization as a result of higher degree of alteration. Exposures of volcanic rocks older than the Eocene are very rare due to erosion. The oldest outcrops belong to Eocene-Oligomiocene age. The outcrops of these old volcanics are observed in Central Anatolia around Ankara. The Miocene aged volcanics exhibit similar properties but with much smaller outcrops in the inner

Aegean region. They are poorly permeable, and therefore they do not constitute aquifers. The rocks of the Mio-Pliocene and the Pliocene are widespread in the northern Aegean, in central Anatolia around Kayseri Province and in Eastern Anatolia. The younger volcanics have higher permeability and form some local aquifers. Boreholes tapping the Mio-Pliocene and Pliocene volcanics are common to supply groundwater to villages and small municipalities. However, the most productive volcanic aquifers are made up of Plio-Quaternary and Quaternary volcanics. The volcanic rocks that are composed of young basaltic lava have relatively high transmissivity values, and they constitute high yield aquifers where their extent and thickness are sufficiently large. Such aquifers supply significant amount of water to municipalities, particularly in Central Anatolia near Kayseri, Southeast Anatolia near Diyarbakır and Southern Anatolia near Iskenderun city. The volcanic rock aquifer in Kayseri is very productive and behaves hydraulically similar to karst due to large lava tubes, where the groundwater flow is conduit flow type. The aquifer is recharged all year long from the snowmelt of the permanent snow cover of the high Erciyes volcano. The volcano-sedimentary units of older age have lower permeability, and therefore, they form low yield aquifers. The older the volcanic unit, the lower the permeability because the openings are filled with clayey material, which are produced by the alteration of the minerals forming the volcanic unit. Similar to karst aquifers, there has been no systematic hydrogeological characterization of the volcanic aquifers. Only individual studies covering some specific portion of the aquifers are available. Thus, it is not reliable to give an estimate of groundwater potential of volcanic rock aquifers. In many areas where the old volcano-sedimentary and volcanic rock units are exposed, the wells are either dry or supply only very limited groundwater (Degirmenci et al. 2011).

6.3.4 Fractured Bedrock Aquifers

Groundwater occurrence in intrusive or metamorphic bedrocks is very limited in Turkey. This is partly because the outcrops of crystalline rocks are not widespread and partly because the permeability of the intrusive rocks is very low and dependent of stress fractures. The fracture frequency decreases significantly with depth. As a consequence, mostly shallow and low yield groundwater systems occur in the intrusive rocks. The upper sections of the outcrops are generally weathered to produce sandy material. This material is thicker particularly at foots of hills where the slope is more gentle. Subsurface waters infiltrated from rain emerge as seepages at topographic depressions. Locals capture these seepages to supply water for their individual use. It is common to observe numerous seepages and fountains in areas covered by intrusive rocks, although no productive boreholes are located. Intrusive rocks crop out mainly in the south of Menderes Massif, in central Anatolia around Yozgat Province, in western part of Eastern Anatolia, and the Marmara (northwest Turkey) region, where they form local, low yield shallow aquifers.

The metamorphic rocks may yield groundwater in very limited rates depending on the rock type. Schists and gneisses are practically impervious; whereas, quartzite and calc-schists may bear groundwater in very limited amounts. In general, hydrogeologically these rocks are not of interest for water supply problems. Instead, they are studied mainly for groundwater control in mines.

6.4 Groundwater Potential of Turkey

According to country-scale water balance calculations, the overall input from direct precipitation is calculated as 450 km^3; 274 km^3 out of this amount is estimated to be lost by evapotranspiration. Another 7 km^3 is calculated to add from neighboring countries in the transboundary river basins. The renewable/annual water potential of Turkey is calculated as about 234 km^3, 41 km^3 of which is the groundwater potential. On the other hand, not all of this potential is considered to be available for development due to technological and economic reasons. The available annual water resource is about 113 km^3, and the available groundwater potential is about 15 km^3. The above given figures are taken from DSI (2009). However, water balance calculations seem to be problematic in several aspects.

Apart from the uncertainty in representativeness of precipitation and evapotranspiration, river flows are assumed to represent surface runoff; whereas, groundwater contributes significantly to the gauged streams and rivers either through springs or as base flow. A reliable figure for the groundwater potential of Turkey cannot be given at the current situation because the hydrogeological appraisals on the basis of individual aquifers or aquifer systems have not been completed. For the year of 2016, the water balance calculation for each basin has revealed that the total recharge is about 23 km^3 and the total exploitable/available groundwater potential is slightly below 18 km^3 (Table 6.2).

Contribution of each basin is depicted on a pie chart in Fig. 6.15. The largest contribution is by Euphrates-Tigris basin with a rate of about 21%. The second largest contribution is found to be by the Konya closed basin with a rate of about 11.5%, followed by Kizilirmak and Sakarya basins with contributions of about 10% and 9%, respectively. This shows that more than 50% of the total groundwater potential is found in the 4 large basins. Unlike the case with surface waters (see Fig. 6.6), the groundwater potential seems to be correlated with the size of the basin (Fig. 6.16). Normally, groundwater potential is indirectly related to size of the basin. Type and extension of the lithological units, precipitation, and the factors that control evapotranspiration and infiltration are more effective in the occurrence of groundwater potential. The Konya closed basin plots are slightly deviated. This closed basin is a large flat plateau in the driest region of Turkey. Drainage network is not well-developed due to the dry climate and the flatness. It is difficult to measure or to make a reliable estimation of surface runoff. This case

Table 6.2 Groundwater potential calculated for individual basin (DSI 2018)

Basin No	Basin	Groundwater recharge ($\times 10^6$ m^3/year)	Exploitable potential
01	Meric – Ergene	507.7	498.2
02	Marmara	241.7	210.7
03	Susurluk	780.4	585.9
04	Northern Aegean	289.4	212.9
05	Gediz	1155.9	866.9
06	Kucuk Menderes	179.2	179.2
07	Buyuk Menderes	1045.4	761.5
08	Western Mediterranean	473.2	316.7
09	Antalya	1164.7	576.3
10	Burdur Lake	106.4	89.5
11	Akarcay	345.4	345.4
12	Sakarya	2197.1	1545.2
13	Western Black Sea	641.2	607.6
14	Yesilirmak	907.2	872.8
15	Kizilirmak	2003.1	1762.9
16	Konya Closed	2597.0	2023.0
17	Eastern Mediterranean	96.5	70.5
18	Seyhan	838.8	749.9
19	Asi	393.2	289.5
20	Ceyhan	985.3	533.5
21	Euphrates-Tigris	4994.8	3763.7
22	Eastern Black Sea	490.9	490.9
23	Coruh	30.0	20.0
24	Aras	388.5	294.4
25	Van Lake	179.2	148.2
	Total	**23032.3**	**17815.3**

also implies that the water balance calculations include significant uncertainties, and a representative observation network is essential for an accurate estimation of the groundwater potential of the country. It can be postulated that the quantity and quality of the existing data and the method of water balance calculations are not adequate for an effective and sustainable management of groundwater resources. The number and the density of groundwater observation wells installed in basins support this postulation (Table 6.3). The total number of observation wells throughout the country is 3448, 2003 of which record the groundwater level on a seasonal basis. The countrywide density of observation wells is less than 5 wells per 1000 km^2. This corresponds to the minimum number in Europe. Equipping all observation wells to record groundwater levels on a monthly basis would be very

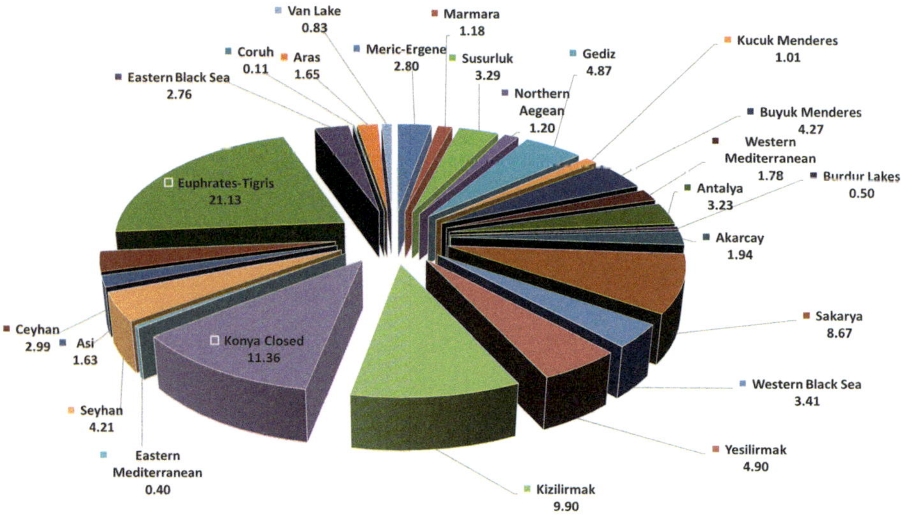

Fig. 6.15 Contribution of basins to total groundwater potential for the year 2016

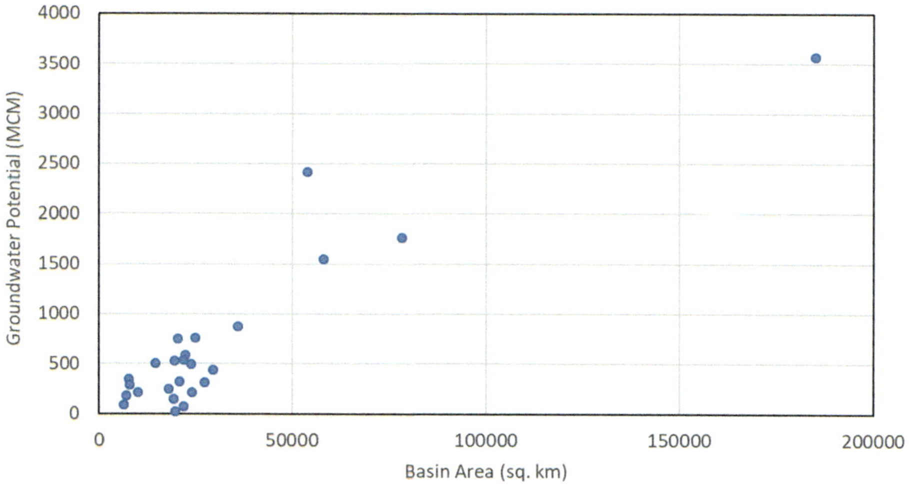

Fig. 6.16 Relationship between basin area and groundwater potential

useful in making the network more efficient. On the other hand, examining Table 6.3 for individual basins, the current situation seems to be quite satisfactory. But what is more important than the number or the density of wells is their distribution and the represented aquifer.

Table 6.3 Number and density of groundwater observation wells as of 2016 (DSI 2018)

Basin No	Basin	Number of groundwater observation wells			Density of wells (Number of wells per 1000 km^2)
		Monthly records	Seasonal records	Total	
01	Meric – Ergene	35	22	57	3.91
02	Marmara	50	13	63	2.61
03	Susurluk	71	95	166	7.41
04	Northern Aegean	42	64	106	10.60
05	Gediz	65	82	147	8.17
06	Kucuk Menderes	57	94	151	21.86
07	Buyuk Menderes	85	46	131	5.25
08	Western Mediterranean	50	56	106	5.06
09	Antalya	52	6	58	2.96
10	Burdur Lake	30	21	51	8.00
11	Akarcay	47	9	56	7.36
12	Sakarya	162	96	258	4.44
13	Western Black Sea	*	33	31	1.05
14	Yesilirmak	56	301	357	9.89
15	Kizilirmak	226	499	725	9.27
16	Konya Closed	239	264	503	9.34
17	Eastern Mediterranean	9	5	14	0.63
18	Seyhan	3	66	69	3.37
19	Asi	26	34	60	7.70
20	Ceyhan	36	9	45	2.05
21	Euphrates-Tigris	92	147	239	1.29
22	Eastern Black Sea	*	*	*	*
23	Coruh	*	3	3	0.15
24	Aras	12	38	50	1.82
25	Van Lake	*	*	*	*
	Total	**1445**	**2003**	**3448**	**4.42**

*No groundwater observation well exists

Table 6.4 Water consumption by sectors for selected years

Year	Total water consumption	Sectors					
		Irrigation		Domestic		Industrial	
	km^3	km^3	%	km^3	%	km^3	%
1990	30.6	22.0	72	5.1	17	3.4	11
2004	40.1	29.6	74	6.2	15	4.3	11
2008	43.0	32.0	74	6.0	15	5.0	11
2023	112.0	72.0	64	18.0	16	22.0	20

Modified after Ayten (2014)

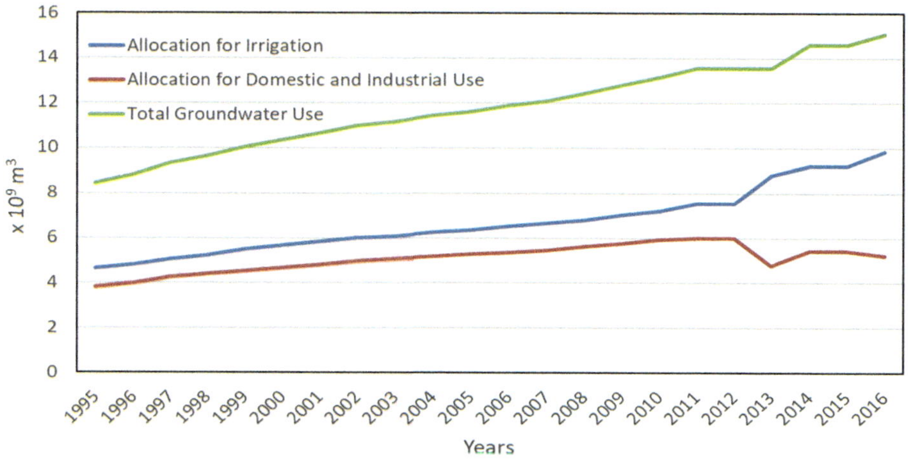

Fig. 6.17 Change in groundwater use in sectoral allocation

6.5 Groundwater Use By Sectors

Sectoral water allocation is one of the important elements of sustainable water management. The water use statistics show that the great majority of water is consumed by three sectors in Turkey: irrigation, domestic and industrial. According to statistics, irrigation has the biggest share with an average of about 73%, followed by domestic use with an average of about 16%, and the industrial use with the smallest share of about 11%. These ratios do not seem to have changed significantly for the last 30 years, although the total annual consumption has increased by about 25%, from about 30 billion m^3 to about 44 billion m^3 (Table 6.4). The total water consumption by the year of 2023 is projected to increase to 112 billion m^3. The sectoral water allocation will change on the advantage of industrial and agricultural use (Fig. 6.17). The domestic use is estimated to remain the same with 16%, the irrigation use to decrease to 64% and the industrial use to increase to 20% (Table 6.4).

Table 6.5 Groundwater use in sectoral water allocation between 1995 and 2016 (DSI 2018)

Year	Allocation for irrigation	Allocation for domestic and industrial use	Total groundwater use
	km^3/year		
1995	4.630	3.820	8.450
1996	4.820	4.000	8.820
1997	5.060	4.270	9.330
1998	5.230	4.420	9.650
1999	5.490	4.560	10.050
2000	5.670	4.680	10.350
2001	5.840	4.830	10.670
2002	5.990	5.000	10.990
2003	6.073	5.103	11.176
2004	6.243	5.200	11.443
2005	6.327	5.295	11.622
2006	6.511	5.371	11.882
2007	6.633	5.463	12.096
2008	6.772	5.647	12.419
2009	7.035	5.776	12.811
2010	7.197	5.941	13.138
2011	7.540	6.016	13.560
2012	7.544	6.016	13.560
2013	8.790	4.772	13.561
2014	9.180	5.420	14.600
2015	9.180	5.420	14.600
2016	9.840	5.220	15.060

The share of the groundwater resources in sectoral water allocation has increased from 8.5 billion m^3/year in 1995 to about 15 billion m^3/year in 2014 (Table 6.5). The figures given in Table 6.5 show that the groundwater resources supply about 25% of the total water use in 2004 and about 30% in 2008. The groundwater allocated to irrigation makes up two-thirds of the total groundwater usage, causing the agricultural sector heavily depending on groundwater resources where surface water resources are limited. The change of groundwater contribution to sectoral water use is depicted in Fig. 6.17. It is clear from the figure that the share of groundwater in irrigation has increased steadily and almost doubled in the last 20 years. The use of groundwater in domestic water supply has increased by half in the last 20 years from about 2 billion m^3/year to about 3 billion m^3/year (Fig. 6.18). However, the share of groundwater has decreased from about 65% to about 45% (Fig. 6.19). It is important to note that the groundwater resources have been developed through wells in general, although the number of springs developed for water supply to municipalities and the total amount of groundwater captured at springs have decreased in the last

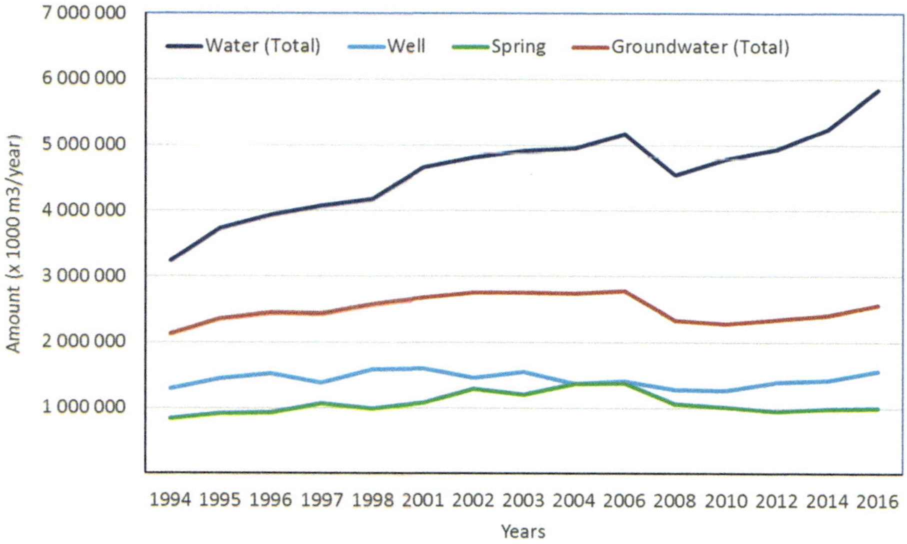

Fig. 6.18 Temporal variation of groundwater contribution to domestic use

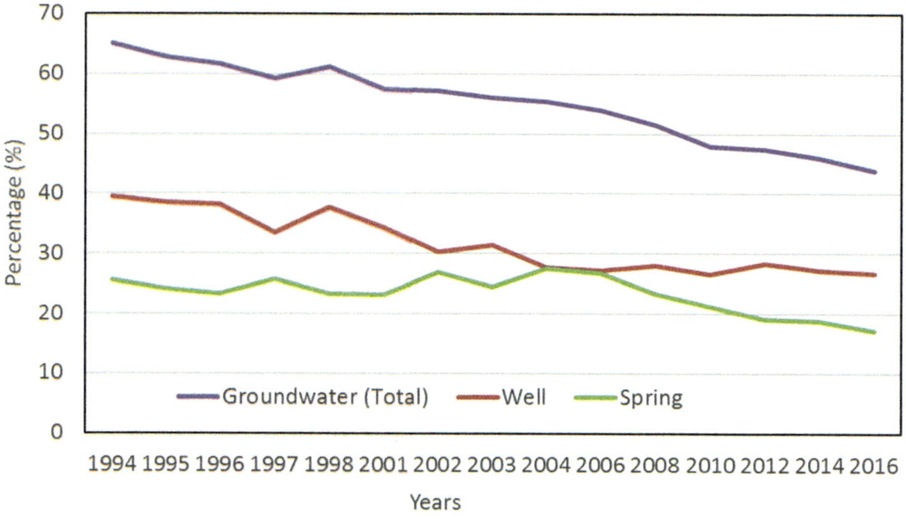

Fig. 6.19 Contribution of groundwater and mode of development to supply water for domestic use

10 years. The drastic decrease of springs and/or spring flows indicates that the groundwater resources are under severe natural and/or anthropogenic stresses. Either climate change or mismanagement of water resources (or both) have adversely affected the spring discharges. Surface waters and groundwater are developed separately without considering that they are interacting subsystems of a single

source. As a consequence, the spring flows have been reduced significantly and several springs have dried up.

6.6 Historical Perspective of Groundwater Resources Development in Turkey

6.6.1 Institutional Infrastructure

Turkey is one of the few countries in the region who has historical roots of institutional infrastructure and water related organizations. In 1914, the General Directorate of Public Works has been established by the Ottoman Government mainly for water supply for irrigation, flood control, and reclamation. Following the foundation of the Republic of Turkey in 1923, the Water Directorate was established in 1925. The Water Directorate, with only five regional offices had been the main authority who had control and responsibility for water related works in the entire country. Lack of sufficient financial capacity and qualified personnel had resulted in inefficient water management, particularly during the severe drought in 1929. Thereupon, the Water Directorate was changed to Water General Directorate, a more comprehensive organization. Recognizing the importance of capacity building, Turkey invited renowned scientists and engineers and sent young people abroad to specialize in several topics including water resources management. As an outturn of this strategy, the Water General Directorate was re-organized in 1939 in Water Works General Directorate to conduct feasibility, planning studies, and hydrological monitoring. The last re-organization of this institution was in 1953 when the General Directorate of State Hydraulic Works (DSI) was established. This organization is less centralized by its 25 regional offices, known as "DSI Districts". Several aspects of water management in terms of agricultural development, hydropower generation, municipal water supply for large cities, and flood control are included in responsibility of the DSI. The responsibility area of the districts was delineated on the basis of administrative boundaries rather than hydrological basin boundaries. Districts are mainly responsible for all sorts of data collection and mapping. DSI is a large governmental organization with about 29,000 personnel, 3700 of which are engineers, and a significant budget. About 1/3 of the annual State's investment budget is allocated to DSI. In 2005, DSI invested approximately 2 billion USD.

In the way of accession to EU, studies on alignment with the acquis have resulted in establishment of two other main water organizations in 2012: the General Directorate of Water Management and the Turkish Water Institute. The major responsibility of the General Directorate of Water Management is the implementation of the EU-Water Framework Directive and related directives. The Turkish Water Institute is to conduct studies on efficient water policy at home and abroad. DSI continues its existence with some changes in its responsibilities. DSI presently implements the water management plans and programs as delivered by the General Directorate of Water Management.

In 2018, Turkey has undergone a constitutional reform and changed its system to a presidential system. A re-organization of the institutions related to water is underway. The new system claims that implementation of the EU-Water Framework Directive will continue and be completed to establish the "good status" of Turkish waters and water dependent ecosystems.

6.6.2 Groundwater Legislation

Water resources development and management has been somehow referred to in several key elements of the Turkish legislation. Provisions related to water use, management and allocation can be found more than 100 laws (acts), bylaws and decrees (Kibaroglu 2006). As an exception, Turkey enacted the Groundwater Law No. 167 in 1960, where the groundwater resources and related concepts of 'safe yield of an aquifer' and 'beneficial use' have been described on the basis of hydrologic budget calculations. The law also defines the ownership of groundwater resources and related rights of inhabitants sharing groundwater resources. According to the law, any kind of action on groundwater resources, including research, exploration, development, use and licensing, is under the responsibility and permission by the DSI. The Groundwater Bylaw issued in 1961 provided a sort of guide for implementation of the Groundwater Law. Implementation of this law and the bylaw has boosted the hydrogeological investigations during the period between 1960 and 1978. Almost all (>150) plain aquifers had been investigated during this period, and 98 reports on hydrogeological appraisal of plain aquifers were published. The reports contained application of contemporary hydrogeological methods and techniques. The potential and safe yield of aquifers were calculated, and the groundwater exploitation reserves were delineated. However, the major missing issue was the studies on delineation of protection zones for the aquifers and for the water supply wells. This is because the law and the bylaw do not include provisions of protection against pollution. To fill this gap, the Bylaw for Water Pollution Control was enacted in 1988, where permissible activities in fixed-radius protection zones for water wells are defined. The issue of protection has been discussed in more detail in the course of implementation of the EU-Water Framework Directive. As a consequence, two bylaws on groundwater, namely Bylaw on Protection of Groundwater against Pollution and Deterioration and the Bylaw on Monitoring of Surface Water and Groundwater were enacted in 2012. In addition to the bylaws, a comprehensive Water Law was drafted and opened to discussion by stakeholders under the coordination of the Ministry of Forestry and Water Affairs. The draft Water Law has been prepared to a great extent, to comply with the Water Framework Directive. Just to mention some key aspects of the draft, the current paradigm of water management is to shift from aquifer-based approach to river basin-scale approach. The Draft Water

Law includes provisions of "river basin management" in all stages from appraisal to planning and to allocation. This provision will also lead to a fundamental change in the organizational structure of management bodies. The current centralized management approach will have to change to a decentralized approach based on individual river basins with new local organizational bodies. The new approach and the associated re-organization of management bodies thus comply with the Water Framework Directive.

6.6.3 Exploration and Resource Assessment

With the implementation of five-year development plans, Turkey's economy has well recovered since 1960s with high growth rates. The share of agriculture in 1960s was about 40% which later, in the early 1980s, declined to about 26% (Cecen et al. 2012; Tanrivermis and Bulbul 2007). The population increased from about 28 million in 1960 to about 45 million in 1980. Similarly, the rural population rate declined from about 65% in the period between 1960 and 1980, and to a post-1980 rate of 40%. Agriculture, furthermore, had a great share as a sector in export with more than 90% in 1960 and 75% in 1980. In 1990, this ratio drastically decreased to 17%, indicating the transformation of the Turkish economy from agrarian to industrial character. The agrarian economy in the period of 1960–1980 boosted the development of large-scale irrigation because the total area of irrigated land increased from 215,000 ha in 1960 to more than 1,000,000 ha in 1980 (DSI 2001). According to the statistics of year 2014, about 700,000 ha of irrigable land is irrigated with groundwater. Consequently, prospecting and development for groundwater was intensified in alluvial plains between the period 1960 and 1980. A total of 465 hydrogeological appraisal studies, 368 of which are at planning stage, was completed. However, hydrogeological reports (also known as the Green Reports) of the 157 alluvial plain aquifers have been published.

Based on the published reports, it can be postulated that, particularly between 1960 and 1975, the contemporary knowledge, techniques and skills were applied at a high level by qualified personnel. On the other hand, it should also be noted that, after 1980, the role of industrialization and privatization model of economic development has had greater importance. Thereupon, DSI started to invest in construction of large dams to produce hydropower. This has resulted in a significant decrease in studies of prospecting for groundwater. Capacity building and institutional and personal improvement in practice of hydrogeological investigations had almost stopped in the 1990s. Publication of hydrogeological investigation reports had also stopped. The knowledge, experience, technical capacity and skills remained at the level of the "golden period" 1960–1980. The methodology of exploration and groundwater resource assessment outlined below depends to a great extent on the published "green" reports.

The hydrogeological appraisal comprised the watershed area of the river draining the plain. However, prospecting for groundwater was focused on the Quaternary alluvial plain aquifer(s), and generally excluded older lithological units and bedrock aquifers. Therefore, the geology, studied at 1:100000 scale in general, was described on the basis of the type of lithology, mostly ignoring the role of the geological setting on the hydrogeological conceptual model. The focus has always been the thickness and extension of the alluvial deposits in the plain area. A resistivity survey was carried out in general to obtain the thickness of the alluvium in the plain, followed by exploratory drillings. The exploratory boreholes, on the other hand, have been drilled and completed as production wells; they have been used as production wells in case they have penetrated a high yield zone. The main reason was to reduce costs. For the same reason, wells have been drilled to penetrate the whole sequence of alluvial deposits; the full lengths of the wells have been screened to obtain the maximum well yield. In addition, almost as a usual application, observation wells have not been drilled and therefore, all pump tests could only be done as single well tests, yielding only transmissivity values for the tested section. As a major drawback of this approach, a reliable and accurate hydraulic characterization of individual aquifer layers has not been possible.

Groundwater potential was estimated on the basis of basin scale hydrological budget calculations. A survey on inventory of hydrographic features, such as rivers and streams, springs, shallow and deep wells, was carried out and used in quantifying the surface runoff and groundwater consumption in the study area. However, it should be mentioned that observations of flows of streams and springs and of groundwater levels in wells were generally done on seasonal intervals and have been limited to the study period which do not exceed 2–3 years. The main input, precipitation, was recorded at quite limited number of meteorological stations, which have not allowed an accurate estimation of the areal average for the basin. In most cases, precipitation and temperature recorded at one major meteorological station have been used in the water budget calculations. The potential and the actual evapotranspiration have been calculated using the Thornthwaite (1948) and Thornthwaite-Mather (1955) methods, respectively. The "actual evapotranspiration" and the "excess water" extracted from the Thornthwaite-Mather budget calculations were the two prominent elements used in groundwater potential estimation. The actual evapotranspiration rate was used to calculate the main loss from the basin. The groundwater potential has been calculated on the basis of the infiltration rate through the lithological units cropping out throughout the watershed. The infiltration rate has been assumed to occur at some percentage of the excess water. The assigned percentages were only rough and subjective estimates and far from being substantial because they did not rely on any measurement. The groundwater inflow or outflow (from or to adjacent basins) has been calculated using the Darcy equation. This is the only part of the reports where a hydraulic characteristic of the aquifer (transmissivity) has been used. The core part of the investigations and the reports was the groundwater balance where the "inputs/recharge" and the "outputs/discharge" were documented as a table. Due to the fact that all calculations were done using long-term averages of the elements of the hydrological cycle, the system was assumed to be under steady-state conditions. As a consequence, the inputs were always equal to the outputs, suggesting that there has been no change in aquifer storage in time.

Perceived partly on the basis of Todd's definition (Todd 1959), the safe yield of the aquifer was defined and calculated as a certain percentage of either the annual recharge or the annual discharge. In general, a percentage of about 80% of the annual recharge was assumed to correspond to the annual safe yield of the aquifer. A typical hydrogeological map and a table documenting the groundwater budget as the major output of the hydrogeological appraisal is shown in Fig. 6.20 and Table 6.6. As can be noticed from Fig. 6.20 the colors used in the hydrogeological map are not those currently used for hydrostratigraphic units.

Although contemporary knowledge and techniques have been used to a great extent in the surveys, uncertainties in almost all components of the water balance reduce the reliability of values of the resulting budget. This approach in the calculation of the safe yield of an aquifer assumes that an amount of groundwater equal or less than the annual recharge can be abstracted without causing an areal depletion in the potentiometric level. Moreover, the location of production wells has no importance in achieving the safe yield. The production wells were proposed to be drilled in areas called as 'groundwater exploitation area'. The hydrogeological appraisal reports delineated areas suitable for groundwater exploitation. The methodology used for this purpose was based upon the specific capacity of the exploration wells. The areas where the specific capacity was relatively higher were declared as groundwater exploitation areas.

The hydrogeological investigations for groundwater resources assessment in green reports also included the evaluation of the groundwater quality. Water samples were collected and analyzed for major ions and in some cases for nutrients. Remembering that the main objective of groundwater resource exploration studies was water supply for irrigation, the water quality evaluation focused on suitability for irrigation. For this purpose, the results of water chemistry analyses have been plotted on the USA Salinity Lab and the Wilcox diagrams.

In an attempt to revise the hydrogeological investigation methodologies and to form a basis for producing technical standards for tendering the hydrogeological investigation projects, DSI has given two projects to Hacettepe and Middle East Technical Universities to study Akarcay (Tezcan et al. 2002) and Kucuk Menderes River Basin Plain aquifers (Yazicigil et al. 2000), respectively. Although these studies have changed the existing view on the hydrogeological investigations, the lack of competent hydrogeologists in the private sector precluded the applications of the proposed methodologies in their projects. Most of the projects conducted by private sector after 2000s still employed the practice followed in the Green Reports of DSI.

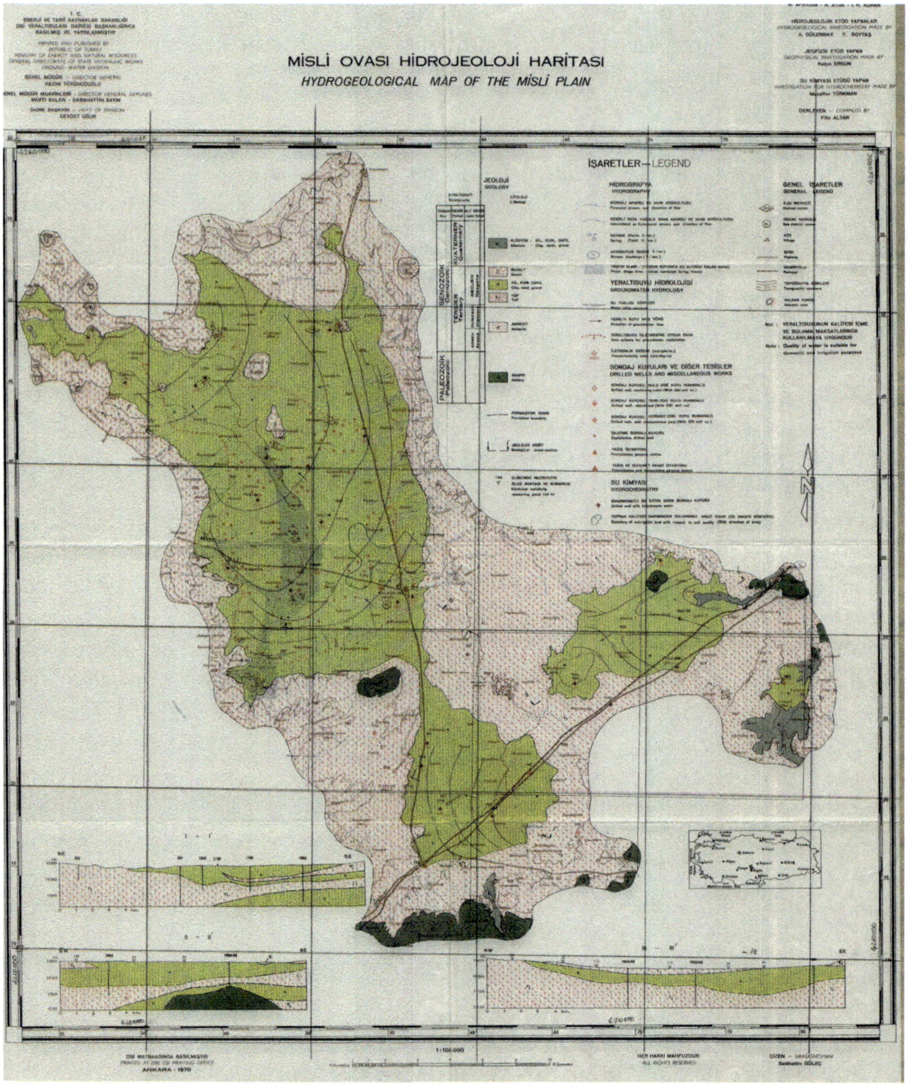

Fig. 6.20 Hydrogeological map of Nigde-Misli plain aquifer. (DSI 1970b)

Table 6.6 Groundwater budget calculated for Nigde-Misli plain aquifer. (DSI 1970b)

Recharge ($\times 10^6$ m^3/year)		Discharge ($\times 10^6$ m^3/year)	
Direct infiltration from precipitation	18.5	Evapotranspiration	3.0
Seepage from runoff	14.0	Discharge through springs	11.0
Indirect infiltration through tuffs	29.0	Withdrawal by wells	24.0
		Groundwater outflow	27.5
Total input	**61.5**	**Total output**	**65.5**

6.7 Groundwater Management: State of Practice

In Turkey, water resources management has always been under the responsibility of the government and until 2011, the main responsible organizations have been four General Directorates; namely, the State Hydraulic Works (DSI), the Electrical Power Survey Administration (EIEI-abolished in 2011), the Iller Bank and the Rural Services (KHGM-abolished in 2005). DSI has been responsible for all types of investments related to water resources development. Construction of multipurpose large dams (i.e. hydropower generation, large scale irrigation schemes, supplying drinking water to cities with a population over 100,000 inhabitants) is among the primary responsibilities of DSI. The EIEI had been responsible for hydrological and geotechnical surveys for hydropower projects. Water supply to cities with a population less than 100,000 was under the responsibility of the General Directorate of Iller Bank. The General Directorate of Rural Services had been responsible for small scale irrigation schemes and water supply for domestic use in villages.

From the standpoint of water resources management, these organizations have been structured with different divisions and departments for surface water development and for groundwater management. This organizational structure can actually be regarded as a reflection of the general approach of the government to water resources management. In addition, water resources "management" and water resources "development" have long been regarded as identical by the authorities. These two major conceptions have influenced the practice in water resources management until the "management" and the "development" were defined under responsibilities of separate organizations. This happened in 2011, when the General Directorate of Water Management has been established and the responsibility of DSI was limited to development projects created on the basis of water management plans developed by the General Directorate of Water Management. It is noteworthy that this transformation was initiated by the implementation of the EU-Water Framework Directive. It is also interesting to analyze water management practices before and after the year 2011, although the change should not be regarded as instantaneous.

The groundwater "management" practice in Turkey until 2011 can be explained on the basis of the organizational structure and the concept of safe yield. The organizations, as mentioned above, have been structured with departments and divisions responsible either for surface waters or for groundwater resources. The main body responsible for groundwater management at department level is the Department of Geotechnical Services and Groundwater of DSI. The department has two divisions: Division of Geotechnical Services and Groundwater Division. Apparently, the main organizational body responsible for development and management of groundwater resources of Turkey has been at a division level, despite the fact that groundwater resources are approximately forming one-third of the total water used in the whole country. This is almost the only department where geologists and hydrogeologists are employed in DSI.

On the other hand, management of surface waters (understood as development of river basins) is under the responsibility of the Investigation, Planning and Allocations Department of DSI. This department is responsible for reconnaissance, master plan and feasibility studies on technically, economically, and environmentally viable projects for the purpose of the integrated land and water development of 25 river basins. The other two powerful departments are the Design & Construction Department, which works on final design, application project, adjudication, and construction of irrigation projects and small dams, and the Dams & Hydroelectric Power Plants Department, which is responsible for the final design, application project, adjudication, and construction of dams and hydroelectric power plants. The great majority of employees in these departments are civil engineers and hydrologists. This type of organizational structure has led to reinforce the misconception that surface waters and groundwater are separate sources and that surface waters are more important than groundwater due to large structures constructed on surface waters. This negligence on surface waters and groundwater forming a single resource became apparent in early 2000s as the base-flow of most streams draining plain aquifers diminished severely as a result of decline in groundwater levels due to overexploitation (Sakiyan and Yazicigil 2004).

Natural systems form continua and in nature there are no sharp boundaries. As also conceptualized in the hydrological cycle, the natural water resources systems should also be regarded as a continuum. This means that the subsystems are not separated nor isolated, but are in continuous interactions (Ekmekci 2015). In terms of water resources, groundwater systems should be expected to respond to any stress on surface waters, and vice versa. The response may not be immediate nor recognizable in short term. But the effect might be significant in the long term. Therefore, even though the organizations may be structured with departments and divisions responsible for surface waters and groundwater separately, they should be in very close cooperation in order to predict the interactions between these subsystems. In practice, the organizational structure did not allow cooperation between the departments. Turkey has been suffering severe management problems such as desiccation of wetlands, drying springs and streams, decline of groundwater levels etc. due to the lack of cooperation.

The practice of groundwater management in Turkey has been regarded as identical to development of groundwater resources. Locating productive wells to make the groundwater resources potential available for irrigation has been the main objective. Almost all plain aquifers have been developed to achieve this objective. Development programs have been planned on the basis of the "calculated safe yield" of the aquifers. In spite of the fact that, in many of the aquifers, groundwater is exploited below the given safe yield, significant depletion of groundwater levels is being recorded. Overexploitation is suggested to be the main reason for this problem, even in plain aquifers where the exploited amount is less than the "calculated safe yield". These examples imply that the current practice of groundwater management in Turkey is somehow problematic. Firstly, the development of groundwater is considered to be identical to management. Development includes production of groundwater to meet the current demand, whereas the concept of management is also associated with sustainability in terms of quantity and quality. Secondly, the "safe yield" defined on the basis of long-term annual recharge does not take into account the hydraulic characteristics of the aquifer. On the contrary, the transmissivity, storativity and hydraulic diffusivity control the drawdown in the aquifer and time rate of propagation of the drawdown cone even when the pumping rate is much below the "calculated safe yield". Thirdly, the boundary conditions are not considered in locating the production wells. Wells located close to positive boundaries would reduce the abstraction from the aquifer storage which, in turn, allow the well attain a steady-state much faster, stopping further decline of groundwater level. And, lastly, the protection of groundwater resources has been ignored during development. Protection has been perceived as safeguarding the wellhead. The protection of aquifers against pollution has not been included in the management plans.

After 2011, the institutional structure has changed in the progress towards the adoption of acquis and in accordance with the EU Water Framework Directive. This change is expected to be accompanied with a paradigm shift in conception of groundwater management from a "source management" to meet the current demand to "basin-scale management" without sacrificing the sustainability of ecosystems. However, although a significant progress in institutional structure was achieved, it is not possible yet to claim that the current practice of groundwater management has evolved to what is described in the Water Framework Directive. This is to a great extent due to the fact that the parliament has not yet approved the Draft Water Law, and partly due to the lack of sufficiently competent and qualified personnel who would comprehend and apply the new paradigm of groundwater management. As a result, almost no change was recorded in the practice of groundwater management, except releasing two bylaws in 2012 as explained in the section devoted to groundwater legislation. Upon the release of these bylaws, studies to delineate wellhead protection zones for wells used to supply drinking water have been undertaken in certain basins. In addition, a national monitoring network has been established to record several quality parameters in groundwater systems. However, it is not yet possible to state that the quality issue is included in the practice of groundwater management.

6.8 Implementation of EU-Water Framework Directive for Sustainable Management of Groundwater Resources in Turkey

The practice of groundwater development in Turkey, based on the simple approach explained above, has resulted in significant declines of groundwater levels, overdraft, quality degradation, dried-out springs and streams, and loss of wetlands. Efforts to adapt the Water Framework Directive of the European Union (WFD) to Turkish legislation were very welcomed by the politicians and the technical water community. The political willingness to adopt the Directive has been high. In the course of complying with the related chapter acquis, three General Directorates have been abolished, the General Directorate of the State Hydraulic Works was re-organized, and the Turkish Water Institute and the General Directorate of Water Management have been established. In addition, more than 20 regulations and bylaws related directly or indirectly to water have been enacted. And finally, more than 266 projects of implementation of different aspects of the Water Framework Directive, including capacity building and twinning projects, have been funded. All these efforts should be regarded as an indication of the high willingness to adopt the Water Framework Directive.

However, the practice has not been as productive as expected. This is due to several reasons. Firstly, the projects have not been implemented in a correct sequence. For instance, the project for sectoral allocation of water preceded hydrogeological modelling. On the contrary, hydrogeological characterization, conceptualization and modelling should precede any development and allocation project. Secondly, lack of competent technical personnel who are not qualified in hydrogeology have been employed as "experts of water resources management". Environmental engineers, chemists, biologists and even aquaculture engineers who lack the basic understanding of hydrogeological systems have been assigned to conduct projects on water resources management. As a consequence, there had been great inconsistencies between the available data and the methods applied in the projects. This situation reduced the reliability of the results significantly. This is also because the projects have been tendered with unrealistically high bids, which commercialized the work. The administration had to overlook this fact because the political priority had been "closing the chapter".

On the other hand, the main logic behind the WFD is that water resources can and should be managed as isolated systems, whereas they are all interacting systems. The WFD defines the concept of "groundwater body" as a management unit, as a distinct volume of groundwater within an aquifer or aquifers. However, in practice, this term is not well understood; it even caused serious confusion among those who are not qualified in hydrogeology. Different methodologies applied for the same basin resulted in different numbers and delineations of groundwater bodies. Furthermore, interactions between groundwater bodies are ignored. On the other hand, the Directive defines the available groundwater resource as "the mean of the long-term annual average rate of overall recharge of the body of groundwater less the long-term annual rate of flow required to achieve the ecological quality objectives for associated surface waters, to avoid any significant diminution in the ecological status of such waters and to avoid any significant damage to associated terrestrial ecosystems". Apparently, application of management at groundwater body level without considering the interactions will not achieve these objectives described for the ecosystems because groundwater systems form spatio-temporal continua. They interact with adjacent systems through cross-formation flow and with surface waters. The magnitude and the direction of the interaction depend on the stress and the hydraulic characteristics of the hydrogeological system. Therefore, delineation of groundwater bodies as isolated individual management units does not comply with the essence of the Water Framework Directive, whose main objective is to maintain the good ecological status of water systems and terrestrial ecosystems.

This objective requires a thorough understanding of the concept of sustainability. The term sustainability in groundwater or more generally in water management studies is actually used to indicate two different concepts: sustainable use and sustainable management. Sustainable use or sustainable yield of a groundwater system is not the same with sustainable management; the former being related to sustainability of the use of the aquifer at the expense of degradation of the interacting adjacent systems, whereas the latter requires to maintain the interacting systems while groundwater is utilized. As a concluding remark, the practice of groundwater management/development in Turkey can be demonstrated as a representative example for most countries, where the problems and challenges limit the benefit expected from implementing the Water Framework Directive. Apparently, the concepts of sustainable yield and sustainable management should be revisited on the basis of hydrogeological and ecohydrological approach in order to achieve the objectives described in the Directive (Ekmekci 2013).

6.9 Groundwater Dependent Ecosystems and Sustainable Management

In comparison to the neighboring countries, Turkey is extremely rich in wetlands. A total of 135 wetlands have been reported as internationally important aquatic ecosystems on the basis of Ramsar Criteria by the Ministry of Forestry and Water. On the other hand, currently only 14 of the wetlands are registered as Ramsar Areas (Fig. 6.21). Owing to its geographical location, Turkey exhibits a great variety of

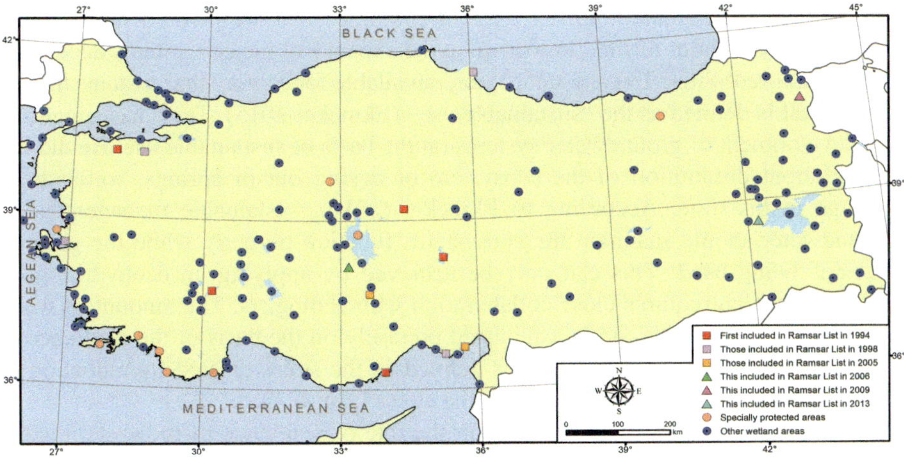

Fig. 6.21 Major wetlands of Turkey. (Erdem et al. 1998)

topographical settings and climatologic conditions. In addition, due to its location on the active tectonic Alpine-Himalayan Zone, Turkey has a very complex geological setting.

Groundwater dependent ecosystems are vulnerable to changes in groundwater systems. Any change in level and/or the quality of groundwater will ultimately affect other systems interacting with the aquifer. Keeping in mind the fact that sustainable management can be achieved only when significant diminution in the ecological status of surface waters and significant damage to associated terrestrial ecosystems are avoided, the amount of groundwater abstraction needs to be precisely calculated, and the production wells should be correctly located. The amount that should be abstracted without causing any undesired consequence on the adjacent ecosystems is not directly related to the annual recharge as commonly believed. The hydraulic characteristics of the aquifer and the boundary conditions control the dynamics of interaction with adjacent systems. It is useful to revisit Theis (1940), where he defined the three sources of water derived from wells: storage of the aquifer, induced recharge, reduced discharge, and a mixture of the last two. In early periods of pumping, water comes from the storage of the aquifer, and this causes drawdown in and around the well, forming the depression cone. As the pumping continues, the depression cone propagates until it intersects with a hydrogeological boundary. The drawdown accelerates when the cone intersects an impermeable (negative) boundary. In case of a positive boundary, such as a stream, spring or a wetland, the well starts to capture some of the rejected recharge or the groundwater outflows. Conceptualization of such a system is depicted in Fig. 6.22. As the depression cone intersects an effluent stream or a marsh, the well starts to capture some of the effluent (reduced discharge) and/or the marsh water (induced recharge). The drawdown rate depends on the pumping rate and the water available for capture.

Theis (1940) suggests to keep the pumping rate equal or less than the water available for capture. This would stop further depletion of groundwater level in the well and in the aquifer because all water derived from the well will come only from the captured sources and not from the aquifer storage. The well would supply water continuously without further drawdown at the expense of the ecosystems dependent on the captured water. The amount of water available for capture that sustains the use of the well is defined as the "sustainable use" (Ekmekci 2015). Theis has suggested the development of groundwater systems on the basis of sustainable use. He did not worry about diminution of the ecosystem or drying out of springs, wetlands or streams in his time. According to Ekmekci (2010), sustainable management of groundwater should maintain the ecosystems function properly while the groundwater is being used. This can only be achieved by applying an ecohydrological approach, which requires close collaboration with ecologists. The amount of water that can be captured should be calculated precisely on the basis of the resilience of the ecosystem. The resilience here is defined as the stress created by reduction of water without causing any response to the stress (Fig. 6.23).

The amount of water captured within the rate of resilience of the system is the "sustainable yield" of the aquifer. Development of groundwater on the basis of sustainable yield is then called as "sustainable management". The objective of the

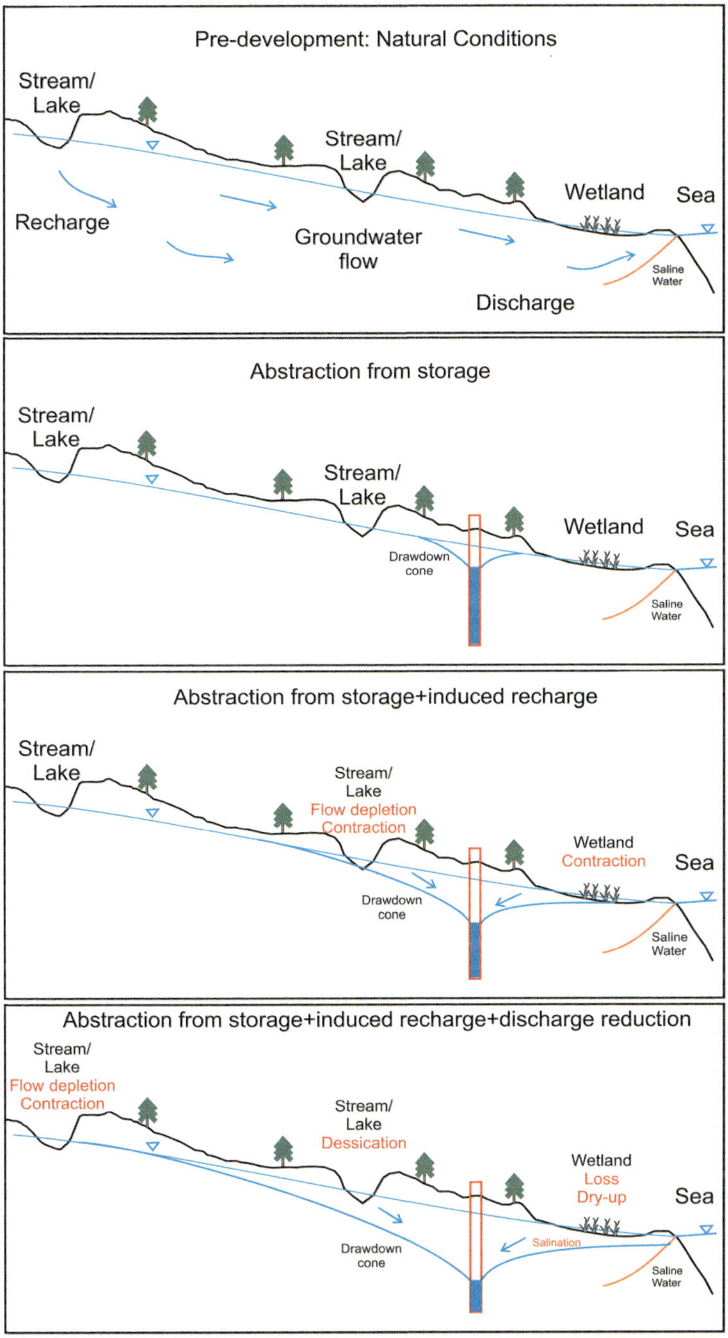

Fig. 6.22 Sources of water derived from wells. (Ekmekci 2010)

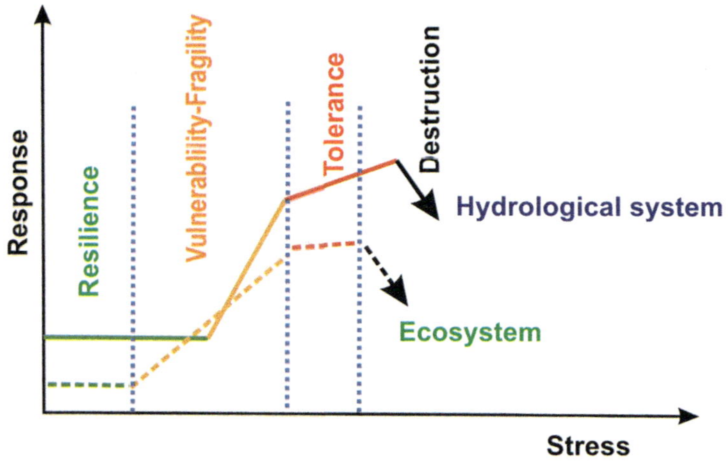

Fig. 6.23 Conceptualization of response of ecohydrological systems to stress. (Ekmekci 2010)

Water Framework Directive can be achieved by applying the sustainable management in the sense described above. Apparently, the sustainable yield does not depend on the annual recharge as the safe yield does. The annual recharge is the amount of water infiltrated and stored in the aquifer, whereas some of the water that can be captured as a part of the sustainable yield constitutes the rejected recharge. Another important difference between the safe yield and the sustainable yield is that the characteristics of the aquifer is the main controlling factor. The hydraulic characteristics play no role in the safe yield. Similarly, the location of wells does not affect the safe yield, while the location of production wells is of major importance in maintaining sustainable yield. Considering the differences between the concepts of safe yield and sustainable yield, the Draft Water Law and the bylaws should be revised accordingly.

Turkey is one of the richest countries in Europe and the Middle East in terms of biodiversity, variety, and number of wetlands. Characteristics such as climatic variability, geographic extent, distinctive characteristics of biological diversity, particular features of the ecosystem services, and the passage of two of the most important bird migration routes in west Palearctic region from Turkey, place the country in global significance in terms of lakes and wetlands. These aspects forced Turkey not only to become a party to Ramsar Convention in 1994, but also to develop targeted policies and produce legislative background for wetland protection (Gul et al. 2014). The Regulation on the Protection of Wetlands, issued in 2002 and amended in 2005, 2014 and 2017, is the primary legislative reference that directly serves for implementation of the rules brought by the Ramsar Convention. As a result of national surveys conducted by the former Ministry of Forestry and Water in Turkey to date, 135 wetlands that conform to the Ramsar definition were identified in Turkey. However, only 14 of these sites have been designated as the Ramsar Site to date (Fig. 6.21). The former Ministry of Forestry and Water issued a circular in 2017, entitled ‘Lakes and Wetlands Action Plan’, to study the existing physical conditions, to analyze the pressures and impacts, to determine the contamination load from point and areal sources, as well as the assimilative capacity, the quality and water budget for 303 lakes and wetlands, including the 14 Ramsar Sites, in 25 river basins. Unfortunately, most of these lakes and wetlands, either did not have any prior hydrogeological characterization and conceptualization study, or some had only an outdated study. In most of the earlier studies, the groundwater component in water budget of the lake is generally determined as the difference between the measured inflows and outflows, ignoring the interaction between the two systems and individual components of the groundwater recharge and discharge. There are few modelling studies which considered the interaction between the groundwater and the lake system to determine the lake budget, to study sustainable management of the lakes (Yagbasan and Yazicigil 2009), and to predict the future response under climate change effects (Yagbasan and Yazicigil 2012; Yagbasan et al. 2017). Despite all these studies, environmental engineers, chemists, biologists, and even aquaculture engineers who are lacking the basic understanding of hydrogeological systems have been assigned to conduct these projects. Furthermore, the lack of competent hydrogeologists in the private sector will be an important constraint in the near future to model the interactions between the two systems for groundwater dependent lakes and wetlands for sustainable management of both resources.

Acknowledgements The authors are thankful to Dr. Levent Tezcan (Hacettepe University) for his support in producing the maps of different rock types. Hatice Kilic (Middle East Technical University) assisted in reproduction of some of the figures.

References

Ayten N (2014) Principles of sectoral water allocation. Expertise thesis. General Directorate of Water Management, Ankara

Bakalowicz M (2005) Karst groundwater: A challenge for new resources. Hydrogeol J 13:148–160

Bener M (1965) Karst in Goksu Valley and Taseli Plateau. Istanbul University, Faculty of Literature. Unpublished Ph.D. thesis

Canik B, Corekcioglu I (1985) The formation of sinkholes (Obruk) between Karapinar and Kizoren-Konya. In: Gunay G, Jhonson AI (eds), Karst water resources, IAHS Publ. No.161, pp 193–205

Cecen AA, Dogruel AS, Dogruel F (2012) Economic growth and structural change in Turkey in 1960–1988. Int J Middle East Stud 26(1):37–56

Degirmenci M, Ekmekci M, Tezcan L, Atmaca E, Namkhai O, Yazici M, Sozudogru O, Aykar E, Akpinar H (2011) Hydrogeological characteristics, contamination risks and protection zones of the Kayseri City Aquifer TUBITAK-CAYDAG Project No 107Y170, Ankara

DSI (1970a) 1/500000 Scale Hydrogeological map sheets of Turkey. Publication of General Directorate of State Hydraulic Works, Ankara

DSI (1970b) Hydrogeological investigation of Nigde-Misli plain. General Directorate of State Hydraulic Works Report, Ankara

DSI (2001) Evaluation report on irrigation schemes operated or transferred by DSI. General Directorate of State Hydraulic Works, Ankara

DSI (2009) Turkey water report. Publication of General Directorate of State Hydraulic Works, Ankara
DSI (2017a) Official bulletin of charted statistics http://www.dsi.gov.tr/yayinlarimiz/haritali-istatistik-bultenleri. Accessed 29 Aug 2018
DSI (2017b) Report of hydrology working group, 2nd. Forestry and Water Council, Ankara
DSI (2018) Official statistics of water potential for the year of 2016. http://www.dsi.gov.tr/dsi-resmi-istatistikler/resmi-i-statistikler-2016/2016-yili-verileri. Accessed 29 Aug 2018
Ekmekci M (2003) Review of Turkish karst with emphasis on tectonic and paleogeographic controls. Acta Carsologica 32:205–218
Ekmekci M (2010) Sustainable management of water resources. Lecture Notes. Hacettepe University, Ankara
Ekmekci M (2013) Ecohydrological approach to protect water dependent ecosystems. Ministry of Forestry and Water, Symp on Water and Biodiversity 22–23 May 2013 Marmaris
Ekmekci M (2015) What we learned from CV Thies: is the safe yield safe? General Directorate of State Hydraulic Works Geotechnical and Groundwater Seminar, 4–18 December 2015 Istanbul
Erdem U, Nurlu E, Akarsu Z (1998) Degradation problems in the coastal zone lakes of the Aegean region, Turkey: In New Tasks for Ecologists After Rio 1992: Proceedings of VII International Congress of Ecology-INTECOL, Florence, Italy, 19–25 July 1998, Florence
Erol O (1984) Geomorphology and neotectonics of the Pluvial Lake Basins in the Taurus Belt and South Central Anatolia. In: Tekeli O, Goncuoglu C (eds) Proceedings of international symposium on geology of the taurus belt, Ankara
Ford DC, Williams PW (2007) Karst geomorphology and hydrology. Wiley, New York
Gul A, Onusluel Gul G, Dikmen AC, Kuzucu A (2014) Investigating hydrologic alterations for wetland systems in terms of river environmental flows. Water Environ J 28:329–339
Gunay G (1985) Karst groundwater studies in Manavgat river basin, Turkey. In: Gunay, G, Jhonson, AI (eds) Karst water resources, IAHS Publ. No.161, pp 333–340
Izbirak R (2001) Turkey I. Publication of Ministry of Education, Istanbul
Kibaroglu A (2006) Legislative framework for water management in Turkey. www.mpfpr.de/.../Water_Law/...Turkey/Kibaroglu_Turkish_Water_Law_English.pdf. Accessed 22 Feb 2018
MTA (1989) Geological map of Turkey scale 1:2.000.000. Publication of General Directorate of Mineral Research and Exploration, Ankara
Nazik L, Poyraz M (2017) A region that characterize the general karst geomorphology of Turkey: central Anatolia Plateau karst zone. Turkish Geogr Rev 68:43–56
Nazik L, Tuncer K (2010) Regional characteristics of Turkish karst morphology. Turkish Speleology J Karst Caves Res J 1:7–19
Okay AI (2008) Geology of Turkey: A Synopsis. Anschnitt 21:19–42
Okay AI, Tuysuz O (1999) Tethyan sutures of northern Turkey. Geol Soc Lond Spec Publ 156:475–515
Rojay B, Heimann A, Toprak V (2001) Neotectonic and volcanic characteristics of the Karasu fault zone (Anatolia, Turkey): The transition zone between the Dead Sea transform and the East Anatolian fault zone. Geodin Acta 14:1–17
Sarikaya MA, Zreda M, Ciner A (2011) Quaternary glaciations of Turkey. In: Ehlers J, Gibbard PL, Hughes PD (eds), Quaternary glaciations – extent and chronology: a closer look. Developments in Quaternary Science, vol 15, pp 393–404
Sahin S, Toker CE, Abdelwahed MF (2017) The determination of the continental borders in Anatolia region from the long-wavelength gravitational data and seismic velocity perturbation. Eur Int J Sci Technol 6(9):1–17
Sakiyan J, Yazicigil H (2004) Sustainable development and management of an aquifer system in western Turkey. Hydrogeol J 12:66–80
Sengor AMC, Gorur N, Saroglu F (1985) Strike-slip faulting and related basin formation in zones of tectonic escape: Turkey as a case study. SEPM Spec Publ 37
Seymen I (1983) Tectonic features of the Kaman group in comparison with those of its neighbouring formations around Tamadag (Kaman-Central Anatolian Crystalline Complex). Bull Turkish Geol Assoc 26:89–98

Tanrivermis H, Bulbul M (2007) The role of agriculture in Turkish economy at the beginning of the European Union accession negotiations. J Appl Sci 7(4):612–625

Tekeli O (1981) Subduction complex of pre-Jurassic age, northern Anatolia, Turkey. Geology 9:68–72

Tezcan L, Meric BT, Dogdu N, Akan B, Atilla AO, Kurttas, T (2002) Hydrogeology and groundwater flow model of Akarçay Basin. Final Report Int Res Center for Karst Water Res, Hacettepe University, Ankara

Theis CV (1940) The source of water derived from wells: essential factors controlling the response of an aquifer to development. Civ Eng 10:277–280

Thornthwaite CW (1948) An approach toward a rational classification of climate. Geogr Rev 38:55–94

Thornthwaite CW, Mather JR (1955) The water balance. Lab Clim Publ in Climatology, Centerton 8(1):1–104

Todd DK (1959) Groundwater hydrology. Wiley, New York

TUIK (2018) Sectoral water and waste water statistics. http://www.tuik.gov.tr. Accessed 24 Feb 2018

Yagbasan O, Yazicigil H (2009) Sustainable management of Mogan and Eymir Lakes in Central Turkey. Environ Geol 56:1029–1040

Yagbasan O, Yazicigil H (2012) Assessing the impacts of climate change on Mogan and Eymir Lakes' levels in Central Turkey. J Environ Earth Sci 66:83–96

Yagbasan O, Yazicigil H, Demir V (2017) Impacts of climatic variables on water-level variations in two shallow Eastern Mediterranean lakes. J Environ Earth Sci 76(575):1–11

Yazicigil H, Doyuran V, Karahanoglu N, Yanmaz M, Camur MZ, Toprak V, Rojay B, Yilmaz KK, Sakiyan J, Suzen L, Yesilnacar E, Gundogdu A, Pusatli T, Tuzcu B (2000) Investigation and management of groundwater resources in K. Menderes River basin under the scope of revised hydrogeological studies, Final Report, Middle East Technical University, Ankara

Yazicigil H, Ekmekci M (2003) Perspectives on Turkish ground water resources. Ground Water 41:290–291

Chapter 7
Water Quality

Orhan Uslu

Abstract Turkey's present population exceeds 80 million. It has increased six-fold in the last 90 years. Her per capita Gross National Income (GNI) increased more than 25 times in the last 50 years. The associated expansion in production and consumption causes serious pressures on resources and the environment. Based on the latest available, and to some extent reliable data, the present situation of water quality of inland and coastal waters (river, lake, groundwater and marine pollution) in Turkey is discussed in this chapter. The wastewater management schemes and efforts to improve the situation using available treatment and disposal technologies are summarized.

Keywords Turkey · Water quality · River basins · River pollution · Lake pollution · Groundwater pollution · Marine pollution · Wastewater management · Wastewater treatment · Deep sea discharges · Sludge management

7.1 Introduction

The population of Turkey has soared from 13.6 million to 80 million in the last 90 years as depicted in Fig. 7.1. It is expected to come to a stagnant state towards the end of this century (TSA 2018). During this period, the Turkish economy has also shown a remarkable expansion despite numerous economic crises and wars. In Fig. 7.2, the development of the annual Turkish gross national income per capita (GNI) is shown for the period 1967–2017 (World Bank 2018).

Turkey is the 17th largest economy in the world, with a GDP (Gross Domestic Product) of $820 billion (World Bank 2017). Turkey has set an ambitious target to become one of the ten largest economies in the world by 2023, the centennary of the foundation of the Turkish Republic. Doing so will require Turkey to triple its economy

O. Uslu (✉)
Emeritus Professor, Environmental Engineering Department, Dokuz Eylul University, Izmir, Turkey

N. B. Harmancioglu, D. Altinbilek (eds.), *Water Resources of Turkey*, World Water Resources 2, https://doi.org/10.1007/978-3-030-11729-0_7

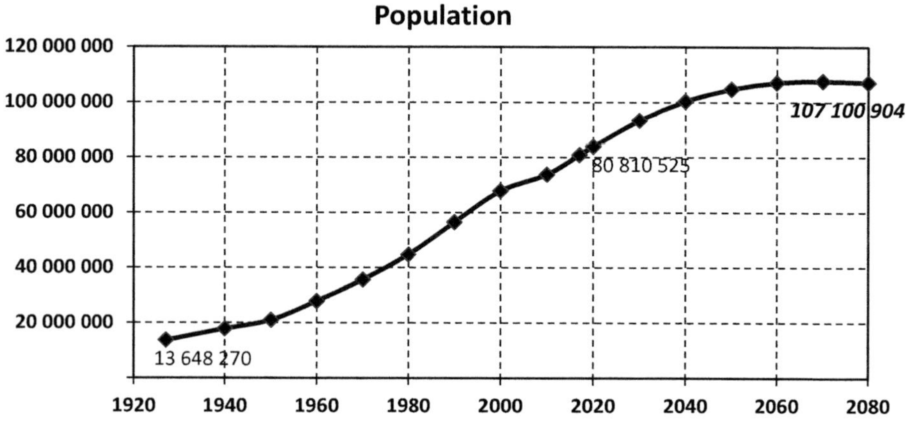

Fig. 7.1 Development of Turkish population (TSA 2018)

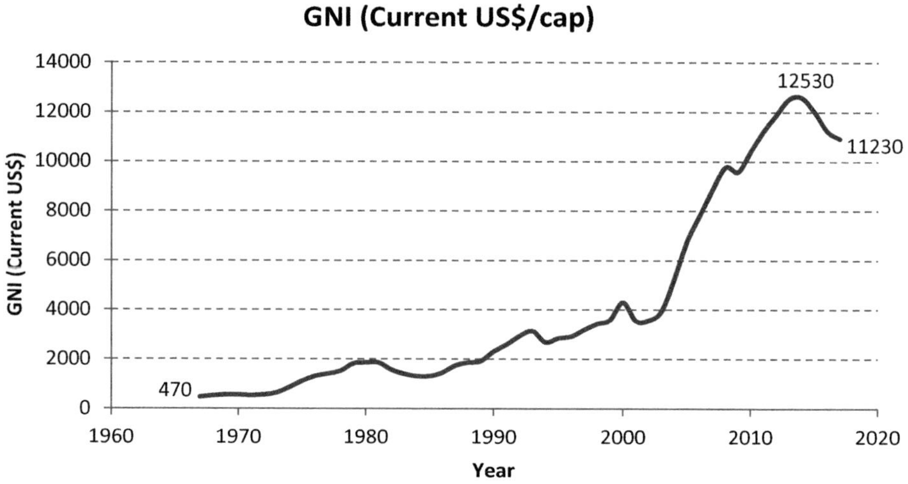

Fig. 7.2 Development of the Turkish Gross National Income per capita and year between 1967 and 2017 in current US$ (World Bank 2018)

to more than $2 trillion and to make significant upgrades to its energy, information technology, finance, and physical infrastructures (US Dept. of State 2018).

This development requires a tremendous increase in natural resource use and results in a parallel swell in waste production. Within this context, the resulting inland and coastal water quality problems will be presented in this chapter.

Water potential of Turkey is summarized in Table 7.1. According to the "Falkenmark Water Stress Indicator" (Falkenmark et al. 1989), the minimum amount of daily domestic water demand per capita is 100 liters. 500–2000 lt/cap/day is

Table 7.1 Water potential of Turkey

Hydrologic component		109 m^3/year	Explanation
1	Average precipitation	501	
2	Evapotranspiration	274	
3	Infiltration	69	
4	Surface runoff	158	1-2-3
5	Net bank inflow into surface waters	28	
6	Surface water potential	186	4 + 5
7	Inflow from bordering countries	7	
8	Gross surface water potential	193	6 + 7
9	Gross groundwater potential	41	
10	Total renewable gross water potential	234	8 + 9
11	Technically and economically usable surface water	98	
12	Technically and economically usable groundwater	14	
13	Total usable water potential of Turkey	112	11 + 12

Table 7.2 Extractions from water resources with regard to usage (10^9 m^3/year)

Years	2008	2010	2012	2014	2016
Municipalities	4.55	4.78	4.94	5.23	5.83
Villages	1.22	1.01	1.04	0.43	0.38
Manufacturing industry activities	1.31	1.56	1.79	2.20	2.12
Thermal power plants (*)	4.54	4.27	6.40	6.53	8.61
Organized industrial zones	0.11	0.11	0.14	0.14	0.15
Mining facilities	(**)	0.05	0.11	0.21	0.23
Irrigation	33.77	38.15	41.55	35.85	43.06
Total		49.95	55.96	50.59	60.38

Note: (*) data includes sea water use. (**) No Information
Source: TURKSTAT, "Sectoral Water and Wastewater Statistics" Press Release
Source For 'Irrigation' Values: Ministry of Forestry and Water Affairs General Directorate of State Hydraulic Works, http://www.turkstat.gov.tr/PreHaberBultenleri.do?id=27672

required for agricultural and industrial activities. The annual threshold value for water scarcity has been determined as 1700 m^3 per capita. It is stated that if the water availability drops below this value, related water problems will occur. According to the calculations by State Hydraulic Works (DSI), Turkey has an annual water potential of 1652 m^3 per capita. It is predicted that the population of Turkey will reach the threshold 100 million in 2040, and as a result, the amount of water potential per capita will decrease to 1120 m^3. Turkey will be among the countries suffering from water stress and will have to follow policies that will enable effective use of her water resources. Actual extractions from water resources with regard to usage between 2008 and 2016 are given in Table 7.2. Withdrawal projections according to the main water consuming sectors for 2023, as given by DSI, are summarized in Table 7.3.

Table 7.3 Water consumption projections for 2023 (10^9 m^3/year)

Sectors	10^9 m^3/year	%
Irrigation	72	64
Domestic water	18	16
Industry	22	20
Total	112	100

Turkey is already among the countries suffering from water stress, and the situation will worsen in the future. The country will have to follow policies that will enable effective use of the water resources.

Water footprint is a parameter that shows the amount of water consumed by humans. "Project of Turkey's Water Footprint" was completed in 2014 (WWF 2014). According to this report, the total virtual water footprint of Turkey was $163 \bullet 10^9$ m^3/year in 2013.

According to Mekonnen and Hoekstra (2011), the average annual per capita water footprint of the world is around 1240 m^3 and Turkey is over this average value with a water footprint of 1610 m^3.

7.2 Water Pollution Legislation

Water related legislation in Turkey is presented extensively in Chap. 15 of this book. Administrative structures related to water management and associated legal regulations are elaborated in Chap. 11. The reader is advised to refer to these chapters for further details.

In this section, some of the water pollution related criteria defined in the By-law on Water Pollution Control that entered into force in 1988 (Official Gazette No. 19919), revised in 2004 (Official Gazette No. 25687), in 2008 (Official Gazette No. 26786), and in 2018 (Official Gazette No. 30332) will be summarized, because they will frequently be used in the subsequent sections.

The author of this chapter had the privilege to draft the initial version of the By-law in 1988. Considering the Turkish accession to EU, the Turkish By-law was initially based on the regulations of the German association ATV (Abwassertechnische Vereinigung – Association for Wastewater Technology) which, at that time, was internationally renowned for more than 50 years. These regulations also influenced the later accepted EU directives concerning water pollution.

ATV later united with DVWK (Deutscher Verband für Wasserwirtschaft und Kulturbau – German Association for Water Resources and Land Improvement) and finally (2004) adopted the short name DWA (Deutsche Vereiniging für Wasserwirtschaft, Abwasser und Abfall – German Association for Water Management, Wastewater and Solid Wastes).

Within the By-law, principles on classification of surface waters including coastal and transitional waters used for different purposes, assessment of pressures and impacts, delineation of the measures towards improving water quality, and identification of trophic levels were designated. Classification of inland waters is carried out according to Quality Criteria of Inland Waters.

The By-law introduced a classification system for inland waters, which is summarized in Table 7.4. The water quality classes in this table indicate:

- Class I:
  High quality water,
- Class II:
  Water with low contamination,
- Class III:
  Contaminated Water,
- Class IV:
  Highly contaminated water.

For lakes, wetlands, marshlands and artificial reservoirs, limit values for dissolved inorganic nitrogen, total phosphorus, secchi disk depth and chlorophyll-α are given in Table 7.5 to determine the trophic level in these water bodies.

Within the context of quality assessment for coastal waters, the target quality criteria for Aegean Sea, Mediterranean Sea, Sea of Marmara and Black Sea, in terms of dissolved oxygen, pH, colour and turbidity, temperature, hazardous substances and floating matters, are stated in the By-law (Table 7.6).

Table 7.7 summarizes the discharge criteria for deep sea discharges as given in the By-law.

7.3 Water Pollution with Respect to Medium

The main reasons why the quality of Turkey's water resources decreases are:

- Overuse of natural resources,
- Untreated industrial and domestic wastewaters mixing into water resources due to unplanned and rapid urbanization,
- Agricultural activities,
- Insufficiency of present waste water treatment facilities in terms of capacity, process selection and operation.

Water quality evaluations should go hand in hand with natural and anthropogenic pollution estimates and the amount of water available in that particular region. In this section, the pollution in inland and coastal waters is briefly examined. Since extensive discussion of water quantities in various media is given in the preceding chapters of this book, it is not intended to replicate these aspects in order to prevent unnecessary repetition.

This section relies extensively on previous publications of the author (Uslu 1985, 1995) and the more recent watershed management plans and publications of relevant

Table 7.4 Classification system for inland water quality

Water quality parameters	Water quality classes			
	I	II	III	IV
A. Physical and inorganic chemical parameters				
1. Temperature (°C)	25	25	30	>30
2. pH	6.5–8.5	6.5–8.5	6.0–9.0	<6.0 or > 9.0
3. Dissolved oxygen (mg O_2/L)[a]	8	6	3	<3
4) Oxygen saturation (%)[a]	90	70	40	<40
5. Chloride (mg Cl^-/L)	25	200	400[b]	>400
6. Sulphate ion (mg $SO_4^{=}$/L)	200	200	400	>400
7. Ammonium-N (mg NH4+-N/L)	0.2[c]	1[c]	2c	>2
8. Nitrite-N (mg NO_2^- N/L)	0.002	0.01	0.05	>0.05
9. Nitrate-N (mg NO_3^- N/L)	5	10	20	>20
10. Total-P (mg P/L)	0.02	0.16	0.65	>0.65
11. Total dissolved matter (mg/L)	500	1500	5000	>5000
12. Colour (Pt-Co units)	5	50	300	>300
13. Sodium (mg Na^+/L)	125	125	250	>250
B. Organic parameters				
1. Chemical oxygen demand (COD) (mg/L)	25	50	70	>70
2. Biochemical oxygen demand (BOD) (mg/L)	4	8	20	>20
3. Total organic carbon (mg/L)	5	8	12	>12
4. Total Kjeldahl nitrogen (mg/L)	0.5	1.5	5	>5
5. Oil and grease (mg/L)	0.02	0.3	0.5	>0.5
6. Methylene-blue surface active agents (mg/L)	0.05	0.2	1	>1.5
7. Phenolic substances (volatile) (mg/L)	0.002	0.01	0.1	>0.1
8. Mineral oils and derivates (mg/L)	0.02	0.1	0.5	>0.5
9. Total pesticide (mg/L)	0.001	0.01	0.1	>0.1
C. Inorganic parameters[d]				
1. Mercury (μg Hg/L)	0.1	0.5	2	>2
2. Cadmium (μg Cd/L)	3	5	10	>10
3. Lead (μg Pb/L)	10	20	50	>50
4. Arsenic (μg As/L)	20	50	100	>100
5. Copper (μg Cu/L)	20	50	200	>200
6. Total chromium (μg Cr/L)	20	50	200	>200
7. Chromium (VI) (μg Cr^{+6}/L)	n.d.	20	50	>50
8. Cobalt (μg Co/L)	10	20	200	>200
9. Nickel (μg Ni/L)	20	50	200	>200
10. Zinc (μg Zn/L)	200	500	2000	>2000
11. Total cyanide (μg CN/L)	10	50	100	>100
12. Fluoride (μg F^-/L)	1000	1500	2000	>2000
13. Free chlorine (μg Cl_2/L)	10	10	50	>50
14. Sulphide (μg $S^{=}$/L)	2	2	10	>10
15. Iron (μg Fe/L)	300	1000	5000	>5000
16. Manganese (μg Mn/L)	100	500	3000	>3000

(continued)

Table 7.4 (continued)

Water quality parameters	Water quality classes			
	I	II	III	IV
17. Boron (μg B/L)	1000[e]	1000[e]	1000[e]	>1000
18. Selenium (μg Se/L)	10	10	20	>20
19. Barium (μg Ba/L)	1000	2000	2000	>2000
20. Aluminium (mg Al/L)	0.3	0.3	1	>1
21. Radioactivity (Bq/L) α-activity	0,5	5	5	>5
β-activity	1	10	10	>10
D. Bacteriological parameters				
1. Faecal coliforms (MPN/100 mL)	10	200	2000	>2000
2. Total coliforms (MPN/100 mL)	100	20,000	100,000	>100,000

[a]Compliance with only one of the parameters, oxygen concentration or saturation is permissible
[b]It may be necessary to lower these limits for the irrigation of chloride sensitive crops
[c]Depending on the pH level the free ammonium nitrogen should not exceed 0.02 mg NH_4-N/L
[d]The parameters in this group indicate the total values of the chemical species indicated
[e]It may be necessary to lower the criterion to 300 μg/L for irrigation of boron sensitive crops

Table 7.5 Eutrophication limiting values for lakes, wetlands, marshlands and artificial reservoirs

Parameters	Lake types	
	Natural protection area and recreation	Other (includes naturally occurring salty, bitter and carbonate lakes
pH	6.5–8.5	6–10.5
COD (mg/L)	3	8
DO (mg/L)	7.5	5
SS (mg/L)	5	15
Total coliforms (MPN/100 mL)	1000	1000
Total-N (mg/L)	0.1	1
Total-P (mg/L)	0.005	0.1
Chlorophyll a (mg/L)	0.008	0.025

authorities MoEU (Ministry of Environment and Urbanization)(2016), Fanack (2016), MoFW (Ministry of Forestry and Water Works) (2016), and MoEU (2018).

The Republic of Turkey will apply the European Union Directives in order to improve the quality of its waters. The key directive in this field is the Water Framework Directive which is under the "EU Water Quality Sector". It aims for a good status of water to be achieved by 2015 or subject to justification by 2027 at the latest. The good status covers both ecological and chemical quality criteria.

Within this context, work on preparation of Watershed Protection Action Plans was started by the former Ministry of Forestry and Water Works (MoFW) of the Turkish Republic. Initially, 25 hydrological watersheds of the country (Fig. 7.3) were rated, considering the water quality, pollutant sources, protection areas and drinking water resources in the catchment area. Based on this prioritization,

Table 7.6 General quality criteria for sea waters

Parameter	Criterion	Remarks
pH	6.0–9.0	–
Colour and turbidity	Natural	90% of natural photosynthetic activities along the depth should be preserved.
Floating matter	–	No floating oils, tar and solid wastes.
Susp. solids (mg/L)	30	–
DO (mg/L)	>90% sat.	DO values are to be observed along the entire depth.
Degradable organic matter	–	DO values after dilution should not be affected exceeding the criterion given in previous line.
Crude oil and derivates (mg/L)	0.003	Should be measured in water, biota and sediments separately.
Radioactivity	–	Natural levels in the examined water body should not be exceeded. Artificial radioactivity should be below detection levels.
Productivity	–	Natural levels of seasonal productivities should be preserved.
Toxicity	None	
Phenols (mg/L)	0.001	
Ammonium (mg/L)	0.02	
Heavy metals		
Copper (mg/L)	0.01	
Cadmium (mg/L)	0.01	
Chromium (mg/L)	0.1	
Lead (mg/L)	0.1	
Nickel (mg/L)	0.1	
Zinc (mg/L)	0.1	
Mercury (mg/L)	0.004	
Arsenic (mg/L)	0.1	

Table 7.7 Discharge criteria for deep sea discharges

Parameter	Limit
pH	6–9
Temperature	35 °C
SS (mg/L)	350
Oil and grease (mg/L)	15
Floating substances	–
BOD_5 (mg/L)	250
COD (mg/L)	400
TN (mg/L)	40
TP (mg/L)	10
Methylene blue surface active agents (MBAS) (mg/L)	10

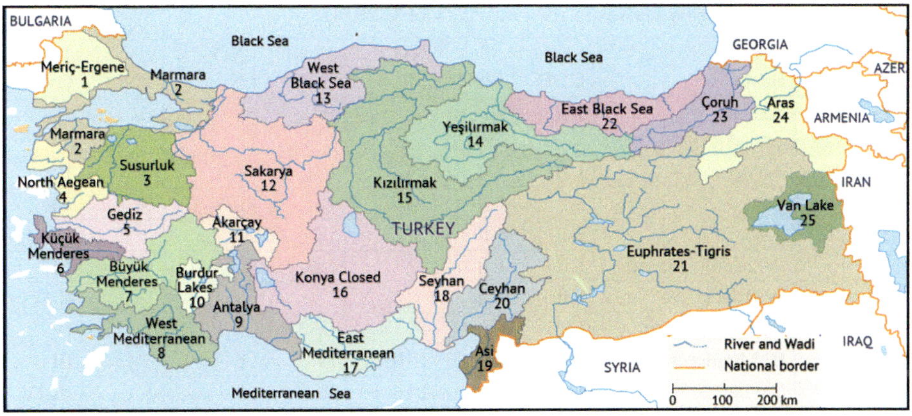

Fig. 7.3 Watershed basins map of Turkey (Fanack 2016)

protection action plans were already completed for 6 watersheds: Meric-Ergene Basin, Sakarya Basin, Akarcay Basin, Gediz Basin, Van Lake Basin, Asi Basin.

Preparation of protection action plans for 11 of the remaining 19 watersheds were undertaken by TUBITAK Marmara Research Centre. Within the scope of the Project, Watershed Protection Action Plans were prepared for the following 11 hydrological watersheds based on the 5th article of Water Pollution Control Act: Marmara Basin, Susurluk Basin, Northern Aegean Basin, Kucuk Menderes Basin, Buyuk Menderes Basin, Burdur Basin, Yesilirmak Basin, Kizilirmak Basin, Konya Closed Basin, Seyhan Basin and Ceyhan Basin. These plans can be accessed through the website of the General Directorate for Environmental Management: http://www.suyonetimi.ormansu.gov.tr/AnaSayfa/eylemplanlari/Havza_koruma_eylem_planlari.aspx?sflang=tr

7.3.1 River Pollution

Water quality monitoring activities in 25 river basins in Turkey have been carried out by the General Directorate of State Hydraulic Works (DSI) since 1970s. As of 2014, monitoring programs covering biological, chemical, physicochemical and hydromorphological quality factors have been prepared in accordance with the European Union Water Framework Directive in order to achieve standardization in monitoring and to obtain long term and effective water quality monitoring data.

The data obtained at the end of monitoring activities is evaluated according to the By-law on Surface Water Quality, and Water Quality reports are prepared for each basin and present water quality conditions are publicized. The results are summarized in Tables 7.8, 7.9 and 7.10.

Table 7.8 Water quality classes in river basins (MoEU 2014)

Basin no	Basin	COD	BOD	NH_4-N	NO_2-N	NO_3-N	Tot. P	Overall
11	Akarcay	I-II	III-IV	IV	IV	I	IV	IV
9	Antalya	I-II	I-II	I-II	I-II	I-II	II-III	II
24	Aras		I-II	I-II	I-II	I-II		I-II
19	Asi	I-II	I-II	III-IV	IV	I		IV
8	W. Mediterranean	I-II	I-II		I-II	I	II	I-II
13	W. Black Sea	I-II	I-II					II-IV
10	Burdur	I,IV		II, IV	IV	I		IV
7	B. Menderes	II-III		II-III	IV	I		III
20	Ceyhan	I		I-III	III-IV	I		III-IV
23	Coruh	I-II	I-II	I-II	II-III	I	II-III	II-III
17	E. Mediterranean	I	I	I-II	I-III	I		I-III
22	E. Black Sea	I	I	I	III	I	III	I-III
1	Ergene	IV	III	III	III-IV	I		III-IV
21	Euphrates & Tigris	I-II		I-IV				I-IV
5	Gediz	I-II	II-III	I-IV	III-IV	I		III-IV
15	Kizilirmak	I-II		II-IV	III-IV	I-II	II-IV	II-IV
16	Konya	III	III	II	IV	II		III-IV
4	N. Aegean	I-IV	I-IV	I-IV	III-IV	I-II		I-IV
6	K. Menderes	IV		IV	IV	I		IV
2	Marmara	I-IV	I-IV	I-IV	III-IV	I-II		II-IV
12	Sakarya	II-IV		II-IV		I	II-IV	III-IV
18	Seyhan	I		II-III	III-IV	I	II-IV	II-IV
3	Susurluk	I-II		II-IV	IV	I, III		III-IV
25	Van Lake		II	II	III-IV	I		I-IV
14	Yesilirmak	III-IV	III-IV	III-IV	III-IV	III-IV	III-IV	III-IV

When the average results of 2014 are evaluated within the frame of the By-law for Water Pollution and Control, the following basins has been recognized as hot-spots (MoEU 2016):

- Ergene River holds Class IV water in terms of general condition, (A) oxygenation Parameters, (B) Nutrient Parameters and (D) Bacteriological Parameters, and it is in Class III in terms of (C) Trace Metals.
- Gediz River water quality is in Class IV in terms of (A) oxygenation Parameters, (B) Nutrient Parameters, (C) Trace Metals and (D) Bacteriological Parameters, and it is in Class III in terms of general condition.
- Bakircay River water quality is in Class IV in terms of (A) oxygenation Parameters, (B) Nutrient Parameters and (D) Bacteriological Parameters, and it is in Class III in terms of (C) Trace Metals and in Class II in terms of general condition.

Table 7.9 Relative distribution of point loads in river basins (MoEU 2014)

		Point loads (%)												Contribution of Pt. loads to Total	
		COD				TN				TP					
Basin no	Basin	Mun.	Ind.	Leak.	Total	Mun.	Ind.	Leak.	Total	Mun.	Ind.	Leak.	Total	TN	TP
11	Akarcay	57	41	2	100	87	8	6	100	93	7	0	100	19	34
9	Antalya	48	42	10	100	68	20	12	100	93	5	2	100	9	22
24	Aras	78	22	0	100	94	6	0	100	95	5	0	100	8	20
19	Asi	74	25	1	100	90	8	2	100	90	10	0	100	27	50
8	W. Mediter.	80	7	13	100	80	3	17	100	97	3	0	100	10	27
13	W. Black Sea	32	67	1	100	61	36	3	100	73	26	1	100	21	42
10	Burdur	57	39	4	100	82	17	1	100	57	39	1		4	4
7	B. Menderes	48	51	1	100	80	18	2	100	83	17	0	100	12	16
20	Ceyhan	72	28	0	100	89	11	0	100	85	15	0	100	16	20
23	Coruh	96	1	3	100	92	8	1	100	99	1	0	100	5	21
17	E. Mediter.	75	19	6	100	88	8	4	100	95	4	1	100	12	30
22	E. Black Sea	90	8	2	100	96	2	2	100	98	1	1	100	17	48
1	Ergene	20	77	3	100	42	50	8	100	39	60	1	100	26	51
21	Euphrates-Tigris	79	20	1	100	89	10	1	100	96	4	0	100	19	39
5	Gediz	50	49	1	100	71	27	2	100	83	17	0	100	26	44
15	Kizilirmak	83	17	0	100	92	8	0	100	95	5	0	100	12	15
16	Konya	69	31	0	100	88	11	1	100	94	6	0	100	4	12

(continued)

Table 7.9 (continued)

		Point loads (%)												Contribution of Pt. loads to Total	
		COD				TN				TP					
Basin no	Basin	Mun.	Ind.	Leak.	Total	Mun.	Ind.	Leak.	Total	Mun.	Ind.	Leak.	Total	TN	TP
4	N. Aegean	79	21	0	100	84	16	0	100	86	14	0	100	17	31
6	K. Menderes	57	43	0	100	82	18	0	100	80	20	0	100	18	27
2	Marmara	68	32	0	100	80	20	0	100	87	13	0	100	26	33
12	Sakarya	67	32	1	100	86	13	1	100	90	10	0	100	36	56
18	Seyhan					98	2	0	100	97	3	0	100	10	11
3	Susurluk	60	40	0	100	61	39	0	100	71	29	0	100	16	23
25	Van Lake	97	3	0	100	98	2	0	100	99	1	0	100	11	31
14	Yesilirmak	78	22	0	100	92	8	0	100	94	6	0	100	10	15

Note: Mun. = municipal; Ind. = industrial; Leak. = uncontrolled leakage from point sources; TN = Total Nitrogen; TP = Total Phosphorus

Table 7.10 Relative distribution of diffused nutrient loads in river basins (MoEU 2014)

		Diffuse loads (%)									
		TN					TP				
Basin no	Basin	Agri.	Breed	Land	Other	Total	Agri.	Breed.	Land	Sept.	Total
11	Akarcay	30	35	28	8	100	52	36	9	3	100
9	Antalya	23	15	41	11		41	19	18		
24	Aras	7	46	14			11	66	16	7	
19	Asi	54	15	16			53	21		17	
8	W. Mediter.	23	17	43	17	100	39	24	19	18	100
13	W. Black Sea	16	30	41	13	100	55	36			
10	Burdur	62	23	12	3	100	86	12	2		100
7	B. Menderes	64	18	13	5	100	82	14	3	1	100
20	Ceyhan	57	20	16	7	100	80	14	3	3	100
23	Coruh	6	24	59	11	100	10	52	36	2	100
17	E. Mediter.	33	13	42	8		55	17	18	10	100
22	E. Black Sea	31	16	32	12		21	32	19	26	
1	Ergene	66	16	11	7	100	72	19	2	1	
21	Euphrates-Tigris	29	23	36	12	100	50	25	14	11	100
5	Gediz	35	29	26	10	100	47	38	10	5	100
15	Kizilirmak	57	14	22	7	100	83	11	5	1	100
16	Konya	25	66	7	2	100	80	13	4	3	100
4	N. Aegean	36	22	30	12	100	50	28	10	11	
6	K. Menderes	31	35	21	13		28	49	23		100
2	Marmara	57	22	14	7	100	64	28	5	3	100
12	Sakarya	37	24	31	8	100	61	27	10	2	100
18	Seyhan	66	15	16	3	100	87	10	3		100
3	Susurluk	63	20	14	3	100	73	23	4		100
25	Van Lake	6	40	37	8		11	46	19	23	
14	Yesilirmak	44	26	23	7		69	25	5	1	100

- K. Menderes water quality is in Class IV in terms of general condition, (A) oxygenation Parameters, (B) Nutrient Parameters and (D) Bacteriological Parameters, in Class II in terms of (C) Trace Metals.
- Susurluk River water quality is in Class III in terms of general condition, in Class IV in terms of (A) oxygenation Parameters, (B) Nutrient Parameters and (D) Bacteriological Parameters and in Class II in terms of (C) Trace Metals.
- Sakarya River water quality is in Class III in terms of general condition, and in Class IV in terms of (A) oxygenation Parameters, (B) Nutrient Parameters, (C) Trace Metals and (D) Bacteriological Parameters.

In an overall assessment of river water quality with regard to dissolved oxygen (DO) and nutrients (TN and TP), the following results have been obtained (MoFW 2017):

- Class I:
  11%
- Class II:
  38%
- Class III:
  30%
- Class IV:
  21%

With regard to biological parameters, the results were as follows:

- Class III:
  29%
- Class IV:
  50%
- Class V:
  21%

7.3.2 Lake Pollution

There are more than 120 natural lakes, 293 dams and about 1000 small dam reservoirs in Turkey (Fanack 2016).

Comprehensive and reliable information about all these water bodies is missing. In this section, an attempt is made to give at least a general idea about the lake water quality in the country. Quality of some of the larger lakes is discussed in the following paragraphs on the basis of available data. The main references to this section are Uslu (1995) and MoFW (2017).

In an overall assessment of river water quality with regard to dissolved oxygen (DO) and nutrients (TN and TP), the following results have been obtained (MoFW 2017):

- Class I:
  11%
- Class II:
  58%
- Class III:
  18%
- Class IV:
  13%

With regard to biological parameters, the results were as follows:

- Very good:
  2%
- Good:
  9%
- Medium:
  7%
- Weak:
  53%
- Bad:
  29%

7.3.2.1 Marmara Region

Lake Sapanca
The catchment area of Lake Sapanca is 251 km^2. It is surrounded by mountains in the south and small hills in the north. Water is taken from the Lake for domestic and industrial needs.

Lake Sapanca is one of the few lakes in Turkey, which provides drinking water, but it is exposed to heavy urbanization because of its natural beauty and proximity to the metropolitan Istanbul. With its charming natural beauty, the region around Sapanca has become very important for day trips and weekend vacations. It is a wetland area where millions of migrating birds reside seasonally.

Until now, heavy pollution has not been observed in the lake. Dumping of wastewater from domestic areas into the lake should be prevented, and no licenses should be given to factories in the future.

Table 7.11 Nutrient loads to Lake Sapanca

Diffuse TN load (t/year)	38
Diffuse TP load (t/year)	5
Point TN load (t/year)	–
Point TP load (t/year)	–

Water quality problems encountered in the area are:

- wastewater from settlement areas and commercial facilities such as motels, restaurants, petrol stations etc.;
- pollution coming from streams;
- agricultural pollution (pesticides, fertilizers);
- land based pollution sources such as highways and roads passing along the north and south shores of the lake;
- erosion problems caused by decline of forests;
- pollution from fish farms along the streams; motor boats on the lake;
- waste from factories surrounding the catchment area of Lake Sapanca.

The lake has mesotrophic characteristics. The nutrient loads are summarized in Table 7.11.

Lake Manyas
Lake Manyas is a rather shallow freshwater lake with a surface area of 162 km^2. It is located south of the Sea of Marmara. The lake is Turkey's only national park possessing a Class A diploma from the Council of Europe. Up to 230 species of birds visit this area, 44 of them also breed in the catchment, which is extremely important for its natural beauty and scientific value. It is also included into the RAMSAR treaty since 1996.

There are agricultural areas and factories around the lake. As a consequence, the water quality is Class III with regard to COD and BOD, Class IV regarding nutrients, and Class II regarding trace elements. General quality is regarded as Class IV.

Lake Iznik
Lake Iznik has a surface area of 308 km^2 and a maximum depth of 65 m. The southern shores of the lake have extensive beaches. It contains freshwater fishes. Lake Iznik is included in the national list of wetlands to be protected.

There are over 60 settlements in the catchment, the residents of which are mostly occupied with irrigated agriculture and olive farming. Presently 9000 ha of land is being irrigated with the water drawn from the lake. It is planned to add 7000 ha new irrigated area to the system. The agricultural residues (fertilizers and pesticides) are washed out into the lake through precipitation. All the settlements around the lake discharge their wastewaters directly or indirectly into the lake.

Although the lake is being used as a drinking water resource, the rivers and creeks feeding it have a "very polluted" (Class IV) quality. As a consequence, the lake itself has also a Class IV quality with regard to COD, BOD, TKN, NH_4-N, TP and DO. With regard to salt content, it is qualified as Class III.

Lake Apolyont (Ulubat)

Lake Apolyont, with an average depth of 1.90 m and a maximum depth of 6.00 m has a surface area of 154 km^2. Its catchment area totals to 10,413 km^2. It is a shallow, turbid and eutrophic freshwater body. The lake was included into the RAMSAR wetlands list in 1998.

The lake waters, together with the unpolluted waters supplied from Mustafakemalpasa Creek, are used to irrigate an approximate area of 26,800 ha. Close to 30% of Turkey's crayfish production comes from Lake Apolyont. It is also planned to be used as a supply of water for drinking and other purposes to the city of Bursa in the future.

The overall water quality is evaluated as Class IV. With respect to COD and BOD, it is of Class II, regarding nutrient (TN and TP) Class III, and as for trace elements Class I.

7.3.2.2 Lakes Region

Lake Aksehir

Lake Aksehir extends over an area of 339 km^2 and has an average depth of 4.5 m. The catchment area is 7340 km^2. It is a closed lake with no outlet and affected by the pollution coming from the creeks in the catchment area, domestic wastewaters from the city of Aksehir, and conserved fruit factories on its shore. To prevent environmental pollution and to achieve an appropriate management, an action plan for Aksehir and the neighbouring Eber Lake has been prepared.

Although a water quality observation station has been established in the middle of the lake, which was planned to take water samples from various depths, it is no more possible to reach it because of the recession of the water level. The shores all around the lake are turned into sump areas which prohibit access. As a result, there are no quality measurements to evaluate the state of the lake.

Lake Eber

Lake Eber is the 12th largest lake of Turkey. It has a surface area of 130 km^2 and, on the average, a depth of 1–3 m. The catchment area is 5000 km^2. It is mostly covered with reeds. The rate of evaporation is high.

The surface waters in the catchment area rank as polluted or extremely polluted. As a consequence, the water in the lake has a water quality class of IV. Both Akarcay and Eber lakes are not suitable as a source of drinking water extraction.

The lake has a water quality Class IV due to pH, BOD and nutrients, and Class IV due to trace elements. It ranks as hypertrophic due to TN, TP and secchi disk depth. The main causes of pollution are wastewater from settlements, industries (sugar and alkaloid factories in Afyon), thermal tourism and agricultural activities.

Lake Karamik

The lake, which resembles Lake Eber with respect to its formation, is located in the Province of Afyon and is eutrophic. Algae production is observed in the summer, and signs of pollution are evident from the colour and odour of the lake water. The lake's eutrophic character is conducive to the growth of reeds and cattails. The SEKA Paper Factory discharges annually 1.5 million m^3 of wastewater into the Lake Karamik.

Lake Beysehir

Lake Beysehir is the largest fresh water lake in Turkey. Its catchment and surface areas are 4052 km^2 and 650 km^2, respectively. It has an average depth of 8.5 m.

The main stresses on the lake quality are domestic and industrial wastewaters, and agricultural activities (use of fertilizers and pesticides).

Lake Beysehir is a protected wetland according to the RAMSAR Treaty. The rivers in the catchment area are of Class IV according to nutrients, organic matter (BOD and COD), and trace elements (Boron).

Lake Burdur

Lake Burdur is one of the largest masses of water in the lakes region. It covers an area of 250 km^2 and has a catchment area of 6150 km^2. It has an average depth of 45 m, and its deepest point is 74 m. The main sources of pollution in the lake are municipal wastewater from the city of Burdur and the wastes of Keciborlu sulphur factory, the sugar factory, the milk factory, and the army base. Because of these discharges, the pollution in the lake has reached critical levels. Its water is not suitable for any kind of use.

Acigol

Acigol has a surface area of 149.5 km^2. It is the second largest salty lake after the Salt Lake in central Anatolia. Intensive agriculture, animal feed, and domestic wastewaters are the main causes of pollution.

Lake Egirdir

Lake Egirdir is located within the Lake District. It has a surface area of 482 km^2 and a catchment area of 3309 km^2. Main uses of the lake are drinking and irrigation water supply, tourism, and fishing. Agricultural, industrial and domestic activities are the main causes of the degradation of water quality. TN, TP and faecal coliform concentrations have a rising tendency.

7.3.2.3 Western Anatolia

Lake Koycegiz

Lake Koycegiz is located in south-western Turkey. It started to form as a result of the clogging of the channel between the lake and the Mediterranean Sea. The lake is among the "Specially Protected" areas because of its natural beauty, and ecological and archaeological importance. The lake is connected to the Mediterranean Sea coast, Iztuzu, where endangered sea turtle species, Caretta caretta, nest and breed. It has a surface area of 58.6 km^2 and a catchment area of 1073 km^2.

Lake Koycegiz is influenced by several external factors such as sulphuric springs, Mediterranean seawater, and a relatively strong changing wind. The complicated layer structure of the lake is determined more by chemical (salt) gradients than by temperature. Its quality is influenced by domestic and industrial wastewaters.

Lake Bafa

Conversion of an Aegean embayment (Latmian Gulf) into a lake took place in the last 6000 years. The propagation of the Menderes River Delta cut off the mouth of the embayment and transformed it into a brackish residual lake in the southern part of the former embayment. Its surface area is 71 km^2.

The main input of the Lake is the Buyuk Menderes River. As a polluted river, it influences the basic water quality. The river contributes substantially to the sediment inputs that form the morphological characteristics. The fish farms around the lake increase the nutrient concentrations. The residues from nutrients and pesticides originating from agricultural activities in the delta plains to the south, olive trees in the slopes of the mountains surrounding the lake, domestic wastewaters from settlements and touristic facilities, and industrial wastewaters from the olive oil factories increment this pollution. These factors characterize the eutrophic and "very polluted" state of the water body.

7.3.2.4 Central Anatolian Closed Basin

Salt Lake

Salt Lake is Turkey's second largest lake following Lake Van. It has a surface area of 1620 km^2 and a maximum depth of 5.0 m. The lake's most significant feature is the salinity of its waters. With evaporation in the summer, its depth decreases to 2.0 m. There is no fish life in the lake.

Since the lake is located in a closed basin, it has no outlet to the sea. All the agricultural, domestic and industrial residues accumulate in the lake.

Because of this increasing pollution, the salt production has declined in the lake, which used to account for approximately 30% of Turkey's salt production. Salt quality has also deteriorated and does no more confirm with the health standards.

Nutrient loads to the Salt Lake are summarized in Table 7.12.

The continuously rising population of Konya and other neighboring settlement areas and increasing industrial activities are the causes of the point pollution. The pollution components originating from these sources are organic matter, detergents, oil and grease, nutrients, sulphur, fluoride and mercury.

The combination of drought and excessive abstraction has also had severe consequences with the Salt Lake in the arid Konya basin. The lake, which in the past was visited by thousands of flamingos each summer, has begun the process of transforming into a dry salt basin. Although the Konya basin has experienced drought conditions since the 1980s, excessive groundwater abstraction, the majority of which is drawn from illegally drilled wells for irrigation, has also played a critical role (Dogdu and Sagnak 2008). The lack of rainfall and excessive abstraction for agriculture has severely depleted groundwater, causing levels to decrease markedly

Table 7.12 Nutrient loads to Salt Lake (t/year)

Diffuse TN load	434
Diffuse TP load	46
Point TN load	381
Point TP load	68

in recent years. In addition, numerous smaller lakes and wetlands in the Konya basin, dependent upon groundwater inputs, have also dried up (EEA 2009).

7.3.2.5 Lake Van Closed Basin

Lake Van

Lake Van, with its surface area of 3574 km^2 and a volume of 607 km^3, is the largest lake in Turkey. It has a maximum depth of 457 m and an average depth of 170 m. It is situated in a closed basin with a catchment area of 12,225 km^2. The average elevation of the surface is 1656.5 m from the sea surface. Lake Van is one of the most thoroughly studied lakes in Turkey. In addition to groundwater inflows, there are several rivers and creeks feeding the lake. Seasonal fluctuations of inflow, rainfall, and evaporation cause annual fluctuations of 0.05–0.60 m of the surface level, which in the last years tends to increase.

It is the first among the soda rich lakes in the world in terms of volume. Its water is bitter, salty, and soda rich with a salt content of 2.24 ppt. The various inorganic species are 42% NaCl, 34% Na_2CO_3, 16% Na_2S0_4, 3% K_2S0_4 and 2.5% $MgC0_3$.

In general, Lake Van ranks as much polluted (Class IV) with respect to pH, salinity and TP; with regard to organic matter (BOD and COD) and NH_4-N, it is of Class III. High values of TP signalize diffuse loads coming from agriculture and animal breeding.

There are municipal wastewater treatment plants in Van, Ahlat, Tatvan, Ercis and Muradiye. But only the one in Ahlat is working satisfactorily. The others cannot function properly due to insufficient capacity and operational problems.

The situation with respect to industrial wastewaters is even worse. The wastewater treatment plant of Van Organized Industrial Zone has been damaged during the great earthquake of 2011. Some of the wastewaters are connected to the municipal system, but this increases the problems encountered in the municipal wastewater treatment plant as mentioned before.

The municipalities and industries around the lake discard their solid wastes into uncontrolled dumping sites. The leakage water carries organic matter, toxic chemicals, and heavy metals to the lake. In the following tables (Tables 7.13, 7.14, 7.15, 7.16 and 7.17) the flows and loads of pollutants are summarized, which contribute to the pollution of the Van Lake (TUBITAK-MAM 2013).

Table 7.13 Municipal wastewater discharges (m^3/day) and loads (tons/year) in Lake Van

Year	Population equivalent	WW discharge	Municipal WW loads				
			Parameter	Produced	Treated	Discharged into the basin	(%) Discharged into the basin
2012	1,225,637	110,329	COD	24,062	15,093	8969	37
			TN	1905	810	1095	57
			TP	306	129	178	58
2040	1,484,866	139,878	COD	35,166	28,133	7033	20
			TN	3210	1540	1670	52
			TP	494	197	297	60

7.3.2.6 General Remarks About Lake Water Quality

Finally, it can be stated that the main problems encountered in Turkey, regarding lakes and wetlands are (MoFW 2016):

- diminishing volumes/surface areas due to excessive water extraction and evaporation;
- deterioration of water quality due to domestic and industrial wastewaters;
- increasing eutrophication risk due to agricultural activities;
- accumulation of toxic compounds (heavy metals, persistent and perdurant organic compounds);
- loss of biological diversity;
- insufficiency of data quantity, quality, and accessibility, concerning the causes and the effects of the above mentioned problems.

7.3.3 Groundwater Pollution

It is estimated that Turkey's total annual groundwater resource is approximately 14 billion m^3. Due to recent rapid technological developments and population increase, water resources such as groundwater are in danger of severe pollution in the world, including Turkey. The following paragraphs are summarized from Baba and Tayfur 2011.

Groundwater contamination can be classified as having either natural or anthropogenic sources. Natural groundwater contamination is mainly due to geological formation with shallow groundwater mass (water-rock interaction in cold waters), infiltration from low-quality surface water bodies (streams, rivers, lakes), or due to the effect of geothermal fluids (water-rock interaction in hot waters).

Table 7.14 Industrial wastewater discharges (m^3/year) and loads (ton/year) produced in the catchment (2012 estimates) of Lake Van

	Industrial WW loads				
WW discharge	COD	BOD	SS	TN	TP
370,986	145	73	66	11	2

Table 7.15 Diffuse TN loads (ton/year) produced in the catchment (2012 estimates) of Lake Van

Natural land cover	Agriculture	Animal breeding	Atmospheric transport	Septic tanks	Leakage from garbage dumps	Total
2078	319	2254	453	406	130	5610

Table 7.16 Diffuse TP loads (ton/year) produced in the catchment (2012 estimates) of Lake Van

Natural land cover	Agriculture	Animal breeding	Septic tanks	Leakage from garbage dumps	Total
48	30	121	59	1.9	360

Table 7.17 Total nutrient loads in the catchment (2012 estimates) of Lake Van

	Loads (tons/year)					
	Total nitrogen (TN)			Total phosphorus (TP)		
Year	Point	Diffused	Total	Point	Diffused	Total
2012	660	5610	6270	114	260	374
2020	1226	4657	5884	223	170	393
2030	1498	4368	5865	260	155	415
2040	1682	4052	5734	299	139	437

The geological sources of contamination, which mostly release arsenic into groundwater, are mostly located around the Aegean region. TDS, calcium, and sulphate contamination due to salt and gypsum are mostly found in Central and Mediterranean regions. Polluted surface waters can also contaminate groundwater through infiltration. Such contamination resulting in high concentrations of Mn, Fe, Zn, etc. is found in Upper Kizilirmak, Gediz, and Buyuk Melen river basins and some drinking water reservoirs of the Metropolitan Municipality of ˙Istanbul. Another major problem is the contamination due to geothermal fluids, especially from the geothermal fields in the Aegean region, releasing heavy metals, especially B, fatal to living beings. It not only contaminates groundwaters but also major rivers such as Buyuk Menderes and Gediz in this region. Young volcanic rocks and geothermal fluids which contain fluoride cause dental and bone problems, especially in the areas of Denizli, Isparta, and Aydin. Excess levels of Al in drinking water as a result of mining sites, cause Alzheimer disease in and around the Aegean region, especially in Canakkale. A mild to severe degree of iodine deficiency, which is fundamental for growth and development, exists in settlements along the west coast of Turkey.

Anthropogenic groundwater contamination is generally ascribed to extreme use of agricultural pesticides and fertilizers, mining wastes, disposal of industrial wastes, seepage from waste disposal sites, seawater intrusion, and imperfect well construction.

Industrial wastewaters pollute surface and groundwater in the industrialized regions, such as Izmit and Izmir bays, Torbali region of Izmir, Gediz, Kizilirmak, and Porsuk rivers. It is important to carry out toxicity tests in wastewater discharge regulations to control this pollution. Surface and groundwater around metropolitan areas, such as Istanbul, Bursa, Izmir, Ankara, Antalya, and Eskisehir, are also contaminated by improperly constructed municipal and industrial waste sites. The uncontrolled seepage from these sites conveys heavy metals into groundwater, thus causing major pollution.

Deterioration of water quality as a result of fertilizers and pesticides is another major problem in Turkey, especially in the regions of Mediterranean, Aegean, Central Anatolia, and Marmara. Due to heavy fertilization, nitrate and nitrite contamination is very common in these regions, where the levels are above the standards.

Abandoned mercury mines, lead mining, colemanite mine, gold mining, and lignite mining in the regions of western Turkey, especially in Canakkale, Izmir, Mugla, Kutahya and Balikesir, cause serious groundwater quality problems by raising concentration levels of arsenic, boron, Ca, Mg, Al, SO_4, Cu, Cr, Co Ni, Zn, and Mn. In addition, coal burning power plants located mostly in the Aegean region cause pollution by their disposal sites and fly ash.

Excessive groundwater abstraction from a coastal aquifer causes the freshwater levels to lower and seawater to flow into the aquifer - a process known as 'saline intrusion'. Seawater intrusion causes problems in coastal regions, especially in the Aegean coast. It raises the concentrations levels of Na, Mg, Ca, and HCO_3 in groundwater. This diminishes the quality of the aquifer and prevents the subsequent use of the groundwater because conventional treatment methods do not remove the salt. Furthermore, the long residence time of groundwater means that the saline contamination may remain for decades. Large areas of the Mediterranean coastline have been affected by saline intrusion driven by abstraction of water for agriculture and public water supply, with demand for the latter being markedly increased by tourism. The regions in Europe where saline intrusion is a problem are shown in Fig. 7.4 (EEA 2009).

7.3.4 *Marine Pollution*

Marine pollution along the Turkish shorelines is mainly due to major land-based sources such as:

- untreated wastewater from domestic and industrial settlements;
- pollutants brought from inland areas by river;

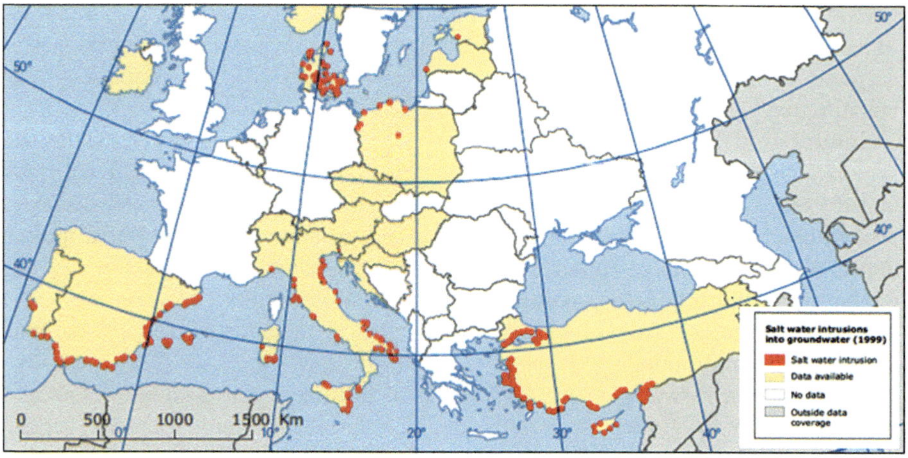

Fig. 7.4 Saline intrusion in Europe (EEA 2009)

- artificial or natural fertilizers and pesticides used in agricultural activities;
- tourism activities;
- extensive concentrations of secondary, holiday homes;
- port and marina establishments;
- cooling water discharged from thermal power plants; and, to some extent,
- aquaculture facilities.

Additionally, transboundary pollution sources from neighbouring countries (such as the pollutants brought by the Danube River and transported to the Turkish coast and into the Marmara Sea by Black Sea surface currents; the litter brought in by sea currents from the eastern Mediterranean countries and transported by surface currents to the Turkish Mediterranean coast), maritime transport, yachting, oil spills, and tanker and pipeline accidents are important sources of marine pollution (PAP/RAC 2005; MoEU 2016).

Bio-accumulative pollutants accumulate especially in the tissues of living organisms and pass from species to species in the food chain with increasing concentrations. Heavy metals such as mercury, cadmium, lead and substances such as pesticides can be included in this group.

80% of marine pollution results from land based and 20% from maritime activities:

1. Land based pollution sources are:
   - Point source pollution; domestic waste, industrial waste, landfills
   - Diffuse source pollution; soil and other pollutants (pesticides) that reach to the seas as a result of agricultural activities

2. Pollution sources from marine transportation:

   - Pollution from ships and other marine vehicles (fuel, oil, toxic liquids, waste water, garbage etc.)
   - Pollution from marine accidents (as a result of the accidents, cargo and fuel may fall and pour into seas)

A number of early industrial facilities that were developed in the 1960s and '70s along the shores of relatively sheltered sea areas, such as the northern Marmara coast, Izmit Bay, Izmir, Aliaga and Nemrut Bays, and Iskenderun Bay, are responsible for the major coastal "hot spots". These areas still suffer from the impacts of water pollution from industries due to the relatively enclosed nature of the basins. As a result of industrialization, population of these coastal areas has increased over-proportionally, causing an increase in domestic wastewater production.

A major increase in the yearly municipal wastewater discharge rate into coastal waters is observed between 1994 and 2016 ($557 \bullet 10^6$ to $1813 \bullet 10^6$ m^3). Whereas only 15.6% of the wastewaters had been treated in 1994, a significant increase in treatment can be observed in 2016 to 95.2% (Table 7.23).

The estimate of the total wastewater discharge from manufacturing coastal industries was $467 \bullet 10^6$ m^3/year in 1997 (PAP/RAC 2005). Of this figure, about 11.22% received some kind of treatment. In 2014, the amount of industrial wastewater discharged into the sea has reached $1559 \bullet 10^6$ m^3/year. However a large portion (at least 80%) of these discharges consists of cooling water from thermal power plants.

Coastal waters of the Mediterranean Sea are generally oligotrophic (very good) and mesotrophic (good water quality). However, a trend from oligotrophic condition to mesotrophic condition was observed in the shallow coastal waters between Mersin provincial wastewater discharge area and delta of Seyhan River on the eastern side of Mersin Gulf, due to land-based influx into the coastal waters. It is observable that, especially in Iskenderun and the interior gulf of Antalya, urban waste- and groundwaters affect coastal waters. Total mercury accumulation in multiple sediment samples from Goksu delta was higher than other coastal sediment results. No remarkable variation was observed in analysis of the trends of other metals (MoEU 2016).

In the Black Sea, the ecological condition of 3 coastal bodies of water was described as average/poor, and only Sile area was described as of good quality. Especially Sakarya action radius and Zonguldak region are under intense land-based pressure. Of the 3 water management units in Central-Western Black Sea, 2 are in average/poor ecological condition, and the easternmost one is in "good" condition. According to the analyses, sediments in Zonguldak region and nearby areas are highly polluted with respect to all groups of pollutants. Sinop region was identified as in very good condition in terms of ecological quality; however, remarkable sediment pollution is present in the west part of Sinop Cape. In Kizilirmak-Yesilirmak impact area, ecological conditions of the Kizilirmak impact area was identified as “average”, Samsun port and it’s adjacent coastal waters as “bad”, and Yesilirmak impact area as “poor”. Sediments in the Samsun area, which is under the influence of domestic wastewater discharge, were determined to be polluted by DDT and its derivatives. Ecological condition of 6 bodies of water in the Eastern Black Sea was determined as average to poor (MoEU 2016).

Sea of Marmara shows average to good ecological qualities, and only in the Gonen Creek impact area, the coastal waters exhibit eutrophic conditions. It was observed that the amount of phytoplankton was relatively low in August, and there was no considerable change in species composition. Eutrophia and hypertrophy were observed in terms of TN and TP, but oligotrophic conditions were present in terms of chlorophyll-α and secchi disc depth. This is due to low levels of chlorophyll on the surface of the Sea of Marmara in the summer season and the low levels of planktonic activities. The eutrophic and hypertrophic conditions in terms of total nitrogen and phosphorous indicate high levels of land-based influx. When the gulfs on the Sea of Marmara are classified according to the By-law on Surface Water Quality, the transitional waters in Gulf of Erdek, Gulf of Bandirma, and the interior Gulf of Izmit are in hypertrophic condition; and Gulf of Gemlik, central and exterior gulfs of Izmit and other regions of Gulf of Erdek are in mesotrophic condition (MoEU 2016).

When total petroleum hydrocarbon content of the sediment samples are compared, the average values of the Black Sea are much lower than the average levels of Marmara and the Aegean Sea, except the Zonguldak area which has the highest level due to coal mining. The concentration levels of polyaromatic hydrocarbon compounds in the Black Sea, Sea of Marmara, and Aegean Sea exhibit levels below limit values, and no concentration was detected in Mediterranean Sea. According to metal enrichment factors, the areas of high metal content (>10) in sediment are Black Sea, Aegean Sea and Sea of Marmara in terms of lead content, Aegean Sea and Black Sea in terms of arsenic content and Sea of Marmara in terms of mercury content (MoEU 2016).

Every year in the swimming season, bacteriological monitoring activities are performed in sea and lake waters in 34 cities with a coast to sea or lake by the Ministry of Health, Public Health Institution of Turkey. The number of swimming areas monitored was 1085 in 2010, and quality monitoring activities has risen to a total of 1239 swimming areas in 2016. According to the results of the monitoring in 2016, among the 1239 swimming zones, 932 zones (75%) were identified as Class A, 267 zones (22%) as Class B, 38 zones (3%) as Class C, and only 2 zones (0.002%) were identified as Class D. The classification of bathing waters is summarized in Table 7.18.

Shipping is one of the most important factors causing marine pollution. The main environmental impacts of shipping operations include air pollution, oil discharges or other hazardous substances/wastes, and transferring invasive alien organisms in global scale. More than 90% of the foreign trade of Turkey, in terms of volume, have been realized with maritime transport. The Turkish Straits System (TSS), which consists of the Bosporus Strait (17 nm), the Canakkale Strait - Dardanelles (37 nm) and the Sea of Marmara proper (110 nm), is the most important ship route in Turkish

Table 7.18 Classification of bathing waters (MoEU 2018)

	2010	2011	2012	2013	2014	2015	2016
Class A	721	776	786	963	932	1021	932
Class B	285	345	284	186	130	183	267
Class C	68	50	105	24	44	40	38
Class D	11	7	8	0	2	5	2

Seas. The narrow straits at Bosporus and Dardanelles with blind turns and dangerous currents (up to 8 knots) have always been potential threats to the passing ships. Every year, more than 40,000 ships cross the TSS. Oil tankers are the most prone vessels to possible accidents especially during storms (Unlu 2016).

According to Unlu (2016) environmental impacts of maritime transportation activities are:

- Ship-generated oil discharges and emissions
  - Accidental spillages (tanker and non-tanker accidents)
  - Shipping emissions
- Operational discharges and environmental impacts
  - Bilge water
  - Ballast water discharges and transfer of alien species
  - Sewage
  - Solid waste
- Physical effects of marine vessels on marine habitats
  - Anchoring
  - Antifouling paints on ships

7.4 Wastewater Management

7.4.1 Development of Wastewater Collection and Treatment in Turkey

The primary objective of all wastewater treatment and disposal schemes is control and minimization of detrimental effects on the environment, caused by the liquid wastes originating from various fields of human activity. To achieve this goal, a great number of technologically feasible treatment alternatives (unit operations and processes) are presently available. However, the application of these wastewater treatment methods in developing countries is limited by a variety of reasons listed below, not necessarily in order of priority:

- Economic constraints,
- Skilled manpower limitations for design, construction and operation,

- Scale of the project, and
- Energy requirements.

Up to 1990s, Turkey's per capita GNI was below 2000 US$. The nation had other priorities (nutrition, health care, education, employment, national security) superseding environmental protection. At thc same time, knowledge and expertise in environmental engineering was very limited. As a consequence, efforts made for environmental protection in general and water quality management in particular were inadequate. In the years following 1990, the situation has started to change, and investments were underway in environmental infrastructure.

While the percentage of population that receives sewerage services was 69% of municipal population in 1994, this rate reached 90% in 2014 (Table 7.19). Although the rate was 92% for 2012, it decreased to 90% in 2014. This was mostly due to the fact that some municipalities were given district/village status with the Municipality Law numbered 6360 (MoEU 2016).

In the last decades, there has been a considerable increase in the number of operating wastewater treatment plants. Treatment of wastewaters is an important application to ensure an effective use of water and protecting the present resources. Significant investments are made in Turkey within this context. While 71 municipalities were providing service with 41 wastewater treatment plants in 1994, at the end of 2014, this number reached to 513 municipalities with 604 wastewater treatment plants. As of the end of 2015, 551 municipalities have been providing services with 653 wastewater treatment plants (Tables 7.20 and 7.22).

As can be seen on Table 7.21, while 13% of municipal population received waste water treatment services in 1994, this rate reached 68% at the end of 2014 (MoEU 2016).

Table 7.23 summarizes the remarkable development of municipal wastewater collection, treatment and disposal systems in the last 24 years. The amounts of wastewaters included to the systems have increased more than 300%. The treated fractions have increased from 10% to 90% in most discharge alternatives. Similarly, Table 7.24 shows the development of industrial wastewater schemes in the years 2010 to 2014.

According to the Wastewater Treatment Action Plan 2015–2023 of the Ministry of Forestry and Water (MoFW 2017), at the end of 2014, wastewaters of 55.8 million people were treated in municipal wastewater treatment plants. This number corresponded to over 90% of the municipal population. 303 of these plants were secondary (biological) treatment, 40 deep sea outfalls, 40 package treatment, 13 physical treatment, 124 natural treatment (waste stabilization ponds), and 77 advanced treatment (biological nutrient removal – BNR). It is planned to upgrade all existing treatment plants to BNR by the end of 2022 to remove N and P loads of the wastewaters. Today (2018), it is estimated that 95% of the municipal population is connected to sewerage systems and 80% is receiving wastewater treatment.

Establishment of Organized Industrial Zones (OIZs) in Turkey dates back to 1960s. The investors in the OIZs could benefit from a pre-installed infrastructure and public structures. They could collectively benefit from information technologies,

Table 7.19 Percentage of population served with a sewerage system in municipal population

Year	1994	1995	1996	1997	1998	2001	2002	2003	2004	2006	2008	2010	2012	2014
%	69	72	72	77	78	81	83	85	86	87	88	88	92	90

Table 7.20 Number of municipalities that provide wastewater treatment services

Year	1994	1995	1996	1997	1998	2001	2002	2003	2004	2006	2008	2010	2012	2014	2015
Number	71	75	82	97	115	238	248	278	319	362	442	438	536	513	551

Table 7.21 Percentage of population served with a wastewater treatment plants in municipal population (%)

Year	1994	1995	1996	1997	1998	2001	2002	2003	2004	2006	2008	2010	2012	2014
%	13	12	14	19	22	35	35	38	45	51	56	62	68	68

Table 7.22 Number of wastewater treatment plants owned by municipalities

Year	1994	1995	1996	1997	1998	2001	2002	2003	2004	2006	2008	2010	2012	2014	2015
Number	41	46	55	68	80	126	145	156	172	184	236	326	460	604	653

Table 7.23 Amount of wastewater discharged from municipal sewage treatment plants into receiving bodies, 1994–2016 (10^6 m^3) (MoEU 2016)

	1994	%	2001	%	2008	%	2016	%
Tot. amount discharged	**1510**	**100.0**	**2301**	**100.0**	**3261**	**100.0**	**4484**	**100.0**
Treated	150	9.9	1194	51.9	2252	69.0	3842	85.7
Untreated	1360	90.1	1107	48.1	1010	31.0	642	14.3
Sea	**557**	**36.9**	**836**	**36.4**	**1458**	**44.7**	**1813**	**40.4**
Treated	87	15.6	658	78.7	1232	84.5	1725	95.2
Untreated	470	84.4	178	21.3	227	15.5	88	4.8
Lake/dam	**53**	**3.5**	**38**	**1.7**	**67**	**2.1**	**78**	**1.8**
Treated	3.5	6.6	19.3	50.8	48	71.9	53	67.8
Untreated	49.5	93.4	18.7	49.2	19	28.1	25	32.2
River	**796**	**52.8**	**1223**	**53.1**	**1404**	**43.1**	**2153**	**48.0**
Treated	48	6.0	461	37.7	778	55.4	1728	80.3
Untreated	749	94.0	762	62.3	626	44.6	425	19.7
Dam	**58**	**3.8**	**89**	**3.9**	**115**	**3.5**	**126**	**2.8**
Treated	10	16.9	23	25.8	84	73.1	76	60.7
Untreated	48	83.1	66	74.2	31	26.9	50	39.3
Land	**41**	**2.7**	**41**	**1.8**	**50**	**1.5**	**20**	**0.4**
Treated	2	4.3	8	19.3	14	28.0	14	70.0
Untreated	39	95.7	33	80.7	36	72.0	6	30.0
Other	**4**	**0.3**	**73**	**3.2**	**166**	**5.1**	**293**	**6.5**
Treated	0	0.0	24	32.7	95	57.1	245	83.7
Untreated	4	100.0	49	67.3	71	42.9	48	16.3

Note: Includes the amount of wastewater treated outside the municipal treatment plants
Figures in table may not add up to totals due to rounding

increasing efficiency with regards to utilization of fuel, energy and water resources, as well as combating with the unemployment.

In 2014, the OIZs have produced 254•10^6 m^3 of industrial wastewater, 72.9% of which was discharged into rivers, 9.1% into municipal sewers, 6.4% into vadis, and 11.6% into other receiving media. 221•10^6 m^3 of the produced wastewater was treated. 50.9% of this treatment took place in advanced plants, whereas 49.1% in chemical and biological wastewater treatment plants (MoEU 2016). Detailed information about OIZ wastewater discharges is compiled in Table 7.25.

In process selection for planned WWTPs, the Turkish legislation gives criteria for process selection (Table 7.26). These criteria were determined, based on the populations and considering the requirements given in Municipal Wastewater Treatment and Water Pollution Control By-laws (drinking water catchment basin, vulnerable and less vulnerable areas). Hence, all the treatment facilities for population above 2000 and located in a drinking water catchment area, and those for population above 10,000 and located in a vulnerable area, were planned to be able to remove nutrients (N, P).

Table 7.24 Data on wastewaters from manufacturing industry (MoEU 2016)

	2010	2012	2014
Amount of discharged wastewater (10^6 m^3/year)	**1256**	**1539**	**1931**
Amount of discharged cooling water	884	1197	1572
Amount of discharged wastewater except cooling water	372	342	359
Total amount of treated and discharged wastewater (10^6 m^3/year)	**164**	**188**	**207**
Amount of treated and discharged cooling water	10	11	10
Total amount of treated and discharged ww. except cooling water	154	178	198
Amount of wastewater discharged by receiving bodies (10^6 m^3/year)	**1256**	**1540**	**1931**
Municipal sewerage	81	59	60
Sea	–	1194	1559
Lake	0.239	–	–
River	245	148	140
Dam	0.731	–	0.486
Septic tank	5	2	2
Organized industrial zone sewerage	109	113	121
Other(*)	28	14	48
Number of wastewater treatment plants	**1825**	**2075**	**2096**
Physical/chemical	656	778	878
Biological	1089	1190	1094
Advanced	80	107	124
Capacity of wastewater treatment plants (10^6 m^3/year)	**490**	**556**	**539**
Physical/chemical	103	160	157
Biological	335	334	318
Advanced	51	62	64
Amount of ww. treated in wastewater treatment plant (10^6 m^3/year)	**244**	**240**	**244**
Physical/chemical	55	58	67
Biological	170	151	150
Advanced	20	30	27

(*) Includes wastewater discharged to village sewerage, tax free zone sewerage, wastewater treatment plants of cooperatives, State Hydraulic Works channels, dry stream bed, mining site, wastewater used for on-site or off-site irrigation, etc.; Figures in table may not add up to totals due to rounding

7.4.2 *Deep Sea Discharge*

In general, the concern for water quality is primarily directed towards the reduction of pollutant concentrations through the application of the well-known technologies. Even after the most intensive and complex treatment processes, however, there remains a residual pollution which has to be returned and dispersed in the environment. Therefore, the most important objective is to prevent the build-up of excessive pollutant concentrations. Independent of the interest in developing more efficient

Table 7.25 Indicators of water, waste water and waste in OIZs (MoEU 2016)

	2010	2012	2014
Number of organized industrial zones in operation	**134**	**181**	**196**
Sewerage network			
Number of OIZ that have their own sewerage system	102	136	162
Number of OIZ benefitting from municipal sewerage system	17	21	12
Number of OIZ without a sewerage system	15	24	22
Amount of ww. discharged through OIZ sewerage (10^6 m^3/year)(*)	**190**	**235**	**254**
Treated	161	192	220
Untreated	29	43	33
Number of waste water treatment plants	**38**	**57**	**76**
Chemical/biological treatment	24	41	54
Advanced treatment	14	16	22
Waste water treatment capacity (106 m^3/year)	**297**	**319**	**369**

(*) The amount of discharged wastewater is greater than the amount of abstracted water as it includes wastewater generated from self supplied water of some establishments
Figures in table may not add up to totals due to rounding

techniques of wastewater treatment, the dilution and dispersion wastes in the environment thus constitute a main problem of interest in water quality management (Uslu 1985).

The deep sea discharges (marine disposal) of untreated or partially treated sewage constitutes a complimentary or, in most cases, even an alternative strategy for the reduction of the detrimental effects of wastewaters on the environment. By this type of disposal, however, the prediction of pollutant concentrations caused by the waste input into the marine environment is very important for the design and operation of such systems. Dispersion models are necessary to estimate the dilution of wastes due to mixing and transport. The degradation of organic matter and die-off of pathogenic bacteria and viruses also contribute to the waste reduction. The efficient mixing through well-designed outfall diffusers is significant for the dilution of sewage effluents in the immediate vicinity of the outfall. The combined effects of mixing and degradation in open bodies of water in a marine environment, where adequate exchange is present, are often sufficient to achieve the receiving water quality standards imposed on such waters.

Main mechanisms involved in these processes are:

- Initial dilution (S_1): Municipal wastewaters normally have a lower density than the salty seawater. Thus, the wastewater discharged through diffusers at the sea bottom rises to the surface. During this rise, the wastewater plume is diluted with the ambient sea water, and the pollutant concentrations decrease.
- Secondary dilution (S_2): The plume which rises to the surface forms a cloud, which is transported through the surface currents. During this transport, the cloud mixes through lateral dispersion with the clean surface waters, and the concentrations decrease once more.

Table 7.26 Criteria for process selection (TUBITAK-MAM 2014)

Population	Location	Process	Treatment level	Pretreatment	Sludge treatment
N < 2000	Drinking water basin	Package treatment	Secondary	CS	Drying beds
	Vulnerable area	Natural treatment/package treatment	Secondary	CS/septic tank	Drying beds
	Others	Natural treatment/package treatment	Secondary	CS/septic tank	Drying beds
2000 < N < 10,000	Drinking water basin	Extended aeration activated sludge	Sec./Adv.	CS + FS + HFGC	Grav. Thickener + mechanical/drying beds
	Vulnerable area	Extended aeration activated sludge	Secondary		
	Others	Extended aeration activated sludge	Secondary		
10,000 < N < 50,000	Drinking water basin	BNR (Carbon+nutrient removal)	Advanced	CS + FS + HFGC	Mechanical
	Vulnerable area	BNR (Carbon+nutrient removal)	Advanced		Mechanical
	Others	Extended aeration activated sludge	Secondary		Grav. Thick. + Mech.
50,000 < N < 100,000	Drinking water basin	BNR (Carbon+nutrient removal)	Advanced	CS + FS + AGC	Mechanical
	Vulnerable area	BNR (Carbon+nutrient removal)	Advanced		
	Others	Extended aeration activated sludge	Secondary		

(continued)

Table 7.26 (continued)

Population	Location	Process	Treatment level	Pretreatment	Sludge treatment
100,000 < N < 250,000	Drinking water basin	BNR(Carbon+nutrient removal)	Advanced	CS + FS + AGC	Mechanical
	Vulnerable area	BNR (Carbon+nutrient removal)	Advanced		
	Others	BNR (Carbon+nutrient removal)	Advanced		
N > 250,000	Drinking water basin	BNR (Carbon+nutrient removal)	Advanced	CS + FS + AGC	Sludge digestion + mechanical
	Vulnerable area	BNR (Carbon+nutrient removal)	Advanced		
	Others	BNR (Carbon+nutrient removal)	Advanced		

CS Coarse Screen, *FS* Fine Screen, *HFGC* Horizontal Flow, *AGC* Aerated Grit Chamber, *BNR* Biological Nutrient Removal

- Tertiary dilution (S_3): The bacteria and viruses die-off in the marine environment mainly through the action of solar radiation and salt content of the sea water, thus decreasing their concentrations.
- The total dilution (S_t) is defined as: $S_t = S_1 \cdot S_2 \cdot S_3$

The typical values for S_1, S_2 and S_3 are 100, 10 and 10, respectively. Thus, a total dilution (S_t) of 10,000 is easily achievable under suitable ambient conditions. For further information about deep sea outfalls, the reader is advised to consult Uslu (1985) and other related sources.

The availability of a highly developed wastewater treatment technology has in most parts of the world obscured the fact that open bodies of sea water have a natural capacity to handle enormous quantities of organic loads. Many of the opponents of sea disposal seem to assume that the sewage remains unchanged in the coastal waters. However, treatment also occurs in the natural environment even when untreated sewage is discharged into the sea.

When sewage is diluted 100 times with clean sea water, it is, from the bacteriological point of view, indistinguishable from a fully treated secondary effluent, and as far as suspended solids and BOD are concerned, it is not only indistinguishable but, in fact, is of superior quality. Subsequent to the dilution, the natural chemical and biological processes of purification occur with the result that almost no change in the character and composition of the receiving seawater can be detected. In an inland river, the prime indicator of pollution is the depletion of the oxygen content, but in the open sea environment, there are few places, except in the immediate vicinity of a large discharge, where any oxygen depletion can be recognized (Calvert 1975).

The addition of nutrients, on the other hand, may change the overall marine food production. When this increase remains below the eutrophication limits, it can only be beneficial to man.

The possible presence of toxic pollutants, such as heavy metals and perdurant organic compounds in the wastewaters discharged into the ocean, seems to be the weakest link in the argument in favour of marine outfalls. However, these substances are also present in the effluents of conventional treatment plants, and their removal requires highly specialized and costly technologies. As mentioned previously, the most feasible solution both from the technological and economic points of view is the control of these pollutants at the source and their segregation from the main waste streams at the earliest possible opportunity.

The effects of these eco-toxic conservative substances depend on the concentration build-up in the biota. Fortunately, these pollutants would, due to the control measures stated above, generally be present in extremely small concentrations, and the time required to bring about distinguishable changes in the vast bodies of open seas would have to be measured in geological time scales.

The characteristics of the recipient water body are important in such cases because of the pertinent detention times of conservative pollutants in these water bodies. The detention time is simply defined as the volume of the recipient divided by the volume of wastewater discharge per unit time. The detention times associated with marine outfalls are in order of 10^6 years for the oceans and 10^{-1} to 10^{-2} years for semi enclosed water bodies such as bays. These time scales constitute a measure for determining the impact of perdurant and conservative substances on a recipient. A river, for instance, reacts almost immediately to such a pollutant discharge, whereas the reaction time in a sea may be significantly longer (Uslu 1985).

To sum up, marine outfalls have been extensively used since 1980's and still are being used in the coastal regions in Turkey with sufficient dilution capacity. In the past, the country did not have sufficient experts, constructors and operators to manage wastewater treatment plants. The funds were also very limited. As a result, it was decided to build marine outfalls to discharge and dilute wastewaters in the coastal cities in an ecologically appropriate manner. Since tourism was rapidly developing especially in the Mediterranean and Aegean coasts, it was a top priority for the country to protect these coastlines against pollution. The properties of wastewaters that are allowed for marine disposal are depicted in Table 7.7, as defined in the By-law for Water Pollution Control.

Deep sea discharge is required for the discharges of cooling waters and concentrated salt waters with a waste water capacity of more than 5000 m^3/day, by considering the above mentioned dilution and modelling criteria.

7.4.3 Sludge Management

During sludge treatment, stabilization is used in a rather smaller number of treatment plants, and dewatering processes are utilized by a larger number of wastewater treatment plants in Turkey. Approximately 25% of sludge generated is stabilized, and aerobic digestion (53%) constitutes the most common type of the stabilization process; anaerobic digestion (29%), lime stabilization (16%), and composting (2%) are applied as other stabilization processes in Turkey.

165 of the plants operate one type of sludge dewatering unit. Sludge is typically dewatered in belt filters (54%), centrifuges (29%), drying beds and lagoons (13%), and plate filter presses (4%).

Although the dried solids contents of the produced sludge vary from plant to plant, daily total sludge production on dry basis in Turkey is 910 tons according to surveys filled out by wastewater treatment plant operators.

Mainly land filling in municipal landfill areas (46%) is used for ultimate disposal. Land filling in special storage areas (30%), land application (10%), usage of sludge as a supplementary fuel in cement factories (6%), and other disposal methods (sewage truck and catch basin) (7%) are other ultimate alternatives. In addition, incineration composes only 1% of ultimate disposal routes (Insel et al. 2013).

Newer information on the topic is given by MoEU (2016) about the final disposal of municipal wastewater treatment plant sludge as in Fig. 7.5.

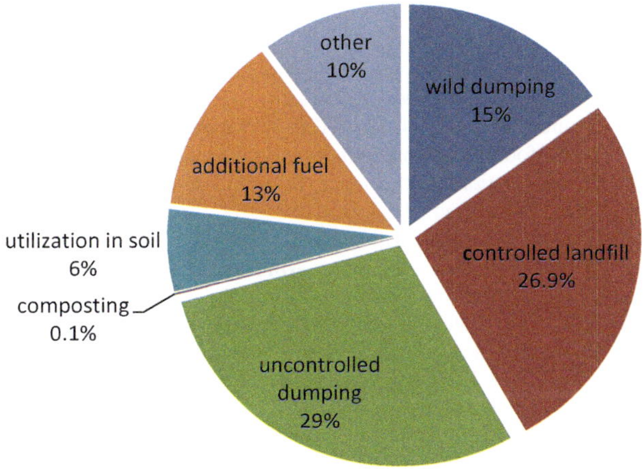

Fig. 7.5 Final disposal routes for municipal wastewater treatment plant sludge

7.5 Conclusion

Due to population increase, inland migration from rural to urban areas, industrialization, agriculture, expansion of tourism, and increases in economic activities and resource depletion, environmental loads have soared several hundred times in the last 50–60 years in Turkey.

This situation had severe effects on the water quality of inland and coastal waters. At the beginning, the country did not have sufficient expertise, funds, regulations and governmental agencies to combat with the emerging problems.

The country has responded quickly and consistently to the situation. Considerable improvement has been accomplished in wastewater management during the last two decades. However, there is still a long way to go especially for the improvement of now established infrastructure, data collection, and research.

References

Baba A, Tayfur G (2011) Groundwater contamination and its effect on health in Turkey. Environ Monit Assess, https://doi.org/10.1007/s10661-011-1907-z; Published online, Springer

Calvert JT (1975) The case against treatment. In: Gameson ALH (ed) Discharge of sewage from sea outfalls. Pergamon Press, London

Dogdu MS, Sagnak C (2008) Climate change, drought and over pumping impacts on groundwaters: Two examples from Turkey. Paper submitted to the Third International BALWOIS Conference on the Balkan Water Observation and Information System, Ohrid, the Former Yugoslav Republic of Macedonia, May 2008

EEA (2009) Water resources across Europe – confronting water scarcity and drought. EEA Report No 2/2009. EEA, Copenhagen

Falkenmark M, Lundqvist J, Witstrand C (1989) Macro-scale water scarcity requires micro-scale approaches. Nat Res Forum 13(4):258–267

Fanack (2016) Surface water and groundwater resources in Turkey. https://water.fanack.com/turkey/turkey-surface-water-groundwater/

Insel G, Kendir E, Ayol A, Erdincler A, Arikan O, Imamoglu I, Alagoz BA, Gencsoy EB, Sanin FD, Buyukkamaci N, Karatas O, Saygili G, Sener G, Cokgor EU, Filibeli A (2013) Current situation and future perspectives in municipal wastewater treatment and sludge management in Turkey. J Res Sci Technol 10(3):133–138

Mekonnen MM, Hoekstra AY (2011) National water footprint accounts: the green, blue and grey water footprint of production and consumption. Value of water research report series no. 50. United Nations Educational, Scientific and Cultural Organization (UNESCO) – Institute for Water Education (IHE), Delft

MoEU (2014) Atiksu Aritimi Eylem Plani (2014–2023) (Wastewater treatment action plan (2014–2023)). Republic of Turkey Ministry of Environment and Urbanisation (MoEU), Ankara, (www.csb.gov.tr)

MoEU (2016) State of the environment report for Republic of Turkey. Republic of Turkey Ministry of Environment and Urbanisation (MoEU), Ankara, (www.csb.gov.tr)

MoEU (2018) Environmental indicators 2016. Republic of Turkey Ministry of Environment, (http://www.csb.gov.tr/gm/ced/)

MoFW (2016) Goller ve Sulak Alanlar Eylem Plani Taslagi (Action plan for lakes and wetlands). T.C. Orman ve Su Isleri Bakanligi, Su Yonetimi Genel Mudurlugu (Republic of Turkey Ministry of Forestry and Water Resources, Generaldirectorate for Water Management (MoFW)), Ankara, (http://www.ormansu.gov.tr)

MoFW (2017) Su Kalitesi Yonetimi Calisma Grubu Raporu (Water Quality Working Group Report), 2. Ormancilik ve Su Surasi. T.C. Orman ve Su Isleri Bakanligi, (Republic of Turkey Ministry of Forestry and Water Resources, General Directorate for Water Management (MoFW)), Ankara

PAP/RAC (2005) Coastal area management in Turkey. Priority Actions Programme Regional Activity Centre, Split

TSA (2018) Results of population censuses, 1935–2000 and results of address based population registration system, 2007–2017; TurkStat, Population Projections, 2018–2080, http://www.tuik.gov.tr

TUBITAK-MAM (2013) Havza Koruma Eylem Planlarinin Hazirlanmasi Projesi Van Golu Havzasi Nihai Raporu, Proje Sahibi Kurulus: T.C. Orman ve Su Isleri Bakanligi (TUBITAK-MAM – Turkish Scientific and Technical Research Council-Marmara Research Centre) (2013) Watershed Protection Action Plan for Van Lake Watershed. Responsible Authority: Turkish Ministry for Forestry and Water Resources)

TUBITAK-MAM (2014) Havza Koruma Eylem Planlarinin Hazirlanmasi Projesi Marmara Havzasi Nihai Raporu, Proje Sahibi Kurulus: T.C. Orman ve Su Isleri Bakanligi (TUBITAK-MAM – Turkish Scientific and Technical Research Council-Marmara Research Centre): Watershed Protection Action Plan for Marmara Watershed. Responsible Authority: Turkish Ministry for Forestries and Water

Unlu S (2016) Marine pollution from ships in the Turkish straits system. In: Ozsoy E, Cagatay MN, Balkis N, Balkis N, Ozturk B (eds) The sea of Marmara; marine biodiversity, fisheries, conservation and governance. Turkish Marine Research Foundation (TUDAV), Istanbul, pp 755–767

US Dept. of State (2018) https://www.state.gov/documents/organization/229296.pdf

Uslu O (1985) Marine outfalls as a wastewater treatment and disposal alternative in developing countries. In: Curi K (ed) Appropriate waste management for developing countries. Plenum Press, New York/London

Uslu O (1995) Water pollution. In: Environmental profile of Turkey. Environmental Foundation of Turkey (EFT) Onder Matbaasi, Ankara

World Bank (2017) https://data.worldbank.org/indicator/NY.GDP.PCAP.CD

World Bank (2018) https://data.worldbank.org/indicator/NY.GNP.PCAP.CD?end=2017&locations=TR&start=1967&view=chart

WWF (2014) Turkiye'nin Su Ayakizi (Water Footprint of Turkey). World Wildlife Fund (WWF) – Turkey. ISBN: 978-605-86596-7-4, Ankara

Chapter 8
Water Resources Potential of Turkey

Bülent Selek and Hakan Aksu

Abstract Assessment of water resources potential plays a key role in effective water resources management. This chapter provides the details of water balance components over Turkey. It presents a compilation and review of the available data on precipitation, evapotranspiration, groundwater and surface flows between 1981 and 2010, and the main drivers affecting Turkey's water potential are explained. The study is based on the most recent reports of the Ministry of Forestry and Water and related institutions (State Hydraulic Works, Meteorological Services, and the Directorate General of Water Management). It gives the details of the country's gross water resources, using a water balance approach. Compiled data on the water resources potential are given separately for each of the 25 large river basins of Turkey and also in total figures for the whole country. Furthermore, an overview of the 1981 and 2010 period is also presented to provide information on maximum and minimum values of the hydrologic and meteorological components. This chapter is intended to quantify Turkey's water resources but does not assess how these resources should be managed or developed.

Keywords Water resources potential · Turkey · Precipitation · Evapotranspiration · Surface water · Groundwater

8.1 Introduction

Turkey is located at the crossroads of Europe and Asia, between the latitudes 36°N and 42°N and longitudes 26 °E and 45°E. The country has an area of 78 million ha and extends 1650 km from the east to the west and 1000 km from north to south.

B. Selek (✉)
Ministry of Agriculture and Forestry, Ankara, Turkey

H. Aksu
Faculty of Aeronautics and Astronautics, Department of Meteorological Engineering, Samsun University, Samsun, Turkey
e-mail: hakan.aksu@omu.edu.tr

N. B. Harmancioglu, D. Altinbilek (eds.), *Water Resources of Turkey*, World Water Resources 2, https://doi.org/10.1007/978-3-030-11729-0_8

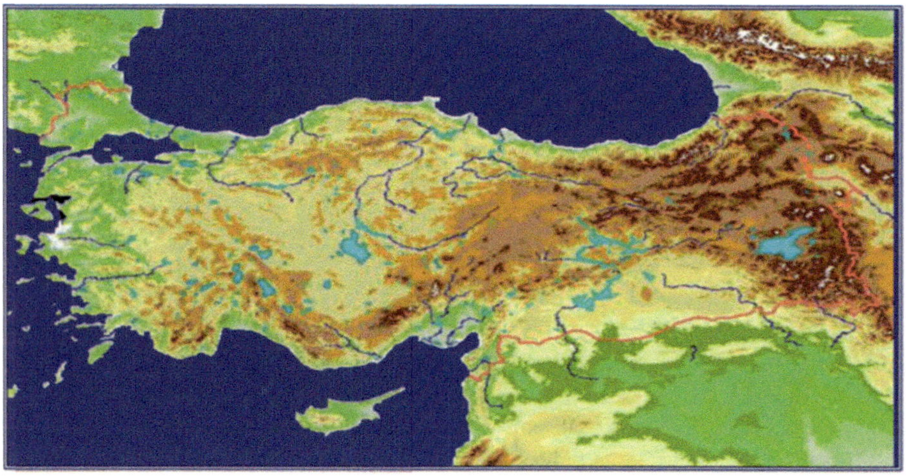

Fig. 8.1 Elevation map of Turkey. (DSI 2017)

Turkey's average altitude (1132 m) is higher than that (1050 m) of Asia and much higher than that of Europe (330 m). The elevation rises from the west to the east, and the highest point is the Agri Mountain (Ararat) (5137 m) in the eastern region (Fig. 8.1) (DSI 2017).

Even though Turkey's climate can be categorized as semi-arid, many parts of the country enjoy the Mediterranean, continental, and sub-tropic climates. Varying distances from the sea and changes in altitude result in marked climatic variance within short distances. While the average annual rainfall for the country as a whole is 574 mm, it reaches 2500 mm in the Black Sea region (in the North) and decreases to 300 mm in the Central Anatolia region.

Drought periods are also a characteristic of Turkey's climate, and over the last 40 years, the longest and the most severe drought events occurred between the years 1971–1974, 1983–1984, 1989–1990, 1996–2001, and 2007–2008 (Kurnaz 2014).

The contribution of snow to water potential is considerable, especially in the eastern regions. The national average temperature for the period of 1981–2010 is calculated as 13.5 °C (Report on Climate Change and Adaptation WG 2017). However, increases in the annual average temperature have been detected at some locations, altering the hydrological regime by increased evapotranspiration and melting of snow earlier in the year.

In addition, according to Intergovernmental Panel on Climate Change Fourth Assessment Report (IPCC 2013), the Mediterranean Basin in which Turkey is situated is one of the most vulnerable regions to global climate change. Annual and seasonal water availability is projected to change significantly in the coming decades, and increased precipitation intensity and variability will lead to greater risks

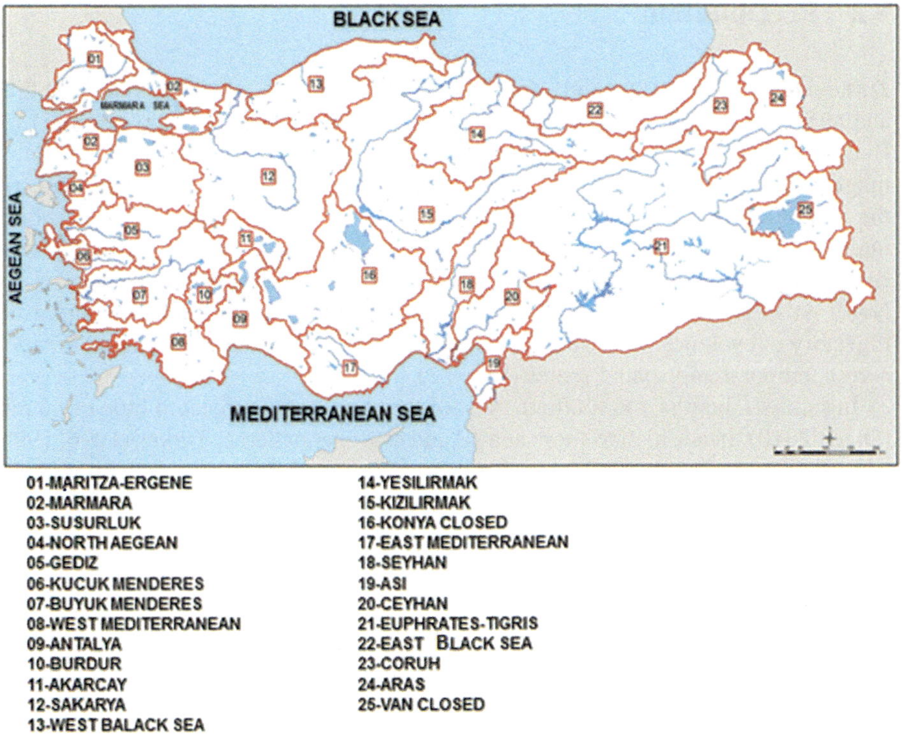

Fig. 8.2 Main hydrological basins of Turkey

of floods and droughts (UK Met Office 2011). Investigation of water resources potential and possible trends over time are of crucial importance for water management. Furthermore, as approximately one third of the country's water resources come from transboundary river basins, the current and potential future variations are also important for hydro-politics of the region.

Given the complex relationships between hydrology, water management, hydro-politics, and hydro-climatology, an accurate determination of water resources potential is of upmost importance. Only then, it is possible to determine water availability for human and environmental needs under the pressure of global changes such as rapid urbanization, climate change, and population growth (WMO 2012).

This chapter provides detailed information on the water resources of Turkey's 25 main river basins (Fig. 8.2). It is intended here to quantify various components of water resources, but not to assess how these resources should be managed or developed. The chapter is presented with the sections on precipitation, evapotranspiration, groundwater, surface water, and finally on an overall assessment of the water budget.

8.2 Precipitation

Turkey's precipitation is mainly characterized by highly variable distribution in time and space. The main physiographic factors affecting Turkey's climate are: (i) the Black Sea and Mediterranean basins; (ii) the west to east oriented high mountain ranges along the northern and southern coasts of the Anatolian Peninsula; and (iii) the Anatolian plateau with a mean elevation of 1130 m (Turkes 1996). The air masses driving Turkey's weather and climate are predominantly maritime and continental tropical and polar air masses. Positive and negative phases of the North Atlantic Oscillation (NAO) affect the precipitation regime by changing the trajectory of cyclones. These complex relations result with variations in precipitation across temporal and spatial scales (Fig. 8.3).

In summer months, the southern parts of the country are subject to little or no rain (Olgen 2010). In addition to these spatial and seasonal patterns, Turkes (1996, 1998) and Turkes and Erlat (2003) found a decreasing trend of winter precipitation for many stations all over Turkey, especially in the Mediterranean region.

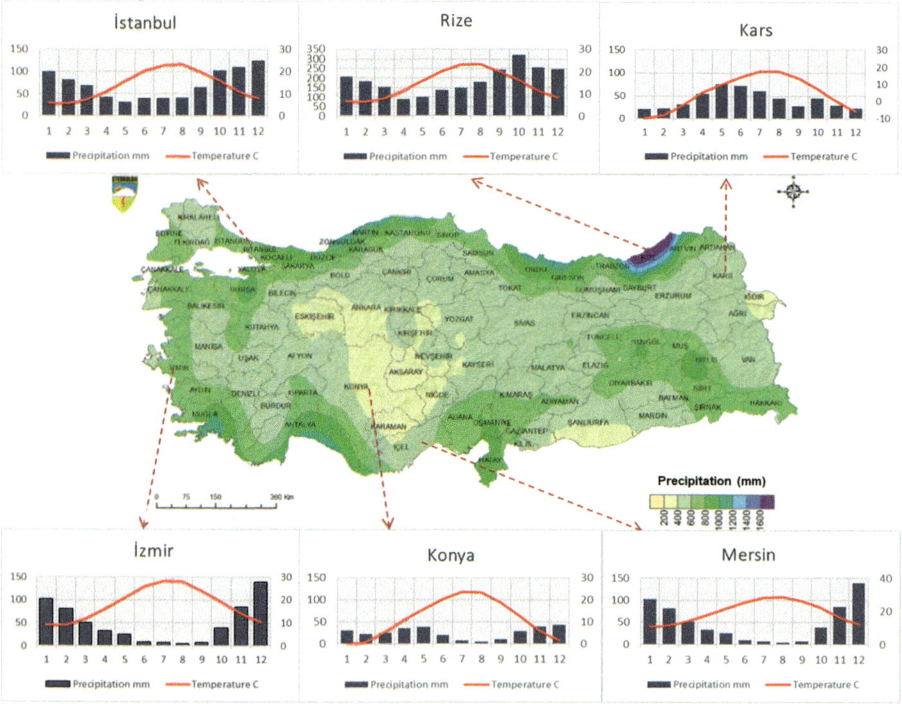

Fig. 8.3 Rainfall and temperature distribution over Turkey. (MGM 2017)

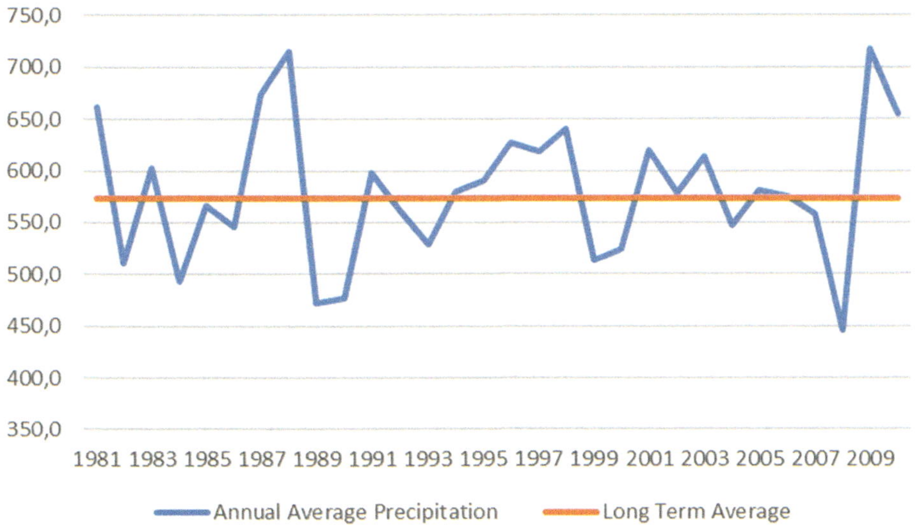

Fig. 8.4 Temporal variation of annual average precipitation in Turkey

In a recent study conducted by the Expert Group on Hydrology of the former Ministry of Forestry and Water of Turkey (responsibilities are transferred to the newly established Ministry of Agriculture and Forestry, MoAF, as of June 2018) a time series of 255 meteorological stations all over the country were used to estimate the mean areal precipitation for the period of 1981–2010, using six different methodologies: Thiessen (1911), Percentage Weighted Polygon (Sen 1998), Inverse Distance (Shepard 1968), arithmetic mean, Isohyetal, and Kriging (Krige 1951). The study period was determined in line with the recommendations for normal climate by the World Meteorological Organization (WMO 2017). Among the methodologies used, the results of the Ordinary Kriging – Exponential technique were taken into account as Turkey's areal precipitation amounts (Ulupinar et al. 2015).

According to results of the above mentioned study, Turkey's annual average precipitation for the 1981–2010 period was calculated as 574 mm. Temporal variation of the annual precipitation values is presented in Fig. 8.4. For the study period, the maximum annual areal total precipitation were estimated as 717.0 mm in 2009, which is 25% higher than normal. The minimum was 445.0 mm in 2008, representing 22% less than normal. The normal seasonal precipitation is 210 mm in the winter, 169 mm in the spring, 141 mm in the autumn, and 60 mm in the summer.

Precipitation normals for the main 25 basins of Turkey are presented in Table 8.1. Maximum normal precipitation is 1000.1 mm in the East Black Sea Basin, and the minimum is 390.1 mm in the Konya Basin.

Table 8.1 Annual average precipitation for large river basins of Turkey

No	Basin name	Average precipitation (mm)
1	Maritza-Ergene	591.7
2	Marmara	693.9
3	Susurluk	649.8
4	North Aegean	606.9
5	Gediz	578.5
6	Kucuk Menderes	611.1
7	Buyuk Menderes	598.7
8	West Mediterranean	739.9
9	Antalya	768.6
10	Burdur	476.0
11	Akarcay	476.3
12	Sakarya	463.8
13	West Black Sea	761.1
14	Yesilirmak	538.7
15	Kizilirmak	451.3
16	Konya Closed	390.1
17	West Mediterranean	582.0
18	Seyhan	576.2
19	Asi	829.5
20	Ceyhan	649.1
21	Euphrates-Tigris	565.3
22	East Black Sea	1000.1
23	Coruh	705.5
24	Aras	483.5
25	Van Closed	518.7
Total Average		**574.0**

Report on Water Resources Development and Hydrology WG (2017)

8.3 Evapotranspiration

Evapotranspiration (ET) is a collective term for all the processes by which water in the liquid or solid phase, at or near the earth's land surfaces, becomes atmospheric water vapor (Dingman 2002). Evaporation occurs from open-water surfaces, whereas transpiration occurs on the bare-soil and soil with vegetation. Evapotranspiration is the combination of these two processes and is controlled by the balance between available energy and accessible water. The estimation of evapotranspiration is not only important for hydrological analyses, but also for energy balance due to the fact that more than half of the solar energy absorbed by land surfaces is currently used to evaporate water (Trenberth et al. 2009).

Unfortunately, there is no systematic measurement network in Turkey for ET, even though it is the second largest component of the water cycle after precipitation. ET varies in time and space considerably. The methods for estimating basin-wide ET

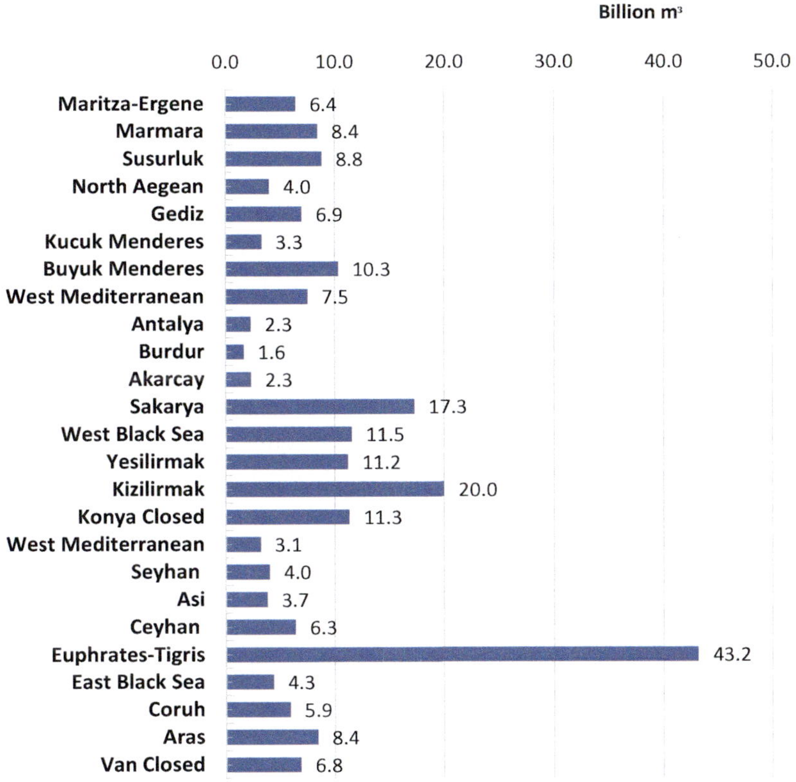

Fig. 8.5 Annual ET for large river basins of Turkey

are generally grouped into three categories: the water budget approach, meteorological estimates, and remote sensing of the land–atmosphere interface (Aksu and Arikan 2017).

In a recent study, the hydrology expert group of the Ministry of Forestry and Water Affairs of Turkey estimated the actual evapotranspiration by using the Thornthwaite – Mather Water Balance Model on semi-distributed and monthly spatial/time scales throughout Turkey for the period of 1981–2010. Conceptual Thornthwaite Water Balance model parameters were precipitation, temperature and potential evapotranspiration (Okkan 2009). Potential evapotranspiration estimations were carried out by using data obtained from the meteorological observation stations scattered in sub-basins. Estimations of open-water surface evaporation were based on Class-A pan evaporation measurements multiplied by pan coefficients at the meteorological stations.

The total average annual evapotranspiration was estimated at 218.6 billion m^3 over the country. Details of the actual evapotranspiration values for large river basins of Turkey are given in Fig. 8.5. Average actual evapotranspiration equivalents were

279.7 mm, with the maximum at 473.1 mm in the Asi Basin and the minimum at 110.9 mm in the Antalya River basin.

8.4 Groundwater

Groundwater is used for domestic and irrigation water purposes in Turkey (Vliegenthart et al. 2007), particularly where the surface waters are limited. General water quality of groundwater is better than surface waters, given its slower response to contamination and droughts. Groundwater potential varies with precipitation, temperature, evapotranspiration, and hydrogeology and, similar to the other components of hydrological cycle, it is affected by climate change and variability. Variation in groundwater can go unnoticed because current aquifer levels are largely invisible to the naked eye. Therefore, detailed and continuous monitoring of groundwater status is fundamental for water allocation strategies.

The geological structure of Turkey is heterogeneous and anisotrop as a result of extended seismic and volcanic activities. The Turkish aquifer system can be categorized mainly in two parts: alluvial plain aquifers and karstic carbonate aquifers (Report on Water Resources Development and Hydrology WG 2017). Coupled with the uneven distribution of precipitation over Turkey, this leads to marked spatial and temporal variations in groundwater recharge.

Groundwater studies in Turkey, including investigation, usage, allocations, and licensing, are conducted by the State Hydraulic Works (DSI) in accordance with the 1960 Groundwater Law number 167. Groundwater levels are monitored monthly by DSI at 1160 wells, and with additional seasonal observation stations, the total number of monitoring points accounts to 2748 monitoring points across the country (Gumus and Toklu 2018). Periodic discharge measurements are carried out at 1402 springs which have greater discharge than 50 l/s. All the monitoring activities are in accordance with the regulation entitled "Monitoring of Surface and Groundwaters" published by the Ministry of Forestry and Water Affairs in 2014.

To determine Turkey's groundwater potential, DSI carries out hydrogeological investigation studies across Turkey, which began in the 1950's and do not yet cover all of Turkey.

The designated groundwater potential of Turkey, based on hydrogeological investigation reports, is 23 billion m^3. The distribution of the potential throughout the large river basins is shown in Table 8.2. The number of springs with a discharge of greater than 50 l/s is 1442, producing a total flow of 27 billion m^3 (Table 8.3).

Table 8.2 Designated groundwater potential of Turkey

Number	Basin name	Designated groundwater potential (hm^3/year)
1	Maritza-Ergene	507.7
2	Marmara	241.7
3	Susurluk	780.4
4	North Aegean	289.4
5	Gediz	1155.9
6	Kucuk Menderes	179.2
7	Buyuk Menderes	1045.4
8	West Mediterranean	473.2
9	Antalya	1164.7
10	Burdur	106.4
11	Akarcay	345.4
12	Sakarya	2197.1
13	West Black Sea	641.2
14	Yesilirmak	907.2
15	Kizilirmak	2003.1
16	Konya Closed	2597.0
17	West Mediterranean	96.5
18	Seyhan	838.8
19	Asi	393.2
20	Ceyhan	985.3
21	Euphrates-Tigris	4994.8
22	East Black Sea	490.9
23	Coruh	30.0
24	Aras	388.5
25	Van Closed	179.2
Total		**23032.2**

Report on Water Resources Development and Hydrology WG (2017)

8.5 Surface Waters

Surface water potential is directly related to hydropower production, irrigation, domestic, industrial, and touristic water supply. Global changes (including climate change/variability and land use changes as a result of rapid urbanization, and other factors) affect surface water potential just as they act on all the other water budget components. Flow regime characteristics, such as the times of maximum and minimum flow over the year, are also subject to change. For these reasons, monitoring of surface waters and periodic assessments of the potential are needed.

In Turkey, a comprehensive stream gauging network is under operation, which includes personnel like hydrographers and data management engineers, as well as modern infrastructure (instruments and communication systems). Surface water potential of Turkey has recently been reviewed by the Hydrology Expert Group under the scope of the Turkey Water Budget Project (Report on Water Resources

Table 8.3 Spring flows of large basins

Number	Basin name	Number of springs	Total spring flows (hm^3)
1	Maritza-Ergene	23	79.3
2	Marmara	5	62.8
3	Susurluk	131	560.3
4	North Aegean	6	71.4
5	Gediz	57	438.8
6	Kucuk Menderes	6	37.7
7	Buyuk Menderes	105	2100.1
8	West Mediterranean	39	1520.7
9	Antalya	177	8664.6
10	Burdur	31	213.7
11	Akarcay	31	349.4
12	Sakarya	57	947.6
13	West Black Sea	21	257.9
14	Yesilirmak	58	153.1
15	Kizilirmak	102	1267.5
16	Konya Closed	121	1950.3
17	West Mediterranean	40	918.6
18	Seyhan	0	0
19	Asi	24	508.5
20	Ceyhan	120	2125.2
21	Euphrates-Tigris	235	4409.8
22	East Black Sea	2	71.8
23	Coruh	0	0
24	Aras	7	55.3
25	Van Closed	44	302
Total		**1442**	**27066.4**

Report on Water Resources Development and Hydrology WG (2017)

Development and Hydrology WG 2017). The study evaluated data from 401 flow gauging stations over the period between 1981 and 2010, analyzing all the water budget components. Known water abstractions for irrigation, either by water user associations or farmers, and domestic water supply amounts were added to measured flows at the stream gauging stations to derive estimates of the natural flows for the 25 river basins of Turkey.

Average surface water potential was estimated at 180,789.3 hm^3 across the 30 year period, with a maximum of 258,772.06 hm^3 in 1988 and minimum of 104,429.40 hm^3 in 1994 (Table 8.4). The Euphrates-Tigris River Basin accounts for 30.7% of the average national surface water potential.

The overall average runoff for the country was calculated as 7.3 l/s/km^2, although this figure is rather biased statistically due to the high differences between the maximum and the minimum runoff, and thus between the standard deviations, among the basins. Flow regimes are strongly influenced by the combined effect of

Table 8.4 Flow characteristics of large river basins

No	Basin name	Annual average flow hm^3/year (1981–2010)	Contribution ratio (%)	Runoff coefficient	Yield (l/s/km^2)
1	Maritza-Ergene	1842.3	1.02	0.22	4.0
2	Marmara	7540.3	4.17	0.47	9.9
3	Susurluk	4226.0	2.34	0.27	6.0
4	North Aegean	1500.6	0.83	0.25	4.8
5	Gediz	1544.8	0.85	0.17	2.7
6	Kucuk Menderes	527.6	0.29	0.12	2.4
7	Buyuk Menderes	2968.7	1.64	0.19	3.8
8	West Mediterranean	6969.1	3.85	0.44	10.5
9	Antalya	13082.8	7.24	0.83	21.2
10	Burdur	255.9	0.14	0.09	1.3
11	Akarcay	325.5	0.18	0.09	1.4
12	Sakarya	5158.2	2.85	0.18	2.8
13	West Black Sea	9914.1	5.48	0.45	10.6
14	Yesilirmak	6582.3	3.64	0.33	5.8
15	Kizilirmak	6120.0	3.39	0.16	2.5
16	Konya Closed	2647.3	1.46	0.14	1.6
17	West Mediterranean	8240.1	4.56	0.65	11.8
18	Seyhan	6785.8	3.75	0.52	10.6
19	Asi	1813.3	1.00	0.28	9.5
20	Ceyhan	7371.5	4.08	0.52	10.6
21	Euphrates-Tigris	55419.3	30.65	0.56	9.5
22	East Black Sea	16461.4	9.11	0.72	21.7
23	Coruh	7047.2	3.90	0.49	11.2
24	Aras	4181.8	2.31	0.31	4.8
25	Van Closed	2263.4	1.25	0.25	3.7
	Total	**180789.3**	**100.0**	**Areal Ave. 0.4**	**7.3**

Report on Water Resources Development and Hydrology WG (2017)

the diversities in precipitation regimes, topography, landscape and geology. Maximum runoff is 22.8 l/s/km^2 in the East Black Sea and 1.3 l/s/km^2 in Akarcay and Burdur Basins. The average rainfall-runoff coefficient for the country is 0.40, but the value reaches 0.83 in the Antalya basin and is as low as 0.09 in the Burdur and Akarcay River basins, which are characterized by their karstic geology.

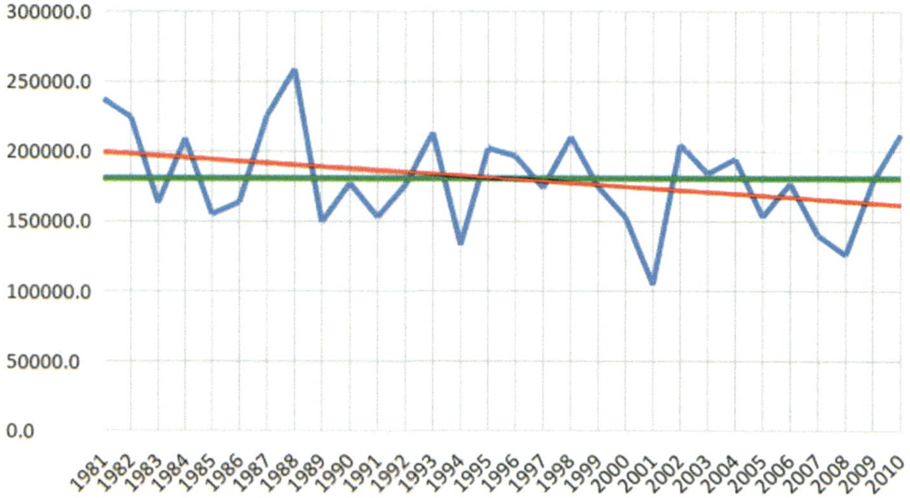

Fig. 8.6 Turkey's annual surface flows (hm^3)

Seasonal analysis of flow regimes shows that nationally 46% of the total flow occurs during the spring season (March, April and May), and the maximum surface flows in April (mainly due to snowmelt). The share of winter flows are 26%, while that in summer and autumn are 17% and 12%, respectively. Considering the sector that uses water the most, irrigation (76% of surface waters in Turkey), the times of peak flows do not match the peak irrigation demand periods.

A significant downward trend has been detected in ten of the river basins according to Mann Kendall Non-parametric trend test with 10% significance level (Buyuk Menderes, Bati Akdeniz, Burdur, Akarcay, Sakarya, Bati Karadeniz, Kizilirmak, Konya, Dogu Akdeniz, and Seyhan) (Ozdemir and Erkus 2017). Only one, Aras, shows an upward trend. A downward trend is also detected in the overall surface water potential for the study period (Fig. 8.6).

The maximum duration of less than average flows is 5 years between 2005 and 2009. Between 1981 and 2010, the second period is during 1989–1992 (4 years), and the third period is 1999–2001 (3 years).

8.6 Water Budget of Turkey

By combining the results of four component studies of areal precipitation, surface water, groundwater and evapotranspiration, the overall water budget of Turkey was recalculated (Fig. 8.7). Water balance components were estimated by

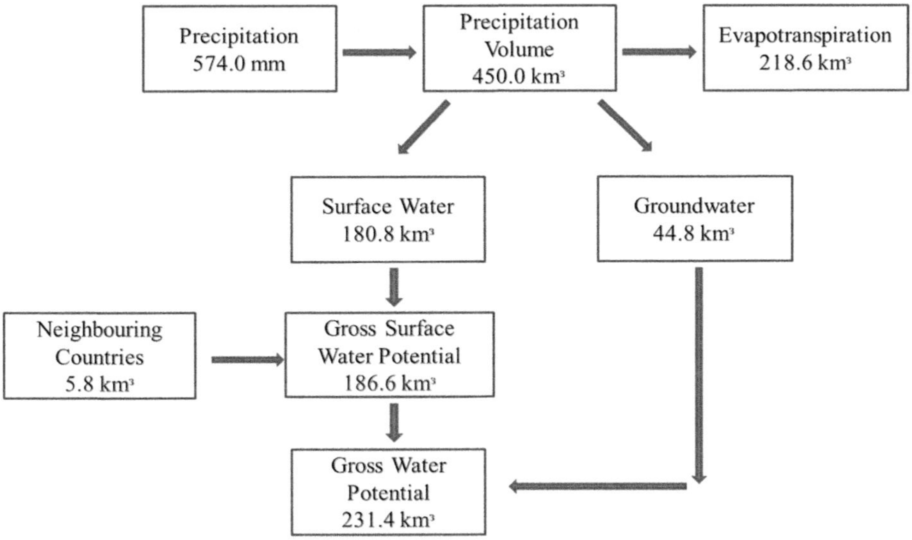

Fig. 8.7 Turkey's water budget and the gross water resources potential

Thornthwaite – Mather model in a monthly time scale for sub-catchments with the recently updated areal precipitation and surface flows. Then, the results were aggregated for each large river basin. The water balance components for large river basins including infiltration are presented in Table 8.5.

In the study mentioned in the previous section, the annual average precipitation for the country is estimated as 574 mm over the study period, which corresponds to 449.6 billion m^3 of water (Fig. 8.6). Evapotranspiration losses account for 218.6 billion m^3, representing approximately 49% of the annual precipitation. The total surface water potential is estimated at 180.8 billion m^3, which is about 40% precipitation, while the groundwater recharges represent around 44.8 billion m^3 (approximately %10 of the precipitation volume). The additional contribution of surface waters from neighboring countries is considered to be around 5.8 billion m^3.

Following from the above, the gross water resources potential of Turkey is calculated as 231.4 billion m^3. Approximately half of this potential (surface and groundwater) is discharged into the seas and to neighboring countries and becomes unavailable for use in technical and economic terms. Exploitable water resources of Turkey is reported as 112 billion m^3 (DSI 2018).

Five of the 25 river basins in Turkey are transboundary, and the surface water potential of these basins corresponds to 40% of the total surface water potential.

Table 8.5 Water budget components of large river basins of Turkey (1981–2010)

No	Basin name	Drainage area (km^2)	Million m^3			
			Precipitation	Flow	ET_a	Infiltration
1	Maritza-Ergene	14444.1	8561.4	1858.9	6382.7	319.8
2	Marmara	23107.1	16186.8	7537.9	8405.5	243.4
3	Susurluk	24332.0	15645.1	4227.2	8776.5	2641.4
4	North Aegean	9973.6	6051.4	1500.5	3969.0	581.8
5	Gediz	17034.0	9002.9	1536.3	6916.7	549.9
6	Kucuk Menderes	7059.7	4323.4	527.1	3260.3	536.1
7	Buyuk Menderes	26133.2	15889.0	2993.3	10279.4	2616.4
8	West Mediterranean	21223.9	15705.5	6965.1	7458.6	1281.7
9	Antalya	20330.8	15670.5	13076.2	2255.4	338.9
10	Burdur	6306.2	3020.3	264.4	1630.0	1125.9
11	Akarcay	7982.6	3805.7	325.6	2290.1	1190.0
12	Sakarya	63357.8	29352.3	5290.3	17254.4	6807.6
13	West Black Sea	28929.8	22017.6	9905.1	11534.0	578.6
14	Yesilirmak	39628.0	20170.9	6584.6	11173.6	2412.7
15	Kizilirmak	82197.3	37126.8	6123.6	19956.7	11046.6
16	Konya Closed	50037.8	19524.8	2649.7	11294.3	5580.7
17	West Mediterranean	21807.0	12709.8	8250.4	3139.4	1319.9
18	Seyhan	22241.6	12935.4	6778.1	3960.8	2196.5
19	Asi	7912.4	6556.9	1825.9	3743.5	987.6
20	Ceyhan	21598.5	14025.6	7349.4	6338.1	338.0
21	Euphrates-Tigris	176657.0	99900.5	55577.3	43168.5	1154.6
22	East Black Sea	22844.6	22844.8	16476.3	4318.3	2050.1
23	Coruh	20248.7	14286.0	7047.1	5858.3	1380.6
24	Aras	28114.6	13593.2	4182.2	8370.2	1040.8
25	Van Closed	17977.0	9164.3	2263.2	6823.7	77.4
	Total	**781479.4**	**448070.8**	**181115.7**	**218558.0**	**48397.0**

8.7 Concluding Remarks

This chapter presents the results of a recent comprehensive study on water resources potential of Turkey, realized by the hydrology expert group of the former Ministry of Forestry and Water. General characteristics of the water budget (precipitation, evapotranspiration, groundwater and surfacewater) are presented for 25 large river basins. Finally, the aggregation of the hydrologic and meteorological components in all basins has led to the estimation of the gross water potential of the country.

Considering the importance of water resources assessment on development strategies and management, a few concluding remarks are worth to reemphasize, regarding Turkey's water potential:

- Water potential changes substantially at temporal and spatial scales as does the climate.
- Turkey has experienced drought periods due to decreasing precipitation within the last years.
- Surface water potential exhibits significant decreasing trends for ten out of 25 basins, and decreasing trends are also observed for the overall water resources potential of the country.
- The change in areal precipitation becomes 25% higher and also 22% lower than normals for the study period (1981–2010).
- Since not all of the perspectives of water budget estimation are included in this study, it is hoped that water professionals and researchers will conduct more comprehensive studies with new methodologies in the future.
- A well designed and operated hydrometeorological monitoring network (flow gauging, rainfall, temperature, monitoring wells, evaporation, snow measurements etc.) is crucial for water balance studies.
- Data of all institutions and organizations producing water resources data should be standardized.

Acknowledgment The authors are grateful to Ministry of Forestry and Water of Turkey and related institutions (State Hydraulic Works, Meteorology Service, DG of Water Management) for their valuable published materials. The authors also thank the chief editor Prof. Dr. Nilgun Harmancioglu for inviting them to prepare this chapter of the book "Water Resources of Turkey".

References

Aksu H, Arikan A (2017) Satellite-based estimation of actual evapotranspiration in the Buyuk Menderes Basin, Turkey. Hydrol Res 48(2):559–570. https://doi.org/10.2166/nh.2016.226

Dingman SL (2002) Physical hydrology, 2nd edn. Prentice Hall, Upper Saddle River, p 646

DSI (2017) Water and DSI. Ankara

DSI (2018) http://en.dsi.gov.tr/land-water-resources. Accessed Sept 2018

Gumus V, Toklu MM (2018) Groundwater activities and developments in recent years in Turkey, National geodesy and geophysics congress, Izmir, Turkey

IPCC (2013) Climate change: 2013 the physical science basis contribution WG-1, WMO and UNEP

Krige DG (1951) A statistical approach to some basic mine valuation problems on the Witwatersand. J Chem Metall Min Soc S Afr 52:119–139

Kurnaz L (2014) Drought in Turkey. Istanbul Policy Center, Istanbul, pp 1–16

MGM (2017) Rainfall assessment report of 2017, Ankara

Okkan U (2009) Impacts of monthly precipitation and temperature on Izmir freshwater basin's runoff, Ms. Sci. thesis, Dokuz Eylul University, Department of hydraulic-hydrology and water resources

Olgen MK (2010) Spatial distribution of annual and seasonal rainfall variability over Turkey. Aegean Geogr J 19/1(2010):85–95, Izmir, Turkey

Ozdemir D, Erkus MK (2017) Trend analysis of total annual flows of Turkey basins, 8. National Hydrology Congress, Diyarbakir

Report on Climate Change and Adaptation WG (2017) Second forestry and water council. Ankara, Turkey

Report on Water Resources Development and Hydrology WG (2017) Second forestry and water council. Ankara, Turkey

Sen Z (1998) Average areal precipitation by percentage weighting polygon method. J Hydrol Eng 1:69–72

Shepard D (1968) A two-dimensional interpolation function for irregularly-spaced data. In: Proceedings of the 23rd ACM national conference. Association for Computing Machinery, New York, pp 517–523

Thiessen AH (1911) Precipitation averages for large areas. Mon Weather Rev 39(7):1082–1084

Trenberth KE, Fasullo JT, Keihl J (2009) Earth's global energy budget. Bull Am Meteorol Soc 90:311–323

Turkes M (1996) Spatial and temporal analysis of annual rainfall variations in Turkey. Int J Climatol 16:1057–1076

Turkes M (1998) Influence of geopotential heights, cyclone frequency and southern oscillation on rainfall variations in Turkey. Int J Climatol 18:649–680

Turkes M, Erlat E (2003) Precipitation changes and variability in Turkey linked to the North Atlantic Oscillation during the period 1930–2000. Int J Climatol 23:1771–1796

UK Met Office (2011) Climate: observations, projections and impacts. Turkey, UK

Ulupinar Y, Kervankiran S, Akbas, AI, Ugurlu A (2015) Areal precipitation assessment of Turkey, VIII. National Hydrology Congress, Sanliurfa

Vliegenthart FJL, Sargin AH, Gorkmen A, Dogdu MS (2007) A new approach for groundwater management in Turkey river basin management congress, Antalya

WMO (2012) Technical material for water resources assessment, WMO No: 1095, Switzerland

WMO (2017) WMO guidelines on the calculation of climate normals, WMO No: 1203, Switzerland

Chapter 9
Agricultural Water Use

Mahmut Cetin

Abstract Turkey is located in a semi-arid region. Due to topography, geological conditions, sea effect and the geographical position of the country, distribution of precipitation has shown temporal and spatial variability. Consequently, soil and water resources in the 25 river basins are not distributed evenly over the country. Therefore, irrigation is a prerequisite in order to develop a highly productive agricultural system, as well as to optimize agricultural production. Turkey has a total of 25.85 million hectares (Mha) irrigable area, of which 22.6 Mha land can be envisaged as economically and technically irrigable under today's conditions. Based on the soil, topography and drainage conditions, the national governmental agency -State Hydraulic Works (DSI)- responsible for development of soil and water resources including irrigation in Turkey set a goal nearly 60 years ago that 8.50 Mha area had been economically and technically irrigable under the available technology. Of the targeted irrigable area, as of 2017, 6.57 Mha land was equipped with irrigation facilities, 65% of which was constructed by DSI. Four major irrigation organizations were emanated for operation and maintenance services of rrigation schemes. As of the end of 2017, 4.28 Mha gross area (net 3.37 Mha) was equipped with irrigation facilities constructed by DSI; 73% of net irrigation area constructed by DSI was transferred to water user associations (WUAs). 15% was operated by irrigation cooperatives (ICs) and the remaining 12% by DSI. In order to increase the sizes of agricultural enterprises and decrease the average parcel numbers of undertakings, implementation of land consolidation (LC) projects in agricultural areas is necessary. Implementing LC projects before irrigation construction renders at least 40% savings in expropriation and construction costs. It also helps irrigation managers to increase the very low irrigation efficiencies (average 37%) and irrigation ratios (42% in DSI operated, 66% in WUAs operated irrigation schemes). As of the end of 2017, totally 54 BCM water was consumed in irrigation, domestic and industrial sectors. Of this, a total of 74% was consumed by irrigation such that 56% (30.2 BCM) was supplied from surface water resources and 18% (9.8 BCM) from

M. Cetin (✉)
Department of Agricultural Structures and Irrigation, Faculty of Agriculture, University of Cukurova, Adana, Turkey
e-mail: mcet64@cu.edu.tr

N. B. Harmancioglu, D. Altinbilek (eds.), *Water Resources of Turkey*, World Water Resources 2, https://doi.org/10.1007/978-3-030-11729-0_9

groundwater resources. Regional development projects regarding irrigation have an important role so as to eliminate in inter-regional differences in terms of development. In this regard, *The Southeastern Anatolia Project*, i.e. *the GAP*, is considered as one of the biggest integrated regional development projects in the world. The GAP is a brand name of Turkey and contributes widely to increase the agricultural production of Turkey. After the completion of the GAP components, 1.79 Mha area will have been equipped with irrigation facilities. Net agricultural income has increased about four-fold in the GAP irrigation areas. However, it was observed that gross domestic agricultural product increased about six-fold in the GAP areas due to the irrigation practices. Irrigation return flows need monitoring in order to take preventive measures on time.

Keywords Irrigation ratio · irrigation efficiency · land consolidation · Water User Associations (WUAs) · Southeastern Anatolia Project (GAP)

9.1 Introduction

Turkey is a Mediterranean country and a peninsula surrounded by the Mediterranean, Aegean, Marmara and Black Seas. Geographically, it is located between 26°–45° east longitudes and 36°–42° north latitudes (DSI 2016a). Thus, there is a 6° difference in latitudes. Topography varies from mean sea level of 0–5137 m, which is the highest elevation at the culmination of Mount Ararat. Mean altitude is 1132 m. Latitude differences, the distance from seas, extent and orientation of mountain chains, and topographical variations cause to emanate primarily four different climate types over the country, i.e. continental climate dominant in southeastern Anatolia, east Anatolia, central Anatolia and Thrace regions; Mediterranean climate, Thrace transition zone climate, and Black Sea climate. Except for the Black Sea region, semi-arid climate conditions prevail in Turkey. No arid zones have been experienced up to the present. However, research results by Selek et al. (2018) indicated clearly that there is an increase in semi-dry and dry humid zones (approximately 14%) and a decrease in the semi-dry-less humid to humid zones of Turkey. Each climate zone has its own characteristic ecological zones for agricultural production. For this reason, Turkey is known as the realm of ecologies. However, spatial and temporal variabilities in temperature and precipitation values hinder year-round production of field and horticultural crops under natural conditions. In this context, precipitation is the most limiting factor for consumptive use. Therefore, sunflower, cereals and pulses are the main field crops, which can be economically grown under rain-fed conditions over the country, except for the Black Sea region. On the other hand, olive and fig in the Mediterranean region and pistachio in the Eastern Anatolia region are horticultural crops grown under rain-fed conditions. Consequently, irrigation is necessary to ensure sustainability of agricultural production and food security in Turkey, except for the eastern Black Sea region.

Thus, for sustainable development, irrigated agriculture is inevitable and a primary key sector to provide goods and finance with the other sectors such as industry, energy, environment etc. The history of irrigated agriculture in Turkey dates back to 3000 B.C. In this regard, the twin rivers, the Tigris and the Euphrates, created the "Fertile Crescent" in ancient Mesopotamia (Loucks and van Beek 2005). So, civilizations of Mesopotamia, Sumer, Akkad, Babylonia, etc. rose in the Tigris-Euphrates valley, and eventually doomed due to the ignorance of shallow water table and salinity development by irrigation applications (Hillel 2004). Nevertheless, agriculture and irrigation, in turn, have had a place in Anatolian rural society since then. Furthermore, irrigated agriculture is directly associated with industry and other sectors by supplying goods and finance. In this regard, it is of vital importance in Turkish economy through providing employment, contributing to the gross domestic product, creating added value, increasing agricultural yield per unit area, improving income distribution in rural areas, and ensuring food security for the society during dry spells etc.

Employment by the agricultural sector is about 60% in developing countries, but less than 7% in developed and/or industrialized ones. In Turkey, agriculture employed around 68% of active employees in 1928, the very first stages of the foundation of Turkish Republic. Agricultural employment showed an increasing trend in the period of 1950–1960 and reached the highest level of 85%. As expected, agricultural sector employment rates have decreased gradually since then (DSI 2004), and it was 46%, 35%, 21%, and 19% in 1990, 2008, 2015 and 2017, respectively. As of the end of 2017, an employee in agricultural sector was able to feed 5 people, himself included, contrary to the fact that an agricultural employee is able to feed almost 47 in the Netherlands, 40 people in Germany, and 83 in UK. On the other hand, the sector of agriculture has contributed to the gross domestic product (GDP) of Turkey. The shares of agriculture, industry and services sectors in GDP by current prices were almost 40%, 17% and 43% in 1968, respectively. As seen in Fig. 9.1, the share of agriculture in GDP has decreased continuously in favor of industry and other sectors, and it was about 7.5% at the end of 2015.

Detailed economical evaluations by TOBB (2018) show that the share of agricultural sector in GDP was somewhere around 6–7% in the 5-year period between 2013 and 2017. As understood, agricultural sector is of importance for the country's economy and food security. Hence, irrigation is a must in the agricultural sector, and water is one of the most important inputs to enable agriculture partly independent on unfavorable climatic conditions. Although existing water use in agriculture has difficulty in competing against other sectors' water demands, irrigation intrinsically creates additional employment, makes technological agricultural inputs such as fertigation, pest management etc., improves income distribution in rural areas, and renders growing more than one crop, depending on the length of the growing season. For this reason, irrigation developments have gained momentum depending on the financial status of the country since the foundation of Turkish Republic in 1923.

At the beginning of the foundation of Turkish Republic, some small scale irrigation projects were realized with very limited budget conditions. Areas opened for irrigation were only 142.6 thousands ha with a 34% overall irrigation ratio in

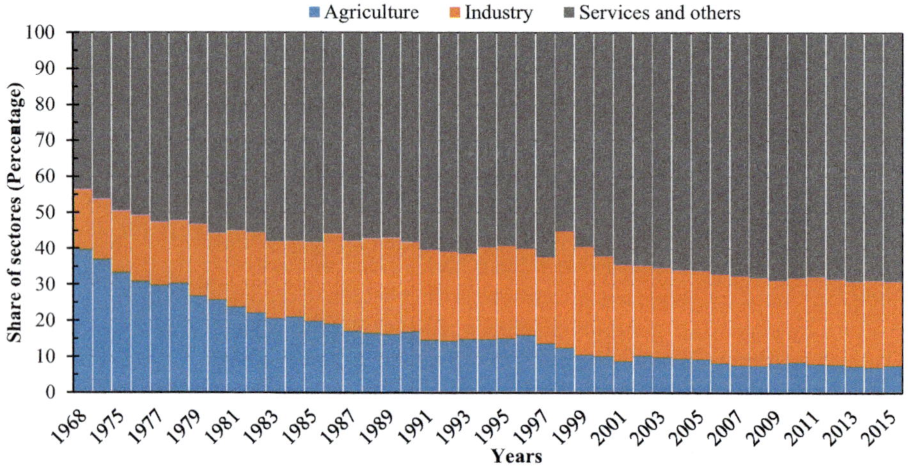

Fig. 9.1 Gross domestic product (GDP) in current prices by sectors

1950s. Turkish irrigation agency -*State Hydraulic Works (DSI),* which is the one and only state institution responsible for the development of soil and water resources in the country, was established in accordance with the Law No 6200 in 1954. Furthermore, DSI was empowered to plan, design, construct and operate dams, hydroelectric power plants, and domestic water and irrigation schemes (Bayazit and Avci 1997). Irrigated areas have started to increase considerably since the establishment of DSI. Consequently, as of the end of 2017, totally 6.57 million hectares (Mha) area was equipped with irrigation and drainage facilities by state institutions of DSI and Rural Services plus the community. Since its establishment to the end of 2016, DSI has invested heavily in irrigated agriculture (US$ 65 billion, (DSI 2016a)). In this sense, the share of irrigation sector is, on the average, 41% of total investments made in sectors of agriculture, energy, environment, and drinking, and domestic water supply. The share of the agricultural sector in the total investment budget of the primary national irrigation agency, DSI, is given in Fig. 9.2.

Data for Fig. 9.2 were collected from DSI (2004), DSI (2016a), annals of DSI, and yearly activity reports released by DSI; then, compiled and necessary calculations were done accordingly. It is observed in the activity reports of DSI that about 1/3 of the State's investment budget has been annually allocated to DSI. As seen in Fig. 9.2, the share of irrigation investments in DSI's investment budget varied between 31% and 77%. It is clear that there has been an increasing trend in the irrigation investments by DSI since 2003 partly due to decrease in hydropower investments after involvement of private sector. At the end of 2017, the share of irrigation investments reached the culmination (77%).

In Turkey, suitable climate, soil, and topographical conditions, coupled with the irrigation water sources, as well as good agricultural practices, have made it possible to cultivate various plants throughout the year. Accordingly, the average rate of yield increase per ha by irrigation is 188% for cereals, particularly in wheat, 328% for

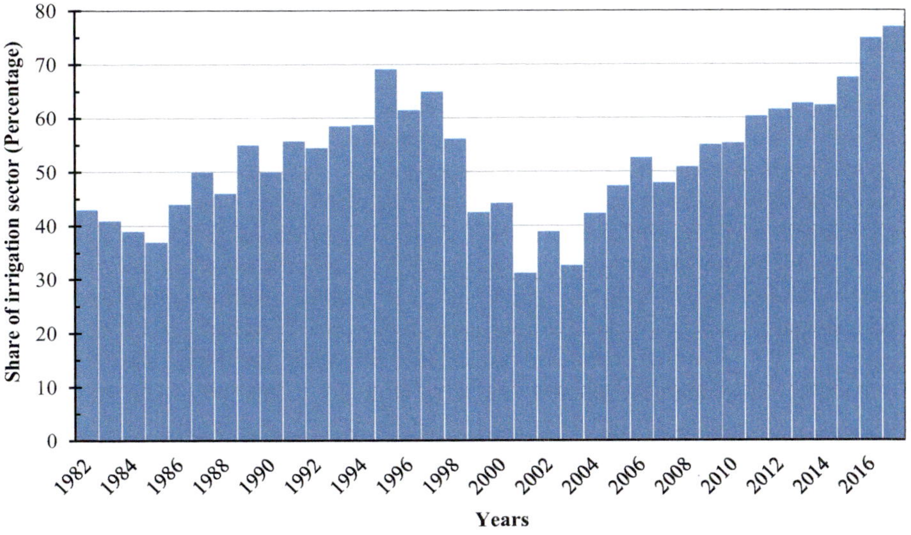

Fig. 9.2 Share of agricultural sector in total investment budget of DSI

cotton, 699% for maize, 184% for forage crops, 190% for vegetables and 116% for mixed-fruits in 2017 (DSI 2018a). Agricultural production increases have resulted in disposable incomes for the farmers. In this context, taking into account production costs and production values in current prices of 2003 (DSI 2004), gross domestic agricultural income (GDAI) per ha, on the average, increased from US$ 350 to US$ 1750 by the irrigation projects, indicating an increase of US$ 1400 per ha in gross agricultural income (GAI). In brief, data provided by DSI (2016a) indicate clearly that irrigation increases agricultural production value as well as GDAI about four-fold to six-fold in Turkey, in comparison with rain-fed conditions.

Although irrigation projects have contributed a lot to the national economy, a great majority of irrigation schemes do not satisfy the expectations. The most conspicuous problems are the low irrigation ratios as well as low irrigation efficiencies (Cetin et al. 2015).

9.2 Soil Resources and Limiting Factors Regarding Irrigated Agriculture

9.2.1 Soil Resources

National irrigation agency, DSI, has been trying to complete land evaluation surveys for irrigated agriculture within the context of "*Master Plans*" for the 25 Turkish river basins. Facts and figures regarding potential soil resources and their suitability for irrigation have not been released yet. Nonetheless, provisional data are available for the river basins. However, except for the very limited detailed soil survey works, there has been no detailed soil survey study in Turkey to figure out soil inventory by land capability classes. In this sense, it has been verified by experts and practitioners (MoD 2014) that, presently, there has been no up-to-date soil inventory data by capability classes, consisting of 8 classes, except for the "*Land Inventory*" document. This document was prepared separately for 67 provinces of Turkey, based on the reconnaissance soil survey results done between 1982 and 1984 by General Directorate of Rural Services (GDRS), formerly known as TOPRAKSU (1961–1984), which was dissolved in 2005. Capability class is the broadest category in the land capability classification system. Class codes from I to VIII are used to represent both irrigated and non-irrigated land capability classes (Atalay 2016). Capability class I to IV soils are considered as suitable for agricultural practices; thus, the remaining soil capability classes are unsuitable for cultivation. The higher the capability class number, the more severe are the limitations that restrict the choice of plants, or they require very careful management, or both.

Planar surface area of Turkey is 77.95 Mha, roughly the size of Zambia or Mozambique. It has been reported by several references, for example by DSI (2004) and MoD (2014) (the **M**inistry **o**f **D**evelopment), that total arable and irrigable lands are about 28.05 Mha and 25.85 Mha, respectively. However, GDRS/TOPRAKSU data given in MoD (2014) shows that a total of 26.51 Mha (34% of surface area) land is in the capability class I-IV, in turn, suitable for agricultural practices, i.e. arable land for field and horticultural crops. On the other hand, 0.2%, 13.9%, and 46.0% of the total land, respectively, consist of flat but not cultivable commercial pasture and forestry and degraded pasture and forestry. The remaining 5.9% area is never suitable for agricultural purposes.

Agricultural lands in Class I, Class II, Class III and Class IV are composed of 5.09 (6.5%), 6.71 (8.6%), 7.28 (9.4%) and 7.43 (9.5%) Mha, respectively. In this regard, it is clear from these data that 1.55 Mha area, which is not capable for agricultural purposes due to the severe limitations regarding soil, topography and drainage, have been under cultivation presently, indicating misuse of precious soil resources. Misuse of lands may trigger unexpected results under irrigated conditions. On the other hand, by considering the capability classes of soils, it may be claimed that at least 11.8 Mha land is readily available for irrigation by applying surface irrigation methods. Additionally, 14.71 Mha areas in Class III and Class IV may be irrigated by introducing modern techniques, namely drip and sprinkler irrigation, in order to ensure sustainability of soil use and crop production. Contrary to these interpretations, economically and technically irrigable area has tenaciously been declared as 8.5 Mha by DSI and others since 1960s. Kanber and Unlu (2008) highlighted that lands with the slope of less than 6% are about 13 Mha and are suitable for irrigation. On the other hand, arable lands with the soil depth greater than 90 cm are somewhere around 11 Mha (MoD 2014). In view of the fact that there have been enormous technological developments in irrigation methods, lands suitable for irrigation are much more than 8.5 Mha. Likewise, Yildiz (2007) argues that this figure needs updating as well. In fact, Kanber and Unlu (2008) concluded that the economically irrigable area in Turkey is not 8.5 Mha, but directly 25.85 Mha.

Lands suitable for irrigated agriculture should be determined by surveying soils for *"irrigated agriculture land classification"*, i.e. *SAT*. There are six irrigation suitability classes in SAT. Lands in the SAT Classes I-IV are considered as suitable for irrigation. Lands in Class V are provisionally non-irrigable under existing conditions because they need reclamation, to some extent, for irrigation. Therefore, project engineers make comprehensive evaluations based on the availability of water, reclamation requirements, and economic considerations during the irrigation project planning stage in order to decide whether Class V soils will be irrigated or not. If the decision is in favor of irrigation, then, based on the limitations, Class V soils are duly transferred into Class I to VI, otherwise into Class VI. Soils in Class VI are never suitable for irrigated agriculture, but for wildlife. Evaluation process of SAT explained is important in order to quantify the total amount of soil resources suitable for irrigation in a river basin. Concordantly, DSI has an "*Investigation, Planning and Allocations Department*", responsible for conducting surveys at reconnaissance, planning and detailed levels, and preparing master plan and feasibility studies on technically, economically, and environmentally viable projects for the purpose of the integrated land and water development in the major 25 river basins of Turkey. "*Soil and Drainage Branch Office*" of the Department is considered as the only technical office to realize soil surveys regarding SAT in Turkey, and it has carried out SAT soil surveys at the River Basin level. Updated SAT inventory data from the on-going works in the context of "River Basin Master Plans" are provided by DSI (Table 9.1). Those data are based on DSI SAT surveys and are the archived data of GDRS/TOPRAKSU. As seen from Table 9.1, the distribution of lands suitable for irrigation (SAT Class I-IV) is spatially uneven and less than the articulated irrigable land of 25.85 Mha. It is accepted that lands potentially available for agriculture have been in a gradually decreasing trend year by year due the misuses of precious soil resources. For example, migration from rural areas to cities has accelerated the loss of agricultural lands in SAT Class I-IV, as well as that of pastures and grasslands previously devoted to extensive animal husbandry near the outskirts of urban centers. Accordingly, the irrigation potential in Turkey has been reckoned with different figures in different sources, i.e. 8.5 Mha in DSI (2017a), 12.5 Mha in DSI (2017b), 25.85 Mha in DSI (2018a), and Kanber and Unlu (2008). As a matter of fact, lands to be irrigated by irrigation projects are technically determined by SAT classes (DSI 2004, 2016a). In this regard, by looking at the data given in Table 9.1, it is reasonable and justifiable to conclude that 22.6 Mha land, out of 28.05 Mha arable lands, is economically and technically irrigable under present conditions. In this regard, the afore-mentioned 8.5 Mha land may be considered as the land readily suitable or priority given areas for irrigated agriculture by introducing moderate irrigation methods, i.e. surface irrigation methods reinforced with some technical measures. Thereby, priority given areas, *i.e. "targeted irrigation area of 8.5 Mha"*, are aimed to be equipped with irrigation schemes by 2023 –the hundredth anniversary of the foundation of Turkish Republic.

Table 9.1 Planar surface area of river basins and potentially suitable areas for irrigated agriculture (Mha)

Catchment Name	No	Surface Area	Plain Area	Suitable for irrigation (SAT Class I-IV)	Actually irrigated (% of SAT Class I-IV)[a]	Provisionally non-irrigable (SAT Class V)	Potentially suitable for irrigated agriculture
Meric-Ergene	1	1.456	1.381	1.310	5.9	0.054	1.364
Marmara	2	2.410	0.319	0.226	25.5	0.011	0.236
Susurluk	3	2.240	0.529	0.389	28.6	0.044	0.433
North Aegean	4	1.000	0.904	0.245	20.0	0.017	0.262
Gediz	5	1.800	0.521	0.446	35.9	0.009	0.455
Kucuk Menderes	6	0.691	0.202	0.151	9.0	0.008	0.159
Buyuk Menderes	7	2.498	0.812	0.950	23.6	0.039	0.989
West Mediterranean	8	2.095	0.322	0.301	20.2	0.025	0.326
Antalya	9	1.958	0.444	0.379	34.4	0.030	0.410
Burdur Goller	10	0.637	0.220	0.192	21.8	0.046	0.238
Akarcay	11	0.761	0.324	0.263	8.4	0.036	0.299
Sakarya	12	5.816	2.075	1.986	7.2	0.131	2.118
West Black Sea	13	2.960	0.392	0.188	17.4	0.366	0.554
Yesilirmak	14	3.611	1.326	0.517	23.7	0.114	0.631
Kizilirmak	15	7.818	3.529	3.421	5.5	0.153	3.574
Konya Closed	16	5.385	2.702	2.257	10.5	0.269	2.526
East Mediterranean	17	2.205	0.212	0.243	19.7	0.325	0.567
Seyhan	18	2.045	0.485	0.315	59.3	0.048	0.363
Asi	19	0.780	0.442	0.233	13.0	0.045	0.277
Ceyhan	20	2.198	0.734	0.523	41.8	0.027	0.550
Euphrates sub-catchment	21	12.730	4.948	4.144	11.1	0.819	4.964
Tigris sub-catchment	21	5.761	1.951	1.252	11.0	0.025	1.276
East Black Sea	22	2.408	0.737	0.297	0.0	0.029	0.327
Coruh	25	1.987	0.158	0.250	6.4	0.990	1.240
Aras	24	2.755	0.811	1.708	5.8	0.262	1.970
Van Closed	25	1.941	0.229	0.416	14.8	0.344	0.759
Total area		**77.945**	**26.712**	**22.602**	–	**4.264**	**26.866**

Source: DSI master plans

[a]0.494 Mha groundwater irrigation areas, which have been managed by "*groundwater irrigation cooperatives*", are left out of the calculation (DSI 2016a).

9.2.2 *Land Consolidation Associated with Irrigation Developments and Legal Status*

By implementing land consolidation (LC), expropriation costs to be incurred may be eliminated in irrigation projects (DSI 2016a). More importantly, LC projects help service providers decrease construction as well as operation, maintenance and repairing costs, to some extent, in irrigation schemes (FAO 2003; van Dijk 2007; DSI 2017b). Based on the information obtained from the Turkish Statistical Institute (TUIK) and Ministry of Food, Agriculture and Livestock (MoFAL), there have been ongoing defects in land ownership in Turkey, and so Turkish agriculture has experienced a number of problems that directly prevent farmers from introducing modern production inputs (TUIK 2004). Furthermore, such inherited problems hinder achieving targeted goals in irrigation projects. Fragmented parcels of agricultural enterprises, the very small parcel sizes, and uneven distribution of landownership are only a few of the inherited problems regarding irrigated agriculture (MoFAL 2013). For example, the average parcel number per enterprise is 5.9 in 2002, but it went up 6.9 in 2011, whereas the average size of enterprises was 7.7 ha in 1950. While the average size of an agricultural enterprise is some 16.0 ha in the European Union countries (DSI 2016a), it is only 6.1 ha -well below the economical operational farm size- in Turkey (MoFAL 2013). Based on the agricultural census results, the number of parcels per enterprise, on average, was over 6 in 2001 (TUIK 2004). More importantly, average parcel size is less than 1.0 ha in Turkey whereas it is between 1.8 and 4.0 ha in EU countries. General characteristics of agricultural enterprises by farm sizes are given in Fig. 9.3.

In 2001, the total number of agricultural enterprises was 3.02 million in Turkey. As seen from Fig. 9.3, agricultural enterprises with the size of ≤4.9 ha constitute 65% of the total enterprises. Contrarily, 65% of enterprises cultivates only 21.4% of

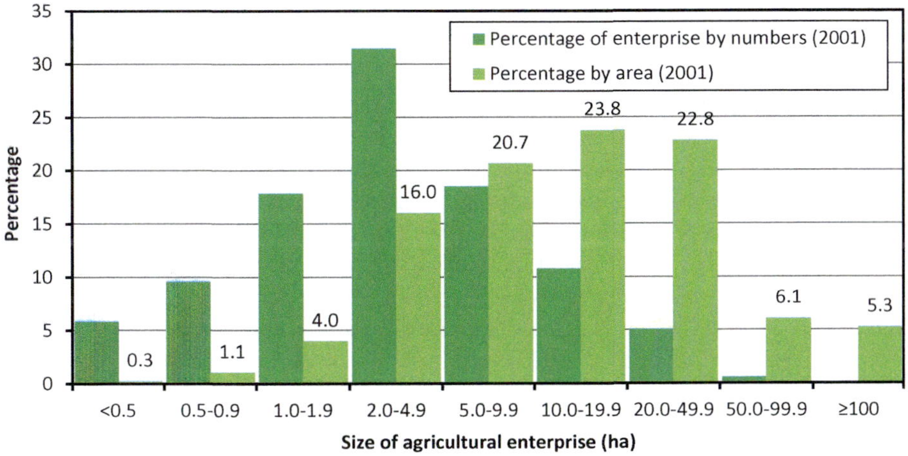

Fig. 9.3 Percentage distribution of agricultural enterprises numbers and their landownerships by farm sizes

3.02 Mha land. On the other hand, only 16% of area belongs to the nearly one third of the enterprises (≈32%) with a farm size of 2.0–4.9 ha area. It is more important to emphasize that very few of the enterprises (5.1%) with a farm size of 20.0–49.9 ha area owns almost a quarter of arable lands (23%). Far worse than this, farmers with greater than 20 ha area are composed of only 6% of total enterprises but cultivate 34.2% of the total cultivated area, indicating an uneven and inequitable land property distribution among the farmers. Inequitable land property distribution is an ill-conditioned and inherited structural defect in land tenure in Turkey. As explained by the statistics above, it is primarily characterized by small holding size, intense land fragmentation, mixed land tenure, such as land held in undivided form and multiple ownerships, lack of farm loads, and irregular shaped plots.

In order to enhance farm productivity and increase irrigation ratios, as well as the overall irrigation efficiency in the irrigation schemes, implementation of land consolidation (LC) has been a remedy in areas under rain-fed or irrigation agriculture, or both. Satana et al. (2017) points out that one of the key structural constraints in the agricultural sector in Turkey is directly linked with land fragmentation, and that the sole remedy for land fragmentation is reversing the process by implementing land consolidation (LC) projects. Apart from the huge economic and social benefits of LC under rain-fed farming, economic and social benefits of LC may be amplified furthermore when implemented in irrigated farming conditions due to the double effect of significant cost savings during infrastructural development (FAO 2003) and enhanced operational efficiency on the farm (MoD 2014). On the other hand, there are also important side benefits of land defragmentation (MoFAL 2013; MoD 2014). For example, Satana et al. (2017) pointed out that defragmented lands improve tenure security and increase agricultural land values.

In order to facilitate modern agricultural practices and irrigation, land consolidation is traditionally used as an instrument for improvement of the business structure of farms through the consolidation of fragmented agricultural parcels (van Dijk 2007; Satana et al. 2017). The need for a better structure has evolved out of the provision of food security for society at low cost and a better cost-benefit ratio for individual farmers. When, in the past, it became obvious that the free land market could not solve the consolidation of fragmented lands, many countries embarked on the development of land consolidation regulations. These regulations provided for mechanisms to consolidate land parcels within a certain area in a coordinated way.

The LC works in Turkey are implemented in compliance with the provisions of two different laws: a) *"Agricultural Reform Law on Land Rearrangement in Irrigated Areas"* Law No 3083, invoked in 1984, b) *"Soil Conservation and Land Use Law"* Law No 5403, promulgated in 2005. Additionally, it is important to highlight that *"Regulation on Implementation of Agricultural Reform Law Regulating Land Regulation in Irrigation Areas"*, and *"Land Consolidation Technical Directive"* and *"Land Consolidation Bylaw"* were issued by virtue of the Law No 3083. According to the laws, bylaws, decrees and regulations regarding LC works, *Ministry of Food, Agriculture and Livestock (MoFAL)* is the sole responsible agency for all LC implementations in Turkey.

General Directorate of Agricultural Reform (GDAR), acting under the auspices of (MoFAL), and the national irrigation agency *State Hydraulic Works (DSI),* acting under the auspices of *Ministry of Forestry and Water (MoFW)* are the two primary state institutions, authorized by laws, to implement LC projects in Turkey. GDAR has implemented LC projects on rain-fed agriculture as well as on irrigated agriculture; while DSI, under GDAR's oversight, has been authorized to undertake LC projects on irrigated lands in conjunction with irrigation development projects. Until the end of 2017, both institutions realized LC projects over Turkey. As of 2018, DSI is authorized as the sole lead state institute to implement LC projects in Turkey.

Land suited for LC is estimated at some 14 million ha (DSI 2016a; Satana et al. 2017). In Turkey, the first LC project was implemented by the GDRS in Konya-Cumra irrigation project area in 1961 (Satana et al. 2017). Efforts for expanding LC implementation have reached a certain level since then. In accordance with the laws, DSI initiated the first LC project in irrigation areas in 2007. Data provided by DSI (2016a) and Satana et al. (2017) indicate that DSI has implemented land consolidation projects in 0.54 Mha of irrigated land in 32 provinces, presently. LC projects are completed in 5.0 Mha areas, of which 3.12 Mha area is in rain-fed farm conditions and the remaining 1.88 Mha area is in irrigable lands. The remaining potentially suitable area for LC project implementation is about 9.0 Mha, of which 5.87 Mha is irrigable. According to the strategic plan of DSI, in the coming years, more efforts will be spent to complete LC projects in irrigable lands in order to eliminate expropriation costs of irrigation projects and to make cost savings in the O&M stages of irrigation schemes. In this regard, it is targeted by DSI that LC projects will have been implemented in 5.88 Mha area by 2023. Consequently, consolidation works done so far fell proportionately well behind the irrigation projects realized by DSI. Therefore, they need some time and finance to be realized completely.

LC project implementation is an important remedy to achieve the expected benefit from irrigation projects as well as from irrigated agriculture. The Turkish experience shows that if LC can be implemented at the designated farming areas prior to the construction of irrigation infrastructures, approximately 40% savings could be provided from expropriation payments, construction and operation expenditures. Furthermore, a proper implementation of LC projects will help farmers to get savings in money, time and labor in agricultural activities. DSI (2004) and Satana et al. (2017) point out that, after the completion of LC projects, irrigation ratios have reached 90% and irrigation efficiencies have attained values as high as 85%, which is over the acceptable levels in large-scale irrigation schemes.

In order to prevent the defragmented agricultural lands from re-fragmentation, a number of technical and legal measures have been taken so far. In this regard, the necessary amendments were made in the laws concerning the issue in order to minimize fragmenting of agricultural parcels. The progress achieved so far is grouped under two categories: a) *"indivisible parcel sizes"* were determined; b) amendments were made in the *Turkish Civil Law* (Law No 4721, came into force in 1926 and amended in 2001) in order to prohibit sales of agricultural parcels less than *indivisible parcel sizes*. Therefore, marginal agricultural lands less than 2.0 ha, cultivated agricultural lands below 0.5 ha and the greenhouse areas below 0.3 ha were considered as *the indivisible parcel sizes* by law in 2007. Over and above the determination of *indivisible parcel sizes*, the *"Law on Soil Protection and Land Use and Revision on Turkish Civil Law"* was adopted in order to prevent the fragmentation of agricultural lands in 2014. FAO (2003) addresses that effectiveness of those measures will depend to a great extent on durability of the consolidation policy. Then, LC implementation will help rural people make the Turkish agriculture in the region more competitive while promoting rural development.

9.3 Institutional Framework in Irrigation Developments and Irrigation Managing Organizations

9.3.1 Institutional Framework in Irrigation Developments

There are two aspects of irrigation in Turkey: *(a) Irrigation developments, (b) Operation and maintenance (O&M) of irrigation facilities.* DSI, Provincial Special Administrations (PSAs), defunct TOPRAKSU and General Directorate of Rural Services (GDRS) are the only governmental institutions in Turkey for developing irrigation schemes of different scales, i.e. small-scale irrigation command areas (from 7 ha to 2600 ha (DSI 2016a, b)) irrigated by small earthen dams and large-scale irrigation schemes of 100,000 ha or over, irrigated by large-dams (DSI 2018a), such as Ataturk, Bozkir, Demirkopru, Menzelet, Seyhan, Aslantas etc. Government institutions have carried out small-, medium- and large-scale irrigation developments in Turkey since the foundation of the Republic. In this regard, there are only two principal government/state organizations concerned with irrigation developments (Cakmak et al. 2010; DSI 2016a): (a) *General Directorate of State Hydraulic Works (DSI)*, (b) *Provincial Special Administrations (PSAs, formerly General Directorate for Soil and Water, i.e. TOPRAKSU and General Directorate of Rural Services (GDRS))*.

9.3.1.1 General Directorate of State Hydraulic Works (DSI)

DSI was established by the Law No 6200 in 1954. Its headquarter is located in Ankara, the capital city of Turkey, and there are 26 regional directorates distributed over Turkey. It has been accepted by authorities that establishment of DSI is the milestone of systematic water resource developments as well as the start of institutional irrigation management in Turkey (Bayazit and Avci 1997). Its mandate is to develop water and land resources in Turkey. Therefore, the country's overall water resources surveying, planning, construction, execution and operation is under the responsibility of DSI. As such, DSI has been active in four sectors: Agriculture, hydroelectric energy, environment, and service. In fact, DSI is the main executive irrigation agency of Turkey. Additionally, DSI is authorized by the law amended to execute land consolidation and on-farm development (land levelling, subsurface drainage and field access roads) projects in the areas equipped with irrigation facilities by DSI. Apart from irrigation developments, it is responsible for flood control, drainage, hydropower development, and supplying water to cities, too. It has further responsibilities related to river basin planning, water quality monitoring and improvement, outdoor recreation, basic studies on stream gaging and soils classification, and research on water-related structural design and construction materials. DSI centralizes most of the state functions involved in planning and developing large-scale water resources schemes (DSI 2016a). However, for the sake of decentralization efforts of the government, it has 26 Regional Directorates over Turkey. For all that, handover initiatives of irrigation schemes to the irrigation managing bodies have been a part of decentralization of services by DSI.

As per Law No 6200, DSI has the responsibility to make irrigated agriculture land classification (SAT) works and construct irrigation projects, and has a mandate to operate as well as to manage irrigation schemes. Since its establishment in 1954 to the end of 1993, the majority of irrigation schemes were operated and managed by DSI branch offices. However, due to financial and political reasons, there had been a paradigm shift in irrigation management policy in the early 1990s. In 1994, services related to operation, maintenance and management of irrigation schemes were started to be transferred to irrigation managing organizations, such as irrigation cooperatives, village legal entities, municipalities, and water user associations (WUAs) by adopting a so-called fast transfer program. Services regarding O&M of irrigation schemes may be taken over to the beneficiaries, but the ownership of the schemes, i.e. the proprietary right of the facilities, remain with the State/DSI. Furthermore, DSI operates the dams and regulators in order to ensure delivery of irrigation water to the irrigation managing bodies.

On the other hand, some small scale irrigation projects on-demand by universities or state farms, such as state hatcheries and the General Directorate of Agricultural Enterprises (TIGEM) acting under the auspices of MoFAL, have been constructed by DSI in exchange of money. DSI takes over proprietary rights as well as O&M services of these small scale irrigations, hereinafter referred to as *"other agency irrigations, OAIs"*, to the demand owners. As of 2016, DSI developed some 16,992 ha area on 29 irrigation units as part of OAIs.

The Law No 6200 orders that the whole operation and maintenance expenditures incurred by DSI are to be paid back by beneficiaries as operation and maintenance charges. Additionally, as per the Law No 6200, total construction costs incurred in the construction of irrigation facilities shall be recovered from beneficiaries after the completion of the irrigation projects. Therefore, a farmer, benefiting from DSI constructed irrigation schemes, has to make payments on management services and payback. Accordingly, financial evaluations show that the recovery payment period of irrigation schemes constructed by DSI, on the average, is 11 years in irrigation schemes.

9.3.1.2 Provincial Special Administrations (PSAs)

In order to ensure rural development, the General Directorate for Soil and Water (TOPRAKSU) was established, under the auspices of Ministry of Rural Affairs and

Cooperatives, in 1960 (DSI 2016a; Satana et al. 2017). After having dissolved TOPRAKSU in 1984, the General Directorate of Rural Services (GDRS) was established under the Ministry of Agriculture, Forestry and Rural Affairs. GDRS was active until 2005. Then, the GDRS was reconstituted in 2005, and the Provincial Special Administrations (PSAs) concurrently substituted for the GDRS (MoD 2014; DSI 2017a). The PSAs, under the Governor's Office, have taken over the responsibilities of the TOPRAKSU and GDRS. Defunct TOPRAKSU and GDRS used to be active at the village level (Cakmak et al. 2010). TOPRAKSU was responsible for soil surveys, developing small-scale groundwater resources for irrigation, developing surface water sources with flows of less than 500 liters per second for irrigation (Svendsen 2001), construction of small earthen dams for irrigation or domestic purposes; and on-farm development tasks including farm irrigation facilities, construction of subsurface drainage in agricultural fields and the construction of rural roads and village water supplies. Because the GDRS/TOPRAKSU did not have an operation and maintenance capacity, minor irrigation schemes constructed by the TOPRAKSU/GDRS were directly transferred to farmers' cooperatives, a group of farmers, village legal entities or local governments (municipalities) to manage, upon completion. PSAs services have been liable to the principles of TOPRAKSU/GDRS. Therefore, there has been no payback schedule for the farmers benefitting from PSAs constructed irrigation schemes.

Irrigation developments by TOPRAKSU, GDRS and PSAs, hereinafter referred to as *"PSAs developed irrigations"*, were not documented accordingly. Therefore, there have been some uncertainties on the amount of PSAs developed irrigations due to the fact that these irrigation practices have been somewhat out of control. Nonetheless, it is estimated by DSI (2017b) that PSAs developed irrigations are about 1.29 Mha in Turkey at present. Municipalities and offices of PSAs, in close collaboration with TUIK (State Institute of Statistics, http://www.tuik.gov.tr/), are expected to update this figure in order to account for either a reliable or representative inventory of PSAs developed irrigations.

It should be kept in mind that mandates of Provincial Special Administrations offices, due to the course of law, have to be taken over to the metropolitan municipalities if established. Therefore, irrigation developments or services may be provided by PSAs acting under the authority of either metropolitan municipalities or the office of the governor in the provinces.

9.3.2 Irrigation Managing Organizations: Operation and Maintenance (O&M)

Based on the operation and maintenance (O&M) characteristics of irrigation facilities, irrigation managing organizations in Turkey may be grouped under four categories: a) *Private Irrigations*, b) *Local Authority Irrigations*, c) *Irrigation Cooperatives and Water User Associations (WUAs)*, d) *Public Irrigation Managements*.

9.3.2.1 Private Irrigations (PIs)

A farmer or a group of farmers can irrigate their own fields by utilizing either surface or groundwater (GW) resources, or both. In order to operate GW wells, farmers have to get a user's license from DSI, pursuant to Law No 6200 and Groundwater Law No 167 enacted on 16th December 1960, if the well depth is more than 10 m. Farmers must utilize their own private money to develop this type of irrigation, and O&M is under their own responsibility. This type of irrigation management is indigenous and known as "*private*" or "*community based irrigations*", i.e. *PIs*. PIs are village-based schemes and have been viable for years. Hence, the head of a village acts as the primary coordinator in operation and maintenance activities as well as in repairing works. In irrigation water management terminology, PIs are expressed as "*small-scale privately owned irrigation schemes*". Although PIs contribute much to the country's economy, it could be postulated that there has been no regular and reliable record on the acreage of PIs. It is reported by DSI (2016a, 2017a, b) that PIs areas are about 1.0 Mha. However, it is useful to note that this figure has not been updated since 1970s due to lack of data. This figure supports this postulation, suggesting that a comprehensive survey should be initiated by the provincial offices of the Ministry of Agriculture in order to get a representative, reliable and updated PIs inventory of Turkey.

9.3.2.2 Local Authority Irrigations (LAIs)

Some of the irrigation projects developed by PSAs and DSI may be taken over to the village legal entities or local governments (municipalities) to manage. Those entities, i.e. LAIs, take the responsibility of O&M services of the irrigation schemes as well as the necessary repairing works. All operation, management and repairing expenses incurred are paid by the managing entities, i.e. LAIs. The source of budget for LAIs consists of the "*collected water fees*" from the farmers. This management type is observed because PSAs have no O&M capacity, and heads of local governments and villages demand the construction of irrigation facilities from DSI by ensuring O&M services for the farmers. No recorded data is available on how much area/acreage have been managed by LAIs due to the fact that PSAs have engaged in informal transfers (Svendsen 2001) to the village legal entities and/or municipalities.

9.3.2.3 Irrigation Cooperatives (ICs) and Water User Associations (WUAs)

Irrigation Cooperatives (ICs) and Water User Associations (WUAs) are commonly and widely adopted irrigation-managing organizations in Turkey. Svendsen and Nott (2000) defined the term *Water Users Associations (WUAs)* as "*a local-level organization based on the active involvement of water users who come together for the purpose of organizing and practicing irrigation system operation and maintenance"*. However, WUAs differ from ICs and LAIs, i.e. village legal entities and municipalities, in that the WUAs is a new paradigm O&M institution with a public legal personality that is structurally different from any existing government body in Turkey. In fact, ICs and WUAs are formal water user organizations (WUOs) established for the purposes of O&M services in the irrigation schemes regardless of whether the water is provided from surface water or groundwater resources. In this regard, ICs and WUAs act as recipients of O&M responsibility in the transfer/takeover of irrigation facilities constructed by the Government institutions. Irrigators and/or farmers involve directly in the O&M services provided by organizations, and participate in decision-making processes to take decisions on their own. Therefore, ICs and WUAs should be considered as the paradigm for the participatory irrigation management (PIM). Most of the industrialized countries such as USA, Japan, Australia and Spain adopted PIM policy a long time ago due to fiscal necessities. In Turkey, WUAs Law No 6172 encourages legally the evolution of PIM institutions in the rural areas in particular.

Irrigation Cooperatives (ICs) were established according to the Cooperatives Law No 1163, issued in 1969. If irrigation water is provided from surface water resources, they are named as surface water irrigation cooperatives (SWICs), otherwise groundwater irrigation cooperatives (GWICs).

Until the end of 2004, WUAs, formerly known as irrigation associations (IAs), were established according to the 47–48 articles of Village Law No 442 issued in 1924, 133–138 articles of the Provincial Administration Law No 5442 issued in 1949, and Municipality Law No 1580 issued in 1930. By amending the Law on Local Administrative Unions (Law No 5355), a regulation about the foundation of unions was put into force in 2005. In doing so, the legal gap regarding irrigation management has been filled indirectly. In practice, due to legal shortcomings in the Law No 5355, a number of problems were experienced in the course of taking over O&M services to WUAs until 2011. Afterwards, the intensive studies done by DSI technical staff from 2008 to 2011 resulted in Irrigation Unions Law No 6172 in 2011. Today, with some legal amendments, WUAs provide farmers with O&M services in irrigation schemes within the scope of this law. By law, WUAs assume the responsibility for O&M services/rights of the irrigation facilities; but they are not able to take the ownership rights of the irrigation facilities.

9.3.2.4 Public Irrigation Management (DSI)

By the Law No 6200, DSI has a task to provide farmers with O&M services in the DSI constructed irrigation facilities in the expense of the participatory management concept. This type of irrigation management has been traditional and known as publicly managed irrigation management. Accordingly, until 1993, irrigation management by the Government, i.e. DSI, was the preferred model. As of 1993, DSI has changed the policy of O&M services in parallel with the cyclical changes in the world, and started an accelerated program of transferring O&M services responsibility for publicly-managed large-scale irrigation schemes to ICs, WUAs, municipalities, village legal entities, universities and/or TIGEM-like state institutions.

9.3.3 *Transfer of O&M Services Responsibility to Local Managing Entities*

National budgetary crisis, high inflation rates -minimum 21% and maximum 113% between 1974 and 1993- rapid growth in the wage costs of unionized labor, overtime payments, and financial needs for new investments to expand irrigation areas etc. in the early 1990s gave a fresh impetus to DSI to accelerate the transfer of O&M services to the WUAs. Hence, the budgetary crisis led to a squeeze on financial allocations to DSI in general and to the O&M Department in particular. Furthermore, financial constraints led to the curtailment of overtime work by DSI field staff. Financial curtailments caused to accelerate deterioration of irrigation schemes because it was impossible to recover all the incurred costs for operation, maintenance and repairing of irrigation facilities, augmenting every year. Hence, there was a decline in the quality of irrigation services, indicating dissatisfaction with the farmers. In order to solve the problem, Turkish authorities tried to find alternative irrigation management models to transfer O&M responsibilities and/or services of irrigation facilities to local managing organizations. However, it was not an easy task, but quite challenging.

By financial support of the World Bank, a *"Drainage and On-Farm Development Project"* was initiated in Turkey in the early 1990s (Svendsen 2001). During the execution of this project, the World Bank superintendents participated in discussions on the financial crisis and poor management issues facing the irrigation management sector in Turkey. DSI authorities were encouraged to explore new ways either to put O&M financing on a sounder footing or to transfer O&M services to local managing entities. In this context, Turkey's previous experiences on irrigation group works dating back to 1940s and the transfer of O&M responsibilities to irrigation cooperatives and village legal entities as well as irrigation associations were seen as a valuable precedent. It was clear from the information given in Svendsen and Murray-Rust (2001) that the funds of the *"Drainage and On-Farm Development Project"* in Turkey were partially made available to DSI authorities to broaden their experiences abroad. Then, a number of study tours were organized to other countries of experience on the devolution of authority (DSI 2016a). As an outturn of those actions, a remarkable shift has been observed in the vision of DSI staff and in the capacity building efforts of DSI regarding the future of irrigation management in Turkey. Before those activities, the transfer of irrigation services to local authorities such as irrigation cooperatives, village legal entities, municipalities or water user associations were not at the expected pace until 1993 (Uskay 1999). As of 1994, the transfer of O&M services to irrigation managing organizations has gained remarkable acceleration in Turkey through implementing a fast transfer program. The core of this fast transfer program is the adaptation of participatory irrigation management (PIM) concept by introducing the Water User Associations (WUAs) model. In doing so, Turkey has gained valuable technical expertise in exemplary takeover of government managed irrigation systems. Even so, Svendsen (2001) addresses that the foundation of Water User Associations (WUAs) has been on a downward reaching link between DSI and local administrations, rather than through the bottom-up organization of village-level associations of irrigators. Therefore, WUAs are local-level and participatory irrigation management organizations, consisting of farmers/irrigators who get together in order to organize and practice operation and maintenance (O&M) of irrigation schemes. On that sense, today, the WUAs model should be considered as a reasonable solution and a remedy for sustainable management of irrigation schemes.

Nonetheless, it is worth mentioning that the Turkish irrigation management transfer program, including the WUAs model, developed by technical staff of DSI, gained worldwide attention as well as celebrity regarding its speed and effectiveness. At the end of 2000, 1.62 Mha (72% of DSI-constructed irrigation areas) irrigation schemes were handed over to WUAs, whereas the transferred area was 2.18 Mha in 2010 (80% of total). As a result, as of 2017, the existing large-scale irrigation area operated by DSI consists of only 11% of the total (3.37 Mha) area equipped with irrigation and drainage facilities by DSI, and 73% of irrigations has been handed over to WUAs. Numerous studies have been undertaken in order to manifest the concrete implications of handover activities and devolution of authority regarding irrigation schemes in Turkey. Concordantly, a number of performance evaluations of WUAs and their level of success were done by Mengu and Akkuzu (2010), Murray-Rust and Svendsen (2001), Svendsen (2001), Nalbantoglu and Cakmak (2007) and the others to figure out whether the transfer reached its objectives. By the transfer of O&M services to WUAs, the financial burden of the State, related to the costs of the irrigation management, have fallen sharply. For example, the number of maintenance and operation staff of DSI in the Aegean region was decreased by 65% in 2006. On the other hand, there are important increases in the quality of services received by the farmers/irrigators and a remarkable decline in the M&O costs of WUAs in the handover irrigation schemes, in comparison with the State managed irrigation schemes. Research results indicate clearly that the targeted or desired benefits from the transfer of O&M services to irrigation managing bodies have not been sufficiently achieved, at the national level, so far due to the fact that there have been legal, technical, physical, financial, social etc. shortcomings of the takeover activity. Infrastructure regarding those issues needs strengthening well enough by: a) Developing strategies related to the O&M, b) Transferring O&M know-how to the managing bodies, c) Rehabilitating and rejuvenating urgently the dilapidated or aged irrigation facilities, d) Establishing the principles of financial framework regarding O&M services, and e) Facilitating social, political and legal framework etc.

9.4 Progress Achieved in Irrigation Developments: Areas and Water Consumption

Hydraulic works regarding irrigation were constructed by Foundations in the Ottoman period. In Anatolia, the first modern irrigation facility, i.e. Cumra Irrigation and Drainage Project, was therefore constructed between 1908 and 1914. Publicly funded irrigation areas have gradually expanded since then. Expansion of irrigated lands has gained momentum since the establishment of DSI in 1954 and TOPRAKSU in 1960s. According to the official records, publicly managed irrigated areas were about 122,000 ha in the beginning of 1950s. As of the end of 2017, a total of 6.57 Mha gross area (DSI 2018a) has been equipped with irrigation infrastructures, i.e. irrigation and drainage canals, access roads, flow gauging facilities, flood protection levees, and hydraulic check structures etc., in Turkey. Of this, 65% (4.28 Mha) was developed by the national irrigation agency (DSI), mainly under large schemes, 20% (1.29 Mha) by PSAs (formerly General Directorate of Rural Service (GDRS) and TOPRAKSU), mainly under medium-to-small schemes, and 15% (1.00 Mha) by individual famers as small-scale privately owned irrigation schemes, viz. PIs.

9.4.1 Areal Distribution of Irrigation Schemes Based on Managing Entities

As explained in this Chapter, there have been six different undertaking organizations to manage irrigation schemes. Figure 9.4 includes those organizations and shows their percentage shares, based on the total gross area in O&M services. DSI (2017c) points out that the total gross area irrigated with surface water (SW) and groundwater (GW) resources is about 6.37 Mha, of which 4.07 Mha (net 3.08 Mha) was constructed by DSI. Of this, 0.35% (net 16,992 ha) was constructed by DSI in exchange of money for TIGEM and/or universities, and operated by those institutes. On the one hand, DSI operated areas constituted only 4.37% of the total area; on the other hand, areas transferred to WUAs and groundwater-irrigated areas operated by groundwater irrigation cooperatives (GWICs) constituted almost 50% the total irrigated lands (net 2.36 Mha), and 10% (net 496,633 ha), respectively, in Turkey.

By 2023, the targeted irrigation area of 8.5 Mha will have been equipped with irrigation infrastructures by the national irrigation agency of DSI. Put another way, almost three quarters of the targeted irrigation area is irrigated presently. The remaining 2.14 Mha land readily suitable for irrigation needs investing in the coming years in order to attain the goals of 2023.

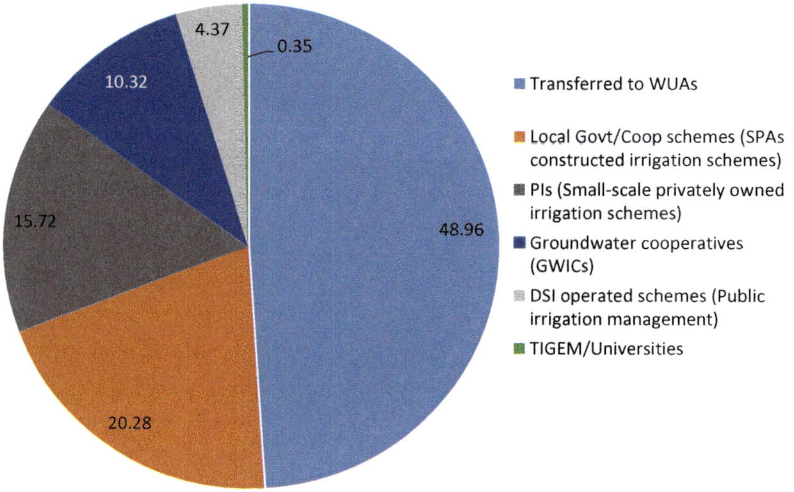

Fig. 9.4 As of the end of 2016, percentage shares of gross 6.36 Mha total irrigated area by managing entities

9.4.2 State Constructed and Operated Irrigation Schemes: Process and Outcomes

Irrigable area inventories and total irrigation developments are usually provided by the national irrigation agency and the others as "*gross irrigation area*", which is the total area supplied with irrigation facilities and commonly documented in the units of hectares (ha) in DSI (2004, 2016a). However, it has become a tradition that "*net irrigation areas*" -which are the portions of the gross irrigation area remaining after deducting land under service roads and operating buildings, water bodies, houses, settlements and so on- have been used in takeover studies, monitoring and evaluation processes of irrigation schemes. Although there are different ways to obtain net irrigation area in an irrigation scheme, DSI simply calculates the net irrigation area by multiplying the gross irrigation area by a factor of 0.864 (DSI 2004; Svendsen and Nott 2000). Figure 9.5 includes net areas opened for irrigation by DSI, and DSI and WUAs operated irrigation acreages, with irrigation ratios by the years.

In early 1960, net irrigation areas constructed by DSI was 184,750 ha. Only 14% of this area was transferred to the beneficiaries, i.e. WUAs. O&M services of the remaining areas were provided by the national irrigation agency. Until the end of 1993, irrigated areas attained totally 1.44 Mha, of which 80% was operated by DSI, and the remaining minor area (only 20%) was transferred to WUAs, ICs, and village legal entities or municipalities. As a matter of financial necessity, the activated transfer program of the Turkish government has caused a sharp decline in DSI-operated irrigation schemes in favor of WUAs operated areas. In 1995, the total net irrigated area by DSI facilities attained 1.90 Mha, of which almost 52% (0.98 Mha area) and 15% (0.28 Mha land) were taken over to WUAs and GWICs,

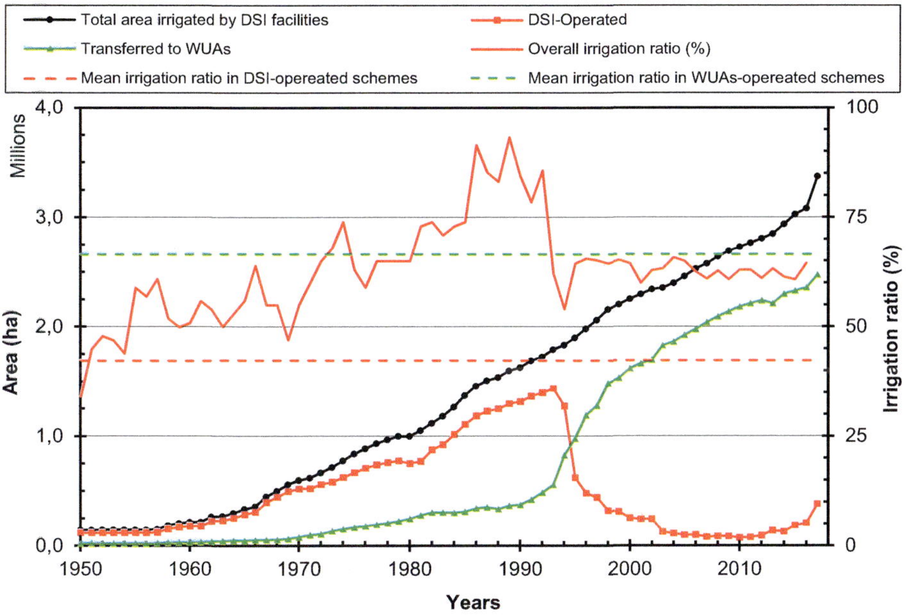

Fig. 9.5 Accumulated areas of DSI constructed irrigations, pace of O&M Services transfer to WUAs, and irrigation ratios (Data source: Yearly activity reports of DSI, archived files, and fact and figures released by DSI since 1960s)

respectively. As of the end of 2017, the total of irrigated areas by DSI facilities reached 3.37 Mha (Fig. 9.5). However, only 11% and 15% of this land were operated by DSI and GWICs, respectively, in order to provide the farmers with O&M services.

Irrigated agriculture is a series of dynamic processes. Therefore, it needs paying attention to O&M services. It is considered that it is not enough to provide only irrigation infrastructures to the farmers. On-farm development investments should be accordingly carried out in order to achieve the targeted goal of irrigation developments. Ignorance may result in low irrigation efficiencies as well as irrigation ratios. As seen in Table 9.1, the actually irrigated areas are much less than the potential command area. Therefore, irrigation ratios, which represent the area irrigated over the command area, are expected to be less than anticipated ones. As seen in Fig. 9.5, irrigation ratios were about 45% in the early 1950s. Between 1985 and 1993, the overall irrigation ratios were over 75% in irrigation schemes without regarding managing entities. It is important to highlight that the mean irrigation ratio is about 42% in DSI operated irrigation schemes, and 66% in WUAs operated irrigation schemes. Data given in DSI (2017a) indicates that irrigation ratios are as low as 30% in some DSI-operated irrigation schemes, which is considered as the threshold or critical value by DSI.

On the other hand, irrigation efficiencies are not at the expected levels in irrigation schemes of Turkey. The overall irrigation efficiency is as low as 36% in

DSI- and WUAs-operated facilities. On the eve of the takeover program in 1992, the overall irrigation efficiency was 43% with 1308 mm actual water usage in DSI operated irrigation schemes. A drastic change was observed in irrigation efficiencies at the end of 1998: 29% with 1308 mm actual water use in DSI operated irrigation schemes and 43% with 1096 mm actual water use in WUAs operated irrigation schemes. In 2000, it was 33% with 1194 mm actual water use and 42% with 1085 mm actual water use in DSI and WUAs operated irrigation schemes, respectively. There is no doubt that the more irrigation water was used in DSI operated irrigation schemes, the less irrigation efficiency was obtained in the irrigation facilities. Put another way, irrigation efficiencies are always higher in WUAs irrigated and/or operated areas compared to DSI operated schemes. As for actual water uses, the reverse is always correct, indicating that WUAs have been using less irrigation water in favor of high irrigation efficiencies. In 2014, Turkey has experienced a stern drought event. Therefore, the actual water use in DSI and WUAs operated areas was 956 mm with the highest irrigation efficiency (51%). On the other hand, in 2016, the actual water use in agricultural areas operated by DSI and WUAs was some 1133 mm and the irrigation efficiency decreased to 46%.

Apart from WUAs, irrigation cooperatives have been active in DSI constructed irrigation schemes to provide O&M services for farmers. DSI (2017b) indicates that, as of 2017, the irrigation units operated by SWICs and GWICs are 257 and 1458, respectively. SWICs have been operating 130,105 ha irrigation area since 2017. During the peak irrigation season, surface waters may not be able to provide sufficient irrigation water to the farmers in some specific regions. If so, the irrigation scheme is reinforced, based on the availability of GW sources, by harnessing GW resources. The areal difference between "*all GW irrigations*" and "*GW cooperatives*" in Fig. 9.6 accounts for the areas of GW-reinforced surface irrigated schemes.

GW irrigated areas shared 17% of the total irrigated area (1.00 Mha) in 1980. Afterwards, the share increased up to 20% in 2007; then decreased to 17%, attaining the total area of 567,550 ha in 2017. However, areas operated by GWICs have gradually increased in parallel with the increases in all GW irrigated areas. Until 2017, there had been 1457 GWICs operating on 496,633 ha area, constructed by DSI in Turkey. The asymptotic convergence of GW irrigated areas less than 0.60 Mha in Fig. 9.6 may help us postulate that areas to be irrigated by GW resources is about to attain the marginal limits in Turkey. Records on GW allocations to the agricultural sector justify this postulation.

Low irrigation ratios, as well as low irrigation efficiencies, are a precursor of poor O&M services, legal gaps regarding irrigated agriculture, structural defects in agricultural enterprises, dilapidated or aged irrigation schemes and the others. In this regard, DSI (2016a) justifies that, as of 2012, DSI has developed 826 irrigation facilities, of which 19% is over 40-year old, 34% is 21–39, 26% is 11–21, 9% is 6–10, and 12% is 1–5 years old. Considering this fact, the state irrigation agency aims at materializing rehabilitation and rejuvenation projects in the coming years. Consequently, it is targeted by DSI that modern irrigation systems will have been constructed in irrigation projects, and all traditional open canals will have been converted into pipeline-closed canals by the year 2023. In doing so, it will be

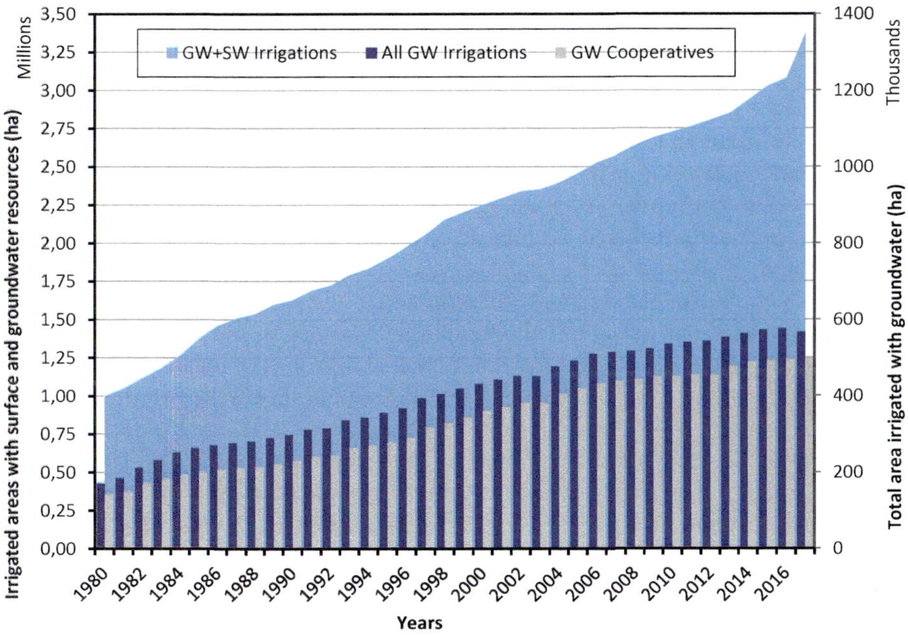

Fig. 9.6 Total irrigated areas by surface and groundwater resources, total groundwater irrigations, and areas operated by *Groundwater Cooperatives* by years

possible to make use of modern irrigation systems/methods such as drip and sprinkler, partial root drying irrigation (PRD), conventional deficit irrigation methods, and the others. Furthermore, on-farm water management practices, improved water distribution by implementing rehabilitation and rejuvenation projects, and modern infrastructures can reduce avoidable water losses in irrigation schemes. Applying efficient irrigation strategies such as the partial root drying, i.e. *PRD* (Kirda et al. 2007), deficit irrigation, subsurface drip and sprinkler irrigation techniques at the farm/plot level will further increase water use efficiency for the sake of saving irrigation water at the district and farm level. The shift from conventional irrigation techniques to more efficient irrigation strategies in irrigation schemes will be a remedy to combat water scarcity problems in the near future.

Regarding the existing water distribution systems in Turkey, they are designed to provide measuring devices at the heads of the main and most of the secondary and tertiary canals. However, because irrigation systems constitute of open canals, turnouts and measuring devices are not accordingly provided for individual farms. Although measuring devices are essential for equitable water distribution among water users and for the control of water application, they are somehow excluded from the water distribution system. Non-existence of these measuring structures/ devices has provoked farmers to use excess irrigation water. Additionally, there are only a few large scale irrigation schemes in Turkey -for example, Kayacik irrigation in Gaziantep, Yaylak and Bozova pumping irrigation schemes located in the southeastern Anatolia project (GAP) area- where water meters may be introduced during irrigation. Because of the shortcomings in the flow rate measuring system, irrigation water cost is charged to farmers based on the crop type and the size of the irrigated area. This type of irrigation water price policy is the major cause of excess irrigation water use, resulting in low irrigation efficiencies and irrigation ratios, too. Due to the problems encountered in irrigation schemes (Cetin and Ozcan 1999), it is observed in recent years that farmers of some sections in irrigation schemes have switched back from profitable irrigated farming to less profitable dry farming. Abandoning irrigated lands is an unexpected situation and causes a great decrease in irrigation ratios. This attitude of farmers may be attributed to the problems in O&M services, i.e. poor management, structural shortcomings in irrigation infrastructure, government's price policy, subsidies, high labor cost, and problems regarding pesticides, insecticides, fertilizers and seeds, lack of crop rotation, insufficient credits, and the others.

In order to increase irrigation efficiencies and irrigation ratios in DSI-constructed irrigation schemes, an *action plan* on *"enabling efficient use of water in agriculture"*, for the period of 2014–2018, was put into action by the Ministry of Development in 2014 (MoD 2014). Within this scope, it was targeted that the share of areas under modern irrigation systems, such as drip and sprinkler, and the others, will be increased from 20% to 25% by the end of 2018. Furthermore, the goals were extended so that irrigation ratios and irrigation efficiencies will have reached up to 68% and 50%, respectively, and the number of modern irrigation units saving water will be increased 10% per annum.

Based on survey results and availability of funds, the Turkish government has envisaged that 8.5 Mha area will have been equipped with irrigation facilities by 2023. Of this amount, some 93% is to be irrigated from surface water resources, and the remaining 7% from groundwater resources. Although there have been great achievements and promising progress in the irrigation works, issues regarding irrigated agriculture are challenging and need spending much more efforts than before in the near future. Efforts on modernization of irrigation schemes will help irrigation managing entities increase both irrigation efficiencies and irrigation ratios, resulting in great amount of water savings. Hence, irrigated areas may be increased by reusing the saved water in order to ensure food security, or the water authorities may supply the saved water to the other water demanding sectors.

9.4.3 Water Use by Agricultural Sector

Precipitation distribution over Turkey is extremely irregular due to topographical and geographical differences, and rivers flow under irregular conditions. Usable or exploitable water potential of river basins therefore shows great variability. Spatial and temporal distribution of usable water in the river basins restrains its use in agriculture. In parallel with water resources, as seen from Table 9.1, the spatial distribution of soils suitable for irrigated agriculture is uneven in the river basins. De facto irrigated areas in the river basins are not proportional to the potentially irrigable soils of the basins because the distribution of irrigable soils does not comply with exploitable water resources in the basins.

In Turkey, rights regarding surface water usage provide that water is a public good, and everyone is entitled to use it, depending on the rights of prior uses. However, pursuant to Groundwater Law No 167, farmers should get a license from the state, based on the depth of GW well and withdrawals.

As is the case in other countries, in Turkey, irrigation is the largest water-consuming sector compared to domestic and industrial sectors. However, accounting for total agricultural withdrawals by irrigation managing organizations is a challenging issue due to lack of data. For example, the very small agricultural withdrawals from springs, watercourses, intermittent creeks etc. have been utilized by small-scale privately owned irrigation schemes, viz. PIs. The problem is that there are no data on how much irrigation water have been consumed by PIs per annum, except for some estimated figures. This is because agricultural withdrawals from different water sources by PIs are off-the-record. Additionally, PSAs developed irrigation schemes and irrigation schemes transferred to the village legal entities and/or municipalities by DSI have no continuous flow recording systems, either. Although manual measuring devices such as staff gauges or orifices may be installed at the heads of main canals or of water diversion cross-sections, the field staff employed in O&M services is not receptive to keep water records.

Irrigation schemes constructed by DSI are equipped with water measuring devices, in contrast to small-scale privately owned irrigation schemes and irrigation schemes operated by village legal entities and/or municipalities. Pursuant to the regulations of DSI, irrigation water diversions to the irrigation schemes must be recorded, without considering whether the irrigation unit is being operated by DSI, WUAs, SWICs and GWICs or not. At the end of each irrigation season, DSI technical staff makes necessary evaluations for each irrigation scheme of the area greater than 1000 ha, and a report titled as "*Evaluation Report of Irrigation Facilities Operated by DSI and Transferred to Water Users*" is released for the year of evaluation. For example, as of the end of 2016, DSI-constructed irrigation schemes consisted of net 3.08 Mha (a total of 2613 irrigation units), of which 2.20 Mha area (296 irrigation units) was monitored and evaluated by DSI. According to the evaluation results, a total of 1133 mm irrigation water was, on the average, consumed in the monitored areas in 2016. It is important to point out that this amount is valid exclusively for monitored irrigation schemes. Total water consumption by sectors can be found in the yearly reports.

Water consumption by sectors for selected years is given in Table 9.2. Although water is a limited resource, agriculture already accounts for over 70% of the water consumed by all three sectors (i.e. agricultural, municipal and industrial). As seen in Table 9.2, in 1990, 30.6 km^3 was consumed by various sectors, 22.0 km^3 (72%) in

Table 9.2 Water consumption by sectors for selected years

Years	Water demanding sectors						Total water consumption	
	Irrigation		Domestic		Industrial			
	km^3	%	km^3	%	km^3	%	km^3	%
1990	22.0	71.9	5.2	17.0	3.4	11.1	30.6	27.3
2000	29.3	74.6	5.8	14.8	4.2	10.7	39.3	35.1
2004	29.6	73.8	6.2	15.5	4.3	10.7	40.1	35.8
2008	32.0	74.4	6.0	14.0	5.0	11.6	43.0	38.4
2014	32.0	72.7	7.0	15.9	5.0	11.4	44.0	39.3
2016	40.0	74.1	7.0	13.0	7.0	13.0	54.0	48.2
2023[a]	**72.0**	**64.3**	**18.0**	**16.1**	**22.0**	**19.6**	**112.0**	**100.0**

Data Source: DSI (2017c), DSI activity reports and archived files
[a]Projected value by Turkish national irrigation agency

agriculture, 5.2 km^3 (17%) in domestic water supply and 3.4 km^3 in the industrial water supply. This 30.6 km^3 consumed water only accounts for 27% of total exploitable water resources, i.e. surface and groundwater resources (112.0 km^3), of the country. The more area is equipped with irrigation facilities, the more is the water consumed by the agricultural sector.

In 2016, although the ranking remained the same, i.e. irrigation>domestic>industrial, the share of agriculture in water consumption attained 74% as in 2000 and 2008. However, the share of the total consumption in the total usable potential increased from 27% in 1990 to 48% in 2016. The national irrigation agency envisaged that all the exploitable water resources will have been developed by 2023, and total amount of water to be allocated to irrigation sector will be around 72.0 km^3. Based on this projection, it is clear from Table 9.2 that the share of irrigation water use will have decreased from 74% to 64% by 2023 in favor of other water demanding sectors. But on the basis of past progress, it would be difficult to achieve this target by 2023.

As of the end of 2017, the total water amount consumed by irrigation was about 40.0 km^3. Of this, 30.20 km^3 (75.5%) was provided by surface water resources, and the remaining 9.80 km^3 (24.5%) from GW. As of the end of 2017, percentage shares of water consumption by sectors are given in Fig. 9.7. Note that the figures with respect to water consumed by agriculture in Table 9.2 account only for water consumed in the irrigation schemes operated by DSI, WUAs, GWICs, and in the fields of private farmers having official license to use GW.

DSI (2017c) cited that the safe yield of groundwater in Turkey was around 18 km^3 at the end of 2017. Of this, 9.8 km^3 was consumed in the irrigation sector; 5.8 km^3 out of 9.8 km^3 groundwater was withdrawn by private GW wells in the context of PIs; and the remaining 4.0 km^3 groundwater was consumed in irrigation schemes operated by DSI, TIGEM/Universities and GWICs. As seen in Fig. 9.7, domestic and industrial sectors consumed 26% of 54 km^3 total consumed water, the same as in 2016 (Table 9.2). It is clear from Fig. 9.7 that surface- and ground-waters accounted for 56% and 18% of the total consumed water, respectively.

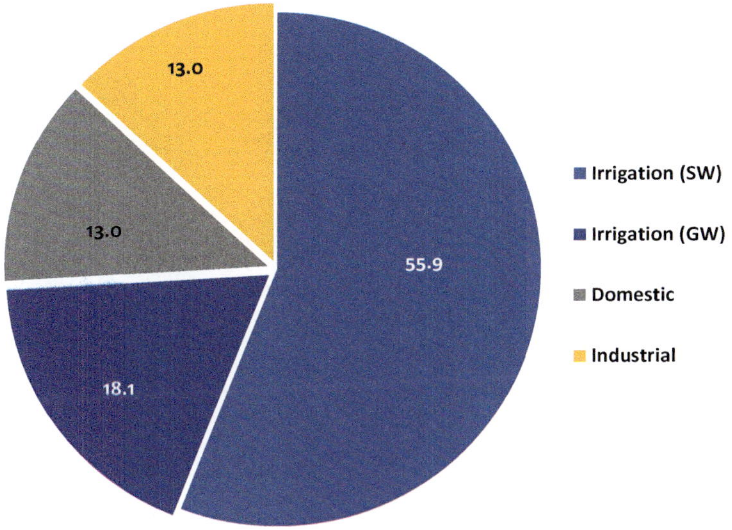

Fig. 9.7 Percentage shares of water consumption by sectors in 2017 (Data was obtained from DSI (2017c) and DSI activity report for 2017)

9.5 Determination of Crop Water Requirements in Practice: The "Two-Step" Approach

The amount of water needed to match crop evapotranspiration (*ETc*), i.e. to compensate the evapotranspiration loss from a cropped area, is known as crop-water requirement (CWR). Determination of *ETc* still relies on the so-called "*two-step*" approach (Steduto 2000): (a) Deriving reference crop evapotranspiration (*ETo*), and (b) Utilizing a crop specific coefficient *Kc* derived experimentally. *Kc* is supposed to account for a specific crop, development stages of the crop, and physical conditions of soil surface wetness as well as crop-atmosphere interactions. Ultimately, *ETc* is estimated by the product of such two parameters *ETo* and *Kc*. As such, CWR data are commonly needed in planning irrigation schemes so as to define irrigation canal capacities, operation and managing irrigation districts, designing farm irrigation system layouts, preparing irrigation scheduling programs, and improving irrigation practices etc.

Considering the importance of *ETc*, soil and water research institutes -located in different climatic regions of Turkey (Fig. 9.8) of the General Directorate of Agricultural Research and Policies (TAGEM, formerly TOPRAKSU), acting under the auspices of MoFAL, have conducted research on determining CWRs of agricultural crops in Turkey since 1954. Hence, a number of CWR data, including *Kc* values based on the SCS Blaney-Criddle reference evapotranspiration, have been obtained

Fig. 9.8 Geographical distribution of TAGEM's Soil and Water Research Institutes (▲) and boundaries of different climatic zones in terms of irrigation scheduling (Modified and improved in GIS platform after Kanber et al. (2017)) [The zones with the same numbers indicate that climatic conditions in the locations shown with capital letters prevail there]

by universities and agricultural research institutes of MoFAL. In 1982, a Guide on *CWRs of crops under irrigated agriculture* in Turkey was released by TOPRAKSU (Kanber 1982). Data on crops, crop developing stages, Blaney-Criddle reference evapotranspiration, CWRs obtained experimentally, and *Kc* values of crops derived experimentally based on the Blaney-Criddle model were provided in the Guide for different locations in Turkey. Until 2013, the SCS Blaney-Criddle method and *Kc* values given in the Guide were utilized by the Turkish national irrigation agency (DSI), defunct TOPRAKSU and GDRS, and offices of MoFAL in order to estimate reference evapotranspiration for agricultural purposes in areas where *ETc* data obtained by research are not available. Accordingly, the "*two-step*" approach had been utilized by using *Kc* values available in the Guide in order to get CWRs in planning irrigation schemes or for operational purposes of irrigation schemes, i.e. management purposes.

Though the "*two-step*" approach has been shown to be robust, practical and effective, experiences have shown that the method has as such been prone to some significant uncertainties in its application, in particular, in the Mediterranean region where a great portion of Turkey is located (Steduto 2000). Accordingly, FAO spent great efforts to adopt a universal method in the estimation of *ETo* for agricultural purposes. To this end, in May 1990, FAO organized a panel of experts and researchers, in collaboration with the *International Commission for Irrigation and Drainage (ICID)* and with the *World Meteorological Organization (WMO)*, so as to review the FAO methodologies on crop water requirements. The result of experts' panel and additional remarks are summarized by Allen et al. (1998) as "The panel of experts recommended the adoption of the Penman-Monteith combination method as a new standard for reference evapotranspiration and advised on procedures for calculation of the various parameters. The FAO Penman-Monteith method was developed by defining the reference crop as *a hypothetical crop with an assumed height of 0.12 m having a surface resistance of 70 s m^{-1} and an albedo of 0.23*, closely resembling the evaporation of an extension surface of green grass of uniform height, actively growing and adequately watered. The method overcomes shortcomings of the previous FAO Penman method and provides values more consistent with actual crop water use data worldwide." With respect to those developments, Steduto (2000) addressed clearly that the FAO Penman-Monteith (FAO56 PM) model can be used confidently as the sole standard method so as to estimate *reference evapotranspiration, ET_o*, at any location evaluated. Hence, ET_o can be computed by using representative meteorological data of solar radiation (sunshine), air temperature, humidity and wind speed.

On the other hand, the FAO56 PM model for the reference crop evapotranspiration, i.e. *ETo,* estimation was standardized by the ASCE-EWRI Task Committee in order to make calculation as easy as possible (ASCE-EWRI 2005). The standard equation to use in the first step (*Step 1*) can be expressed by the following equation (Eq. 9.1):

$$\mathrm{ET_o} = \frac{0.408\,\Delta\,(\mathrm{R_n} - \mathrm{G}) + \gamma\,\frac{900}{\mathrm{T}+273}\,\mathrm{u_2}\,(\mathrm{e_s} - \mathrm{e_a})}{\Delta + \gamma\,(1 + 0.34\,\mathrm{u_2})} \tag{9.1}$$

where, ET_o is reference crop evapotranspiration (mm day^{-1}), Δ is the slope of vapor pressure curve (kPa °C^{-1}), R_n and G are net radiation at the crop surface and soil heat flux density (MJ m^{-2} day^{-1}), respectively, γ is the psychrometric constant (kPa °C^{-1}), T is mean daily air temperature at 2 m height (°*C*), u_2 is wind speed at 2 m-height (m s^{-1}), and $e_s - e_a$ is saturation vapor pressure deficit (*kPa*) between saturation vapor pressure (e_s) and actual vapor pressure (e_a). It should be noted that the derivation of Eq. 9.1 was purely based on SI units. As explained before, after obtaining a representative *ETo* (Eq. 9.1) for the area evaluated, CWR (*ETc*) for a specific crop can be estimated by utilizing the dimensionless crop coefficient *Kc* as the following (Eq. 9.2):

$$\mathrm{ET_c} = \mathrm{ET_o}\,\mathrm{K_c} \tag{9.2}$$

In parallel with the global developments in adopting the standard *ETo* estimation method, institutions of the Turkish government developed a new attitude towards irrigation development strategy in 2012 (DSI 2016a). By 2023, it is aimed to construct modern irrigation systems in irrigation projects, to convert all traditional open canal irrigation schemes into pipelined and closed systems, and to encourage the use of modern irrigation methods/systems at the farm level by revitalizing some incentives. In this regard, rehabilitation of irrigation projects have already started, and Ministry of Development (MoD) has given some priorities to irrigation developments in the "Tenth Five-year Development Plan (2014–2018)". One of the most important priorities is to materialize an immediate program so as to make an effective and efficient use of water in agriculture. In 2013, TAGEM established a Task Committee consisting of experts and researchers, in collaboration with DSI, to review the *ETo* estimation methods and to revise the Guide on *CWRs of crops under irrigated agriculture* in Turkey. The committee concluded that the standardized ASCE-EWRI model gives more reliable results than the Blaney-Criddle model under the present conditions due to the fact that the Turkish meteorological services has strengthened its observation network by investing automatic meteorological stations since the beginning of 2000. Therefore, the use of the standardized ASCE-EWRI model for *ETo* determination is recommended for the national irrigation agency, practitioners, farmers, water managers, and private farms etc. Additionally, the committee reviewed the existing research results published by universities, TAGEM's soil and water research institutes shown in Fig. 9.8.

As a result, based on the FAO56 PM model, reference evapotranspiration (*ETo*) series of 259 stations -distributed geographically over Turkey- were generated and made ready for the users by using meteorological data with record lengths of more than 30-years. Additionally, the Committee compiled tables on growing periods, crop growing stages and corresponding *Kc* values, the real measured crop water requirements obtained from field studies, and other useful information for the 85 specific crops under irrigation. The compiled document regarding water requirements of crops under irrigated agriculture in Turkey is published by TAGEM (2017), superseding the guide by Kanber (1982). In order to completely achieve the goal of *"making an effective and efficient use of water in agriculture"* in the Development Plan designated in MoD (2014), another comprehensive work has been materialized by preparing irrigation scheduling programs for a number of crops grown in the climatic zones shown in Fig. 9.8. Apart from TAGEM (2017), details of the work were published in Kanber et al. (2017).

Regional and climatic differences greatly influence *ETc* values, and then the required irrigation water volume is affected by the length of the growing period of the crops. As seen in Table 9.3, both the growing period and the measured seasonal *ETc* of a specific crop differs remarkably from region to region. Note that *ETc* is not proportional to the length of the growing period due to the preponderant impacts of climatic factors on crop water requirements. This should not be overlooked in the planning of irrigation schemes or in "*general irrigation planning*" works realized before irrigation season, for the operation and management (O&M) purposes.

For example, in Table 9.3, the length of the growing period (LGP) for maize is 120 days in Adana/Mersin and Antalya research institutes located in the Mediterranean region (Fig. 9.8). However, seasonal *ETc* of maize is 503 and 738 mm in Mersin and Antalya, respectively, indicating 47% difference in the same climatic region due to the local variabilities in climate. More importantly, although the LGP is less (115 days) in Sanliurfa research station located in the Southeastern Anatolian Region, the seasonal *ETc* of maize attained the value of 938 mm, indicating 86% increase due to the fact that the temperature is higher and the relative humidity much less than it is in Adana. These interpretations show that great differences are observed in measured *ETc* values. Consequently, specific attention should be paid

Table 9.3 Experimentally measured seasonal crop water requirements (*ETc*, mm) of some selected crops in different climatic zones shown in Fig. 9.8, and the length of the growing period (*LGP*, day) (amended after Kanber et al. (2017) and TAGEM (2017))

Research station		Kırklareli	Ankara	Eskisehir	Konya	Erzurum	Tokat	Samsun	Sanlıurfa	Antalya	Mersin	İzmir
Elevation (m)		203	750	781	1072	1650	550	20	410	28	11.6	10
Wheat	*LGP*⇒	248	304	277	270	165	232		210			
	ETc⇒	786	690	548	571	398	616		769			
Quinoa											90	
											497	
Sunflower		150	120–140		133		145		100–120		125	100
		628	670		648		650		962		504	479
Rice		121						141				140
		1674						2137				1768
Maize		122	155	150	180		153	128	115	120	120	130
		762	808	659	660		637	371	938	738	503	420
Cotton									190			130
									1190			421
Soybean							146	163	118		103	100
							1049	726	1050		439	445
Sugar beet		166	200	218		170	170	173				
		1177	997	1018		726	925	909				
Alfalfa							231	245	247			
							1040	1313	1830			
Apple			170		217		210					
			725		714		845					
Citrus											365	365
											852	1099
Peach							170				245	225
							609				893	783

(continued)

Table 9.3 (continued)

Research station	Kırklareli	Ankara	Eskisehir	Konya	Erzurum	Tokat	Samsun	Sanlıurfa	Antalya	Mersin	İzmir
Olive								365			365
								834			673
Pepper		143				122	130	225		203	
		560				825	656	1441		817	
Tomato		152	150	120		150		192			180
		1376	577	570		659		1742			995
Green bean		70			125	120				80	
		784			604	517				403	

to the calculated crop water requirements (Eqs. 9.1 and 9.2) by utilizing meteorological data of the station considered, and *Kc* values, LGPs and crop development stages given in TAGEM (2017) if there is no measured data on *ETc* required for the location evaluated. Comparisons of *ETc* values with the nearby station results will always be helpful to the practitioners in order to prevent likely errors.

9.6 Evaluation of the Implications of Large-Scale Irrigation Schemes

Irrigation enhances crop diversity and it is one of the most important factors in increasing and stabilizing agricultural production in any climate. However, after implementing irrigation projects, experiences have shown that a number of problems associated with irrigation practices (Cetin and Ozcan 1999) have emerged over time. Therefore, monitoring and evaluation of irrigation schemes is necessary to take preventive actions for ensuring sustainability of agriculture and saving water in favor of other water demanding sectors. In Turkey, the national irrigation agency, DSI, has assumed the responsibility of monitoring irrigation schemes since 1954. Evaluation reports released yearly by the Agency provide valuable information on land utilization patterns, irrigation ratios as well as irrigation efficiencies, problems such as salinity, waterlogging, land ownership, and others associated with irrigation, irrigation systems and methods etc. to relevant people. Land utilization patterns have been the most problematic issue in the irrigation schemes. For example, during the planning stages of the irrigation projects in the Lower Seyhan and Ceyhan Plains located in the eastern Mediterranean region, where the irrigated areas reached well over 250,000 ha, the share of cotton in the land utilization pattern was envisaged as 35%, no maize and 20% forage crops (alfalfa) in 1960s. However, the proposed optimum land utilization pattern had not been realized or stabilized up to this time. Although cotton cultivated areas change from 1 year to another, the cotton cultivated area reached well over 96% in 1990s, and it was less than 9% in 2016. Surprisingly, maize became one of the major crops in Seyhan and Ceyhan irrigation schemes, located in Adana and Osmaniye provinces in Fig. 9.8, respectively, and its share attained over 38% in Seyhan irrigation and nearly 80% in Ceyhan irrigation in 2016. Notwithstanding that cropping patterns vary in the irrigation districts, cropping pattern, on average, in an irrigation scheme should be known in order to make "*general irrigation planning*" for management purposes. Concordantly, DSI (2017a) performed a comprehensive monitoring and evaluation task in 2.2 Mha irrigation areas consisting of 296 irrigation units greater than 1000 ha in 2016. Some information excerpted from DSI (2017a) are given in Table 9.4. As seen in the table, DSI and WUAs operated irrigation, and *de facto* irrigated areas have gradually increased by years. However, irrigation ratios have remained more or less constant (63%). Note that, in 2006, the irrigation ratios were as low as 10% in some irrigation schemes, which is rather below the DSI defined critical level of 30%. This figure indicates that 90% of the irrigation schemes were not irrigated although water was available for the farmers.

In Turkey, as stated before, information on irrigated agriculture is available only for DSI and WUAs operated irrigation schemes. Cropping patterns in DSI-constructed irrigation schemes for over 20-year period (1995–2016) are given in Table 9.4. As seen from the table, major crops are cereals, sugar beet, cotton, maize, forage crops (i.e. alfalfa), citrus, fruits and vegetables, and the others. However, land use types and their share in the cropping pattern vary from region to region because each region has its own special climatic characteristics. For example, the DSI-constructed irrigation schemes in Konya (see Fig. 9.8) located in the Central Anatolian Plateau, produce only grains/cereals (69%) and sugar beets (16%) on their irrigated area. Cotton (49%) and grapes (41%) are dominant crop types in Izmir located in the Aegean zone of Anatolia. Adana and Antalya represent the Mediterranean zone, suitable for growing varying proportions of cotton, maize, vegetables, citrus, and the "Others". In this context, dry beans, melons, and sesame are the major crops of the "Others" category in Table 9.4. If Turkey is considered as a whole, land utilization or cropping pattern identified in DSI-constructed irrigations are 17% cereals, 6% sugar beet, 17% cotton, 20% maize, 4% forage crops, i.e. alfalfa, 3% citrus, 12% fruits and vegetables, and 21% others.

In terms of irrigation water savings and salt leaching, irrigation methods are of great importance in the management of irrigation schemes. In 2012, the overall evaluation of irrigation schemes showed that 77% of total area was irrigated by surface irrigation methods such as furrow, border etc. The remaining part (23%) was under high-frequency irrigation methods, such as sprinkler (15%) and drip/trickle (8%). DSI (2016a) states that some 0.85 Mha area is equipped with sprinkler irrigation system of hand-carried pipes, which is widely used among farmers, and that 0.45 Mha area is irrigated by drip irrigation. Note that mainly fruit trees are irrigated by drip and mainly vegetables by using sprinkler irrigation methods. In order to increase irrigation efficiency to some extent and save irrigation water, DSI has already commenced rehabilitating the existing irrigation projects. On the other hand, initiation of rehabilitation projects has caused a remarkable decrease in surface irrigated areas and an increase in the areas irrigated with pressurized ones as seen in Fig. 9.9.

On the other hand, realization of rehabilitation and rejuvenation projects has resulted in an increase in irrigation schemes with traditional open canal and canalette systems. Therefore, as targeted by DSI, areas with pressurized pipeline irrigation schemes have shown an increasing trend by the years (Fig. 9.10).

As seen in Fig. 9.10, it is targeted that irrigation schemes with pipelines will have reached 48% by the year of 2023. Nevertheless, one fourth of the irrigated areas will have traditional open-channel networks in 2023. It is worth stating that intensification of the drainage network is of prime importance so as to preclude negative effects of irrigated agriculture on crops, soils and irrigation infrastructure. In this regard, based on the evaluated 2.28 Mha area operated by either DSI or WUAs, main, secondary and tertiary drainage canal densities are about 3.2, 3.5 and 4.5 m ha^{-1}, respectively. In 2016, shallow water table was monitored in 1.1 Mha irrigation areas

Table 9.4 Irrigation areas operated by DSI and WUAs, irrigation ratios, major irrigated crops and their share by years (DSI 2017a, 2018a)

Years	Number of irrigation units	Irrigation area (ha)	De facto irrigated area (ha)	Irrigation ratio (%)	Irrigated crop types and their share (percentage)							
					Cereals	Sugar beet	Cotton	Maize	Forage crops	Citrus	Fruits and vegetables	Others
1995	495	1,522,226	981,098	64.5	15.0	7.4	28.7	8.3	2.8	3.3	10.9	23.6
2000	588	1,833,579	1,183,382	64.5	18.9	9.7	19.5	16.3	2.8	3.1	10.7	19.0
2001	596	1,862,448	1,116,682	60.0	14.4	8.0	24.5	15.6	3.2	3.2	10.7	20.4
2002	609	1,894,750	1,191,566	62.9	14.8	7.6	22.8	14.3	3.1	3.3	11.3	22.8
2003	656	1,928,510	1,220,231	63.3	19.2	6.4	19.1	17.2	3.1	3.2	11.2	20.6
2004	631	1,933,343	1,272,149	65.8	21.1	6.3	18.6	20.1	3.2	3.2	10.4	17.1
2005	649	1,979,429	1,286,768	65.0	19.0	6.7	14.6	22.7	3.7	3.3	11.3	18.7
2006	670	2,024,884	1,268,562	62.6	16.8	6.0	18.2	18.8	4.3	3.5	12.0	20.4
2007	683	2,074,063	1,265,171	61.0	20.2	5.2	14.3	22.0	4.6	3.5	11.6	18.6
2008	698	2,129,271	1,337,628	62.8	21.6	5.3	11.8	22.3	4.8	3.3	12.1	18.8
2009	717	2,173,042	1,163,553	60.8	21.1	5.4	10.1	21.5	5.0	3.6	12.7	20.6
2010	733	2,213,258	1,393,946	63.0	18.5	5.9	12.8	22.2	4.6	3.4	11.6	21.0
2011	755	2,243,241	1,416,480	63.0	14.0	5.1	18.9	21.6	4.4	3.5	11.8	20.7
2012	774	2,269,542	1,386,954	61.1	14.5	4.9	15.7	23.5	5.1	3.4	11.8	21.1
2013	795	2,256,521	1,428,155	63.3	13.2	5.0	13.5	25.4	5.9	3.3	11.7	22.1
2014	793	2,334,270	1,433,226	61.4	14.8	4.6	15.1	22.4	5.9	3.6	12.3	21.3
2015	833	2,403,753	1,461,444	60.8	13.5	4.3	13.0	26.0	5.9	3.6	12.2	21.5
2016	888	2,428,264	1,577,095	64.6	14.2	4.8	13.4	24.2	6.0	3.2	12.1	22.1

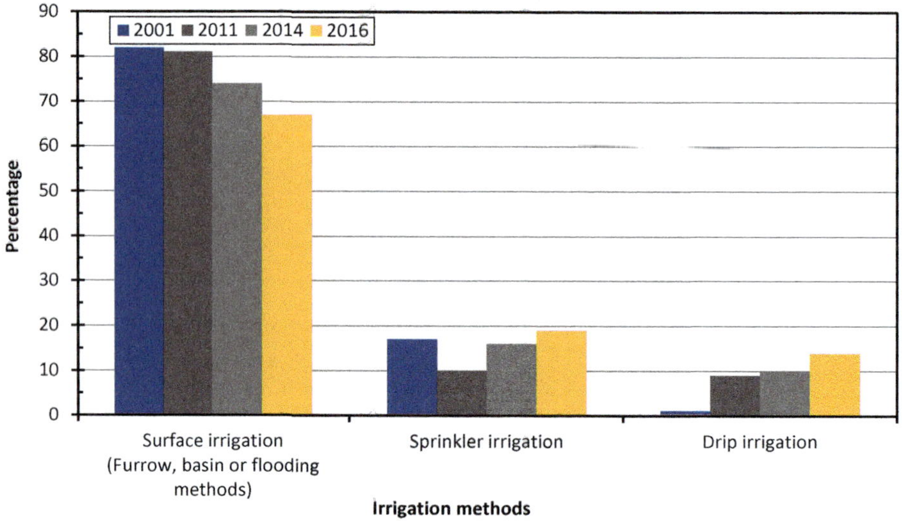

Fig. 9.9 Percentage change in the areas under different irrigation methods by years

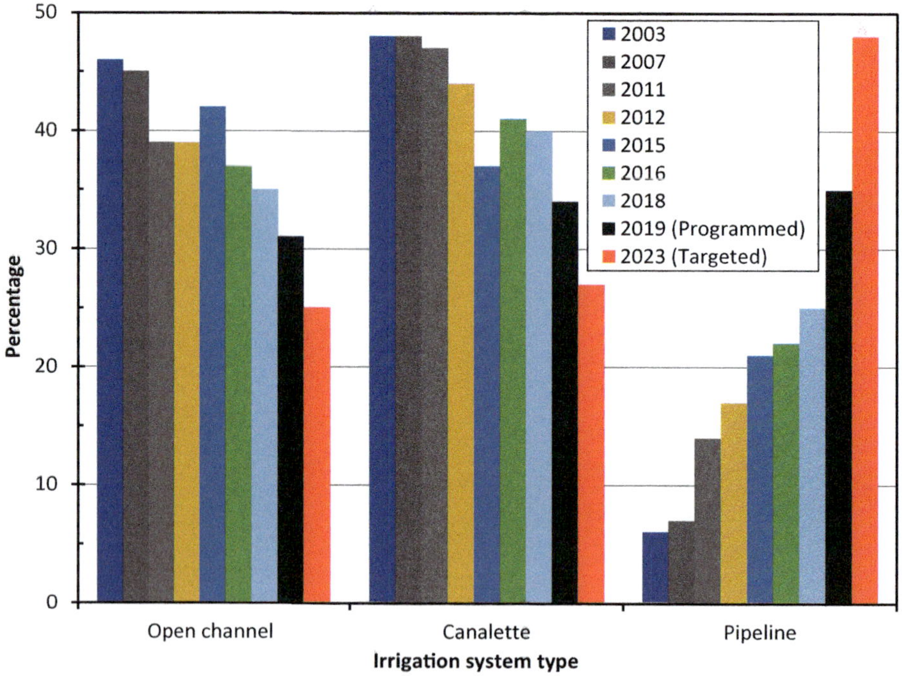

Fig. 9.10 Effects of rehabilitation and rejuvenation projects on irrigation schemes with pressurized ones

by using 10,082-drainage observation well data. Results indicated that 0.25 Mha area (23% of control area) of shallow water table less than 1.5 m depth suffered lack of drainage. Proper O&M services, accelerating rejuvenation of aged irrigation schemes in favor of pressurized irrigation systems, implementation of land consolidation projects and increased drainage densities will be a remedy to eliminate drainage problems in the irrigation schemes.

9.7 Regional Development Projects and Key Aspects of the *Southeastern Anatolia Project (GAP)* Regarding Irrigation

9.7.1 Regional Development Projects Regarding Irrigation

As discussed before, soil and water resources are not evenly distributed over Turkey because geological features, topographical conditions and climate are subject to change from place to place. Thus, living standards in rural areas depend largely on rain-fed agriculture and livestock. Irrigation is a remedy for diversifying production and ensuring food security in the rural areas of Turkey. Furthermore, irrigation developments provide far more than employment opportunities to the local people, compared to rain-fed agriculture and supply diversified goods and fiber to the agro-industries. Therefore, introducing irrigation practices accounts for inter-regional differences in terms of development. So as to eliminate inter-regional development differences, the Turkish government has embarked on ambitious regional development projects including irrigation investments over Turkey for a long time. Detailed information on those projects can be found in DSI (2018a). Konya Plain Project (KOP), East Anatolia Project (DAP) and Eastern Black Sea Project (DOKAP) are the most conspicuous ones (Fig. 9.11) to address regarding irrigation developments. DSI (2016a), DSI (2017b) and DSI's activity report for the year of 2017 (DSI 2018b) show that, after the completion of those projects, 1.36 Mha in KOP area, 1.30 Mha in DAP area and 0.28 Mha in DOKAP area will have been opened for irrigation by 2023. As of the end of 2017, 74% of total area in KOP, 34% in DAP and 39% in DOKAP were equipped with irrigation infrastructures.

Although the *Southeastern Anatolia Project (GAP)* is among the regional development projects in Turkey, it is considered as one of the largest integrated development projects in the world. Besides its physical qualities as well as its dimensions, the GAP draws attention throughout the world due to its social approach, geopolitical position of the area, and hydropolitics. The GAP has become a brand name for Turkey; it has already been placed in irrigation nomenclature today. Note that the acronym of the Project, *i.e. GAP*, is due to the Turkish name of the Project "***G****uneydogu* ***A****nadolu* ***P****rojesi*" (Southeastern Anatolia Project), but not an abbreviation of the made-up name "***G****reat* ***A****natolia* ***P****roject*" in Loucks and van Beek (2005).

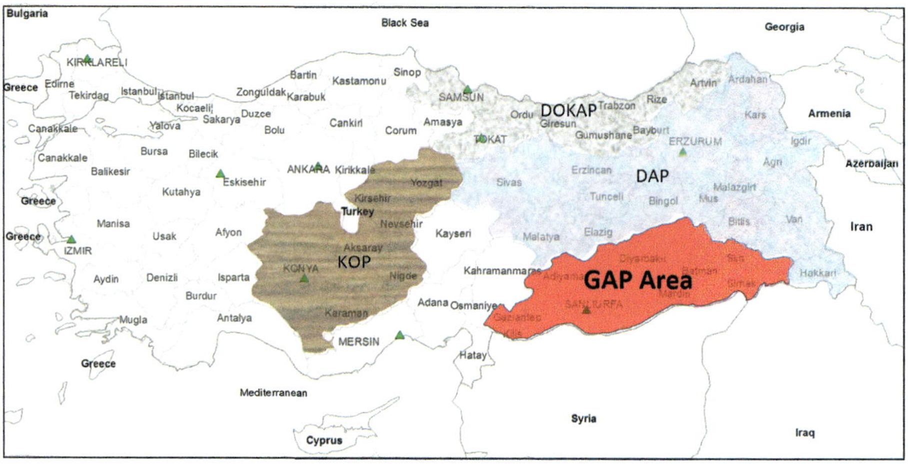

Fig. 9.11 The extent of some of the regional development projects (KOP, DAP and DOKAP), and the location of the *Southeastern Anatolia Project (GAP)* in Turkey

9.7.2 Integrated Regional Development Project: Southeastern Anatolia Project (GAP)

The history of *the Southeastern Anatolia Project* dates back to 1930s (DSI 2014). GAPRDA (2012) addresses that directives of Mustafa Kemal Atatürk, the founder of the Turkish Republic, started the hydrometric surveys in the Euphrates river basin in 1936. The works in the Euphrates and Tigris river basins were accelerated by the foundation of the national irrigation agency (DSI), by the Law No 6200, in 1954. Hence, a number of accomplishments regarding water works and irrigation have been done, and satisfactory results have been achieved since then. Until 1980s, water resources development projects and planning works in the twin river basin had been completed under the name of Upper Euphrates and/or Lower Euphrates Projects. For the first time, DSI enunciated the name of *"Southeastern Anatolia Project"*, i.e. *GAP*, in 1980. Since that time, the projects are well known by the name of "the GAP".

Although DSI is the primary government institution investing in irrigation in the GAP area, a number of state organizations and institutions have been contributing to different components of the GAP. The GAP is substantially an integrated regional development project aiming at leading different sectors such as agriculture, rural-urban infrastructure, transportation, industry, tourism, housing, education, health etc., which have an effect on the economy and the social life in the region.

The GAP area, consisting of 7.52 Mha (nearly one tenth of Turkey's surface area of 77.95 Mha), covers the whole or some parts of 9 provinces, i.e. Kilis, Gaziantep, Adiyaman, Sanliurfa, Diyarbakir, Mardin, Batman, Siirt and Sirnak, located in the Southeastern Anatolia Region (Fig. 9.11). According to the population census results performed in recent years, the percentage of the population in the GAP area is more or less stable and is nearly one tenth of the population of Turkey. However, the population growth rate in the region (GAPRDA 2012) is about 1.804%, which is pretty high in comparison with the national average of 1.201%.

Regarding the natural resources potential in the region, the GAP area includes 21% of hydroelectric energy potential and 21% (presently 1.78 Mha irrigable soils (Table 9.5)) of the targeted irrigable soil potential, i.e. of 8.5 Mha, of Turkey. More importantly, 42% of the total area (7.52 Mha) in the GAP region is arable (≈3.15 Mha) and under cultivation of the crops grown in the region (GAPRDA 2012). Nevertheless, approximately 83% of 3.15 Mha arable lands is presently under rain-fed agriculture due to the lack of irrigation investments. On the other hand, the area comprises 100% of petroleum and phosphate resources of Turkey; furthermore, about one quarter of the surface and groundwater resources of the country is in the GAP area. It is clear from those figures that the area is rich in soil and water resources, indicating a worth for investment.

As emphasized before, the GAP is a world-renowned integrated development project, aiming at socioeconomic development of the region predominantly through irrigation and energy investments in the Tigris-Euphrates twin river basin located in the Eastern Anatolia Region of Turkey (Fig. 9.11). From this point forth, the primary objective of the GAP is to provide goods and services to the urban and rural dwellers in order to increase the living standards in the GAP area by eliminating inter-regional development differences. In order to achieve the two important goals, regarding irrigation developments and energy production, the Project was formulated so that it constituted of primarily 13 subprojects; 7 of which took place in the Euphrates sub-basin and 6 of which took place in the Tigris sub-basin (GAPRDA 2014; DSI 2016a). The national irrigation agency DSI is the sole responsible state institution to implement the major parts of the project, particularly irrigation and drainage works. General characteristics of the Project, regarding irrigation components, are given in Table 9.5. As seen in the table, within the scope of these 13 sub-projects, DSI has foreseen the construction of 22 dams and 19 Hydroelectric Power Plants (HEPPs).

The largest components of this irrigation development and energy production system are Ataturk Dam with a reservoir capacity of 48.7 billion m^3, its HEPP with the energy production of 8.1 TWh per annum, Sanliurfa tunnels, Sanliurfa-Harran and Mardin-Ceylanpinar irrigation-drainage projects. As seen from Table 9.5, all of these components of Ataturk Dam are completed and are under operation. Sanliurfa twin-parallel irrigation tunnel system of 7.62 m inner diameters, which is considered as one of the longest ones of its kind in the world, has a length of 26.4 km. Irrigation water runs through the Sanliurfa tunnels by gravity, i.e. free flow, at a discharge rate of 328 $m^3 s^{-1}$ in order to irrigate a total of 376,689 ha agricultural lands in Sanliurfa-Harran and Mardin-Ceylanpinar plains. The capacities of the tunnels are equal to the 1.4 fold of Ceyhan River's average discharge capacity.

Because the GAP area is located in a semi-arid region, the major inhibiting factor in agricultural production is the lack of rainfall during the growing period of the crops. Thus, irrigation seems to be a prerequisite in order to develop a highly productive agricultural system, as well as optimizing agricultural production. As seen from Table 9.5, after the completion of the Project, 1.78 Mha area (21% of the

Table 9.5 Characteristics of the Southeastern Anatolia Project (GAP) and the development status of projects on irrigation (GAPRDA 2014; DSI 2016a)

TIGRIS AND EUPHRATES BASINS					
ESTABLISHED POWER: 7,389 MW (including individual projects)					
ENERGY PRODUCTION: 26,853 GWh (including individual projects) NUMBER OF DAMS: 22					
IRRIGATION AREA: 1,779,884 ha (including individual projects) NUMBER OF HPPs: 19					
EUPHRATES BASIN			TIGRIS BASIN		
Project and its Units	Irrigation Area (ha)	At the stage of	Project and its Units	Irrigation Area (ha)	At the stage of
1. LOWER EUPHRATES PROJECT	**691,858**		**1. TIGRIS-KRALKIZI PROJECT**	**119,755**	
Şanliurfa Tunnel and Irrigation	**376,689**		Tigris Right Bank Irrigation	50,743	Construction
(a) Sanliurfa-Harran Plain Irrigation	151,419	Operation	Right Bank Pumped Irrigation (P2-P5)	23,085	Operation
(b) Mardin-Ceylanpinar Irrigation	109,184	Operation	Tigris Right Bank Pumped Irrigation (P6)	8100	Operation
(c) Mardin-Ceylanpinari Pumped Irrigation	116,086	Project	Tigris Right Bank Pumped Irrigation (P3-P4)	37,827	Construction
Mardin-Ceylanpinar Groundwater (GW) Irrigation	111,939	Operation	**2. BATMAN PROJECT**	**34,421**	
Siverek-Hilvan Pumped Irrigation	158,053	Project	Batman Left Bank Irrigation	15,828	Operation
Kaya Pumped Irrigation	45,167	Operation	Batman Right Bank Irrigation	18,593	Construction
2. SURUC YAYLAK PROJECT	**113,419**		**3. BATMAN-SILVAN PROJECT**	**235,048**	
Yaylak Plain Irrigation	18,322	Operation	Tigris Left Bank Irrigation	169,321	Project
Suruc Pumped Irrigation	95,097	Operation	Tigris Left Bank Pumped Irrigation	65,727	Project
3. ADIYAMAN-KAHTA PROJECT	**77,631**		**4. GARZAN PROJECT**	**39,164**	
Camgazi Dam Irrigation	8000	Operation	Garzan Irrigation	39,164	Operation
Gomikan Dam and Irrigation	7243	Project	**5. CIZRE PROJECT**	**121,000**	
Kocali Dam and HEPP, Irrigation	17,761	Project	Nusaybin-Cizre-Idil Irrigation	89,000	Operation

(continued)

Table 9.5 (continued)

TIGRIS AND EUPHRATES BASINS					
ESTABLISHED POWER: 7,389 MW (including individual projects)					
ENERGY PRODUCTION: 26,853 GWh (including individual projects) NUMBER OF DAMS: 22					
IRRIGATION AREA: 1,779,884 ha (including individual projects) NUMBER OF HPPs: 19					
EUPHRATES BASIN			TIGRIS BASIN		
Project and its Units	Irrigation Area (ha)	At the stage of	Project and its Units	Irrigation Area (ha)	At the stage of
Buyukcay Dam and HEPP, Irrigation	12,322	Operation	Silopi Plain Irrigation	32,000	Operation
Pumped Irrigation from Ataturk Dam Lake	23,998	Operation	**SUB-TOTAL**	**549,388**	
Samsat Pumped Irrigation	8307	Operation	**6. INDIVIDUAL PROJECTS**	**36,548**	Operation
4. ADIYAMAN-GOKSU-ARABAN	**70,968**		**7. SMALL SCALE IRRIGATION PROJECTS**	**3258**	Operation
Adiyaman-Goksu Araban Irrigation	70,968	Operation	**GRAND TOTAL**	**589,194**	
5. GAZIANTEP PROJECT	**140,903**				
Hancagiz Dam and Irrigation	6945	Operation			
Kayacik Dam and Irrigation	20,000	Operation			
Kemlim Dam and Irrigation	3088	Operation			
Belkis-Nizip Pumped Irrigation	10,164	Operation			
Birecik Dam Pumped Irrigation	95,976	Operation			
Bayramli Regulator and Irrigation	3783	Operation			
SUB-TOTAL	**1,094,779**				
6. INDIVIDUAL PROJECTS	**90,072**	Operation			
7. SMALL SCALE IRRIGATION PROJECTS	**5839**	Operation			
GRAND TOTAL	**1,190,690**				

targeted irrigation potential of Turkey) will be equipped with irrigation facilities, 1.19 Mha of which is in the Euphrates sub-basin and 0.59 Mha in the Tigris sub-basin. Irrigation projects are supposed to fulfill the two primary prerequisites. In this regard, it is worth addressing that irrigated agriculture is the key leading sector in the

GAP area to trigger regional development. As of the end of 2017, 0.53 Mha (31% of potential irrigation area in the GAP) area was put under irrigation; irrigation construction works have been continuing in 11% of the area, and the remaining 58% of the area has been presently excluded from tender.

Provided that the prospective modern agricultural system in the GAP area is supported in terms of research, technology and industry, a 10- to 20-fold production increase is expected in the region. Irrigation practices will directly increase agricultural production and, consequently, the farmers' income will increase because of not only irrigation developments but also of other technological inputs. Accordingly, irrigation development projects play a crucial role in the socioeconomic development of the region. To exemplify, under present conditions, the total cotton production in the region accounts for 55% of the national production, red-lentil 98%, pistachios 87%, barley 23%, wheat 18%, and maize 12%. Sanliurfa-Harran Plain (151,419 ha) has been under irrigation since 1994. Economic analysis indicates that the net agricultural income has increased from US$ 190 per ha to US$ 960 per ha in the Sanliurfa-Harran Plain (in the southern parts of Sanliurfa province depicted in Fig. 9.11) due to the contribution of irrigation. Thus, irrigation investments have proven fruitful by increasing the average net income at least four-fold. According to DSI (2004), on the other hand, the gross domestic agricultural income (GDAI) per ha, on the average, increased from US$ 280 to US$ 1510 by irrigation practices in the GAP area, indicating an increase of US$ 1230 per ha in gross agricultural income (GAI).

Of course, irrigated agriculture has on-site and off-site effects on the aquatic environment. In this regard, the quality of irrigation return flows (IRFs) in the GAP area concerns water demanding sectors, and it has an international dimension, too. The main drainage canal has transported all the IRFs generated by the Sanliurfa-Harran plain. The drainage canal runs southerly and leaves Turkey at the Syrian border. Therefore, the quality and quantity of IRFs need continuous monitoring, and alternatives for the re-use of the IRFs should be in the agenda in order to lower the demand for fresh water.

9.8 Concluding Remarks

Turkey, located in the Mediterranean region, is known as the realm of ecologies for it is a peninsula surrounded by the Mediterranean, Aegean, Marmara and the Black Sea. Latitude differences, the distance from seas, extent and orientation of mountain chains, and topographical variations cause to emanate primarily four different climate types over the country. Semi-arid conditions generally prevail over the country. Spatial and temporal variabilities in temperature and precipitation values hinder year-round production of field and horticultural crops under natural conditions. In this context, precipitation is the most limiting factor for consumptive use. Consequently, irrigation is necessary to ensure sustainability of agricultural production and food security in Turkey, except for the eastern Black Sea region.

Although agriculture has had an important role from the point of view of economic development and food security since the foundation of the Republic, the share of agriculture in GDP has decreased continuously in favor of industry and other sectors, and it was about 7.5% at the end of 2015. However, inequitable land property distribution is an ill-conditioned and inherited structural defect in land tenure in Turkey. Fragmented parcels of agricultural enterprises, the very small parcel sizes, and uneven distribution of landownership are only a few of the inherited problems regarding irrigated agriculture. Therefore, implementation of land consolidation (LC) projects should be given priority in irrigation developments. Implementing LC projects before irrigation construction renders at least 40% savings in expropriation and construction costs.

Considering the preliminary results of irrigated agriculture land classification surveys, it is concluded that 22.6 Mha land, out of 28.05 Mha arable lands, is economically and technically irrigable under present conditions in Turkey. However, only 8.5 Mha area is given priority by DSI and aimed to be equipped with irrigation schemes by 2023 –the hundredth anniversary of the foundation of Turkish Republic.

Irrigation investments need huge amount of financial support and subsidies. Therefore, there have been only two principal government/state organizations concerned with irrigation developments in Turkey: a) *General Directorate of State Hydraulic Works (DSI)*, b) *Provincial Special Administrations (PSAs, formerly General Directorate for Soil and Water, i.e. TOPRAKSU and General Directorate of Rural Services (GDRS))*.

As of the end of 2017, a total of 6.57 Mha gross area is equipped with irrigation infrastructures in Turkey. Of this, 65% (4.28 Mha) is developed by the national irrigation agency (DSI), mainly under large schemes, 20% (1.29 Mha) by PSAs (formerly General Directorate of Rural Service (GDRS) and TOPRAKSU), mainly under medium-to-small schemes, and 15% (1.00 Mha) by individual framers as small-scale privately owned irrigation schemes, viz. PIs.

Based on the operation and maintenance (O&M) characteristics of irrigation facilities, irrigation managing organizations in Turkey are grouped under four categories: a) *Private Irrigations*, b) *Local Authority Irrigations*, c) *Irrigation Cooperatives and Water User Associations (WUAs)*, d) *Public Irrigation Managements*. Turkish irrigation management transfer program, including the WUAs model developed by technical staff of DSI, gained worldwide attention as well as celebrity regarding its speed and effectiveness. As of 2017, the existing large-scale irrigation areas operated by DSI consist of only 11% of the total (3.37 Mha) area equipped with irrigation and drainage facilities by DSI, and 73% of irrigations are handed over to WUAs.

Although water is a limited resource, agriculture already accounts for the major portion (at least 70%) of the water consumed by all three sectors (i.e. agricultural, municipal and industrial). As of the end of 2017, the total water amount consumed by irrigation was about 40.0 km^3. Of this, 30.20 km^3 (75.5%) was provided by surface water resources and the remaining 9.80 km^3 (24.5) from GW. Note that the figures with respect to water consumed by agriculture account only for water consumed in the irrigation schemes operated by DSI, WUAs, GWICs, and in the fields of private farmers having official license to use GW. The water used in agriculture by *small-scale privately owned irrigation schemes* (PIs) and local authority irrigations have been off-the-record yet.

In Turkey, suitable climate, soil, and topographical conditions, coupled with the irrigation water sources, as well as good agricultural practices, have made it possible to cultivate various plants (up to four crops) throughout the year. Accordingly, the average rate of yield increase per ha by irrigation is 188% for cereals, particularly in wheat, 328% for cotton, 699% for maize, 184% for forage crops, 190% for vegetables and 116% for mixed-fruits in 2017. Agricultural production increases have resulted in disposable incomes for the farmers.

Although irrigation has contributed a lot to Turkish economy, the proposed optimum land utilization pattern has not been realized or stabilized in irrigation schemes up to this time. Additionally, the very low irrigation efficiencies (average 37%) and irrigation ratios (42% in DSI operated, 66% in WUAs operated irrigation schemes) indicate poor irrigation management and inherited problems in irrigated agriculture in Turkey.

Introducing irrigation practices accounts for inter-regional differences in terms of development. So as to eliminate inter-regional development differences, the Turkish government has embarked on ambitious regional development projects including irrigation investments over Turkey for a long time. Although the *Southeastern Anatolia Project (GAP)* is among the regional development projects in Turkey, it is considered as one of the biggest integrated regional development projects with various aspects such as irrigation, flood control, environment, hydropower generation, education, health, industry etc. in the world. The GAP has become a brand name for Turkey. The gross domestic agricultural income per ha, on the average, increased from US$ 280 to US$ 1510 by irrigation practices in the GAP areas, indicating an increase of US$ 1230 per ha in gross agricultural income. Despite the remarkable income increase by irrigation, irrigated agriculture has on-site and off-site negative impacts on the fragile environment, particularly in the GAP area of Turkey.

References

Allen RG, Pereira LS, Raes D, Smith M (1998) Crop evapotranspiration: guidelines for computing crop requirements. FAO Irrigation and Drainage Paper No. 56, FAO, Rome, p 300

ASCE-EWRI (2005) The ASCE standardized reference evapotranspiration equation. ASCE-EWRI Task Committee Final Report, January, 2005, p 59

Atalay I (2016) A new approach to the land capability classification: Case study of Turkey. Procedia Environ Sci 32:264–274

Bayazit M, Avci I (1997) Water resources of Turkey: Potential, planning, development and management. International Journal of Water Resources Development 13(4):443–452

Cakmak B, Kibaroglu A, Kendirli B, Gokalp Z (2010) Assessment of the irrigation performance of transferred schemes in Turkey: A case study analysis. Irrig Drain 59:138–149

Cetin M, Ozcan H (1999) Asagi Seyhan Ovasinda sulanan ve sulanmayan alanlarda meydana gelen sorunlar ve cozum onerileri: Ornek bir calisma. TUBITAK Tr. J Agric For, Supplemental Issue 23 (1):207–217

Cetin M, Ibrikci H, Berberoglu S, Fink M, Nagano T, Golpinar MS, Kubota J, Gohman H (2015) Sustainability of agricultural water management in large scale irrigation schemes: A case study in Turkey. The 9th World Congress of the European Water Resources Association (EWRA2015), Water Resources Management in a Changing World: Challenges and Opportunities, Istanbul, Turkey, 10–13 June 2015, Book of Abstracts (http://www.ewra2015istanbul.org/), pp 313–314

DSI (2004) Dunden bugune DSI (1954–2004). TSHOF Plate and Pressing LTD, Ankara, p 170

DSI (2014) GAP: Southeast Anatolia Project. General Directorate of State Hydraulic Works (DSI) Printing House, Ankara, p 28

DSI (2016a) Water and DSI (1954–2016). General Directorate of State Hydraulic Works (DSI) Printing House, Ankara, p 239

DSI (2016b) DSI'ce insa edilerek isletmeye acilan sulama ve bataklik islahi tesisleri (2015–2016). General Directorate of State Hydraulic Works (DSI), Department of Operation and Maintenance, DSI Printing House, Ankara, p 192

DSI (2017a) 2016 yili DSI'ce isletilen ve devredilen sulama tesisleri degerlendirme raporu. General Directorate of State Hydraulic Works (DSI). Department of Operation and Maintenance, DSI Printing House, Ankara, p 543

DSI (2017b) Sulama calisma grubu calisma belgesi. Republic of Turkey, Ministry of Forestry and Water, the 2nd Counsel of Forestry and Water, General Directorate of State Hydraulic Works (DSI) Printing House, Ankara, p 64

DSI (2017c) Su kaynaklarinin gelistirilmesi ve hidroloji calisma grubu raporu. Republic of Turkey, Ministry of Forestry and Water, the 2nd Counsel of Forestry and Water, General Directorate of State Hydraulic Works (DSI) Printing House, Ankara, p 61

DSI (2018a) Turkiye'nin su ve enerji gucu DSI – Ozet bilgiler 2018. General Directorate of State Hydraulic Works (DSI) Printing House, Ankara, p 72

DSI (2018b) Devlet Su Isleri Genel Mudurlugu 2017 yılı faaliyet raporu. Available as pdf file at http://www.dsi.gov.tr/stratejik-planlama/faaliyetraporlari, Accessed 24 July 2018, p 263

FAO (2003) The design of land consolidation pilot projects in Central and Eastern Europe. FAO Land Tenure Studies 6, Rome, ISBN 92-5-105001-5

GAPRDA (2012) GAP eylem plani: GAP atlasi. Republic of Turkey, Ministry of Development, South-Eastern Anatolia Project (GAP) Regional Development Administration (RDA), December 2012, p 96

GAPRDA (2014) GAP action plan: An overview of the South-Eastern Anatolia Project. Republic of Turkey, Ministry of Development, South-Eastern Anatolia Project (GAP) Regional Development Administration (RDA), June 2014, p 43

Hillel D (2004) Introduction to environmental soil physics. Elsevier Academic Press, Boston, p 994

Kanber R (1982) Turkiye'de sulanan bitkilerin su tuketimleri rehberi. 2nd edn. Publication by the General Directorate of TOPRAKSU, Publication No 718, Ankara, p 630

Kanber R, Unlu M (2008) Turkiye'de sulama ve drenaj sorunlari: Genel bakis. Irrigation and Drainage Conference, Regional DSI Meetings on Water for the 5th World Water Forum, Adana, Turkey, 10–11 April 2008, Book of Proceedings, pp 1–45

Kanber R, Unlu M, Aydinsakir K, Aydin O, Kucukcoskun S (2017) Turkiye'de yetistirilen kultur bitkilerinin sulama teknikleri rehberi (Guncellestirilmis baski). Republic of Turkey, Ministry of Food, Agriculture and Livestock (MoFAL), General Directorate of Agricultural Research and Policies (TAGEM), Ankara, ISBN 978-605-9175-88-3, p 321

Kirda C, Topcu S, Cetin M, Dasgan HY, Kaman H, Topaloglu F, Derici MR, Ekici B (2007) Prospects of partial root zone irrigation for increasing irrigation water use efficiency of major crops in the Mediterranean Region. Ann Appl Biol 150:281–291

Loucks DP, van Beek E (2005) Water resources systems planning and management: An introduction to methods, models and applications. Studies and Reports in Hydrology, UNESCO Publishing, p 680
Mengu GP, Akkuzu E (2010) Impact of irrigation management transfer on land and water productivity and water supply in the Gediz Basin, Turkey. Journal of Irrigation and Drainage Engineering, ASCE 136:300–308
MoD (2014) Onuncu bes yillik kalkinma plani (2014–2018): Tarim ozel ihtisas komisyonu "Tarim arazilerinin surdurulebilir kullanimi" calisma grubu raporu. Republic of Turkey, Ministry of Development (MoD), Specialization Commission on Agriculture, Ankara, ISBN 978-605-4667-58-1, p 79
MoFAL (2013) Turkiye tarimsal kuraklikla mucadele stratejisi ve eylem plani (2013–2017). Republic of Turkey, Ministry of Food, Agriculture and Livestock (MoFAL), Ankara, p 59
Murray-Rust DH, Svendsen M (2001) Performance of locally managed irrigation in Turkey: Gediz case study. Irrig Drain Syst 15:373–388
Nalbantoglu G, Cakmak B (2007) Akinci Sulama Birliginde sulama performansinin karsilastirmali degerlendirilmesi. A U Ziraat Fakultesi Tarim Bilimleri Dergisi 13(3):213–223
Satana S, Ceylan AR, Sert A (2017) The Turkish experience in consolidation of irrigated land: productivity and efficiency implications. Presented at the 17th World Bank Conference on Land and Poverty, Washington, DC, March 20–24, 2017, http://www.worldbank.org/en/events/2016/08/22/landand-poverty-conference-2017-responsible-land-governance-towards-an-evidence-based-approach, Accessed 7 July 2018, p 29
Selek B, Tuncok IK, Selek Z (2018) Changes in climate zones across Turkey. Journal of Water and Climate Change 9(1):178–195
Steduto P (2000) Plant nutrient management under pressurized irrigation systems in the Mediterranean Region: crop water requirements. In: Ryan J (ed) Plant nutrient management under pressurized irrigation systems in the Mediterranean Region. Proceedings of the International Fertigation Workshop organized by the World Phosphate Institute (IMPHOS), 25–27 April 1999. ICARDA, Aleppo, p 400
Svendsen M (2001) Irriegation management transfer in Turkey. International E-mail Conference on Irrigation Management Transfer (IMT Case Study), FAO, INPIM
Svendsen M, Murray-Rust DH (2001) Creating and consolidating locally managed irrigation in Turkey: the national perspective. Irrig Drain Syst 15:355–371
Svendsen M, Nott G (2000) Irrigation management transfer in Turkey: process and outcomes. In: Groenfeldt D, Svendsen M (eds) Case studies in participatory irrigation management, World Bank Institute, Washington, DC, pp 27–88
TAGEM (2017) Turkiye'de sulanan bitkilerin bitki su tuketimleri. Republic of Turkey, Ministry of Food, Agriculture and Livestock (MoFAL), General Directorate of Agricultural Research and Policies (TAGEM), Ankara, ISBN 978-605-9175-93-7, p 590
TOBB (2018) Ekonomik rapor – 2017. The Union of Chambers and Commodity Exchanges of Turkey (TOBB), Publication Number: 2018/306, Ozyurt Press, Ankara, Turkey, p 222
TUIK (2004) The 2001 general agricultural census. Republic of Turkey, Turkish Statistical Institute (TUIK), Ankara, ISBN 975-19-3557-1, p 109
Uskay S (1999) Participatory irrigation management practices in Turkey. Paper presented to the 11th Seminar in the Development of Appropriate Technology, 12–19 December, Tokyo
van Dijk T (2007) Complications for traditional land consolidation in Central Europe. Geoforum 38 (3):505–511. https://doi.org/10.1016/j.geoforum.2006.11.010
Yildiz D (2007) Su raporu – Ulusal su politikasi ihtiyacimiz. Industrialists and Businessmen's Association (USIAD), June 2007, Istanbul, ISBN 978-975-98399-1-8, p 159

Chapter 10
Water and Energy

Unal Ozis, Ertugrul Benzeden, and Ahmet Alkan

Abstract The primary energy resources of Turkey are quite limited; therefore Turkey has to develop her water power potential, a renewable and clean energy resource, to produce electrical energy. The hydroelectric energy production of Turkey was barely 1 TWh/y in 1960; the production capacity increased to around 75 TWh/y in 2017. The economically feasible hydroelectric potential of Turkey is in the order of 150 TWh/y. Turkey anticipates to harness the remaining part of this potential, so that several hundreds of major hydroelectric power plants (HEPP) have still to be constructed in the near future. The discharge rates of the water courses in Turkey show significant seasonal variations as well as large annual fluctuations; hence, the discharge regulation by means of large reservoirs, created by dams, becomes a necessity. The additional evaporation from the reservoirs of the dams should not be regarded as an unnecessary loss, but has to be conceived as an in-kind operational cost. Hydroelectric schemes, especially those associated with dams, are the most versatile power plant types to cover the peak power demands. The reservoirs of dams serve often, directly or indirectly, to mitigate the floods. They may also serve to regulate the discharges for irrigation purposes in the context of multipurpose projects. Nevertheless, any hydraulic scheme is an intervention to the nature, and special care should be given not to unnecessarily harm the people and the environment, during the construction as well as the operation stages of water power schemes. The harnessing of water power in Turkey is primarily based on hydroelectric plants associated with large dams and on high-head diversion plants. Embankment dams dominate as dam type, but interesting concrete dams do also exist. The water power plant at the toe of Ataturk dam, with 2400 MW capacity, is the largest hydroelectric scheme in Turkey.

Keywords Water · Power plant · HEPP · Hydroelectric · Energy · Capacity · Dam

U. Ozis · E. Benzeden (✉) · A. Alkan
Faculty of Engineering, Civil Engineering Department, Dokuz Eylul University, Izmir, Turkey
e-mail: ertugrul.benzeden@deu.edu.tr; ahmet.alkan@deu.edu.tr

N. B. Harmancioglu, D. Altinbilek (eds.), *Water Resources of Turkey*, World Water Resources 2, https://doi.org/10.1007/978-3-030-11729-0_10

10.1 Introduction

Hydropower has a special place in covering the electrical energy demands of Turkey. It is a 'renewable' primary energy resource due to the hydrological cycle, driven by the solar energy reaching the earth surface. It is an almost 'clean' energy resource with regard to environmental issues. Hydropower is a 'domestic' primary energy resource in covering the needs of Turkey. Hydroelectric schemes are often 'multipurpose', serving also to other purposes such as flood mitigation, irrigation, water supply. Reservoirs created by dams serve partial flood mitigation even without allocating special volumes for this purpose.

The production of the hydroelectric energy in Turkey was less than 0.1 TWh/y in 1955, reached barely 1 TWh/y in 1960, and increased to 67 TWh/y in 2015; the actual production capacity at present is in the order of 75 TWh/y.

The discharges of the water courses in Turkey show large seasonal as well as annual variations (EIE 1955–1993; DSI 1961–1995), so that dams proved to be necessary for discharge regulation; at the same time, dams create the entire or a significant part of the total head in hydroelectric schemes.

The hydropower potential of Turkey is basically harnessed by power plants located at the toe of dams along the middle stretches of the rivers and by high-head diversion plants at the higher stretches of the rivers. Several diversion schemes are also associated with dams. Hence dams are often a primary component of major water power schemes in Turkey (Fig. 10.1) (Ozis 1961, 1977, 1991; DSI 1965, 1995a; Ozis et al. 1978, 1983, 2006a, b; Ozis and Harmancioglu 1982).

Fig. 10.1 Location of Major Water Power Schemes in Turkey. (U. Ozis, E. Benzeden and A. Alkan)

Fig. 10.2 The Ataturk dam and power plant on Euphrates, with a maximal reservoir surface of 817 km^2. (Photo by U. Ozis)

The reservoirs of dams create, compared to the natural state, larger water surfaces (Fig. 10.2), and, thus, additional evaporation losses. However, these losses should not be regarded as an unnecessary loss, but have to be conceived rather as an in-kind operational cost.

Hydroelectric units can be taken into service within a couple of minutes, and they are taken out of service with a quite small loss of water, so that hydroelectric schemes associated with dams or other reservoirs are the best suited to cover the peak power demands of the interconnected electricity systems.

Besides hydroelectric and other renewable energy based power plants, Turkey will continue to use thermal plants and begins to use nuclear plants for producing electricity. In this context, the availability of hydroelectric plants associated with dams or other kind of reservoirs will be of great importance. Such hydroelectric schemes, especially those close to large consumption centers, should be designed with low operation load (capacity) factor, hence with the largest possible installed capacity. The share of the hydroelectric schemes in Turkey is around one third of the total installed power capacity and one fourth of the total electrical energy demands, reflecting partially the above mentioned characteristics of the hydroelectric schemes.

10.2 Renewable Energy and Hydropower

Hydropower is a function of two basic elements: namely, the ‘discharge’ and the ‘head’. The hydrological cycle, ‘evaporation – precipitation – water flow’, of the earth-atmosphere system, driven by the solar energy, creates the flows in water courses or the ‘discharge’. The topographical elevation difference between two points of the water course creates the ‘head’. The multiplication of these two elements equals the ‘gross power’ of the water course. Hence, the hydraulic energy is a ‘renewable’ form of energy, as long as the solar energy reaches the earth.

The most widespread form to make use of hydraulic energy as a primary energy resource is actually to transform it to electrical energy as a secondary energy resource. This transformation occurs through the 'hydroelectric power plant' (HEPP). The work of this power (capacity) in time, capacity multiplied by time, results in energy: the 'hydroelectric energy'.

Electrical energy can be generated from several primary energy resources. However, Turkey's energy resources are quite limited, and she has to import them in significant quantities from various foreign countries. Fossil fuels like oil, gas, and coal, and nuclear energy resources like uranium and thorium, are not renewable; they may cause significant environmental pollution. Solar energy, wind energy, and water power are renewable and may cause far less, often tolerable, pollution in the environment.

Turkey has to give priority to renewable energy resources. Moreover, in the context of the actual technology and availability, the best suitable domestic energy resource of the country is water power (Ozis 1961, 1991; DSI 1965, 1995b; Ozis et al. 1999; Keloglu 2006; Tutus 2006; Yuksek et al. 2006; Akpinar et al. 2008; Sen 2011; Gokdemir et al. 2012; Akpinar 2013). The total installed capacity of all types of electricity generating plants in Turkey was around 73 GW in 2015. The total installed capacity of hydroelectric power plants was 26 GW, and its share was 35% (Table 10.1).

The total capacity of hydroelectric power plants associated with dams and reservoirs is around 19 GW and that of the hydroelectric power plants without reservoir regulation is 7 GW. It should be noted that this subdivision is not quite definitive, since several downstream plants benefit indirectly from upstream regulations. The total installed capacity of other environment friendly power plants (based

Table 10.1 Installed capacity of hydroelectric schemes in Turkey

Year	Installed capacity (GW)	Share in total installed capacity (%)
1955	0.04	6
1960	0.4	32
1965	0.5	34
1970	0.7	32
1975	1.8	43
1980	2.1	42
1985	3.9	42
1990	6.8	41
1995	9.9	47
2000	11.2	41
2005	12.9	33
2010	15.8	32
2015	25.9	35
2016	26.7	34
2017[a]	27.3	32

[a]Provisional data

Table 10.2 Hydroelectric energy production in Turkey

Year	Annual hydroelectric production (TWh/y)	Share in total electrical energy production (%)
1955	0.1	6
1960	1.0	35
1965	2.2	44
1970	3.0	35
1975	5.9	37
1980	11.3	46
1985	12.0	33
1990	23.1	41
1995	35.5	42
2000	30.9	25
2005	39.6	24
2010	51.8	25
2015	67.1	26
2016	67.2	25
2017[a]	58	20

[a]Provisional data

on wind, geothermal, and solar energy) in Turkey is 5.4 GW, corresponding to a modest share of 7%, but with an increasing trend.

The total installed capacity of fossile fuel based power plants (coal, oil, gas; mostly imported) is 42 GW, with a dominating share of 57% in the total installed capacity. It is aimed, in the long-term, to decrease this share in favor of the increase of renewable sources, like water power. The first nuclear power plant (1 GW) is actually under construction, and the second is in initial stages of realization.

The total electricity generation in Turkey was around 262 TWh/y in 2015. The total energy generation of hydroelectric plants was 67 TWh/y, and its share in the total was 26% (Table 10.2). This share was in the order of 35–45% in the period of 1960–1995, but declined to around 25% thereafter. The main cause appears to be the continuous terrorist activities, especially in the eastern and south-eastern regions of the country, where the bulk of the hydroelectric potential exists.

The total electricity generation of other renewable power plants (based on wind, geothermal, and solar energy) is 15 TWh/y, corresponding to a modest share of 6%, but shows an increasing trend. On the other hand, the total electricity generation of fossil fuel based power plants (coal, oil, gas; mostly imported) is 180 TWh/y, with a dominating share of 68%. It is aimed, in the long-term, to decrease this share in favor of the increase of renewable sources.

The difference in the share of water power, as 35% in installed capacity and 26% in energy production, reflects the suitability of hydroelectric schemes in covering peak demands. Concurrently, the difference in the share of fossil fuel plants, as 57% in installed capacity and 68% in energy production, reflects the efficiency in a rather continuous operation of these plants. This is especially true for nuclear power plants, also due to technical reasons.

10.3 Hydroelectric Potential

10.3.1 Gross Hydroelectric Potential

The gross hydroelectric potential of Turkey has been intermittently computed since 1955 by using the method of hypsographic curves as well as the method of flow-altitude diagrams along each water course, the latter yielding more accurate results parallel to the areal development of hydrometric gauging stations.

Some of the earlier studies yielded figures above 500 TWh/y, but most of the later studies gave results around 430–440 TWh/y (UN/ECE 1953, 1955; Ozis 1961, 1971, 1985, 1991; DSI 1965; Conturk and Bayar 1968, 1969; Erke 1978, 1985; Baran 1987; Baran et al. 1987, 2006; Doluca et al. 1987; Durnabas 1987; Ozis et al. 1997, 2006b). Besides the stochasticity in the runoff, the gross hydroelectric potential should be considered as virtually invariant with time in the regions where the topography and the hydrology of the region can be satisfactorily covered.

The gross hydroelectric potential of Turkey corresponds to 0.56 GWh/km^2 and 2.4 kWh/m^3; the first figure is close to the European average, the second reflects the dominant effect of the topographic head (Ozis 1985).

10.3.2 Technically Exploitable Hydroelectric Potential

The ratio of the effective net head to the gross head of water power schemes is in the order of 0.5–0.9, so that an average ratio of 0.7 can be taken as 'head efficiency'. The ratio of the effective discharge used in the turbine to the natural discharge at the site of the scheme can be assumed to be in the order of 0.9 as 'discharge efficiency'. The ratio of the electrical power obtained at the switchyard to the mechanical power exerted onto the turbine by the water is in the order of 0.8 as the 'power plant efficiency'.

Hence, the overall ratio of the technically exploitable hydroelectric potential to the gross hydroelectric potential is around 0.5, by multiplication of the three separate 'efficiencies' mentioned above. This corresponds to roughly 220 TWh/y, as half of the gross hydroelectric potential, indicating a theoretical upper limit of water power development. This limit would further be reduced by excluding schemes which may cause impoundment of large settlements, industrial estates, vast agricultural lands, and environmentally important areas.

10.3.3 Economically Feasible Hydroelectric Potential

The estimates for the economically feasible hydroelectric potential of Turkey during 1955–1965 varied in the range of 50–100 TWh/y. It concentrated around 65–70 TWh/y during 1965–1975, in accordance with the areal coverage extension of basin development plans. After the 1973 oil crisis, this value first increased to 110 TWh/y, and towards the end of the century to 125 TWh/y. The current estimates are in the order of 140–150 TWh/y (Ozis 1977, 1985, 1991; Ozis et al. 1983, 2006b; Turfan and Bozkus 1992; Pasin and Altinbilek 1997; Basmaci 2004; Baran et al. 2006; Alashan et al. 2018).

The hydroelectric potential of Turkey ranks in Europe only after that of the Russian Federation, west of Ural mountains, and of Norway. A large number of dams associated with hydroelectric power plants serve also other purposes, such as flood mitigation, irrigation, and water supply. Thus, the development of the hydroelectric potential should have a special priority in the supply of electrical energy in Turkey (Ozis 1961, 1977, 1985, 1986, 1998, 2015; DSI 1965; Ozis et al. 1978, 1983, 1999, 2012, 2013a).

10.4 Clean Energy and Hydroelectric Power Plants

Mankind benefited from the mechanical energy of water power through water mills since roughly two thousand years. The electricity producing hydroelectric schemes were first used towards the end of the nineteenth century. They consist of three basic elements: turbine, generator, and transformer (Fig. 10.3).

The 'turbine' is a hydraulic engine to transform the potential energy of water into the mechanical energy of a rotating axis. Generally, the 'Kaplan' type turbines are used with low heads, the 'Francis' type turbines at medium heads, and the 'Pelton' type turbines at high heads, depending on the combination of heads and discharges. Single 'pump-turbines', suitable to work in both directions, are also used in certain pumped-storage hydroelectric plants.

The 'generator' converts the mechanical energy of a 'rotor' revolving around the axis into electrical energy through the magnetic field created by the winding of the outer 'stator'. The turbine and the generator are usually coupled to the same axis, either vertical or horizontal.

The 'transformer' is a non-rotating electrical device used to increase the tension of the electricity, in order to convey it through longer distances with less losses. The interim transformers are situated in or near the powerhouse; the high tension transformers in a suitable switchyard, apart from the powerhouse.

The quality of water changes neither along the flow through the turbine, nor through any conduits; hence hydropower schemes do not cause environmental pollution. On the contrary, the reservoirs of dams act often as large settling basins,

Fig. 10.3 Ataturk power plant of 2400 MW during construction of eight units at various phases. (Photo by U. Ozis)

so that the outflowing water from the power plant appears much cleaner than the inflowing river water. Thus, water power is basically a quite 'clean' energy resource.

10.5 Types of Hydropower Schemes

10.5.1 Classification of Hydropower Schemes

Hydropower schemes are often classified in different forms in the relevant literature (Creager and Justin 1950; Donmezer 1951; Doland 1954; Ludin and Borkenstein 1955–1958; Guthrie-Brown 1958; Ginocchio 1959; Varlet 1962a, b, 1964, 1965; Mosonyi 1963, 1965; Guner 1967; Press 1967; Schleiermacher 1967; Balman and Guven 1973; Cecen 1976; Basesme 1980; Ozis 1983c, 1991; Yildiz 1992; Ozis et al. 2006a, b).

Civil engineers dealing with water power, including the authors of this chapter, adopted three basic groups of water power schemes, especially suitable for Turkey

(Ozis 1973, 1974a, b, 1983c, 1986, 1991, 1998; Benzeden et al. 2006; Ozis et al. 2006a, b, 2012, 2013a):

(a) power plants at the toe of dams (with two major subdivisions: at the toe of embankment dams, and at the toe of concrete dams);
(b) high-head diversion power plants (with two major subdivisions as: free flow diversion plants, and pressure flow diversion plants);
(c) low-head run-of-river power plants.

Major hydroelectric power plants (above 50 MW installed capacity) in Turkey are listed in Table 10.3 (at the toe of embankment dams), Table 10.4 (at the toe of concrete dams), Table 10.5 (diversion by dams), and Table 10.6 (diversion by weirs). The approximate locations of water power plants with installed capacity above 65 MW were also shown in Fig. 10.1.

10.5.2 Power Plants at the Toe of Dams

The discharges of the water courses in Turkey show significant seasonal (periodic) as well as large annual (stochastic) variations, so that the reservoirs created by dams are of paramount importance in discharge regulation. Tables 10.3 and 10.4 show that the majority of large power plants are situated at the toe of dams, especially of embankment dams. However, there are also several major power plants at the toe of different concrete dam types (Ozis and Kocak 1977; Ozis and Yanar 1984; Ozis and Ozel 1989; Ozis et al. 1990b, c, 1992).

The height (above foundation) of related embankment dams is 207 m at Keban and at Akkopru, 195 m at Altinkaya, 175 m at Hasan Ugurlu, and 169 m at Ataturk Dam (Fig. 10.2). The height (above foundation) of the concrete arch dams Deriner is 249 m, Berke is 201 m, Oymapinar is 195 m, of the concrete gravity dam Boyabat is 185 m, and of the concrete arched gravity dam Karakaya (Fig. 10.4) is 173 m.

10.5.3 High-Head Diversion Power Plants with Dams

Topographical conditions may favor, in some cases, to locate the power plant not at the toe of the dam, but at a much lower elevation, in order to increase the total head of the scheme.

The power scheme Ermenek, with a 211 m high arch dam, is a prominent example to this type. The power scheme Demirkopru, with 2.4 km long power tunnel, is another typical example (Fig. 10.5).

Table 10.3 Major hydroelectric schemes at the toe of embankment dams

Scheme/water course	MW	GWh/y
Ataturk/Firat	2400	8900
Keban/Firat	1330	6000
Altinkaya/Kizilirmak	700	1630
Birecik/Firat	672	2520
Hasan Ugurlu/Yesilirmak	500	1220
Yedigoze/Seyhan	320	1000
Borcka/Coruh	300	1040
Alkumru/Botan	266	1000
Obruk/Kizilirmak	212	475
Kandil/Ceyhan	208	590
Batman/Batman	198	480
Kavsak/Goksu	191	770
Karkamis/Firat	189	650
Ozluce/Peri	170	415
Catalan/Seyhan	169	595
Alpaslan/Murat	160	490
Bagistas/Karasu	141	480
Kigi/Peri	140	420
Aslantas/Ceyhan	138	570
Tatar/Peri	128	420
Hirfanli/Kizilirmak	128	400
Menzelet/Ceyhan	124	515
Kilickaya/Kelkit	120	330
Akkopru/Dalaman	116	340
Muratli/Coruh	115	445
Dicle/Dicle	110	300
Kargi/Kizilirmak	102	290
Yamula/Kizilirmak	100	420
Kralkizi/Dicle	94	145
Kurtun/Harsit	85	200
Uzuncayir/Munzur	84	320
Kesikkopru/Kizilirmak	76	250
Tepekisla/Kelkit	70	240
Suat Ugurlu/Yesilirmak	69	270
Koyulhisar/Kelkit	63	175
Adiguzel/Buyuk Menderes	62	280
Topcam/Melet	60	200
Derbent/Kizilirmak	56	255
Seyhan/Seyhan	54	350
Kapulukaya/Kizilirmak	54	190
Kilavuzlu/Ceyhan	54	100
Seyrantepe/Peri	53	220
Garzan/Garzan	52	165

Table 10.4 Major hydroelectric schemes at the toe of concrete dams

Scheme/water course	MW	GWh/y
4.1. At the toe of concrete gravity dams		
Boyabat/Kizilirmak	513	1470
Goktas-II/Zamanti	153	280
Pembelik/Peri	127	405
Torul/Harsit	106	320
4.2. At the toe of arched gravity dams		
Karakaya/Firat	1800	7350
4.3. At the toe of arch dams		
Deriner/Coruh	670	2120
Oymapinar/Manavgat	540	1600
Berke/Ceyhan	510	1670
Artvin/Coruh	332	1030
Gokcekaya/Sakarya	300	560
Sir/Ceyhan	273	725
4.4. At the toe of roller compacted concrete dams		
Beyhan-I/Murat	293	1290
Kopru/Goksu	156	385
Kargi/Sakarya	97	255
Menge/Goksu	89	200
Feke-II/Goksu	70	225

10.5.4 High-Head Diversion Power Plants with Weirs

High-head diversion power plants with weirs, of modest capacities, were dominant in the 50s and 60s. They were constructed especially in remote areas before the national interconnected power transmission grid covered the entire country.

These types of power plants are still applied in areas of high precipitation (north-east/Black Sea) or on water courses with significant (karst) spring contributions (south-west/Mediterranean). Dogankent on Harsit (Fig. 10.6) and Kepez on Duden are interesting examples (Cecen 1957; Cecen and Bayazit 1967; Arisoy 1988; Alkan 1996; Benzeden et al. 1999; Ozis et al. 2006b).

10.5.5 Low-Head Plants with Weirs

Under consideration of the topography and the hydrology, river basin development plans in Turkey include cascades of high-head diversion plants in the upper part of the river, with or without a reservoir for regulation. In the middle part of the river, they often include multipurpose (energy, flood control, irrigation) dams with high capacity power plants.

Table 10.5 Major hydroelectric schemes with diversion by dams

Scheme/water course	MW	GWh/y
5.1. Diversion by embankment dams		
Arkun/Coruh	245	780
Cinarcik/Orhaneli	110	425
Sariguzel/Ceyhan	103	310
Demirkopru/Gediz	69	190
Esen-I/Esencay	60	190
5.2. Diversion by concrete gravity dams		
Akkoy-II/Harsit	230	890
Sariyar/Sakarya	160	400
Goktas-I/Zamanti	122	220
Akkoy-I/Harsit	103	260
Karakuz/Korkun	76	275
Kadincik-I/Kadincik	70	345
5.3. Diversion by arched gravity dams		
Gullubag/Coruh	96	310
5.4. Diversion by concrete arch dams		
Ermenek/Ermenek	302	1100
Gezende/Ermenek	150	525
5.5. Diversion by concrete buttress dams		
Yamanli-II/Goksu	82	235
5.6. Diversion by roller compacted concrete dams		
Ayvali/Oltu	122	310
Kayabeyi/Kura	85	200

Low-head power plants (Fig. 10.7), anticipated in the lower parts of the rivers with relatively large plains, are hard to encounter in Turkey because of the irrigation needs of the agricultural lands. The diversions of the large part of the water to the irrigation systems, in the lower stretches of the watercourses, leave limited water for eventual low-head power plants and significantly restrict their feasibility.

10.5.6 Pumped-Storage Plants

Pumped-storage plants do not actually exist in Turkey, based neither on daily nor on seasonally operation principle. Several feasibility investigations carried out on schemes located in different parts of the country proved unsatisfactory until now, because a capacity increase of the hydroelectric power plants associated with dams proved to be more efficient. In this regard, it is very important that hydroelectric plants associated with dams in areas of dense electricity consumption should be designed with a low plant load (capacity) factor, hence with high unit capacity, to effectively meet the peak load demand of the system.

Table 10.6 Major hydroelectric schemes with diversion by weirs

Scheme/water course	MW	GWh/y
6.1. Diversion conveyance under pressure flow		
Hacininoglu/Ceyhan	360	130
Aslancik/Harsit	98	350
Kokluce/Kelkit	90	590
Darica-II/Turna	74	245
Buyukduzu/Harsit	69	190
Dogancay/Seyhan	62	170
Uzundere/Uzundere	62	135
Kadincik-II/Kadincik	56	320
6.2. Diversion conveyance mainly under free flow		
Karica/Melet	110	400
Cevizlik/Iyidere	90	395
Camlica-I/Zamanti	84	430
Kozbuku/Melet	81	260
Akocak/Karadere	81	255
Kasimlar-II/Ayvali	75	200
Dogankent A&B/Harsit	73	335
Kovada-II/Aksu	53	220
Sanliurfa/S.Urfa-kanal	52	125

Fig. 10.4 Karakaya arched gravity dam and power plant of 1800 MW capacity on Euphrates. (Photo by U. Ozis)

Fig. 10.5 Demirkopru diversion scheme, with 107 m head (fed from the embankment dam on Gediz river by the 2.4 km long pressure tunnel), surge tank, penstocks, and powerhouse. (Photo by U. Ozis)

The fact that the center of gravity of the electricity consumption lies in the north-west of the country, but the center of gravity of the hydroelectric production lies in the south-east of the country, which will still be the case in the near future, may eventually lead to construct certain pumped-storage plants in the vicinity of dense electricity consuming regions in Turkey (Ozis 1968; Harmancioglu et al. 1978; Melikoglu 1980; WPDC 2000; Ozis et al. 2006b; Sarac 2009a; Arisoy and Ilker 2015). Moreover, the use of floating solar energy collectors on the reservoirs of large dams in Turkey, with hydroelectric power plants converted partially to pumped-storage plants, was already investigated in the seventies (Ozis et al. 1977).

Fig. 10.6 Dogankent-A power plant, penstocks of 1.85 m diameter, on Harsit river. (Photo by U. Ozis)

Fig. 10.7 Low-head, run-of-river Maras hydroelectric plant, of 3.6 MW, on Ceyhan river. (Photo by U. Ozis)

10.6 Elements of Hydropower Schemes

10.6.1 Dams and Intakes

The majority of large and middle range hydroelectric power schemes in Turkey are associated with embankment dams (Tables 10.3 and 10.5). The predominant embankment type is rock-fill with central clay core (Fig. 10.8). Earth-fill dams, and rock-fill dams with inclined clay cores or concrete covered upstream faces, are less frequent. Intakes at the abutment, to a lesser extent tower intakes, are used for power plants at the toe of embankment dams as well as for high-head diversion plants associated with dams.

Concrete dams are less frequently used in hydroelectric power schemes in Turkey (Tables 10.4 and 10.5). Examples of almost all types can be encountered: concrete gravity, arched gravity, concrete arch (Fig. 10.9), buttressed concrete, and roller compacted concrete dams. Direct intakes on the dam body are often used for power plants at the toe of concrete dams.

The highest concrete dam in Turkey is actually Deriner on Chorokhi river, with a height of 209 m above thalweg and 249 m above foundation (Kocaker 2010). The Yusufeli dam on the same river, with a height of 270 m above foundation, is under construction (Sarac 2009b).

High-head diversion plants to divert the water and the few low-head run-of-the-river plants to create the head are associated by weirs. In the case of diversion plants with weirs, water is taken by lateral, bottom (tyrolian), or frontal intakes. The latter was developed by Kazim Cecen at the Istanbul Technical University, in order to cope with the high sediment load of Turkish water courses. It was first used at the Ikizdere scheme in 1961 (Fig. 10.10) together with bottom intake, and then at the Dogankent-A scheme in 1971, alone (Cecen 1957, 1967; Cecen and Bayazit 1966).

The run-of-the-river type high-head diversion plants are often equipped with a desilting basin at the beginning of the diversion conveyance (Fig. 10.11). A circular

Fig. 10.8 Altinkaya dam and power plant of 700 MW on Kizilirmak (intakes at the right abutment). (Photo by U. Ozis)

Fig. 10.9 Gokcekaya dam and power plant of 300 MW on Sakarya (intakes on the dam body). (Photo by U. Ozis)

Fig. 10.10 Ikizdere high-head diversion scheme on Iyidere, diversion weir and frontal intake (Photo by U. Ozis)

desilting basin for central discharges and large grain sizes was developed by Neset Akmandor, the former general director of DSI, and implemented at the existing Sizir hydroelectric scheme in 1980 (Akmandor 1973; Cecen 1977, 1981; Bayazit 1998).

Fig. 10.11 The desilting basin of the Dogankent high-head diversion scheme on Harsit (Photo by U. Ozis)

10.6.2 Conduits

Pressure tunnels excavated underground, and/or steel penstocks embedded or visible, convey the water from the intake to the turbine in power plants at the toe of dams. The units of powerhouses at the foot of concrete dams are usually served by individual penstocks. The units of powerhouses at the toe of embankment dams are usually served by conduits bifurcating from the main conduit.

The inner diameter of the pressure tunnels are 9.8 m at Altinkaya, 9.0 m at Kesikkopru, 8.8 m at Aslantas, and 8.4 m at Hasan Ugurlu schemes at the toe of dams. The inner diameter of the penstocks are 8.4–6.9 m at Birecik, 7.5–6.6 m at Ataturk (Fig. 10.12), 7.0 m at Karakaya, 6.4 m at Kesikkopru, 5.5 m at Derbent, 5.2 m at Keban, and 5.0 m at Altinkaya schemes at the toe of dams. The inner diameters of the shaft penstocks are 6.0–5.3 m at Hasan Ugurlu and 5.2 m at Oymapinar underground powerhouses. The inner height of the free flow outlet tunnels is 18 m at Oymapinar and 10 m at Hasan Ugurlu underground powerhouses.

Depending on the relation of head and discharge, a surge tank or tanks (mostly of differential type) are also located towards the end of the penstocks (Fig. 10.13).

In the case of high-head diversion plants, the main conduit is generally a pressure tunnel of low pressure, or eventually a free flow canal, for the diversion stretch, and penstocks or pressure shafts for the steep conduit. A steel surge tank, or tanks, are located at the transition from the low-gradient pressure tunnel to the steep-gradient penstocks. Similarly, a masonry surge pond is located at the transition from the free-flow open channel to the steep-gradient penstocks.

The inner diameter of the pressure tunnels are 2.4–2.1 m at Kepez and 1.85 m at Dogankent-A diversion power schemes associated with weirs. The inner diameter of the penstocks are 4.2 m at Sariyar, 3.3–3.1 m at Kadincik-I, 3.0 m at Gezende, and 2.8 m at Demirkopru diversion schemes associated with dams.

Fig. 10.12 Penstocks of the Ataturk power plant with diameters of 7.5–6.6 m. (Photo by U. Ozis)

Fig. 10.13 Seyhan dam on Seyhan river, surge tanks and power plant. (Photo by U. Ozis)

The height of the surge tank is 66 m at Kadincik-II, 61 m at Demirkopru, and 57 m at Kadincik-I and Kepez-I (Fig. 10.14). The inner diameter of the surge tank is 28 m at Sariyar, 16 m at Dogankent, and 14.9 m at Demirkopru high-head diversion plants. The surge tank diameter is 23 m at Aslantas and 19.5 m each at Seyhan power plants at the toe of dams.

The high-head diversion plants have rarely individual penstocks for each unit. More often, it is a main penstock of a larger diameter, from which penstocks of lesser diameters bifurcate.

Fig. 10.14 Kepez I high-head diversion scheme and penstock in the direction of Antalya Bay. (Photo by U. Ozis)

10.6.3 Powerhouse

The majority of the powerhouses in Turkey are constructed as indoor type powerhouses (Ataturk, Karakaya, Birecik, Boyabat, Borcka). Some are of semi-outdoor type (Keban, Gokcekaya) with portal cranes on top of the powerhouse (Figs. 10.15 and 10.16).

There are also several large underground powerhouses (Fig. 10.17) at the toe of dams (Oymapinar, Hasan Ugurlu, Berke) or at the diversion plants (Ermenek, Dogankent-B).

Large powerhouses display significant dimensions (length of 258 m at Ataturk, 184 m at Karakaya, 177 m at Altinkaya, and 171 m at Keban; height of 60 m at Karakaya, 55 m at Ataturk and Sir, 43 m at Oymapinar, and 35 m at Altinkaya; width of 81 m at Karakaya, 49 m at Ataturk, 46 m by Altinkaya, and 41 m by Keban).

The majority of the power plants at the toe of dams or high-head diversion plants are equipped with Francis-turbines. The unit capacity is 300 MW at Ataturk (8 units) and Karakaya (6 units), 175 MW at Keban (8 units) and Altinkaya (4 units), and 170 MW at Berke (3 units). Some unit discharges exceed 200 m^3/s (230 m^3/s at Karakaya and 220 m^3/s at Ataturk and Birecik).

Fig. 10.15 Ataturk indoor type powerhouse. (Photo by U. Ozis)

Fig. 10.16 Keban semi-outdoor type powerhouse. (Photo by U. Ozis)

Certain diversion plants with very high heads are equipped with Pelton-turbines (1220 m at Akkoy-II, 1010 m at Birkapili, 676 m at Esen-I, 384 m at Kovada-II, and 340 m at Lamas-IV).

Fig. 10.17 Oymapinar underground powerhouse. (Photo by U. Ozis)

Besides modest run-of-river type power plants, some powerhouses at the toe of low dams are equipped with Kaplan-turbines (Manavgat, Suat Ugurlu) or tubular turbines (Karkamis).

10.7 Discharge Regulation and Dams

As noted earlier, the discharges of the water courses in Turkey show significant seasonal (periodic) variations as well as year-to-year (stochastic) fluctuations. This is still valid for river basins which are largely fed, besides rain, from snowmelt and karst spring discharges (EIE 1955–1993; DSI 1961–1995; 1965; Ozis 1971; Ozis et al. 1997). Therefore, large reservoirs have to be created by dams on almost all water courses in order to regulate the discharges for water power, irrigation, water supply, flood control (Fig. 10.18) (DSI 1967; Harmancioglu and Ozis 1981, 1983; Ozis 1982a, b, c, 1983a, b, 1986, 1994, 2015; Ozis et al. 1992, 2004, 2006b, 2013b;

Fig. 10.18 Keban embankment dam on Euphrates and the 1330 MW power plant. (Photo by U. Ozis)

Fistikoglu et al. 2008; Ozis and Ozdemir 2009; Sekkeli and Kececioglu 2011; Ozdemir et al. 2013; Kankal et al. 2014).

As an example, the discharges of Euphrates at the location of Dutluca gaging station, just downstream of the Ataturk Dam, which represents about five sixths of the Euphrates Subbasin flows originating from Turkey, are considered. They showed an average value of around 870 m^3/s during 1937–1980 water years, the annual averages varying from 460 to 1500 m^3/s. Monthly mean discharges, rising up to 4360 m^3/s in May 1969, were usually in the order of 300–500 m^3/s during 8 months from July to February. The latter figures are well below the average and much less during months like August and September, even decreasing to around 170 m^3/s (Ozis et al. 2000).

Discharge fluctuations are somewhat dampened in basins with intense precipitation, as encountered in the eastern Blacksea region. This is an advantageous feature with regard to the difficulty of dam construction in the topography of this region (Ozis et al. 1990a; Alkan 1996). Furthermore, groundwater issuing as karst springs from soluble limestone formations, dampens also the discharge fluctuations, as encountered along a belt in the south and southeast Turkey, extending from western Mediterranean basins up to the Euphrates-Tigris basin (Figs. 10.19 and 10.20) (Ozis and Yanar 1984; Ozis et al. 1985, 1993; Ozis 1989). These dampening effects result primarily in the implementation of high-head diversion power plants in such basins.

Fig. 10.19 Dumanli spring with 50 m^3/s mean discharge, submerged later in the reservoir of the Oymapinar dam in Manavgat river basin. (Photo by U. Ozis)

10.8 Peak Capacity Supply

The demand on electrical energy, therefore the load demand from the system, changes substantially by seasons as well as within the day. Typical load curve shapes are shown on Fig. 10.21. These curves show that some of the power plants used to cover the peak load demand are not to be used during several hours of the day. Nuclear power plants have to be operated continuously because of technical reasons. The discontinuation of thermal plants results in energy losses due to cooling and time losses in restarting the units, even in combined schemes. Hydroelectric plants are taken out of production with quite small water losses and restarted within minutes, so that they are the best suited power plant types for covering the peak demands. In this context, it is very important that hydroelectric plants, associated with dams in areas of dense electricity consumption, should be designed with low

Fig. 10.20 The 185 m high (above foundation) Oymapinar concrete arch dam. (Photo by U. Ozis)

plant load (capacity) factor, hence with high unit capacities, to effectively meet the peak load demand of the system.

Pumped-storage hydroelectric plants, though not implemented yet in Turkey, are especially of interest in this regard. The center of gravity of the electricity consumption lies in the north-west of the country, but the center of gravity of the hydroelectric production lies in the south-east of the country; and this will still be the case in the near future. It might be interesting one day in the future to construct certain pumped-storage plants in the vicinity of dense electricity consuming regions of the country.

10.9 Development of Hydropower Resources

10.9.1 Past Steps

The first electricity was generated in Turkey in 1902 at the low-head, run-of-river Tarsus hydroelectric power plant in Tarsus, which made use of the head created by the falls of the Berdan river. The few hydroelectric schemes of the first half of the

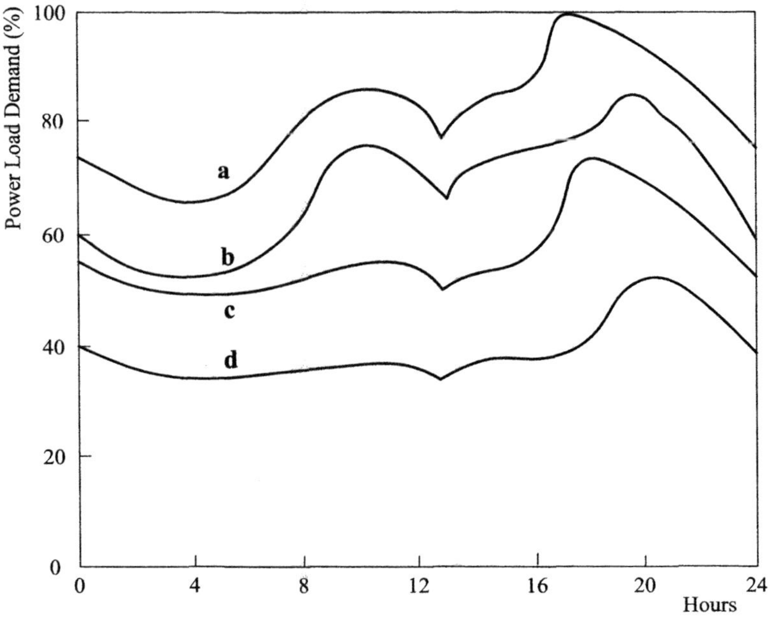

Fig. 10.21 Typical load curves of the demand in an interconnected system: (**a**) winter working day; (**b**) summer working day; (**c**) winter holiday; (**d**) summer holiday

twentieth century were isolated plants of small capacity, developed mostly as diversion plants.

Systematic discharge gauging began in 1936 by the establishment of EIE, Elektrik Isleri Etud Idaresi (Electrical Works Survey Administration) (Erten 1970). Systematic planning of river basin development activities began in 1954 by the establishment of DSI, Devlet Su Isleri (State Hydraulic Works Administration) (Demir 2001). The first two major hydroelectric power plants associated with dams, the high-head diversion scheme Sariyar with the concrete gravity dam (Fig. 10.22), and the power plant at the toe of the Seyhan embankment dam, began operation in 1956.

Dozens of major hydroelectric schemes associated with large dams have been constructed in the following decades. The interconnected transmission system began to develop in the fifties and covered the entire country in subsequent decades (Fig. 10.23).

The TEK, Turkiye Elektrik Kurumu (Turkey's Electricity Administration) was established in 1971 and split in 1993 as TEAS, Turkiye Elektrik Uretim ve Iletim AS (Turkey's Electricity Generation and Transportation Corporation) and TEDAS, Turkiye Elektrik Dagitim Anonim Sirketi (Turkey's Electricity Distribution Corporation).

The 207 m high Keban dam and its 1330 MW hydroelectric power plant began operation in the mid-seventies (Noyan 1965). The feasibility report of the large river

Fig. 10.22 Sariyar concrete gravity dam on Sakarya. (Photo by U. Ozis)

Fig. 10.23 The main switchyard of the Karakaya hydroelectric power plant, which is the link to the interconnected transmission network of Turkey. (Photo by U. Ozis)

basin development plan for Lower Euphrates was prepared at the end of sixties, but had to undergo, after the oil crisis of 1973, certain very important changes in the seventies, rarely encountered in water resources development history. This plan of 1970 (Electrowatt et al. 1970) foresaw two dams (Golkoy and Middle Karababa) downstream of Karakaya, and the bulk of the irrigation water had to be pumped. Their next best alternative foresaw a single dam (High Karababa) and to convey the large part of the irrigation water by gravity.

Comparisons based on price increases after the oil crisis showed first that the difference between these two alternatives faded away (Harmancioglu 1976). Then, detailed analyses led to the selection of the high dam alternative (High Karababa, from 1978 onwards named Ataturk dam), producing 600 GWh/y more energy, irrigating the Urfa region by gravity, and the Mardin region with less pumping energy (DSI 1978; Harmancioglu and Ozis 1978, 1981, 1983; Ozis and Harmancioglu 1994). The Lower Euphrates planning induced the largest basin development plan in Turkey, GAP, Guneydogu Anadolu Projesi (Southeastern Anatolia Project) (DSI 1980, 2000; Ozis 1982a, b, c, 1983a, b; Ozbek 1989; Altinbilek 1997, 2004). More details on GAP can be found in Sect. 12.6 of this book.

The re-encouragement of the private initiative in water power development, with legal context and indirect financial subsidies for the private sector, began in 1991 by the YID, Yap-Islet-Devret (Build-Operate-Transfer) model. This model was enhanced by the Energy Market Law and the creation of EPDK, Enerji Piyasasi Duzenleme Kurumu (Energy Market Regulation Authority) established in 2001 (Basmaci 2004, 2005, 2006; Keloglu 2006; Tutus 2006; Baskan 2011; Sen 2011; Kentel and Alp 2013).

The first decade of the twentieth century witnessed a boom of interest towards hydroelectric projects, dampened somewhat by terrorism effects and environmental opposition during the next decade.

10.9.2 Comparative Evaluations

The hydroelectric power plants of Turkey are perhaps not as gigantic as some famous schemes like Sanxia on Chang-Jiang in China with 22,400 MW capacity or Itaipu on Parana between Brazil and Paraguay with 14,000 MW capacity, but several schemes in Turkey deserve international interest.

The hydroelectric energy production of Turkey was only 1 TWh/y in 1960, but the production potential increased to around 75 TWh/y in 2017, which is equal to half of the ecomically feasible hydroelectric potential of the country. The total installed capacity of hydroelectric power plants was 0.4 GW in 1960 and increased to 27.3 GW in 2017, thus nearly 70 times in 57 years.

The largest hydroelectric power plant was Sariyar with 160 MW in 1960 and reached 2,400 MW at the Ataturk power plant in 1992, corresponding to an increase of 15 times. The most powerful unit was in Sariyar with 40 MW in 1960 and reached 300 MW at the Ataturk power plant in 1992, corresponding to an increase of 7.5 times.

Major hydroelectric schemes in Turkey, of installed capacities exceeding 50 MW, were listed in Tables 10.3, 10.4, 10.5 and 10.6. Roughly 80% of them are associated with dams, reflecting the regulation and storage requirements of Turkey's water courses. Cascades of water power schemes exist and/or are planned on several water courses in Turkey, such as Euphrates (Fig. 10.24), Tigris, Chorokhi (Fig. 10.25), Kizilirmak, Ceyhan, Seyhan, Lamas (Fig. 10.26), and some others. Two thirds of the

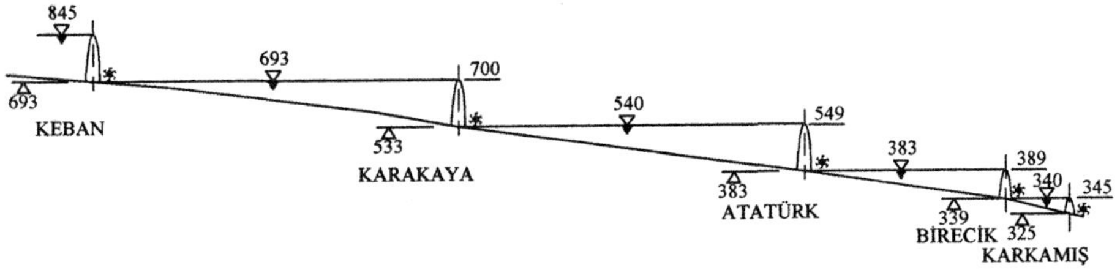

Fig. 10.24 Schematic longitudinal section of Lower Euphrates, with five existing dams and the power plants at the toe of the dams

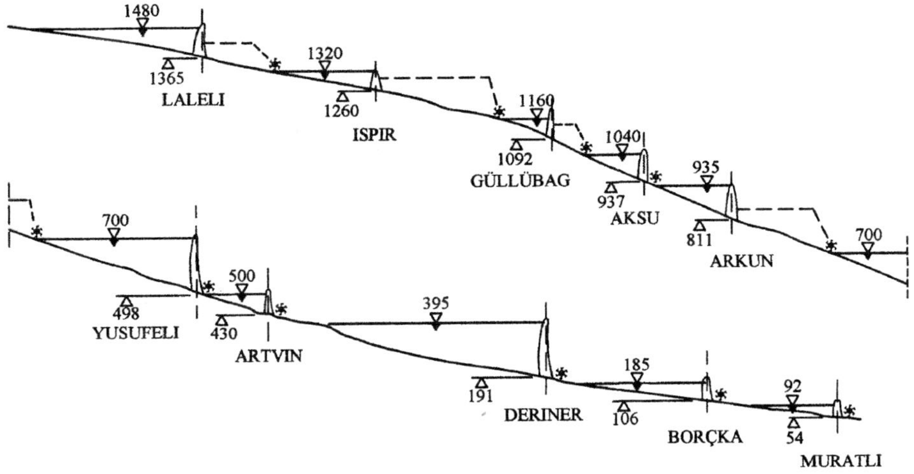

Fig. 10.25 Schematic longitudinal section of Chorokhi River, with ten dams, the power plants of the upper five are high-head diversion plants with dams and the lower five are power plants at the toe of dams

major hydroelectric schemes have power plants at the toe of the dams; around 3/4 of them at the toe of embankment and 1/4 at the toe of concrete dams. Arch dams were used in higher gravity dams and in less higher concrete dams, with a recent trend on roller compacted concrete dams. Roughly 15% of the major hydroelectric schemes, equal to 20% of the schemes associated with dams, are high-head diversion schemes, where part of the head is created by the dam, part of it with diversion to lower elevations when suitable topographic conditions favor such a formulation.

High-head diversion plants with weirs make roughly one fifth of the major hydroelectric schemes, a few with heads over 1000 m elevation difference. These type of schemes are suitable in basins with very high precipitation and moderate seasonal variation, like the northeastern Anatolia, or in basins with significant karst springs contribution, like the Mediterranean region of Turkey.

The number of low-head run-of-river hydroelectric schemes is quite limited and their installed capacities quite low in Turkey. In fact, at the lower reaches of water

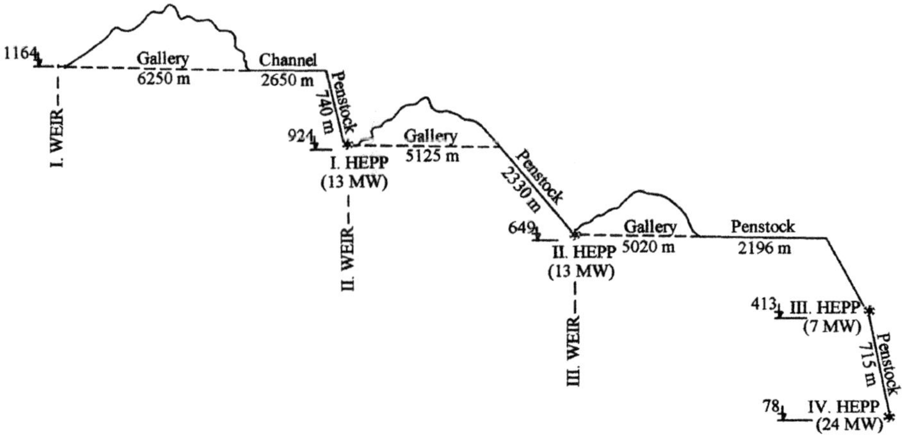

Fig. 10.26 Schematic longitudinal section of Lamas River, encompassing four high-head diversion plants by weirs

courses, where such schemes would likely be installed, the bulk of the river discharge has to be diverted for irrigation purposes, and the remaining part is often far from being feasible for the implementation of such schemes.

10.9.3 Environmental Considerations

An increasing resistance against power plants in general, against hydroelectric schemes in particular, is being observed in Turkey. In fact, certain persons and/or non governmental organizations claim that the hydroelectric plants:

(a) significantly alter the environment with regard to morphology, plant cover, and wildlife;
(b) disturb the local population during construction activities;
(c) cause migration because of the reservoirs of dams;
(d) reduce the surface water by diversion, storage, and evaporation;
(e) reduce and divert the groundwater flow;
(f) increase the price of the water.

On the other hand, in covering the power and energy needs, it is of paramount importance to note that the hydroelectric energy is:

(a) renewable due to the hydrological cycle;
(b) quite clean with regard to environmental pollution;
(c) often economically more feasible compared to other power plants;
(d) especially versatile in covering peak demands;
(e) an important domestic energy resource compared to other resources;

(f) serving other purposes like flood mitigation, irrigation, and water supply, especially when associated with dams.

Therefore, in order to prevent environmental conflicts, emphasis should be put on causing minimal disturbance to the nature and to the population in the project area during construction, and on strictly respecting the local water allocation discharges during operation.

10.9.4 Trends in the Development of Hydroelectric Energy

The trends in the increase rate of annual hydroelectric energy generation of countries with considerable water power potential have been investigated in the late seventies (Ozis et al. 1978, 1983; Ozis and Harmancioglu 1982). The annual hydroelectric energy generation values between 1951 and 1974 of 55 countries around the world, whose 1974 generation exceeded 1 TWh/y, were evaluated. The goodness of fit of various relations were analyzed. Power and logistic models appeared to be more appropriate.

The initial years of development to reach a few percent of the economically feasible potential display irregularities in general. The same is valid for the last few percent. Thus, the length of the development period to be estimated from the trend models was set as the time required to increase the hydroelectric energy production from 2.5% to 97.5% of the economically feasible potential of the country under consideration. The investigations were set forth with 20 European countries, where the data for the economically feasible hydroelectric potential were available. The resulting estimated durations varied from 30 to 100 years. The average of estimated durations was 63 years for the linear, 53 years for the power, and 67 years for the logistic model.

The starting point for the evaluations related to Turkey was taken as 1970, with an annual production of 3 TWh/y. A meaningful end target was the year 2023, the centennial of the foundation of the Republic of Turkey, corresponding to a development duration of 53 years. The annual hydroelectric energy generation in Turkey from 1970 onwards is shown on Fig. 10.27. The increase rate until the late years of the nineties appeared to be in conformity with this target with a power model (Ozis et al. 2008, 2012; Ozis 2015; Melikoglu 2013).

However, the relentless struggle against terrorism, and several dry years slowed down this increase rate. An exponential model, as given in Eq. 10.1:

$$\hat{E} = 4.74^{*} \exp\left(0.06^{*}\, t\right) \qquad (10.1)$$

appears now to be more suitable, where $\hat{E}$ is the expected annual hydroelectric energy generation in TWh/y, and t is the number of years elapsed since 1969 (thus for 1970 t is 1).

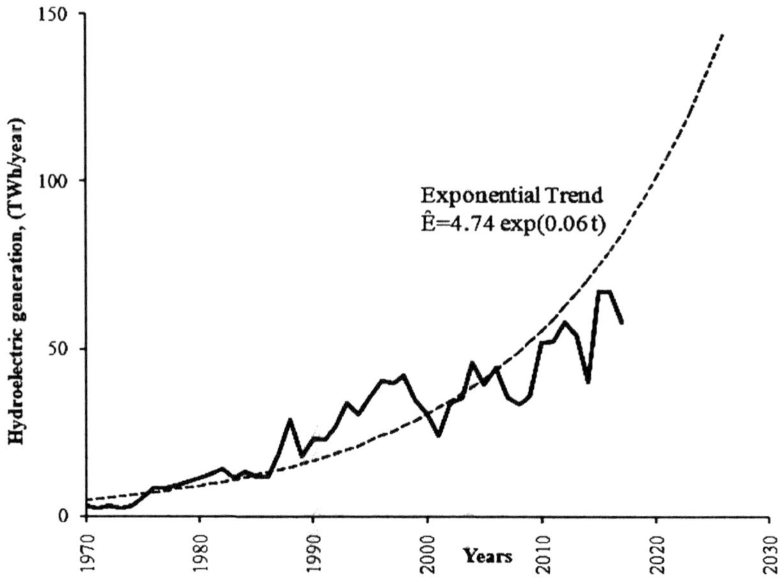

Fig. 10.27 Annual hydroelectric energy generation in Turkey and the corresponding exponential model (t is the number of years elapsed since 1969)

This exponential model in Eq. 10.1 gives 129 TWh/y for the year 2023. The model gives the year 2026 for reaching a production of 146 TWh/y, which will roughly equal 97.5% of 150 TWh/y, the ecomically feasible hydroelectric potential of Turkey.

10.10 Conclusion

Water power, being a renewable and clean energy resource, is one of the most important domestic energy resources in Turkey. Until 2017, with a production capability of 75 TWh/y, only half of the economically feasible hydroelectric potential of the country has been implemented. The development of the entire economically feasible hydroelectric potential is of paramount importance for the country.

Following the discussions in the preceding sections of this chapter, the major conclusions regarding water power in Turkey can be summarized as in the following:

- The hydroelectric energy production of Turkey was only 1 TWh/y in 1960, but the production potential increased to around 75 TWh/y in 2017, which is equal to half of the ecomically feasible hydroelectric potential of the country. The total installed capacity of hydroelectric power plants was 0.4 GW in 1960 and increased to 27.3 GW in 2017, thus nearly 70 times in 57 years. The economically feasible hydroelectric potential of Turkey is in the order of 150 TWh/y. Turkey anticipates to harness the remaining part of this potential, so that several hundreds of major hydroelectric power plants (HEPP) have still to be constructed in the future.

- Roughly 80% of the major hydroelectric schemes in Turkey are associated with dams, either at the toe of the dams (65%) or high-head diversion plants (15%). Dams create thus the necessary head of the power scheme entirely (at the toe) or partially (diversion). Two thirds of the major hydroelectric schemes have power plants at the toe of the dams; around 3/4 of them at the toe of embankment and the remaining 1/4 are concrete dams. Roughly 15% of the major hydroelectric schemes, equal to 20% of the schemes associated with dams, are high-head diversion schemes, where part of the head is created by the dam, part of it with diversion to lower elevations when suitable topographic conditions favour such a formulation.
- The discharges of the water courses in Turkey show significant seasonal (periodic) variations as well as year-to-year (stochastic) fluctuations; even in river basins which are largely fed, besides rain, from snowmelt and karst spring discharges. Therefore, the regulation of discharges through the reservoirs created by dams becomes a necessity. Furthermore, many of the dams serve also other purposes, like flood mitigation, irrigation, and eventually urban & industrial water supply.
- High-head diversion plants with weirs make roughly one fifth of the major hydroelectric schemes, a few with heads over 1000 m elevation difference. These schemes are suitable in basins with high precipitation and moderate seasonal variation. Low-head run-of-river hydroelectric schemes are quite limited and their installed capacities quite low in Turkey.
- Last but not the least, in order to prevent environmental conflicts, emphasis should be put on causing minimal disturbance to the nature and to the population in the project area during construction, and on respecting strictly the local water discharge allocations during operation.

Acknowledgment The authors are grateful for the support of their work on water power and hydroelectric schemes to the State Hydraulic Works Authority (DSI), Electrical Works Survey Authority (EIE), Turkey's Electricity Administration (TEK), Turkey's Electricity Generation and Transportation Corporation (TEAS), Turkey's Scientific and Technical Research Institution (TUBITAK), Iller Bankasi, Cukurova Elektrik AS, Kepez Elektrik AS, Ata Insaat AS, and last but not the least to Ege and Dokuz Eylul Universities. The first author is indebted to Prof. Dr. Fritz Hartung supervising his Ph.D.-Thesis on water power in Turkey in the late 50s and early 60s; to Prof. Dr. Kazim Cecen, former Minister Dr. Eng. Neset Akmandor, Prof. Dr. Istemi Unsal, M.S.C. E. Hikmet Yanar, M.S.C.E. Erdogan Basmaci, M.S.C.E. Ilker Ozel, C.E. Abidin Dursun for their encouragement. He appreciates the support and contributions of Prof. Dr. Yalcin Arisoy and other colleagues in his investigations, most of them being his former students and whose names often appear as co-authors in the lists of references.

The authors extend their thanks to the Springer Institution and to the editors Prof. Dr. Nilgun Harmancioglu and Prof. Dr. Dogan Altinbilek for inviting them to prepare this chapter for the book "Water Resources of Turkey".

References

Akmandor N (1973) Tabani yatay daire bicimli cokeltme havuzunda cokeltme olayi. Dissertation, Istanbul Teknik Universitesi

Akpinar A (2013) The contribution of hydropower in meeting electric energy needs: the case of Turkey. Renew Energy 51:206–219

Akpinar A, Komurcu M, Kankal M (2008) Energy situation and renewables in Turkey and environmental effects of energy use. Renew Sust Energ Rev 12:2013–2039

Alashan S, Sen Z, Toprak ZF (2018) Hydroelectric potential of Turkey: a refined calculation method. Arab J Sci Eng 4:1511–1520

Alkan A (1996) Mathematical modelling of flow duration curves to be used for energy computations in high-head free-flow diversion type hydroelectric schemes, with special emphasis on karst effluents. Dissertation, Dokuz Eylul Universitesi

Altinbilek D (1997) Water and land resources development in Southeastern Turkey. Int J Water Resour Dev 3:311–332

Altinbilek D (2004) Development and management of the Euphrates-Tigris basin. Int J Water Resour Dev 1:15–33

Arisoy Y (1988) Lamas havzasi, Sariaydin (17-12) ve Kizilgecit (1717) akislarinin ustel matematik modelleri ve karst pinar katkilarinin belirlenmesi. DSI Teknik Bulteni 67:49–54

Arisoy Y, Ilker PB (2015) Determining the location of the center of gravity of the energy production and consumption of Turkey. In: Fifteenth International Multidisciplinary Scientific Geoconference, 2015 Albena, vol. 4/1, SGEM, Sofia, pp 161–166

Balman V. Guven, S (1973) Su kuvveti muhendisligine giris. Guven, Ankara

Baran T (1987) Turkiye'nin guneyindeki akarsu havzalarinin brut su kuvveti potansiyeli. M.S.-Thesis, Dokuz Eylul Universitesi

Baran T, Durnabas I, Ozis U (1987) Turkiye'nin su kuvveti potansiyeli. In: IX. Teknik Kongre, vol. II. Insaat Muhendisleri Odasi, Ankara, pp 49–83

Baran T, Dalkilic Y, Ozdemir Y, Ozis U (2006) Turkiye'nin su kuvveti potansiyelinin belirlenmesindeki asamalar ve gelismeler. In: Turkiye 10. Enerji Kongresi, Dunya Enerji Konseyi Turk Milli Komitesi, Istanbul, pp 211–222

Basesme H (1980) Hidroelektrik santrallar ve hidroelektrik tesisler. Turkiye Elektrik Kurumu, Ankara

Baskan A (2011) Liberalization of Turkey's hydroelectricity sector. In: Kibaroglu A, Scheumann W, Kramer A (eds) Turkey's water policy, national frameworks and international cooperation. Springer, Berlin, pp 83–91

Basmaci E (2004) Enerji darbogazi ve hidroelektrik santrallarimiz. Devlet Su Isleri Vakfi, Ankara

Basmaci E (2005) Yeni donemde hidroelektrik santrallarin ozel sektorce gelistirilmesi. Devlet Su Isleri Vakfi, Ankara

Basmaci E (2006) Su politikalari ve hidroelektrik santrallar. In: TMMOB Su politikalari Kongresi. Insaat Muhendisleri Odasi, Ankara, pp 296–307

Bayazit M (1998) Vortex settling basins. In: Terzioglu A, Bayazit M (eds) In Memoriam – Prof. Dr. Kazim Cecen – Anma Kitabi – Gedenkschrift. Renk Ajans, Istanbul, pp 171–176

Benzeden E, Baran T, Alkan A, Arisoy Y, Fistikoglu O, Ozis U, Tuncay S, Gul A, Dalkilic Y, Ozdemir Y (1999) Mersin-Limonlu havzasi Lamas hidroelektrik santrallari. In: Turkiye Insaat Muhendisligi 15. Teknik Kongresi. Insaat Muhendisleri Odasi, Ankara, pp 697–712

Benzeden E, Ozis U, Turkman F, Cetinkaya C, Ozdemir Y (2006; 3.pr. 2014) Su yapilari sistemleri. In: Ozis U, Turkman F, Baran T (eds) Su kaynaklarinin gelistirilmesi. Dokuz Eylul Universitesi, Izmir, p 2.1–2.26

Cecen K (1957) Wasserkraftanlage Ikizdere. Der Bauingenieur 8:300–303

Cecen K (1967) Die Wasserfassung bei Gebirgsflüssen. Die Wasserwirtschaft 10:359–363

Cecen K (1976) Su kuvvetleri. Istanbul Teknik Universitesi, Istanbul

Cecen K (1977) Hydraulic criteria of settling basins for water treatment, hydropower and irrigation. In: IAHR XVII. Congress, vol 4. International Association for Hydraulic Reearch, Baden-Baden, pp 275–294
Cecen K (1981) Sizir hidroelektrik tesisinde yeni bir cokeltme sistemi. Doga Bilim 5:137–149
Cecen K, Bayazit M (1966) Harsit baglamasi su alma yeri model deneyleri. Turkiye Muhendislik Haberleri 131:20–24
Cecen K, Bayazit M (1967) Isletmedeki bir hidroelektrik tesisi uzerinde inceleme. Istanbul Teknik Universitesi, Istanbul
Conturk H, Bayar O (1968) Turkiye'nin su potansiyeli. Turkiye Muhendislik Haberleri 164:9–12
Conturk H, Bayar O (1969) Turkiye'nin brut hidroelektrik potansiyeli. EIE Bulteni 23:11–16
Creager WP, Justin JD (1950) Hydroelectric handbook. Wiley, New York
Demir A (2001) Su ve DSI tarihi. Devlet Su Isleri Vakfi, Ankara
Doland JJ (1954) Hydropower engineering. Ronald, New York
Doluca K, Kulen M, Elbir S (1987) Hydroelectric potential of Turkey. In: Water for peace, 1967 Washington. United Nations, New-York
Donmezer H (1951) Su kuvveti tesisleri. Teknik Okul, Istanbul
DSI (1961–1995) Rasat yilliklari (1959 ila 1989). Devlet Su Isleri, Ankara
DSI (Ozis U, Kargi Y, Akinci V) (1965) DSI Hidroelektrik yilligi 1963. Devlet Su Isleri, Ankara
DSI (Ural OM, Ungan U) (1967) Large dams in Turkey. Devlet Su Isleri, Ankara
DSI (Melikoglu O, Sahcihan S, Tekdal I, Cetinbas Y, Yuksel M) (1978) Yuksek Karababa Baraji ve HES ozet planlama raporu. Devlet Su Isleri, Ankara
DSI (Uskay S, Taskapilioglu E, Gurkan N, Palali Y, Toprakoglu H, Abidoglu A, Koknel B) (1980) Guneydogu Anadolu Projesi. Devlet Su Isleri, Ankara
DSI (1995a) Haritali istatistik bulteni. Devlet Su Isleri, Ankara
DSI (1995b) Turkiye'deki barajlar ve hidroelektrik santrallar. Devlet Su Isleri, Ankara
DSI (2000) GAP 2000, Guneydogu Anadolu Projesi. Devlet Su Isleri, Ankara
Durnabas I (1987) Turkiye'nin kuzeyindeki akarsu havzalarinin brut su kuvveti potansiyeli. M.S.-Thesis, Dokuz Eylul Universitesi
EIE (1955–1993) Akim neticeleri (1936 ila 1989). Elektrik Isleri Etud Idaresi, Ankara
Electrowatt – Tipton and Kalmbach – Société Générale pour l'Industrie – Gizbili (1970) Lower Firat project feasibility report. Devlet Su Isleri, Ankara
Erke H (1978) Turkiye'nin hidroelektrik enerji potansiyeli. In: Turkiye 3. Genel Enerji Kongresi. Dunya Enerji Konferansi Turk Milli Komitesi, Ankara, pp 292–375
Erke H (1985) Ulkemizdeki hidroelektrik enerji potansiyelinin durumu. DSI Aylik Haber Bulteni 283:29–56
Erten S (1970) EIE 33 yil: 1935–1967. Elektrik Isleri Etud Idaresi, Ankara
Fistikoglu O, Ozdemir Y, Ozis U (2008) Firat-Dicle havzasinda akislarin barajla duzenlenmesi. In: Su ve enerji konferansi, vol. II, Devlet Su Isleri XXVI. Bolge Mudurlugu, Artvin, pp 33–42
Ginocchio R (1959) Aménagements hydroélectriques. Eyrolles, Paris
Gokdemir M, Komurcu MI, Evcimen TU (2012) Turkiye'de hidroelektrik enerji ve HES uygulamalarina genel bakis. Turkiye Muhendislik Haberleri 471:18–26
Guner E (1967) Hidroelektrik santrallarin projelendirilmesinde kullanilacak teknik malumat ve abaklar. Devlet Su Isleri, Ankara
Guthrie-Brown J (ed) (1958) Hydroelectric engineering practice: (I) Civil engineering; (II) Mechanical and electrical engineering; (III) Economics, operation and maintenance. Blackie, London
Harmancioglu N (1976) Asagi Firat planlamasinin enerji bedellerinin degisimine duyarligi. M.S.-Thesis, Ege University
Harmancioglu N, Ozis U (1978) Enerji bedellerinin degisimine Asagi Firat planlamasinin duyarligi. DSI Teknik Bulteni 42:3–13
Harmancioglu, N, Ozis U (1981) Enerji buhrani ve Ataturk Baraji. Ege Universitesi Insaat Fakultesi Dergisi, Ataturk'un 100.dogum yili ozel sayisi:451–474

Harmancioglu N, Ozis U (1983) Dynamics of water resources development: lower Euphrates case in Turkey. In: Bayazit M (ed) Application of system analysis on water resources development. NATO Workshop, Istanbul, R. 8
Harmancioglu N, Turan T, Ozis U (1978) Turkiye'deki hidroelektrik enerji uretiminin agirlik merkezi. In: Insaat Muhendisligi VII. Teknik Kongresi, Insaat Muhendisleri Odasi, Ankara, pp 313–330
Kankal M, Bayram A, Uzlu E (2014) Assessment of hydropower and multi-dam power projects in Turkey. Renew Energy 68:118–133
Keloglu N (2006) Turkiye su kaynaklarinin gelecekteki kullanimi ve hidroelektrik enerji uretimi. In: TMMOB Su politikalari Kongresi, Insaat Muhendisleri Odasi, Ankara, pp 13–25
Kentel E, Alp E (2013) Hydropower in Turkey: economical, social and environmental aspects and legal challenges. Environ Sci Policy 31:34–43
Kocaker H (2010) Deriner baraji ve hidroelektrik santrali. Yapi Dunyasi 172:33–35
Ludin A, Borkenstein W (1955–1958) Wasserkraftanlagen: (I) Planung, Grundlagen und Grundzüge; (II) Anordnung und Ausbildung der Hauptbauwerke. De Gruyter-Göschen, Berlin
Melikoglu O (1980) Turkiye elektrik enerjisi uretimi ve tuketiminde pompa-turbin donusebilir santralin yeri ve ekonomisi. Devlet Su Isleri, Ankara
Melikoglu M (2013) Hydropower in Turkey, analysis in the vision of 2023. Energ Rev 25:503–510
Mosonyi E (1963, 2. ed. 1990) Water power development, V. I: Low-head power plants. Akademiai Kiado, Budapest
Mosonyi E (1965, 2. ed. 1991) Water power development, V. II.: High-head power plants. Akademiai Kiado, Budapest
Noyan K (1965) Keban baraji ve hidroelektrik santrali. Turkiye Muhendislik Haberleri 122:4–30
Ozbek T (1989) Das Südost-Anatolien-Projekt-GAP. In: V.Internationales Weiterbildungsprogramm, vol 16. Technische Universität, Berlin, pp 307–317
Ozdemir Y, Ozis U, Baran T (2013) Water resources development in the Euphrates-Tigris basin. In: International perspective on water resources and the environment. American Society of Civil Engineers & Dokuz Eylul University, Izmir, R. 153
Ozis U (1961) Die Hydroenergie in der Entwicklung der Türkei. Dissertation, Technische Hochschule München
Ozis U (1968) Pompajli hidroelektrik santrallar. In: Insaat Muhendisligi IV.Teknik Kongresi, Insaat Muhendisleri Odasi, Ankara, R. I-6
Ozis U (1971) Akislilik katsayilari vasitasiyla ortalama ve dusuk akimlarin tayini, guvenilir su kuvveti potansiyelinin hesaplanmasi. Turkiye Bilimsel ve Teknik Arastirma Kurumu, Ankara
Ozis U (1973) Hydroelectric schemes in Turkey. Water Power 9:325–329
Ozis U (1974a) Der Ausbau der Wasserkräfte in der Türkei. Die Wasserwirtschaft 4:102–106
Ozis U (1974b) Considérations analytiques sur les aménagements hydroélectriques en Turquie. Travaux 470:34–39
Ozis U (1977) Maximal development of water power resources. In: X. World Energy Conference. World Energy Council, Istanbul, R. 3.1–7
Ozis U (1982a) Aménagement de la basse Euphrate en Turquie. Travaux 565:68–72
Ozis U (1982b) Ausbau des unteren Euphrat-Beckens in der Türkei. Die Wasserwirtschaft 5:207–210
Ozis U (1982c) The development plan for the lower Euphrates Basin in Turkey. Nat Resour Dev 16:73–82
Ozis U (1983a) Ausbau des westlichen Tigris-Beckens in der Türkei. Die Wasserwirtschaft 4:109–112
Ozis U (1983b) Development plan of the Western Tigris Basin in Turkey. Int J Water Resour Dev 4:343–352
Ozis U (1983c) Su yapilari. Dokuz Eylul Universitesi, Izmir
Ozis U (1985) Turkiye'nin hidroelektrik potansiyeli ve enerji uretimi. In: EIE 50.Kurulus Yildonumu – Hidroelektrik Enerji Sempozyumu. Elektrik Isleri Etud Idaresi, Ankara, pp 1–30
Ozis U (1986) Hydro-power development in Turkey. Indian J Power River Val Dev 11/12:297–304

Ozis U (1989) Su kuvveti potansiyelinin gelistirilmesinde karst ortamin etkileri. In: Erguvanli Muhendislik Jeolojisi Sempozyumu Tebligleri, 1988 Istanbul, Muhendislik Jeolojisi Turk Milli Komitesi Bulteni 11:20–24

Ozis U (1991) Su kuvveti tesislerinin planlama esaslari. Dokuz Eylul Universitesi, Izmir

Ozis U (1994) La gestion des besoins et d'approvisionnements en eau dans le bassin Euphrate-Tigre. In: IWRA VIII. World Congress on Water Resources, vol.2. International Water Resources Association, Cairo, R. T5–S2

Ozis U (1998) Turkiye'de su kuvvetinin gelisimi. In: Prof. Dr. Kazim Cecen Anisina – Turkiye'nin Hidroelektrik Potansiyelinin Gelistirilmesi Sempozyumu. Istanbul Teknik Universitesi, Istanbul, pp 47–64

Ozis U (2015) Su kuvveti ve Turkiye. In: 4. Su Yapilari Sempozyumu. Insaat Muhendisleri Odasi, Antalya, pp 2–21

Ozis U, Harmancioglu N (1982) Su kuvveti kaynaklarinin gelistirilme sureci. In: VII. Bilim Kongresi, 1980 Kusadasi, Turkiye Bilimsel ve Teknik Arastirma Kurumu, Ankara, pp 245–264

Ozis U, Harmancioglu N (1994) Muhendislik ekonomisi acisindan Ataturk barajinin planlanmasi. In: DSI 40'inci Kurulus Yili, Su ve Toprak Kaynaklarinin Gelistirilmesi Konferansi Bildirileri, vol. 3. Devlet Su Isleri, Ankara, pp 1269–1278

Ozis U, Kocak Y (1977) The first arch dam in Turkey. Int Water Power Dam Constr 4:30–36

Ozis U, Ozdemir Y (2009) Turkey's transboundary watercourses and the Euphrates-Tigris basin. In: Karpuzcu M, Gurol MD, Bayar S (eds) Transboundary waters and Turkey. Institute of Technology, Gebze, pp 21–57

Ozis U, Ozel I (1989) Karakaya dam and power plant. Int Water Power Dam Constr 7:20–24

Ozis U, Yanar H (1984) Turkish dam impounds world's largest karst spring. Int Water Power Dam Constr 12:44–49

Ozis U, Uslu O, Harmancioglu N (1977) Solar energy storage and water power. In: Izmir International Symposium-I on Solar Energy Fundamentals and Applications. University of Miami & Ege University, Izmir, Preprint

Ozis U, Harmancioglu N, Benzeden E, Alpaslan N (1978) Hidroelektrik enerji uretiminin gelisme olanaklari. In: Turkiye III. Genel Enerji Kongresi, vol. 1. Dunya Enerji Konferansi Turk Milli Komitesi, Ankara, pp 280–291

Ozis U, Benzeden E, Harmancioglu N, Turkman F (1983) Opportunities in water power development. In: Veziroglu TN (ed) Alternative energy sources III, vol 4. Hemisphere, Washington, DC, pp 437–452

Ozis U, Benzeden E, Keloglu N, Harmancioglu N (1985) Assessment of karst water resources. In: IWRA V. World Congress on Water Resources Proceedings, vol 1. International Water Resources Association, Bruxelles, pp 95–102

Ozis U, Alkan A, Tatlioglu E (1990a) Yukari havza su kaynaklarinin degerlendirilmesi. In: Turkiye 5. Enerji Kongresi, vol 2. World Energy Conference Turk Milli Komitesi, Ankara, pp 61–75

Ozis U, Basmaci E, Harmancioglu N (1990b) Ataturk nears completion. Int Water Power Dam Constr 9:12–16

Ozis U, Yanar H, Erdem M (1990c) Sir dam and powerplant. Int Water Power Dam Constr 9:17–18, 20–21

Ozis U, Basmaci E, Harmancioglu N (1992) Wasserkraftanlage Ataturk am Euphrat. In: Königsberger A, Matthias HB, Varga T (eds) 7. Internationales Seminar – Wasserkraftanlagen. Technische Universität, Wien, pp 17–31. (p 15–29 in english ed.)

Ozis U, Alkan A, Tatlioglu E (1993) Effects of karst springs on flow duration curves of rivers. In: Gunay G, Johnson I (eds) Hydrogeological processes in Karst Terranes. 1990 Antalya. International Association of Hydrological Science, Gentbrügge, pp 209–222

Ozis U, Baran T, Durnabas I, Seker S, Ozdemir Y (1997) Turkiye akarsularinin su ve su kuvveti potansiyeli. Turkiye Muhendislik Haberleri 391:17–26

Ozis U, Baran T, Harmancioglu N, Benzeden E, Turkman F, Dalkilic Y, Seker S, Ozdemir Y (1999) Turkiye'de su kuvvetinden enerji uretimi. In: Izmir Su Kongresi, Muhendislik ve Diger Meslek Odalari Izmir Subeleri, Izmir, pp 425–441

Ozis U, Fistikoglu O, Canga R (2000) Keban, Karakaya, Ataturk baraj yerlerinde aylik akislar. Insaat Muhendisleri Odasi, Teknik Dergi, 1:2101–2119. & Monthly discharges at Keban, Karakaya, Ataturk dam sites. Chamber of Civil Engineers, Digest 2000:627–630

Ozis U, Ozdemir Y, Dalkilic Y, Turkman F, Baran T (2004) Development of the Euphrates-Tigris basin. In: Harmancioglu NB, Fistikoglu O, Dalkilic Y, Gul A (eds) EWRA Symposium on water resources management, vol 1. European Water Resources Association & Dokuz Eylul University – SUMER, Izmir, pp 85–95

Ozis U, Turkman F, Baran T (eds.) (2006a; 4. pr. 2016) Su kaynaklarinin gelistirilmesi. Dokuz Eylul Universitesi, Izmir

Ozis U, Keloglu N, Alkan A (eds) (2006b; 4. pr. 2016) Su kuvveti. Dokuz Eylul Universitesi, Izmir

Ozis U, Baran T, Dalkilic Y (2008) Hidroelektrik enerjiyi gelistirme hizlari. In: Su ve enerji konferansi, vol. I, Devlet Su Isleri XXVI. Bolge Mudurlugu, Artvin, p 1–14

Ozis U, Alkan A, Ozdemir Y (2012) Ausbau der Wasserkraft in der Turkei. Wasserwirtschaft, 7–8:53–58

Ozis U, Alkan A, Ozdemir Y (2013a) Ausbau der Wasserkraft in der Turkei. In: Heimerl S (ed) Wasserkraftprojekte – Ausgewählte Beiträge aus der Fachzeitschrift Wasserwirtschaft. Springer, Wiesbaden, pp 176–184

Ozis U, Ozdemir Y, Baran T (2013b) Das grenzüberschreitende Euphrat-Tigris Becken. WasserWirtschaft 7/8:74–78

Pasin S, Altinbilek D (1997) Turkiye hidroelektrik potansiyeli ve gelisme durumu. In: 7. Enerji Kongresi, vol. 3. Dunya Enerji Konseyi Turk Milli Komitesi, Ankara, pp 1–26

Press H (1967) Stauanlagen und Wasserkraftwerke: (III) Wasserkraftwerke. Springer, Berlin

Sarac M (2009a) Pompaj depolamali hidroelektrik santrallar. In: Forum 2009, Elektrik Muhendisleri Odasi, Trabzon, pp 90–98

Sarac M (2009b) EIE'nin Coruh havzasi projeleri. In: Forum 2009, Elektrik Muhendisleri Odasi, Trabzon, pp 99–111

Schleiermacher E (transl. Unsal I) (1967) Su kuvveti tesisleri. Istanbul Teknik Universitesi, Istanbul

Sekkeli M, Kececioglu F (2011) Hidroelektrik santrallarin Turkiye'deki gelisimi ve Kahramanmaras bolgesi ornek calismasi. KSU J Eng Sci 2:19–26

Sen Z (2011) Water for energy: hydropower is vital for Turkey. In: Kibaroglu A, Scheumann W, Kramer A (eds) Turkey's water policy, national frameworks and international cooperation. Springer, Berlin, pp 69–81

Turfan M, Bozkus Z (1992) Hydropower potential in Turkey. Int Water Power Dam Constr 12:11–12

Tutus A (2006) Turkiye'de elektrik enerjisinin tarihsel gelisimi ve yeni piyasa duzeni icerisinde hidroelektrik enerjinin yeri. In: TMMOB Su politikalari Kongresi. Insaat Muhendisleri Odasi, Ankara, pp 318–320

UN/ECE (1953) Hydroelectric potential in Europe. United Nations, Economic Commission for Europe, Committee on Electrical Power, Genève

UN/ECE (1955) Hydroelectric potential in Turkey. United Nations, Economic Commission for Europe, Committee on Electrical Power, Genève

Varlet H (1962a) Aménagement, utilisation et prix de revient des usines hydrauliques. Eyrolles, Paris

Varlet H (1962b) Usines de retenue, usines de plaine. Eyrolles, Paris

Varlet H (1964) Turbines hydrauliques et groupes hydroélectriques. Eyrolles, Paris

Varlet H (1965) Usines de dérivation: (I) Captage des eaux, canalisations à écoulement libre; (II) Conduites forcées, tunnels d'amené en charge, usines souterraines. Eyrolles, Paris

WPDC (2000) Pumped storage plants. In: International Water Power & Dam Construction, Yearbook, pp 118–124

Yildiz K (1992) Hidroelektrik santraller hesap esaslari ve projelendirilmesi. Devlet Su Isleri Vakfi, Ankara

Yuksek O, Komurcu MI, Yuksel I, Kaygusuz K (2006) The role of hydropower in meeting Turkey's electric energy demand. Energy Policy 34:3093–3103

Chapter 11
Urban and Industrial Water Uses

Selmin Burak and Hakan Mat

Abstract This chapter intends to reflect urban and industrial water use and management in Turkey, based on an overview of available data related to sectoral uses, constraints and a realization of set targets up to the present. Through Turkey's EU accession process, the harmonization of the existing regulatory system with that of the EU has influenced the development of the water supply and sanitation sector, and consequently, there has been a gradual improvement in service quality and protection of the environment. Therefore, this issue deserves an in-depth analysis with regard to the objectives to be met and the corresponding finances to be allocated by the Turkish government. In line with this, a sectoral assessment giving the cross-cutting issues, followed by drinking water production and its use by households and the public sector, industries connected to the public water supply system and self-supplied industries, as well as pollution generated by municipal and industrial water consumption and disposal are summarized prior to this analysis. Water uses encompassing water supply and treatment, sewerage, wastewater treatment, and disposal will be given in the first section. A general description of the existing institutional structure related to water and sewerage administrations in Turkey will be presented in the second section, which intends to enable a better understanding of the prevailing municipal water management system that has been subject to various revisions in the past.

Keywords Municipal water use · Wastewater treatment · Reuse · Industrial water use · Metropolitan municipalities · Non-metropolitan municipalities · WSAs · NRW

S. Burak (✉)
Institute of Marine Sciences and Management, Istanbul University, Istanbul, Turkey
e-mail: sburak@istanbul.edu.tr

H. Mat
MAT Consult Ltd. Co., Istanbul, Turkey
e-mail: hakanmat@superonline.com

N. B. Harmancioglu, D. Altinbilek (eds.), *Water Resources of Turkey*, World Water Resources 2, https://doi.org/10.1007/978-3-030-11729-0_11

11.1 Introduction

Understanding the respective weight of diverse human water uses and establishing a natural accounting for water resources requires disaggregating information into its smallest elements. Data on sectoral water demand relate to the classical basic sectors of beneficial uses of water, that are mainly: (1) Public (municipal/household) use, which comprises water uses of urban and rural communities/households, as well as public or private sector water uses including industrial water supply. This covers drinking water production and its use by households and public/private sector; (2) Self-supplied industries that are the ones not connected to the public drinking water distribution network; (3) Thermal power stations that use water for cooling purposes. In this chapter, urban (municipal) and industrial water uses encompassing water supply, sewerage, wastewater treatment, and disposal will be given in the first section, followed by the administrative and institutional structure of municipal water management and related regulations in the second section.

11.2 Cross-Cutting Issues

11.2.1 Urbanization and Coastal Environment

Turkey has been affected by a high urbanization ratio since its very first years of development, with a rate of 18.5% in 1950 and increasing to about 62% in 2000 and further to 88% as of 2017. Furthermore, urban development plans do not keep pace with the rapid population growth and changes in settlement patterns, which make them irrelevant and thus subject to revision. These plans are not generally integrated with projections and targets of other plans. Cities with already inadequate infrastructure facilities have had to face congested population problems, coupled with illegal settlements due to migration from the eastern part of Turkey to the western large metropolises. This has been creating severe environmental problems due to urban and industrial pollution of water bodies. Onerous and ambitious investments are planned to remedy environmental pollution and its abatement progressively. An in-depth analysis was carried out in order to calculate the investment needs for a better environment covering the period of 2007–2023 within the context of the EU Integrated Environmental Approximation Strategy (2007–2023) (Republic of Turkey Ministry of Environment and Forestry 2006).

Strengthening the local level capacity to support development of environmental protection and management activities operated by municipalities has become an important management policy. This could include urban pollution abatement strategies and local environmental action plans, monitoring and evaluation of environmental conditions, granting permission and enforcement, and also development of financing mechanisms for environmental management. The latter is especially important; as municipal revenue-generation capacity is rather weak in non-metropolitan municipalities.

The Aegean-Mediterranean littoral of Turkey with a length of 4170 km, having appropriate climatic and environmental conditions, constitutes diverse and unique ecosystems. These coasts, where development is concentrated mainly between Canakkale (Dardanelles) and Mersin Provinces, have been the cradle of many ancient civilizations whose remains make Turkey one of the most attractive Mediterranean countries to visit.

Tourism makes seasonal demands on local resources (water supply and sewerage facilities) and the environment. During the tourist season from May to September, the resident population increases by more than five-fold in coastal settlements located on the Aegean and Mediterranean coasts in particular (The World Bank 1998; Burak et al. 2004). The problems are exacerbated by the concentration of tourist activity in a relatively short holiday season and in specific villages and holiday resort areas. These areas, which are comparatively small, are often subject to environmental pressure from other economic activities such as agriculture, industrial development, and resident population (UNEP/MAP/BLUE PLAN 2007).

Turkey's three largest most industrialized and fastest growing metropolises, namely Istanbul, Izmir and Antalya, are located on the coasts of the Sea of Marmara, the Istanbul Strait (Bosphorus), Aegean and the Mediterranean. Smaller coastal settlements have become increasingly urbanized as a result of legislative and institutional incentives to encourage tourism investment. Construction of hotels and summerhouses has exploded as a result of unearned and real income expectations to the detriment of fertile land (e.g. Izmir, Antalya, Mersin and their surroundings), creating aesthetic pollution and loss of tangerine and olive orchards (Burak et al. 2004). Turkish coastal zones constitute approximately 30% of the total land, whereas the coastal and related hinterland population constituted 43% of the total population of around 80 million in 2015 (UNEP/MAP 2015).

Water has been supplied mostly from groundwater and in excessive amounts to satisfy the demand of the newly developed settlements, lowering the water table and resulting in sea-water intrusion in most of the coastal aquifers. The sewage generated by congested population has caused pollution of bathing waters to exceed the standards relating to human health and environmental protection. Comprehensive rehabilitation and pollution prevention measures have progressively been implemented since more than two decades in coastal large cities, in particular.[1] In general, infrastructure investment mainly targeted to ensure pollution prevention issues, i.e. sewerage, wastewater treatment and solid waste management, has fallen behind schedule due to the lack of adequate urban and environmental planning and financing. Furthermore, the fact that more than twenty acts and decrees are enforced, regarding coastal management issues, resulted in the involvement of more than fifteen institutions, generating biased solutions due to plurality and fragmentation in the decision-making process.

[1]Surface water supply schemes together with extended sewerage network have been developed in Antalya, Adana, Cesme (Izmir) in order to reduce the overexploitation on groundwater and pollution prevention of the fresh water resources (Burak et al. 2004).

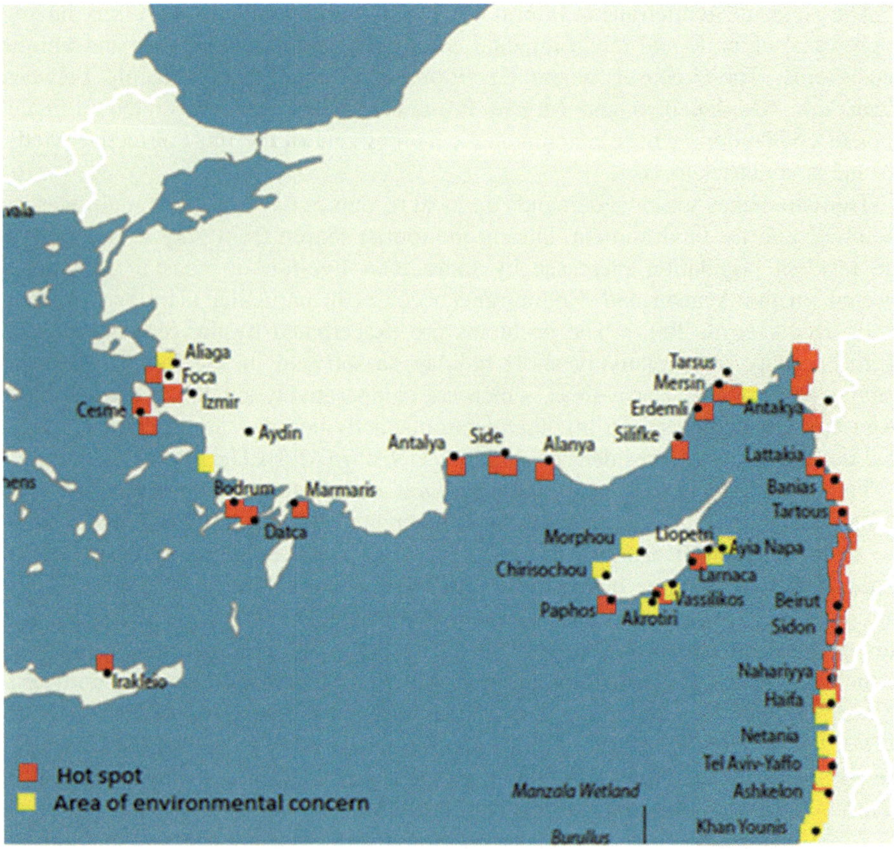

Fig. 11.1 Environmental hot spots in Turkish Mediterranean Coasts. (UNEP/MAP 2012)

Localized pollution problems caused by domestic and/or industrial pollution have been identified as 'hot-spots' within the scope of studies initiated by the Mediterranean Action Plan, as shown on Fig. 11.1 (UNEP/MAP 2012).

It can be highlighted that, particularly in coastal waters, no comprehensive pollution prevention measures have been implemented with strict enforcement. Instead, measures have generally been taken after the medium has been polluted, which has led to onerous investments for a pollution abatement program. The exception is the positive impact gained by the identification of 'Special Environmental Protection Areas' where effective preventive measures are being implemented in sensitive zones with the objective of protection-usage equilibrium.

The Environment Act No 2872 (Official Gazette 18132 1983) dictates very stringent measures for the wastewater treatment plant of the hotels and facilities performing in the tourism sector and their operation. Turkey has been participating since 1994 in the 'Blue Flag Campaign', which has been a driving force to motivate various stakeholders and the local population to reach the guidelines for better beach water quality, as shown on Fig. 11.2.

Fig. 11.2 Beaches awarded with Blue Flag. (Blue Flag 2018)

Starting with the accession process with EU, comprehensive institutional and legal reforms are being introduced in Turkey. Some positive results are noticed, like the reduction of non-revenue-water (NRW), implementation of enhanced sewerage network, and wastewater treatment plants, whereas some other areas have not been improved yet despite such strict enforcement of discharge standards and good water governance.

The case study related to the "Greater Istanbul Sewerage Project" presents a typical example of enhanced sewerage, wastewater treatment and disposal that were started to be implemented in the late 1980s, following the establishment of "metropolitan municipalities" and their affiliated "water and sewerage administrations" (Burak and Demir 2016). The location plan of wastewater treatment plants in Istanbul, treatment levels, and flows discharged in the marine environment are presented in Annex 11.1.

11.2.2 EU Accession Process

Turkey is a negotiating candidate country with the EU (The Council of the European Union 2008). In the Accession Partnership Council Decision of 2008 for Turkey, Chap. 27 related to the environment, where it is stated that Turkey will continue to transpose and implement the acquis related to the framework legislation, international environmental conventions and legislation on nature protection, water quality, chemicals, industrial pollution and risk management and waste management, and pursue integration of environmental requirements into other sectoral policies.

The Draft Framework Water Act was submitted to national authorities, agencies and any other stakeholders concerned, for comments and review in October 2012.

The By-Law on the Protection and Planning of Water Basins (Watersheds) was promulgated on the 17th of October, 2012, which transposed the Water Framework Directive as a closing benchmark.

Turkey has made good progress with the adoption of the legislation on river basin management, groundwater and drinking water, as also reported in the Turkey 2012 Progress Report published by the European Commission (European Commission 2012). In the report, it was stipulated that, in the following stage, "River Basin Protection Action Plans (RBPAP)" will be converted into "River Basin Management Plans (RBMP)". In line with this decision, RBPAPs were carried out in 20 national basins out of 25, excluding transboundary rivers. This was followed by RBMPs in seven river basins, namely Buyuk Menderes, Susurluk, Konya, Maritza/Ergene, Yesilirmak, Bati Akdeniz and Akarcay.

11.3 Definitions Used in Urban Water Supply

Urban (municipal/domestic) water in Turkey stands for potable water supplied by the water authority (i.e. water and sewerage administration of the metropolitan municipality/water department of the non-metropolitan municipality) via the municipal network to various customers (e.g. housing areas, commercial districts, institutional, and recreational facilities) that may use this water for drinking, washing, bathing, culinary, waste removal, yard, cemetery and garden watering purposes. The customers are divided into three main groups as domestic (household), commercial/industrial, and public according to the purpose of their water utilization. Water consumption also includes technical (leakage) and administrative system losses (UNEP/MAP 2012).

The water use efficiency index indicates how to measure progress in water savings through demand management, by reducing losses and wasteful use, mainly during its transmission and distribution (Blinda 2012). It covers total and sectoral efficiency in domestic (municipal), agricultural and industrial water use. The municipal water use efficiency index is defined as the ratio of the 'total drinking water volume billed' to the 'total volume supplied (abstracted/treated and distributed)' to the customers by municipalities, as formulated below Eqs. (11.1) and (11.2):

$$Emun = \frac{\text{Vb}}{\text{Vs}} \tag{11.1}$$

where *Emun* (%): Municipal water use efficiency index, Vb: volume billed to the customers by the municipalities (m^3/year), and Vs: volume supplied to the customers (m^3/year) by the municipalities. Monitoring the water use efficiency index is the commonly applied method for assessing the performance of municipal water management that depends on physical (real) and non-physical (apparent) losses constituting the non-revenue water (NRW), as given by the formula below:

$$NRW = \frac{(\text{Vs-Vb})}{\text{Vs}} \tag{11.2}$$

The sectoral efficiency index indicates both the physical efficiency of municipal drinking water network in terms of physical (technical) losses and commercial efficiency, reflecting the institutional capacity of the water authority concerned to recover operation & maintenance (O&M) costs through water bills. Revenue collection performance is defined as revenues collected divided by revenues billed in a specific year (Burak 2011).

The sources of NRW are mainly physical (real) and non-physical (apparent or commercial). The physical losses constitute that portion of water that is produced but not consumed, whereas the non-physical portion is consumed, but it is either unbilled metered or unbilled unmetered as a result of unauthorized consumption. This implies illegal connection to the municipal network. NRW components are similar to water loss definitions and classifications, as given in Table 11.1.

The case study carried out in three non-metropolitan municipalities gives the detailed components of NRW on Fig. 11.3 (Burak and Mat 2010).

As a common terminology, wastewater treatment incorporates several stages, and they are classified as follows: (i) primary (mechanical/physical) treatment which is a natural process; (ii) secondary (biological) where carbon removal takes place; (iii) tertiary (advanced) treatment where nitrogen and phosphorus compounds are

Table 11.1 Water loss definitions and classifications (Alegre et al. 2006)

System input volume (corrected for known errors) (Water produced + Water imported)	Authorized consumption	Billed authorized consumption	Billed metered consumption (including water exported)	Revenue water
			Billed unmetered consumption	
		Unbilled authorized consumption	Unbilled metered consumption	Non-revenue water (NRW)
			Unbilled un-metered consumption	
	Water losses	Apparent losses	Unauthorized consumption	
			Customer metering/ billing inaccuracies	
		Real losses	Leakage on transmission and/or distribution mains	
			Leakage and overflows at utility's storage tanks	
			Leakage on service connections up to point of customer metering	

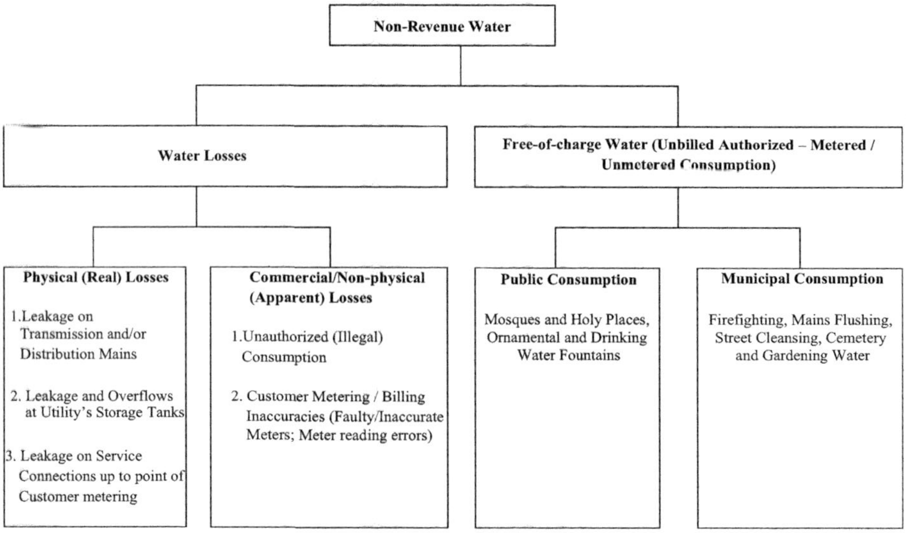

Fig. 11.3 Sources of non-revenue water in the selected municipalities of a case study in Turkey, namely in the cities of Ordu, Carsamba and Ceyhan. (Burak and Mat 2010)
Ordu, Carsamba and Ceyhan are medium size cities, the two first are located on the Black Sea coast; Ceyhan is located on the Eastern Mediterranean coast

removed following a secondary treatment. When industrial wastewater undergoes treatment to the level of domestic sewage, then this process is called a pre-treatment.

In coastal cities, treated wastewater is disposed of via deep sea outfalls, and generally, the treatment level applied is primary when dilution is strong. In case of inland locations, treated wastewater is discharged into the surface water bodies like rivers and lakes. The standards applied to the discharges depend on the effluent flow and the oceanographic properties of the receiving media. This is further discussed in this chapter.

11.4 Water Supply Resources

Water is supplied from surface and groundwater resources for all purposes. Supply from surface water has the largest share with more than 60%, and the rest is withdrawn from groundwater. Domestic water supply with a ratio of 15% and industrial supply with 10% of the total, respectively, follow the irrigated agriculture, which is the highest water-consuming sector with an average rate of 75% (Burak and Margat 2016).

At present, almost the totality of the urban population is connected to the water supply network, with a ratio of 98.2%, and, in rural areas, this ratio is close to 90%. The overall population connected to safe drinking water network is 92.2%, considering rural non-municipal population. A daily allocation of 170 l/ca/d, made

Table 11.2 Water supply and consumption ratio in municipalities between 2012 and 2016 (TURKSTAT 2017a, c)

Water supply and consumption ratio in Municipalities	2012	2014	2016
Water supplied (withdrawn) (l/ca-d)	216	203	217
Water consumed (l/ca-d)	123	132	139
Water lossed (l/ca-d)	93	71	78
NRW (Water lossed/supplied) (%)	43%	35%	36%

available for drinking water supply in the late 1990s, reached more than 200 l/ca/d after 2000 (Burak 2007). Table 11.2 shows water supply and consumption for municipal use. NRW has a high average varying from 35 to 65%, depending on the network and location. However, this ratio has been decreasing, following the enhanced water leak detection and rehabilitation measures supported by legal provisions at central level.

One of the main duties and responsibilities of the public water sector (i.e. water and sewerage administrations (WSAs) at metropolitan municipalities and water departments of non-metropolitan municipalities) is to satisfy the demand for water and to ensure that water quality at consumers' taps consistently meets recognized standards. Large cities like Istanbul, where no perennial water resources exist within the municipal boundaries, water has been transferred from neighboring basins (e.g. Istranca Creeks, Melen River) in order to satisfy the demand of a population exceeding 15 million at present. Until today, water transfer from adjacent basins, accompanied by large water infrastructures, has been the preferred solution in order to provide big metropolises with water (e.g. Istanbul, Ankara). The Istanbul Water Supply Scheme is a good example, putting forward challenges for meeting the demand of an ever-expanding metropolis. Although water transfer is not considered to be an environmentally-friendly approach, this solution is practiced for such cities which do not have perennial water resources within their boundaries. The schematic diagram of the Istanbul Water Supply Scheme is shown on Fig. 11.4, and the details of water resources made available for Istanbul in Table 11.3.

Information on quality, reliability, and continuity of service provision does not exist at centralized level. However, TurkStat, as the National Statistics Institute, is entitled to gather data from both central institutions and municipalities concerned. These data are the national ones that can be used and disseminated. Additionally, TurkStat conducts surveys on population well-being, based on the so-called "household surveys", which give the statistics and identify, inter alia, the level of satisfaction of the piped water customers. Therefore, these surveys can be used as a proxy for the services provided. For example, water supply services were found to be satisfactory by 79% of the customers in 2012 as the result of the latest well-being survey carried out during the period of 2004 and 2012.

In line with the above mentioned practice, TurkStat has the information about the total water abstracted from surface water resources (dams, lakes, rivers) or groundwater through records provided by the State Hydraulic Works (DSI), which meters withdrawals for all uses. In theory, distributed water to customers equals sold water;

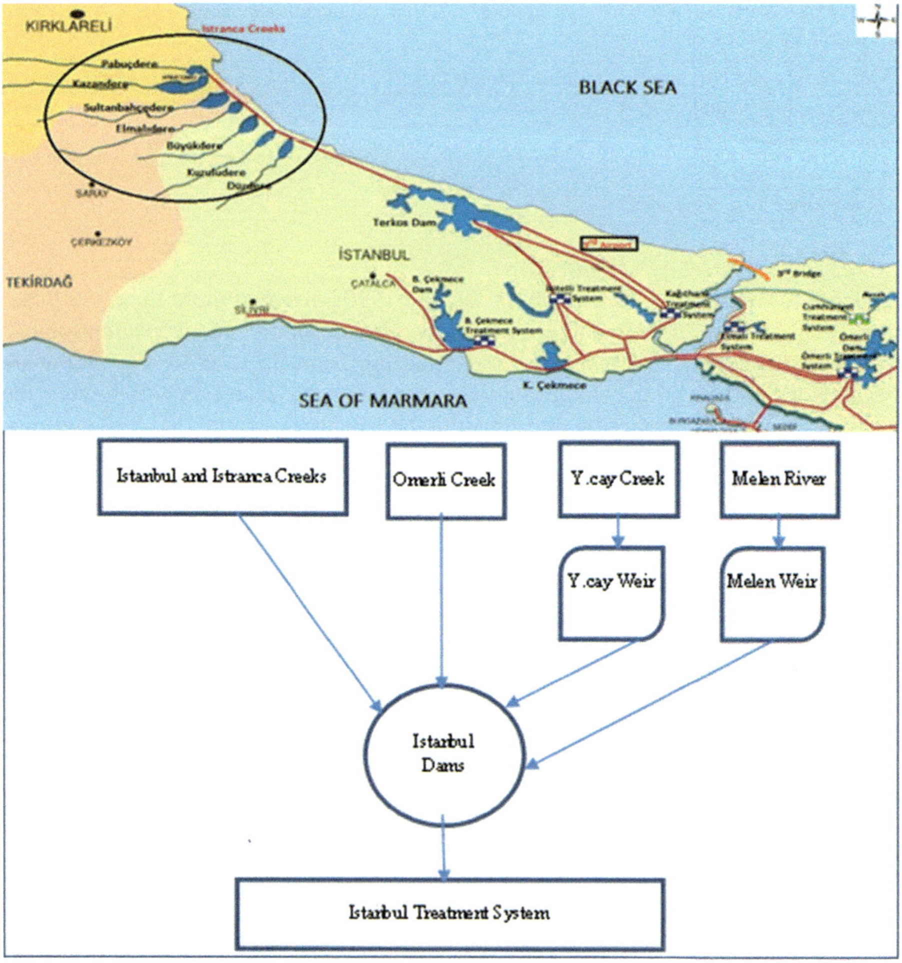

Fig. 11.4 Istanbul water resources and water supply system. (Burak et al. 2017)

but, in practice, there is a discrepancy between the two values due to the difference of water distributed and water sold, which equals to the sum of free-of-charge public and municipal consumption and leak losses. Municipal parks and gardens utilize water free-of-charge in general (some municipalities started billing for this type of use), but it is not clear whether this amount of water is metered or not in some cases. A detailed analysis is given in the case study in three non-metropolitan municipalities (Burak and Mat 2010). Water accounting would be a useful tool to be used at municipal level.

Water supply from non-conventional water resources like desalinated water and wastewater reuse have not been adopted for large cities and at large scale until recently. One of the prevailing arguments of the central planning administrations is that all the national water resources are not fully developed; therefore, there is still

Table 11.3 Details of existing water resources in Istanbul (Burak et al. 2017)

Characteristics of catchment areas and reservoirs	European side			Asian side			
	Terkos	Alibeykoy	B. Cekmece	Sazlidere	Omerli	Darlik	Elmali
1. Drainage area (km^2)	619	160	622	165	621	207	76
2. Mean annual rainfall (mm)	750	837	700	627	880	880	–
3. Mean annual inflow (Mm^3/a)	163	54	219	49,2	236	108	32
4. Average runoff coefficient (%)	35	40	50	47	45	59	–
5. Reservoir							
5.1 Gross capacity (Mm^3)	187	36	182	–	357	113	11.7
5.2 Dead storage (Mm^3)	42	1	20	–	122	6	0.2
5.3 Effective capacity (Mm^3)	162,241	34,143	148,943	88,730	235,371	107,500	9600
5.4 Fully supply level (E.M)	4.5	30	6.3	48	62	52	67.5
5.5 Min. operating level (E.M)[a]	−1.00	11	0.8	–	46	21	37.5
5.6 Reservoir area at FSL[b] (km^2)	39.0	3.0	36	–	22.4	7.0	1.2
6. Nominal annual draught (safe yield) Mm^3/a	142	36	70	–	220	97	14.7

[a]*E.M* elevation in meter
[b]*FSL* Flooding surface level

groundwater and surface water that can be exploited (Burak 2007). Indeed, water supply via the exploitation of conventional resources is the preferred solution compared to non-conventional resources provided that water is available. Furthermore, the use of conventional water resources has better acceptance than any other non-conventional resources for practical and social reasons. However, sea water supply has been considered as an alternative since 2010, as shown in Table 11.4, with low but incremental quantities for water supply in arid and coastal regions. Wastewater reuse is increasingly encouraged and introduced in the irrigation of touristic gulf resorts, gardening, and street washing in municipal areas. Cooling water for industrial supply is also provided through non-conventional resources from the sea, in particular.

Table 11.4 Water abstraction statistics for municipal water supply network, 1994–2016 (TURKSTAT 2017e)

Year		Water abstraction for municipal water supply network (Total)	Dam	Well	Spring	River	Lake – artificial lake/sea[a]
1994	Amount	3,242,733	899,707	1,295,410	837,622	101,270	108,724
	(%)	100,0	28,5	39,5	25,5	3,1	3,3
1995	Amount	3,732,608	976,763	1,451,466	912,014	108,439	283,925
	(%)	100,0	26,9	38,5	24,2	2,9	7,5
1996	Amount	3,938,678	1,033,072	1,521,606	930,533	167,152	286,314
	(%)	100,0	26,9	38,3	23,4	4,2	7,2
1997	Amount	4,080,963	1,077,831	1,378,715	1,060,344	273,452	290,620
	(%)	100,0	27,1	33,5	25,8	6,6	7,1
1998	Amount	4,175,011	1,174,198	1,591,410	984,734	135,606	289,064
	(%)	100,0	28,7	37,8	23,4	3,2	6,9
2001	Amount	4,664,411	1,389,239	1,598,865	1,082,992	131,754	461,562
	(%)	100,0	29,8	34,3	23,2	2,8	9,9
2002	Amount	4,813,097	1,795,963	1,455,114	1,294,660	131,295	136,065
	(%)	100,0	37,3	30,2	26,9	2,7	2,8
2003	Amount	4,918,477	1,925,653	1,547,717	1,206,396	141,194	97,517
	(%)	100,0	39,2	31,5	24,5	2,9	2,0
2004	Amount	4,954,292	1,984,739	1,375,738	1,363,360	143,062	87,392
	(%)	100,0	40,1	27,8	27,5	2,9	1,8
2006	Amount	5,163,500	1,843,736	1,401,815	1,380,057	305,271	232,621
	(%)	100,0	35,7	27,1	26,7	5,9	4,5
2008	Amount	4,546,574	1,810,188	1,275,691	1,060,963	173,928	225,805
	(%)	100,0	39,8	28,1	23,3	3,8	5,0
2010	Amount	4,784,734	2,252,421	1,273,822	1,015,865	159,472	83,154
	(%)	100,0	47,1	26,6	21,2	3,3	1,7
2012	Amount	4,936,342	2,416,018	1,395,957	948,133	78,282	97,953
	(%)	100,0	48,9	28,3	19,2	1,6	2,0
2014	Amount	5,237,407	1,886,617	1,423,751	984,869	652,370	289,800
	(%)	100,0	36,0	27,2	18,8	12,5	5,5
2016	Amount	5,838,561	2,618,225	1,563,154	1,000,205	552,624	104,354
	(%)	100,0	44,8	26,8	17,1	9,5	1,8

[a]Water abstracted from sea is included since 2010

11.4.1 Urban and Industrial Water Demand and Supply

Urban and industrial water demand is mainly generated by domestic and industrial uses. Domestic water use is composed of residential, commercial and institutional (public) uses in urban areas (municipalities) and rural areas (villages). This covers drinking and household uses mainly, whereas industrial water use comprises process

Table 11.5 Evolution of total and municipal population of Turkey between 2012 and 2016 (TURKSTAT 2017b)

Population	31/12/2012	31/12/2014	31/12/2016
Total population of Turkey (Inhabitants)	75,627,384	77,695,904	79,814,871
Municipal population served (Inhabitants)	62,607,813	70,679,533	73,639,909
Municipal population/Turkish population (%)	82.78	90.97	92.26

water, washing and cooling in manufacturing plants. Major industrial water demand is generated by industries that comprise steel, chemical, paper manufacturing, leather, textile, and petroleum refineries.

As one of the first Millennium Development Goals (MDGs) targets to be met, up-to-the standard drinking water demand and supply deserves an in-depth analysis with regard to national statistics. Water supply sources for municipal use and withdrawals to meet the demand are given in Table 11.4, based on the statistics compiled by TurkStat for the period between 1994 and 2016 (TURKSTAT 2017e).

Industrial water demand is mostly met by the individual sources like wells, rivers, lakes and water treatment plants developed by the industrial enterprises (e.g. manufacturing plants, thermal energy generation plants, mining enterprises and Organized Industrial Zones (OIZs).

Within urban areas, industrial water demand is also met partially by the municipal water. Exceptionally, some industrial enterprises, and OIZs purchase water in bulk from municipalities.

The evolution of the municipal population served over the total population is given in Table 11.5 for the period between 2012 and 2016. Based on these population figures, it can be deducted that the water service coverage ratio with piped distribution network in the municipalities was satisfactory, compared to international indicators on population connected to safe drinking water network.

The total water quantity withdrawn by municipalities grew cumulatively by 18.17%, from 4936 million m^3 in 2012 to 5833 million m^3 in 2016. The total water quantity consumed by municipal customers increased cumulatively by 33.23%, from 2802 million m^3 in 2012 to 3733 million m^3 in 2016, as shown in Table 11.6. These data show the increased pressure on water resources.

As being the main water source of the municipalities, the total water quantity withdrawn from dams to provide water to the municipal consumers decreased from 48.95% in 2012 to 44.89% in 2016. Likewise, water withdrawn from wells and springs decreased from 47.49% in 2012 to 43.94% in 2016.

The total water quantity consumed as percentage of the total water quantity withdrawn increased from 56.77% in 2012 to 64.00% in 2016 as a result of decreasing water losses. Consequently, water losses, as percentage of water quantity withdrawn by municipalities, improved from 43.23% in 2012 to 36.0% in 2016 as a result of decreasing technical (pipe leaks) and administrative (unauthorized consumption, metering errors) losses as given in Table 11.7. Nevertheless, the volume of water losses estimated at about 1.84 billion m^3/year is equivalent to the water need of 38 million inhabitants on an average consumption of 133 l/ca/d. Obviously, water

Table 11.6 Municipal water supply, consumption and losses between 2012 and 2016 (million m^3/year) (TURKSTAT 2017c)

Municipal water resources	2012 (million m^3)	2014 (million m^3)	2016 (million m^3)
Dams	2416	1887	2618
Wells	1396	1424	1563
Springs	948	985	1000
Rivers	78	652	553
Lakes/ponds/seas	98	289	99
Total water quantity withdrawn	4936	5237	5833
Total water quantity consumed	2802	3395	3733
Total water quantity treated at WTPs	2729	2995	3350
Water losses	2134	1842	2100

Table 11.7 Evolution of municipal water supply between 2012 and 2016 (% of total water quantity withdrawn) (TURKSTAT 2017c)

Municipal water sources	2012 (%)	2014 (%)	2016 (%)
Dams	48.95	36.03	44.89
Wells	28.28	27.19	26.80
Springs	19.21	18.81	17.14
Rivers	1.58	12.45	9.48
Lakes/ponds/seas	1.99	5.52	1.69
Total water quantity withdrawn	100.00	100.00	100.00
Total water quantity consumed	56.77	64.83	64.00
Total water quantity treated at WTPs	55.29	57.19	57.44
Water losses	43.23	35.17	36.00

demand increases in parallel with demographic and economic growth, whilst resource availability is decreasing due to the impacts of climate change.

11.4.2 Treated Water Quality Objectives

Various standards, guidelines and recommendations consider water quality comprehensively, but the prime requirement is for microbiological quality to ensure that coliform bacteria are absent. This is universally pursued by a combination of appropriate filtration and disinfection, supported by additional treatments as necessary. This is reflected in the WHO Guideline for reliable disinfection, which states that water may be considered to be adequately treated when the following are met:

- turbidity of 1 NTU or less is achieved;
- disinfection of the water with at least 0.5 mg/l of free residual chlorine after a contact period of at least 30 min at a pH below 8.0.

This guideline is taken as the basic starting point for defining "good" quality treated water and, therefore, for assessing the performance of treatment plants. In addition to improving disinfection, taste and odor control, the introduction of ozonation will also prove beneficial in facilitating algae removal, should this become a problem in the future. This method is already being practiced in most of the water supply schemes in Istanbul, in particular. Further details are given in Sects. 11.4.3 and 11.5.

11.4.3 Water Treatment

The quality of drinking water in Turkey is defined by the "Turkish Drinking Water Standards (TS 266 2005)". These standards are generally based on the WHO guidelines. However, since Turkey decided to work towards a comprehensive adoption of the EU Directive relating to the quality of water, drinking water standards in Turkey are set by the Act on "Water Intended for Human Consumption", promulgated in the Official Gazette 25730, published on 17 February 2005. National drinking water standards are consistent with that of the EU set for the Drinking Water Directive (DWD), since these were incorporated in the Turkish regulations with the exception of three parameters, namely, "bromates", "lead" and "trihalomethanes" as given in Annex 11.2 on EU and Turkish Drinking Water Standards (The World Bank 2016).

The percentage of municipal population served by water treatment plants was equivalent to 55% of the total Turkish population and 58.6% of the total municipal population in 2016 (TURKSTAT 2017c).

In Turkey, water supply is almost universal with uneven performance. Differences exist in treatment levels, depending on the raw water quality. Out of the total water treated at 57.4%, 92.2% has undergone conventional treatment, 6.1% advanced treatment, and 1% only physical treatment in 2016 (TURKSTAT 2017c). Considering the rate of customer satisfaction with water supply services conducted yearly by TurkStat, it is assumed that the rest of the supplied water is mostly in "good enough condition", and therefore it can be distributed after a simple disinfection (The World Bank 2016). Disinfection at municipal treatment plants is done generally with chlorination. This conventional method has been replaced with ozonation in a few of the large scale treatment plants during their extension/rehabilitation stage (e.g. Istanbul water treatment plants).

11.5 Institutions Responsible for Water Supply and Provision of Sanitation Service

Water supply and sanitation services are ensured mainly by two institutions at central level for the whole country. Where water and sewerage administrations are established in metropolitan municipalities, this duty is left to them at municipal level.

The General Directorate of State Hydraulic Works (DSI) is the main institution responsible for the development of water and soil resources in general. DSI is in charge of water supply for drinking, utility and industrial water to the cities with a population of more than hundred thousand and providing the related infrastructure. DSI is responsible for the long-term development of water resources for the purpose of drinking water systems (e.g. dams, regulators) together with transmission mains, pumping stations, water treatment plants, main reservoirs and transmission lines feeding these reservoirs. Once the construction works are completed by the contractors and performance bond is released, DSI transfers these facilities, excluding the dams, to the municipalities and undertakes the O&M of the dams only.

IL Bank (formerly Bank of Provinces) is a joint-stock company in which municipalities are shareholders. The duties and responsibilities of the Bank are to provide loans to municipalities in order to realize public works and to assist the municipalities in the design and construction of water and sewerage systems according to the investment programs. While construction of drinking water supply and treatment works are handled by DSI and IL Bank, sewerage, wastewater treatment and marine disposal works are only under the responsibility of IL Bank with the exception of those undertaken by metropolitan municipalities. The infrastructural facilities are handed over to the municipalities after completion of construction for operation and maintenance. DSI and IL Bank are also responsible for providing technical control and supervision of works funded by their loans.

As underlined by the five priority indicators[2] adopted by Turkey as a Contracting Party to the Barcelona Convention, a substantial progress has been experienced during the last decade, after the accession process to the EU, in particular. This progress is mainly pursuant to water supply and sewerage infrastructure projects as put forward by sectoral reports (Blue Plan 2008). However further significant efforts are required for the implementation of wastewater treatment projects.

11.6 Sewerage, Wastewater Treatment and Disposal

Construction of up-to-the-standards sewerage facilities began in the late 1960's, initiated by the IL Bank. New sewerage projects have been designed as separate systems (sewerage and storm water collection), taking into account land development projections. Due to high investment costs, storm water collection systems have been constructed only in limited flood prone areas of big cities (e.g. Istanbul, Izmir).

Approximately 90% of the municipal population is connected to the sewerage system, based on 2016 data (TURKSTAT 2017a). Out of this ratio, more than 90% of the municipal population is served with sewerage facilities, whereas only about 50% of the rural population has this facility. It is estimated that the wastewater of

[2]Five priority indicators were adopted by the Contracting Parties to the Barcelona Convention in 2005, in order to measure the progress made in water management (Blue Plan 2008).

more than 65% of the total population is treated. Out of this ratio, 70% is the share of the municipal population, and 15% belongs to the rural population, on the average. Details of wastewater flows discharged from municipal sewerage systems into receiving bodies between 2002 and 2016 are given in Table 11.8.

So far, in coastal settlements, the final disposal by deep-sea outfall of collected wastewater after preliminary treatment has been a common practice. The treatment level of domestic wastewater to be discharged into the receiving media is assessed under three categories based on the population figures. The regulations prescribe a comprehensive list of effluent standards particular to domestic wastewater treatment works discharging directly to watercourses and the sea. These regulations also apply to individual industries. Areas of high ecological importance and sensitive to environmental pollution must be given special importance as stipulated in the related clause of the Environment Act.

Advance treatment is gradually being introduced to the wastewater treatment plant design located in touristic coastal areas, specially protected areas and water protection basins (e.g. Alanya, Fethiye, and Omerli Watershed in Istanbul, respectively), where nitrogen and phosphorus removal as tertiary treatment after biological treatment is being applied (EU funded projects also have nitrogen and phosphorus removal processes). However, there are some controversial issues related to the overlaps and conflicts between regulations, which hinder the implementation of EU Directives on wastewater in Turkey. There are two regulations in force, namely, the By-Law on Water Pollution Control and By-Law on Urban Wastewater Treatment, whereas other standards, all more stringent than that of EU requirements, are applied in practice. These by-laws set inconsistent treatment standards which create confusion. So far, the common practice has been to take from each by-law the more stringent standard for each parameter and to request municipalities to comply with these new sets of standards, which do not comply with either of the regulations in force (The World Bank 2016).

Consequently, the conservative approach related to wastewater treatment levels has direct and indirect implications in practice. Direct implications are high costs of O&M due to increased technical and personnel expenditures, which may not be affordable in most of the examples, and the advanced level of know-how for O&M which requires on-the-job training and continuity of dedicated staff. Indirect implications are the non-compliance with the standards, which may be perceived as usual, whilst this laxity or neglect is expected to be harmful to the receiving environment in the long-run (Burak and Demir 2016; IME 2017).

11.6.1 Assessment of Wastewater Treatment and Disposal Strategies

Selection of the wastewater disposal strategy depends on several issues including, but not limited to the following: (1) the topography of the site; (2) oceanographic

Table 11.8 Wastewater flow discharged from municipal sewerage systems into receiving bodies (1000 m^3/year), 2002–2016 (TURKSTAT 2017b)

	2002		2004		2006		2008		2010		2012		2014		2016	
	Amount	%	Amount	%	Amount	%	Amount	%	Amount	%	Amount	%	Amount	%	Amount	%
Amount of wastewater discharged	**2,497,657**	**100.00**	**2,922,783**	**100.00**	**3,366,894**	**100.00**	**3,261,455**	**100.00**	**3,582,131**	**100.00**	**4,072,563**	**100.00**	**4,296,851**	**100.00**	**4,484,075**	**100.00**
Treated	1,312,380	52.54	1,901,040	65.04	2,140,494	63.57	2,251,581	69.04	2,719,151	75.91	3,260,396	80.06	3,483,846	81.08	3,842,350	85.69
Untreated	1,185,277	47.46	1,021,743	34.96	1,226,400	36.43	1,009,874	30.96	862,979	24.09	812,167	19.94	813,005	18.92	641,724	14.31
Sea	**885,981**	**74.80**	**1,178,001**	**40.30**	**1,522,695**	**45.23**	**1,458,461**	**44.72**	**1,498,728**	**41.84**	**1,843,115**	**44.95**	**1,915,294**	**44.57**	**1,812,650**	**40.42**
Treated	709,816	80.10	1,003,736	85.21	1,215,440	79.82	1,231,880	84.47	1,347,977	89.94	1,718,588	93.24	1,759,461	91.86	1,724,792	95.15
Untreated	176,165	19.90	174,265	14.79	307,255	20.18	226,581	15.53	150,751	10.06	124,528	6.76	155,833	8.14	87,858	4.85
Lake/ Artifical lake	**38,403**	**1.54**	**43,006**	**1.47**	**46,415**	**1.38**	**67,193**	**2.06**	**76,024**	**2.12**	**75,116**	**1.83**	**93,596**	**2.18**	**78,551**	**1.75**
Treated	23,634	61.54	25,283	58.79	28,166	60.68	48,295	71.87	37,881	49.83	36,748	48.92	47,893	51.17	53,262	67.81
Untreated	14,769	38.46	17,723	41.21	18,249	39.32	18,899	28.13	38,143	50.17	38,368	51.08	45,703	48.83	25,289	32.19
River	**1,356,297**	**54.30**	**1,380,516**	**47.23**	**1,410,614**	**41.90**	**1,404,164**	**43.05**	**1,741,078**	**48.60**	**1,817,352**	**44.32**	**1,898,895**	**44.19**	**2,153,123**	**48.02**
Treated	514,165	37.86	713,395	51.68	705,561	50.02	778,293	55.41	1,180,630	67.81	1,276,456	70.24	1,409,633	74.23	1,728,000	80.26
Untreated	841,402	62.00	667,121	48.32	705,054	49.98	625,871	44.59	560,448	32.19	540,896	29.76	489,262	25.77	425,122	19.74
Dam	**96,434**	**3.86**	**99,551**	**3.41**	**121,532**	**3.61**	**115,405**	**3.54**	**130,224**	**3.64**	**114,199**	**3.46**	**120,781**	**2.81**	**126,325**	**2.82**
Treated	23,849	24.73	52,563	52.80	84,015	69.13	84,375	73.11	83,409	64.05	63,296	55.26	61,843	51.20	76,660	60.68
Untreated	72,586	75.27	46,988	47.20	37,517	30.87	31,030	26.89	46,816	35.95	50,903	44.74	58,938	48.80	49,665	39.32
Land	**37,013**	**1.48**	**40,007**	**1.37**	**120,525**	**3.58**	**50,374**	**1.54**	**35,091**	**0.98**	**35,770**	**0.87**	**17,954**	**0.42**	**20,063**	**0.45**
Treated	5834	15.76	6420	16.05	12,011	9.97	14,108	28.01	9166	26.12	8999	25.16	8367	46.60	14,036	69.96
Untreated	31,179	84.24	33,586	83.95	108,514	90.03	36,266	71.99	25,925	73.88	26,771	74.84	9587	53.40	6027	30.04
Other	**83,528**	**3.34**	**181,702**	**6.22**	**145,113**	**4.31**	**165,857**	**5.09**	**100,985**	**2.82**	**187,011**	**4.48**	**250,332**	**5.83**	**293,363**	**6.54**
Treated	35,082	42.00	99,642	54.84	95,301	65.67	94,631	57.06	60,088	59.50	156,309	83.28	196,649	78.56	245,601	83.72
Untreated	48,446	58.00	82,059	45.16	49,813	34.33	71,226	42.94	40,897	40.50	30,701	16.72	53,683	21.44	47,762	16.28

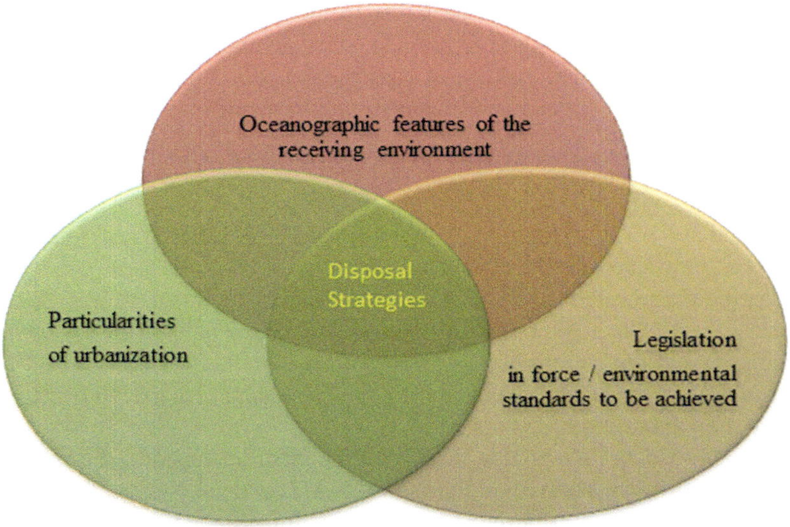

Fig. 11.5 Assessment of disposal strategies. (Burak and Demir 2016)

features of the receiving media in case of a coastal city; (3) the nature and the degree of the pollution and its fluctuation; (3) environmental conditions of the surroundings and receiving waters; (4) ruling pollution abatement criteria and ruling treatment mandatory standards and guidelines; (5) transboundary pollution conditions etc., as presented in Fig. 11.5.

It will be appropriate to remember some points related to ocean disposal strategy experienced in other World's great coastal cities in the recent past:

1. Use of the sea: where a city is near the sea, it can be considered as part of its wastewater disposal system, to take strategic advantage of its assimilation capacity and its great spaces
2. Treatment: wastewater must be treated to some extent. Industrial wastes bearing concentrated toxic chemicals, such as heavy metals, should be treated at their source
3. Discharge away from the shoreline: Treated wastewater discharged to the sea should be released into the optimum currents' location that is capable of assimilation and dispersion far from the coast. Discharge at the shoreline, or to poorly-flushed bays, will be much less effective, because natural dispersion of wastes from the shoreline is very slow compared to offshore regions
4. Environmental impact: Furthermore, the shoreline is where people swim, walk and generally to enjoy the marine environment. The coastal environment is also a very important region for aquatic life and the related food chain. The sustainability of the marine environment is of utmost importance as the coastal waters are shelter to aquatic species

5. Treatment plant location: It is desirable to locate a treatment plant in an area where there is sufficient space to provide adequate treatment to all wastewater to be handled there, both at the time of construction and in the future for extension, when flows may be greater and treatment requirements may be stricter.

The following overall wastewater treatment and disposal strategies and targets have been adopted in Turkey:

1. Sewerage system is separate from the storm water collection system in order to decrease the wastewater flow to be treated
2. Industrial wastewater must undergo preliminary treatment to the degree of domestic wastewater prior to its discharge into the municipal sewerage system
3. Discharge into the bottom of the receiving body
4. Wastewater is treated mechanically, biologically, and tertiary where and when required and possible.
5. Phased approach with regard to the size of the plant (extension over time to cope with the equivalent population load) and treatment level (start with mechanical treatment and deep sea outfall as an urgent implementation and upgrade the treatment level over time)
6. In case of "sensitive zones", tertiary treatment is compulsory.

11.6.2 Industrial Wastewater

Industrial pollution is caused by polluters of different characteristics from various sources generated by concentrated industrialization. The pollution of receiving media by industrial discharges has been a great concern in recent years. The worst affected marine environments are the bays of Izmit, Izmir, and Iskenderun located on the shoreline of the Marmara, Aegean, and the Mediterranean Sea, respectively, where industrial facilities of various categories and sizes are being operated. The treatment approach applied by industries in Turkey is an end-of-pipe treatment, which does not appear to be the optimum solution for the production technologies currently being used. Even the best technologies now being implemented in Turkey only rank moderately when compared to worldwide classifications. As dictated by the Environment Act, industries are obliged to treat their effluent to the standards specified for the category to which they belong. Therefore, under existing conditions, compliance between industrial production and wastewater technologies must be ensured. Where industries discharge into the municipal sewer in metropolitan municipalities, they have to pre-treat their sewage in compliance with the domestic wastewater standards before discharging it into the municipal sewer.

The lack of reliable data about actual operational conditions prevents sound conclusions to be reached about the depolluting rate at industrial treatment plants. Although the best practical technology (BPT) is applied as a standard definition in industry, due to these reasons, it is not clear exactly what this standard is. Under these circumstances, it is not possible for many industries to apply the standards

Table 11.9 Wastewater discharges to receiving water bodies 1000 m^3/yr. (TURKSTAT 2018)

	Water withdrawal	Discharged wastewater	Discharged wastewater (with the exception of cooling water)	Rate of discharged wastewater which is treated (with the exception of cooling water) (%)
Total	17,313,191	14,884,363	5,004,986	80.7
Municipalities	5,832,649	4,250,463	4,250,463	85.3
Villages	375,758	117,040	117,040	16.7
Manufacturing industry	2,115,642	1,677,279	177,969	88.3
Thermal power plants	8,608,370	8,474,339	94,271	10.5
Organized industrial zones	150,359	220,991	220,991	96.2
Mining enterprises	230,412	144,251	144,251	9.6

defined in the regulations, and this leads to their violation. In order to draw up regulations which are applicable in practice, the first step is to lay down a strict definition of the treatment technology. Revision of water quality categories for the receiving media as defined in the regulations and taking into account the feasibility of industrial discharge standards in practice would facilitate the compulsory enforcement of pollution control legislation. Large state-run plants in the textile, sugar, fertilizer and metal industries in particular, were the worst offenders where pollution is concerned. During recent years, with emerging governmental strategies, most of these industries have been privatized. However, although general improvement is foreseen for industrial wastewater discharge, there is no comparative assessment exhibiting polluting discharges between before and after privatization of these large industries. The industrial polluting load generated in each river basin was computed by TUBITAK-MAM within the scope of RBPAPs) (TUBITAK MAM 2010–2013), as given in Annex 11.3. Table 11.9 gives sectoral wastewater discharges with related withdrawals.

11.6.3 Organized Industrial Zone (OIZ) Water Supply, Sewerage and Wastewater Disposal

Organized Industrial Zones (OIZs) in Turkey are established as per Law No. 4562 on Organized Industrial Zones, approved by the National Assembly on April 12th, 2000 and issued in the Official Gazette No. 24021 dated April 15th, 2000.

OIZs are conceived with the objective of developing industries in suitable locations by preventing irregular and unplanned industrialization and urbanization, protecting the environmental pollution by constructing centralized industrial wastewater treatment plants and associated infrastructure schemes (water distribution networks, sewerage and storm water drainage systems), utilizing resources rationally, and improving technical infrastructure effectively.

OIZs can be considered as municipalities providing utility services (water, wastewater, storm water, electricity, natural gas, and telecommunications) to all industrial enterprises situated in the OIZ responsibility area. Principles of "user pays" and "polluter pays" are applied by the OIZs. Water service costs are allocated to the individual industries, based on measured water consumption by water meters; whereas wastewater treatment costs are reflected to the wastewater generators, based on pollution loads and quantities. Wastewater service tariffs are not charged to OIZs by municipalities if these have their own wastewater treatment plants in operation.

There are 308 officially established OIZs in Turkey, (2018); some of them are not active and fully deployed yet. Total employment in OIZs is about 1.7 million.

The total number of active OIZs in Turkey increased from 181 in 2012 to 217 in 2016, as shown in Table 11.10. About 88% of active OIZs have their own water and sewerage network, and 41% (90 OIZs) provide centralized wastewater treatment services.

Out of the total OIZ water demand of 175 million m^3 in 2016, 14.3% (25 million m^3) was provided by the municipalities as bulk water supply, the rest being produced by OIZ managements themselves from different water sources, as shown in Table 11.11.

The total quantity of wastewater discharged by the OIZs in 2016 (263 million m^3) was 50.3% higher than the total quantity of water supplied by OIZ managements, as shown in Table 11.12, since some industries inside OIZs have developed their own water sources (wells) and do not purchase water from the OIZ managements but discharge their wastewater to the OIZ sewers.

Table 11.10 Evolution of the OIZs in Turkey between 2012 and 2016 (TURKSTAT 2017d)

a. Number of OIZs			
Organized Industrial Zone (OIZ)	**2012 (No.)**	**2014 (No.)**	**2016 (No.)**
Active	181	196	217
Having water network	155	172	192
Having sewerage network	157	174	189
Providing wastewater treatment services	57	76	90
b. Active OIZs (%)			
Organized Industrial Zone	**2012 (% of Active OIZs)**	**2014(% of Active OIZs)**	**2016 (% of Active OIZs)**
Active	100.00	100.00	100.00
Having water network	85.64	87.76	88.48
Having sewerage network	86.74	88.78	87.10
Providing wastewater treatment services	31.49	38.78	41.47

Table 11.11 Water supply sources of the OIZs in Turkey in 2016 (TURKSTAT 2017d)

Water supply sources of OIZs	2016 (million m^3)	2016 (%)
Wells	84.4	48.2
Springs	20.1	11.5
Municipalities	25.0	14.3
Dams and rivers	45.5	26.0
Total	**175.0**	**100.0**

Table 11.12 Wastewater discharges of the OIZs in Turkey in 2016 (TURKSTAT 2017d)

Wastewater discharges of OIZs	2016 (million m^3)	2016 (%)
Rivers	190.7	72.5
Municipal sewerage network	31.6	12.0
Dry creek beds	19.7	7.5
Other receiving environment	21.0	8.0
Total	**263.0**	**100.0**

Out of the total quantity of wastewater discharged by the OIZs in 2016 (263 million m^3), 87.0% (229 million m^3) was treated, of which 56.2% with advanced treatment processes and 43.8% with chemical or biological treatment processes.

11.7 Wastewater Reuse

Reuse is an ecological approach that protects water resources and ecosystems as part of a circular economy. It reduces the pressure on scarce water resources, wastage of treated wastewater, and pollution of the receiving environment, providing that wastewater is treated properly and comply with the corresponding usage. In fact, with growing environmental awareness, wastewater is increasingly considered as a "resource" like others in planning.

11.7.1 Agriculture

Wastewater reuse in agriculture is an old practice. Land disposal of wastewater was the first treatment system. The ground operates as a quite efficient filter, and microorganisms present in wastewater supply the soil with nutrients.[3] At present, the main reason for wastewater reuse in irrigated agriculture is to save water in arid regions rather than to treat it on-land.

[3]One hectare comprises one or two tons of microorganisms according to some researches (Degrémont 1991).

In order to ensure safe reuse, measures must be taken to avoid deposits and corrosion in the distribution system in general. It is recommended that, in any case, wastewater undergoes preliminary settling and preliminary biological treatment to reduce odors.

Two categories of risks associated with reuse of wastewater are the main constraints of reused water management. These are: (1) Health risks for close-lying neighborhoods and for consumers of the products irrigated with the reused water. The risks vary greatly depending on the local state of sanitation in the area, farming methods, customs and climate. However, wastewater should not be used on or near vegetables that are eaten raw. Arboriculture, cereals, beets, and oleaginous crops are the types of cultivation most suited. Surface irrigation is preferred to the sprinkler method. (2) Risks to the soil and crops resulting in clogging the soil, increased salinity, and introduction of toxins. Land disposal of wastewater for any reason can alter the physical properties of the soil. Excessive amount of sodium and the absence of leaching can destroy the soil structure in arid regions, in particular (Degrémont 1991).

11.7.2 Reuse of Treated Domestic Wastewater

A reuse project is dependent on an integrated wastewater network comprising sewerage, wastewater treatment, and the needed infrastructure to use the effluent. Therefore, wastewater treatment plants have to be equipped with process technologies like advance (tertiary) treatment. Additionally, the end-use determines the quality necessitated of the effluent, thus the treatment level. The key issue is the question “which quality of effluent for which reuse?” Sectors of reuse that must be treated at tertiary level is given in Fig. 11.6.

11.7.3 Wastewater Reuse in Industry

According to the EUROSTAT Water Statistics Manual, reuse (wastewater treated to some extent and reused) is specified in countries according to the nature of the wastewater. Internal industrial recycle is not the subject of the questionnaire, as commented by TurkStat. Differentiation between definitions of recycled water and reused water is made as the following: (a) if used water remains within the factory fence, it is called recycled water, which depends on the production technology, methods used, raw materials, and substances used during the process; (b) if the factory treats the wastewater with the objective of reusing it outside the factory fence, then it is considered in the reuse statistics.

Municipal wastewater treated up-to-the standards level may be a source of water that is completely suitable for industrial needs, especially for cooling and washing

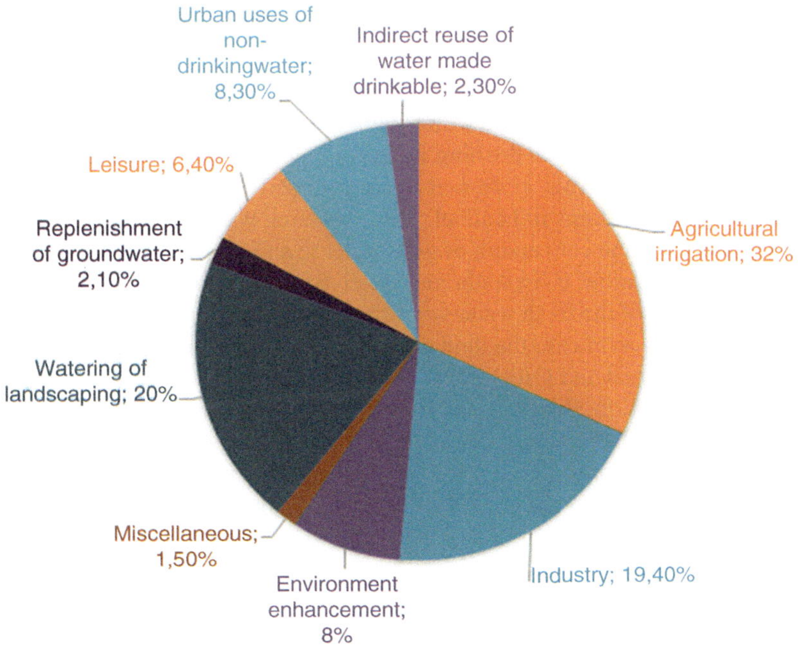

Fig. 11.6 Sectors in which (tertiary) treated waste water can be reused. (IPEMED 2018)

water. Such reuse has a large number of practice in arid regions in general. Very often, enhanced removal of organic pollution is necessary, and biological treatment is then followed by a tertiary treatment. However, this practice requires a sound and integrated wastewater management planning, starting from the very first conceptual level which is yet inadequate in Turkey at this stage. Furthermore, although this practice is environmentally-friendly, a detailed economic analysis is needed to compare the raw water and the treated water investment and operating costs in the case of Turkey. Only 12% of treated municipal wastewater is reused in industries (Demir et al. 2017).

As part of the industrial revolution, WATER 4.0 puts digitization and automation at the center of a strategy for resource-efficient, flexible and competitive water management. In doing this, Water 4.0 incorporates the same main features and terms of the industrial revolution Industry 4.0, such as "networking of machines, processes, storage systems and resources" (German Water Partnership 2016). According to experts from industry and research, the upcoming industrial revolution will be triggered by the Internet, which allows communication between humans as well as machines in Cyber-Physical-Systems (CPS) throughout large networks (Brettel et al. 2014).

Large industrial sectors in Turkey are keen on industrial reforms, (e.g. green production, smart systems) in order to comply with international standards.

11.7.4 Wastewater Reuse for Household Purpose

The reuse of treated wastewater at home or at municipal level is possible for various levels of quality and in accordance with a number of working plans. Partial recycling inside the buildings comprises flushing water for toilets from recirculated wastewater that has been treated. This system is recommended to be introduced in newly constructed smart buildings in Istanbul. Some research was carried out in residences in order to work out the conversion of grey water to a flush water source in a student residence hall in Istanbul (Giresunlu and Beler-Baykal 2016). Some other research stipulates that 25% of the fresh water would be saved out of the total water conveyed from the Melen River, in case the light grey water is reused for household purpose in Istanbul (Beler-Baykal and Oructut 2017). Figure 11.7 gives all uses as a percent of household consumption. Based on these figures, it can be concluded that although it would be unrealistic to expect to use all of the grey water generated in the city, there is a significant opportunity to reuse it as the city is in the phase of urban reconstruction, estimated to reach to about 50% of the existing buildings.

11.7.5 Wastewater Reuse for Municipal Purpose

In arid regions and on Mediterranean coastal areas, wastewater reuse for irrigation and gardening has been strongly recommended by the former Ministry of Environment and Forestry through their regional directorates, encouraging wastewater reuse at secondary houses, hotels and gulf resorts, in particular (Burak 2007). Also, supplying municipal systems with wash water like gardening or street cleaning is a common practice in metropolitan municipalities, in particular. Tourist resorts use exclusively reused water for gardening and golf courses all over the country, especially in the Mediterranean tourist resorts. This reuse practice is enhanced by economic deterrence and incentives related to water tariffs.

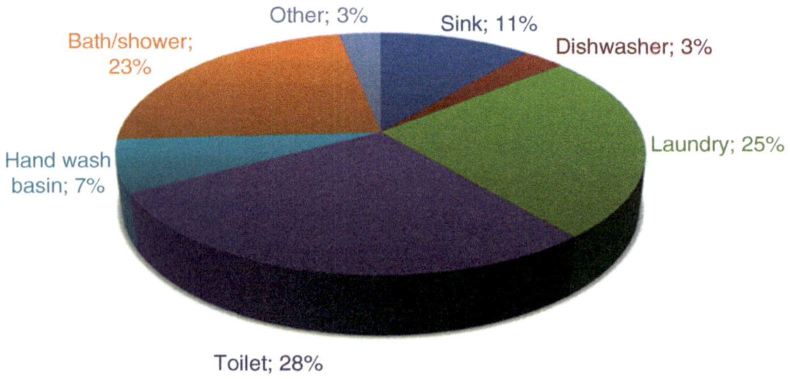

Fig. 11.7 Typical water use in households in Istanbul. (Giresunlu and Beler-Baykal 2015)

11.7.6 Sludge Disposal

"Regulations related to general practices for waste management" were issued by the Ministry of Environment and Urbanization in 2008, and sludge generated by wastewater treatment, water treatment, and industrial treatment plants were incorporated respectively under the Clause 19.08 of the same regulations in 02.04.2015 (Ozturk et al. 2015). In these regulations, wastes are classified with respect to the hazard they represent for the environment and treatment methods they must undergo followed by corresponding disposal systems (The Council of the European Union 2008).

However, advanced sludge disposal does not yet have expanded implementation. It is mostly case-specific as part of an integrated project or for research purposes at pilot scales. Some municipalities have implemented complete schemes of wastewater disposal including sludge (e.g. Marmaris, Antalya, Konya municipalities), where biogas digestion, composting and reuse are all realized. The common practice for sludge disposal is generally dewatering in order to dump it at municipal solid waste landfill sites. Incineration of municipal sludge is not practiced as this is an expensive solution. Furthermore, difficulties in reaching the emission standards related to dioxine is another reason for reluctance with regard to incineration. The use of sludge in agriculture is not an adopted practice yet in Turkey although this is very likely to find large-scale implementations. Here again, sludge reuse remains anecdotal, limited for research purposes (e.g. universities' agriculture departments).

11.8 Cultural Barriers to Reuse

Generally, conventional systems are better accepted culturally and socially with regard to household water use. It is obvious that treated wastewater is not a preferred resource in case other fresh water resources can be made available, like spring water and surface water. However, wastewater reuse is an applied practice in water-scarce regions worldwide, no matter what cultural and religious beliefs are.

With respect to purely religious concerns in Muslim countries, "*... To make use of modern technology in order to recycle wastewater effluents after treatment seems quite in keeping with the spirit and letter of the Islamic teachings, though of course considerations of health, cost and public acceptance are always bound to be taken into consideration...*" according to Shaukat Farooq (Farooq and Ansari 1981).

11.9 Investment Needs of the Water Sector

In line with the accession partnership requirements, a significant improvement in the water sector has been realized in order to comply with the on-going process of the "Acquis communautaire" (The full body of EU Law). Within the scope of the

"Approximation Partnership Document" (APD) accepted as a road map document, Turkey prepared national programs comprising the requested reforms as well as the economic and environmental improvements for the adoption of the Acquis Communautaire. These national programs were launched by the Turkish Government as recommended by the EU, and financial aid was provided to Turkey within the "accession budget" in order to intensify the harmonization work. The preparation of an environmental strategy document was launched with short, medium and long-term objectives to comply with the Accession Partnership and to implement the Acquis Communautaire with respect to the environment.

The Directives that require the highest amount of investments among the regulations are the Urban Waste Water Treatment Directive, Drinking Water Directive, Directive on the quality of the surface waters used for drinking water, Water Framework Directive, Dangerous Substances Directive, Nitrate Directive, and Bathing Water Directive.

The highest and the lowest investment amounts for the implementation of the Directives related with the water sector were determined within the context of the "Environmental Heavy-Cost Investment Planning in Turkey" project (Envest Planners 2004). The Urban Wastewater Treatment Directive requires different treatment types for wastewater that is discharged into sensitive and less sensitive water sources.

The high, medium and low cost scenarios were developed in case of implementation of the Directives, and the total of the costs associated with the above mentioned seven directives were determined to be as follows: 33,969 million Euros, the lowest cost; 35,874 billion Euros, the medium cost, and 37,867 billion Euros, the highest cost scenario, respectively. This revealed that investment discrepancy between the highest and the lowest cost scenarios was approximately 4 billion Euros (Republic of Turkey Ministry of Environment and Forestry 2006).

11.10 Administrative and Legal Framework of Municipalities in Turkey

11.10.1 General Overview

A general description of the existing administrative structure in Turkey will enable to better understand the prevailing municipal water management system that has been subject to various revisions in the past.

In the current administrative structure, there are 81 provinces covering the whole geographic area of the country, each having a governor appointed by the decision of the President. The governor is the highest ranking official in the province representing the Presidency and is responsible for the general public administration of the province.

In each province, there are districts covering the whole geographic area of the province, headed by an appointed district governor. There are 919 districts currently in Turkey.

A municipality, which has a mayor and municipal council members elected by the local population, can be established in urban areas having a population exceeding 5000 inhabitants. Municipalities must be established in provincial and district centers as stipulated in Article 4 of the Act No. 5393 on municipalities, enacted by the National Parliament on 03/07/2005 and published in the Official Gazette No. 25874 dated 13/07/2005. If the population of a municipality falls down below 2000 inhabitants for some reasons, its municipal status becomes discontinued, and it turns into a village status. A village is represented by a "village headman" (Muhtar) and members of the "Village Elderly Board" elected by the resident population. There are 18,333 villages at present.

The country is administered at public services level under two types of municipalities, these are:

1. Metropolitan municipalities established in compliance with the Act No. 5216 on metropolitan municipalities, enacted by the National Parliament on 10/07/2004 and published in the Official Gazette No. 25531 dated 23/07/2004 (Official Gazette 2004).[4]

2. Non-metropolitan municipalities established in compliance with the Act No. 5393 on municipalities, enacted by the National Parliament on 03/07/2005 and published in the Official Gazette No. 25874 dated 13/07/2005.

At present, 30 Metropolitan Municipalities exist altogether as listed in Table 11.13 below.[5]

Fourteen new metropolitan municipalities were established after the municipal elections held on March 30, 2014 as per Act No. 6360 on the "Establishment of New Metropolitan Municipalities", accepted by the National Parliament on 12/11/2012 (Official Gazette 28489, 2004). As per Act No. 6360, responsibility areas of all metropolitan municipalities were extended to cover the whole provincial areas. Out of the total population of Turkey as of 31/12/2017 (80,810,525 inhabitants), 77.61% (62,717,604 inhabitants) live within the responsibility areas of 30 metropolitan municipalities.

According to the information disclosed by the Ministry of Interior Affairs of Turkey, there are in total 1397 municipalities in Turkey as of 2017, whose breakdown with respect to municipality status is given in Table 11.14 below. Out of the 81 provinces covering the whole area of Turkey, 30 provinces have the metropolitan municipality status. In metropolitan municipalities, there are 519 district municipalities.

[4]Metropolitan municipalities are established in certain provinces having a population figure exceeding 750,000 inhabitants with economic, social and cultural importance.

[5]There were 16 Metropolitan Municipalities before the municipal elections held on March 30, 2014 whose responsibility areas were limited to a few central districts of the provinces with the exception of Istanbul Metropolitan Municipality and Izmit Metropolitan Municipality whose responsibility areas covered the whole provincial boundary since 2004.

Table 11.13 Population of metropolitan municipalities as of 31/12/2017 established before and after municipal elections held on March 30, 2014[a]

Metropolitan Municipalities established before March 30, 2014 (16 nos.)	Population (Inh.) as of 31/12/2017	As % of Turkish population as of 31/12/2017 (%)	Metropolitan Municipalities established after March 30, 2014 (14 nos.)	Population as of 31/12/2017 (Inh.)	As % of Turkish population as of 31/12/2017 (%)
Adana	2,216,475	2.74	Aydin	1,080,839	1.34
Ankara	5,445,026	6.74	Balikesir	1,204,824	1.49
Antalya	2,364,396	2.93	Denizli	1,018,735	1.26
Bursa	2,936,803	3.63	Hatay	1,575,226	1.95
Diyarbakir	1,699,901	2.10	Malatya	786,676	0.97
Eskisehir	860,620	1.06	Manisa	1,413,041	1.75
Erzurum	760,476	0.94	Kahramanmaras	1,127,623	1.40
Gaziantep	2,005,515	2.48	Mardin	809,719	1.00
Istanbul	15,029,231	18.60	Mugla	938,751	1.16
Izmir	4,279,677	5.30	Tekirdag	1,005,463	1.24
Izmit	1,883,270	2.33	Trabzon	786,326	0.97
Kayseri	1,376,722	1.70	Sanliurfa	1,985,753	2.46
Konya	2,180,149	2.70	Van	1,106,891	1.37
Mersin	1,793,931	2.22	Ordu	742,341	0.92
Sakarya	990,214	1.23			
Samsun	1,312,990	1.62			
TOTAL	**47,135,396**	**58.33**		**15,582,208**	**19.28**

[a]Total population of Turkey on 31/12/2017: 80,810,525 inhabitants; total population of 30 metropolitan municipalities: 62,717,604 inhabitants (77.61% of total Turkish population live in metropolitan municipalities)

The overall regulatory responsibilities of the municipalities for municipal water and environmental management include mainly the protection of public health and environmental pollution. The duties and responsibilities for service provision of the non-metropolitan municipalities are specified in detail in Article 15 of the **"Municipal Act No. 5393"** and cover water supply, sewerage and drainage services.

The specific duties and responsibilities of the metropolitan municipalities, as well as sharing of them with the district municipalities, are specified in detail in Article 7 of the "Metropolitan Municipalities Act No. 5216".

11.10.2 Administrative and Legal Framework of Water and Sewerage Administrations (WSAs) in Metropolitan Municipalities

Municipalities' Act No. 5393, promulgated on 03.07.2005, states that municipalities have the responsibilities for the provision of water and sewerage services, including setting tariffs and service charges for the said services as stipulated in Article 14 and

Table 11.14 Number of municipalities with respect to municipality status as of 2017

Municipality status	No.	% of total
Metropolitan municipalities	30	2.15
Non-metropolitan provincial center municipalities	51	3.65
Metropolitan district municipalities	519	37.15
Non-metropolitan central district municipalities	400	28.63
Non-metropolitan non-central district municipalities	397	28.42
Total number of municipalities	1397	100.00

15 of the Act, respectively. Article 18 outlines the tasks and the authority of municipal councils and enumerates the requirements for the determination of tariffs and approval of municipal budgets.

As stated under the responsibilities and privileges of non-metropolitan municipalities in Article 15, municipalities are empowered to: (1) billing and collection of water and wastewater tariffs and service charges; (2) supplying of drinking and industrial water, collection and disposal of wastewater and storm water, construction and operation of necessary facilities for the provision of water and wastewater services, either by its own resources or through contracting-out (public-private partnership models), and (3) operation of spring water resources.

As per "Metropolitan Municipalities Act No. 5216 dated 10.07.2004", metropolitan municipalities have the responsibilities for the provision of water and sewerage services; construction and operation of dams and other facilities for this purpose; rehabilitation of creeks; and marketing of spring waters or treated waters produced.

There are two main administrative organizations responsible for the provision of water and sewerage services carried out by metropolitan and non-metropolitan municipalities, respectively:

1. Metropolitan Municipalities: Water and Sewerage Administrations (WSAs) having their own budget, physical and financial resources, and legal entity separate from those of the Metropolitan Municipality Administrations (currently 30 WSAs exist in Metropolitan Municipalities of Turkey)
2. Non-metropolitan Municipalities: (i) Water and Sewerage Departments (an ordinary municipal department within the municipal budget); (ii) Water and Wastewater Operating Enterprises (a separate operating unit responsible for the collection of revenues and spending of O & M expenses within the municipal budget).

11.10.3 Legal framework of Water and Sewerage Administrations (WSAs) in Metropolitan Municipalities

In each metropolitan municipality, a separate public legal entity, named Water and Sewerage Administration (WSA), affiliated to the related metropolitan municipality must be established as per Act No. 2560,[6] the so-called "ISKI Act" (Official Gazette 17523, 1981). WSA has its own budget, accounts, physical/financial resources and staff, and is responsible for the provision of all water, sewerage and storm water services within the responsibility area of the metropolitan municipality.[7]

WSAs established in all metropolitan municipalities are "ring-fenced" public legal entities. WSAs generate their revenues only from the water/wastewater tariffs and service charges and have the discretion of spending them exclusively for the provision of water/wastewater services within the responsibility area of the metropolitan municipalities. WSAs have their own budgets, accounts, staff, physical and financial resources completely separate from those of the metropolitan municipalities.

The list of the main existing Turkish Acts and legislation governing the tasks, responsibilities, and operations of WSAs are presented in Annex 11.4. The duties and responsibilities of the WSA related to water services are itemized in Article 2 of the Act No. 2560, as given in Annex 11.5.

The administration of the WSA is carried out by the General Assembly, Board of Directors and General Manager, as stipulated in the Article 3 of the Act No. 2560. The General Manager is proposed by the Mayor of Metropolitan Municipality and approved by the Minister of Interior Affairs. The General Assembly of the WSA convenes ordinarily twice a year. Its duties and responsibilities are itemized in Article 6, the ones of the Board of Directors in Article 9, and finally the duties and responsibilities of the General Manager are itemized in Article 11 of the same Act No. 2560.

11.10.4 Advantages of having WSAs in Metropolitan Municipalities

Having WSAs in metropolitan municipalities has the following advantages:

- Application of "user pays", "polluter pays" and "full cost recovery" principles are guaranteed in the WSAs since all water/sewerage/storm water service provision and investment and financing costs must be covered by water/wastewater tariffs and service charges to be collected from the beneficiaries, as stipulated in the so-called ISKI Act.

[6]Act for the Establishment of Istanbul Water and Sewerage Administration General Directorate (ISKI) dated 20/11/1981 issued in the Official Gazette No. 17523 on 23/11/1981 (amended by Act No. 3009 dated 23/05/1984 (became affiliated to the Istanbul Metropolitan Municipality from being under the auspices of the Istanbul Governorate) and Act No. 3305 dated 05/06/1986) through additional Article 4 stating that "This Act shall also be applied in other metropolitan municipalities."

[7]Provincial geographic boundaries as regulated in the Act No. 6360 dated November 12, 2012 which came into force after the municipal elections held on March 30, 2014.

- Since the service coverage area and physical and financial resources of the WSAs are considerably higher than the individual non-metropolitan municipalities, WSAs capitalize on the economies of scale and provide higher service levels and quality at lower costs;
- Due to economies of scale unit service provision costs are considerably lower in the WSAs, which makes it easier to apply affordable water/wastewater tariffs to cover the full costs of service provision. This is also supported by differentiating the tariffs charged between urban areas (relatively higher tariffs) and rural areas (relatively lower tariffs) as well as between summer houses (relatively higher tariffs) and main residential houses (relatively lower tariffs);
- WSAs carry out project planning and development activities on a provincial basis, which is much more efficient and effective compared to those on non-metropolitan municipality basis;
- WSAs apply water conservation measures for the whole catchment zones in the province, making public health and environmental protection much more effective;
- WSAs materialize project implementation required in remote rural areas of the province easily, due to revenues generated in condensed urban areas (cross-subsidization of poor rural area population by the wealthier urban area population);
- WSAs attract highly qualified staff for the benefit of all the population in the province by paying market rates, which is generally not possible by the non-metropolitan municipalities;
- WSAs get easy access to financial and money markets (e.g. International Financing Institutions (IFIs)–like the World Bank) to raise loans/funds/grants for investment financing due to their higher financial credibility.

11.11 Administrative and Legal Framework of Water and Sewerage Services in Non-metropolitan Municipalities of Turkey

Water and wastewater services are provided by the following public organizations in non-metropolitan municipalities of Turkey:

- Non-metropolitan Municipalities: Water and Sewerage Departments/Operating Enterprises/Municipal Unions (51 provinces);
- Rural Settlements (Villages): Special Provincial Administrations (18,332 villages in non-metropolitan municipalities outside municipal responsibility areas).

11.11.1 Water and Sewerage Services Departments in Non-metropolitan Municipalities

In most of the non-metropolitan municipalities, municipal services are organized on a departmental basis. Water and Wastewater Services Department in a non-metropolitan municipality is typically one of the operational departments having the same status with the others. Water and Wastewater Services Department has an expenditure budget to execute but does not have a separate revenue budget. Thus, it competes with other municipal departments to get its share from all municipal revenues included in a unified pool of revenues. Therefore, Water and Wastewater Services Department in a non-metropolitan municipality is a "cost center" not a "profit center". Water and Wastewater Services Department does not operate as a semi-autonomous Water Utility Department due to the following reasons:

- All administrative and support units related to the provision of water and wastewater services are not organized under a separate and fully accountable (semi-autonomous) Department;
- Some services and support functions are provided by other municipal departments, including accounting, finance, budgeting, planning, human resources management, communication, press and public relations, and legal services, which result in sharing of staff, vehicles and other resources;
- All water and wastewater service provision costs are not accounted for under a separate water/wastewater cost accounting system, but as part of the overall municipal accounting system with the other municipal departments;
- All water and wastewater tariff and service revenues collected are transferred to the overall municipal revenues budget and shared by different municipal departments based on daily needs. Thus, water/wastewater revenues collected cannot be used for the provision of water/wastewater services and the funding of related investments, primarily at the discretion management of the Water and Wastewater Services Department (ring-fencing principle);
- Annual budgets, tariff levels and structure, and long-term strategic (business) plans are not prepared by the management of the Water and Wastewater Services Department to improve administrative and operational efficiency and effectiveness of the water and wastewater services, which should ideally be prepared at the discretion of the management;
- Integrated and reliable cost accounting systems are missing, and financial statements (e.g. profit -loss statements, cash flow statements, and balance sheets segregated from those of the municipality) exclusively for water/wastewater services cannot be generated.[8]

Therefore, unlike the WSA in a Metropolitan Municipality, which uses water and wastewater revenues that it generates exclusively to provide water and wastewater services, Water and Wastewater Services Department in a non-metropolitan municipality does not have the financial autonomy and ring-fenced financial and accounting systems, which reduces service levels and quality.

[8]Water utility departments lack the ability to assess service based unit costs, cost recovery, profitability analysis, asset utilization efficiency, financial credibility.

11.11.2 Water and Sewerage Operating Enterprises in Non-metropolitan Municipalities

In some of the non-metropolitan municipalities, water and wastewater services are organized under an "Operating Enterprise", which is a proxy of a semi-autonomous water utility department, mostly required to be established by the IFIs and EU to ensure accountability, transparency, and ring-fencing.

In Article 71 of the Act No. 5393 on Municipalities, it is stated that: "The Municipality can establish operating enterprises within its budget to provide services having their own special revenues and expenses with the permission of the Ministry of Interior Affairs".

Consequently, it is possible to establish an "operating enterprise within the municipal budget to provide water and wastewater services", by taking a municipal council decision and getting the permission of the Ministry of Interior Affairs.

Thus, a semi-autonomous "Water and Wastewater Operating Enterprise" can be established legally, having its own revenues, expenses, budgets, chart of accounts, financial statements and physical resources, by compiling together all water, wastewater, storm-water related operations and maintenance, customer management, accounting/finance/budgeting/planning, communication, press and public relations, and all other related services under the organization of this operating enterprise. Such an operating enterprise will be having the right-to-use of the all the water/wastewater/storm water assets under the municipality's ownership and will have the full responsibility for the operation of the facilities and associated maintenance activities.

The benefits generated by the establishment of such a water and wastewater operating enterprise in a normal (non-metropolitan) municipality are the following:

- Running a semi-autonomous operating enterprise within the municipal budget for the provision of all water and wastewater services under one fully responsible organization structure by emphasizing the importance of water and wastewater services through increasing its status;
- Giving management the full power of decision making by reporting directly to the Mayor, but not to another ordinary municipal department which has no right in using revenues generated from its services;
- Materialization of transparency and accountability principles adopted by modern management principles and the use of water and wastewater revenues to cover the expenditures as a fully responsible body (facilitating decision making and execution in repair, maintenance and investment activities);
- Ensuring commitment of the municipal administration for spending of the water/wastewater tariff and service revenues generated to cover water/wastewater service provision costs and financing of investment expenditures, primarily in conformity with the Turkish environmental legislation and full cost recovery principles;

- Determination and follow-up of water/wastewater service provision costs and cost recovery regularly, since the operating enterprise will have its separate chart of accounts, financial statements, and cost accounting systems;[9]
- Calculation of full costs of service provision, including depreciation costs, and determination of full cost recovery tariffs by considering pollution loads and quantities, as well as socially-acceptable tariffs;
- Determination and follow-up of water/wastewater assets and liabilities (i.e. determining the current value of assets, annual depreciation costs and accumulated depreciation costs, which are not available in most of the non-metropolitan municipalities, and developing investment, repair and maintenance plans accordingly; assessment of the effectiveness of asset utilization and evaluation of financial credibility), as well as development of separate financial statements (i. e. profit-loss statements, cash flow statements, balance sheets) for the operating enterprise aside from those of the non-metropolitan municipality;
- Improving effectiveness and efficiency, without incurring any additional costs, by pooling all water/wastewater/storm water services under one fully responsible body, managing all water and wastewater assets (i.e. infrastructure facilities, accounts receivable, liquid funds), and liabilities through a semi-autonomous operating enterprise and its budget;
- Measurement and assessment of operating and financial performance of the departments of the operating enterprise and key personnel, based on pre-determined criteria, taking corrective measures in due time and, thus, ensuring accountability (i.e. non-revenue water (NRW), billing and collection performance, customer complaints, repair and maintenance performance);
- Ensuring visibility of the water and wastewater services and improving public awareness by training the "Press and Public Relations Department" to be established within the organization structure of the operating enterprise.

11.11.3 Water and Sewerage Services in Rural Settlements (Villages) of Non-metropolitan Provinces

Water and sewerage service provision responsibility outside municipal boundaries belongs to special provincial administrations, which exist in 51 non-metropolitan provinces in Turkey as per "Act No. 5302 on Special Provincial Administrations", passed from the National Parliament on February 22, 2005 and published in the Official Gazette No. 25745 dated March 4, 2005.

Special provincial administrations are public legal entities, having administrative and financial autonomy, established in non-metropolitan provinces to meet the common needs of the local population whose ultimate management organ, the

[9]Ensuring spending of tariff and service revenues for the coverage of water/wastewater service provision costs in accordance with the polluter pays / user pays principles – "ring-fencing".

Provincial General Council, is elected by the local electorate once in every 5 years. The Head and Representative of the Special Provincial Administration is the governor of the province appointed by the President.

Special Provincial Administrations are responsible for and authorized to provide the water and sewerage services to 18,332 villages which are located outside the municipal responsibility areas in 51 non-metropolitan provinces, provided that such services have a common local nature as per Article 6 of the Act No. 5302. In addition to the provision of water and sewerage services, Special Provincial Administrations are also responsible for development planning, environmental protection, solid waste management and other public services.

In general, in the organization scheme of special provincial administrations, there is a separate "Water and Sewerage Services Directorate", reporting to one of the deputy general secretaries, whose specific tasks and responsibilities related to water and sewerage services are mainly: (1) implementing the necessary infrastructure works, such as drilling boreholes, construction of pumping stations, storage tanks, and networks in all villages and rural settlements in order to provide drinkable water; (2) extracting groundwater from wells and providing laboratory services to monitor and control the water quality; (3) ensuring safe disposal of wastewaters from all villages and rural settlements by constructing sewerage networks and proving septic collection services via septic tankers, and (4) developing agricultural infrastructure in all villages and rural settlements to provide water for irrigation purposes in the most economic ways, in addition to and as part of water services.

11.12 Legislation on Full Cost Recovery in Turkey

According to Article 14 (Duties and Responsibilities of Municipalities) of the "Act No. 5393 on Municipalities" dated July 3, 2005, municipalities are responsible for the provision of water and wastewater services. Additionally, the municipalities are authorized to collect tariffs for water and wastewater services as per the Article 18 "Duties and Responsibilities of Municipal Councils" of the Act No. 5393, which states that municipal councils have duties and responsibilities on the establishment of tariffs for municipal services, not otherwise governed by other acts as taxes, duties, fees, or contribution charges.

For the efficient and effective management of the water and wastewater services, the following main principles must be taken into account:

- **User Pays principle:** Since all registered water/wastewater customers are obliged to install water meters by legislation, whose consumption is measured and bills are issued on a monthly or bi-monthly basis, the enforcement of this principle is guaranteed in Turkey.

- **Polluter Pays principle:** Wastewater tariffs based on pollution loads (exceeding domestic pollution load level) are applied by a limited number of big WSAs, but neither by the most of the recently established WSAs (14 nos. established in 2014), nor by the non-metropolitan municipalities.
- **Full Cost Recovery:** The enforcement of this principle is guaranteed in WSAs since they have a separate budget, which must be prepared on a "balanced budget" basis to ensure covering all the investment, O&M and financing costs by their own water/wastewater tariff, and service revenues stipulated in their Act of Establishment. However, non-metropolitan municipalities face difficulties in charging full cost recovery tariffs always due to socio-political reasons, and thus, deficits generated from the provision of water/wastewater services need to be subsidized from other municipal budget sources.[10]

In metropolitan municipalities in which WSAs are responsible for the provision of water/wastewater services, the full cost recovery tariff setting principle is stipulated explicitly in Article 23 of the Act No. 2560 on the establishment of ISKI as follows:

"Separate tariffs shall be determined for water sales, wastewater disposal services in locations where there are sewerage networks and emptying of septic tanks. While setting these tariffs, operations and administrative expenses, as well as replacement/renewal/rehabilitation expenditures not amortized but directly expensed and a reasonable profit margin is taken into account."[11]

For non-metropolitan municipalities, there is no legislation or methodology for the calculation and setting of water tariffs "neither on average nor with respect to customer groups". This also gives discretion to municipal councils to set water tariffs without any restrictions in the absence of any regulatory body. Thus, municipalities are allowed to charge full cost recovery and even higher water tariffs if they want to cross-subsidize any other municipal services. However, at the same time, they can also charge water tariffs lower than the full cost recovery (e.g. sufficient to cover cash O&M costs only, without considering depreciation (investment) and financing costs) due to political concerns.

On the contrary, there is a comprehensive legislation for the setting of wastewater tariffs by all municipalities, both metropolitan and non-metropolitan, as per Article 11 of the Act No. 5491 dated April 26, 2006, (Act No. 5491, 2006), amending the Act No. 2872 on Environment dated August 9, 1983, which stipulates the legal basis for the recovery of wastewater services costs as follows:

---

[10]A new institutional system called "Union of Municipalities" was established in the late 1990's in order to overcome this administrative constraint in small coastal municipalities (e.g. CALBIR Union formed with Cesme and Alacati municipalities) recommended by the World Bank mission in Turkey, based on the feasibility study prepared by Hyder Consulting Ltd. (Hyder Consulting Ltd. 1997). This system has had success as it attracted loans for water and sewerage infrastructure (The World Bank 1998; Burak et al. 2004).

[11]After an appeal to the Constitutional Court the phrase "profit margin of at least 10%" was deleted as of January 26, 2012 and replaced by "the Board of Directors shall decide on a *"reasonable profit margin"* over real costs of service provision".

> Beneficiaries or future beneficiaries of wastewater infrastructures have to contribute into all kinds of investment, operation, maintenance, rehabilitation or cleansing costs with respect to the pollution load and quantity of their wastewater generation, notwithstanding whether they are connected to the wastewater infrastructure or not. Those who benefit from wastewater services are charged with wastewater collection, treatment and disposal fees with respect to the tariffs announced by the municipal councils or other authorized institutions (WSAs).

Thus, "polluter pays" and "full cost recovery" principles are reflected in the legislation for the setting of wastewater tariffs, but not for water tariffs.

Based on the requirements in the Act, a detailed Regulation ("Regulation for the Establishment of Principles for Setting Tariffs for Wastewater Infrastructure and Municipal Solid Waste Disposal Services") was published in the Official Gazette No. 27742 dated October 27, 2010. The Regulation is applied to all public administrations involved in the provision of water and wastewater services (e.g. municipalities, WSAs, municipality unions, and organized industrial zones).

The Regulation sets the basic principles for wastewater tariff in Article 5 as follows:

- Full recovery of total system costs by tariffs;
- Establishment of tariffs based on "polluters pays principle"; and
- Utilization of wastewater related revenues for wastewater related expenditures exclusively.

Therefore, in addition to applying the "polluter pays" principle based on total (full) system costs, wastewater revenues collected shall be earmarked for the coverage of wastewater services costs only, implying that surpluses generated, if any, shall not be used for the financing of any other municipal services but for the improvement of wastewater services quality and quantity.

To ensure the application of the "user pays" principle, the Regulation stipulates that all real and legal persons receiving wastewater services shall be registered as official customers *(Article 10)* and that the installation of water-meters for the beneficiaries of both water and wastewater services and wastewater meters for the wastewater-only customers (i.e. private well owners who discharge their wastewater to the municipal sewerage systems) *(Article 11)* is compulsory.

The Regulation defines the "total system cost" as the sum of the following cost items as per Article 13 of the Regulation, that covers (1) Operation and Maintenance (O&M) costs; (2) Financing costs; (3) Depreciation costs of fixed assets; (4) Management and monitoring costs; (5) Related taxes; (6) Expropriation costs; and (7) A return on capital invested to ensure the financial sustainability of the system.

It is also stated in the same Article that the wastewater tariffs should be calculated after deducting "contributions to sewerage investments" collected from the beneficiaries from the total system costs, in accordance with Article 87 of the Act on Municipal Revenues dated May 29, 1981.

The Regulation requires taking into account the following wastewater related service components in the calculation of total wastewater system costs (*Article 14*):

"collection, transportation, pumping, treatment, discharge, sewage sludge disposal, recycling and sales revenue (e.g. use of treated wastewater as irrigation water, or use of sewage sludge as fertilizer, or generation of energy from sewage sludge "as negative costs").

Non-cash cost items are generally disregarded in the Turkish municipal accounting systems. However, being an "investing sector", depreciation cost is one of the biggest cost items in the cost structure of water/wastewater utilities. Having observed this fact, the Regulation stipulates adding depreciation costs to the total system costs: "*Annual depreciation rates to be used in the calculation of tariffs shall be based on useful lives of the assets. Asset values shall be determined based on replacement cost principle* (*Article 16*)".

Therefore, in order to reflect the real depreciation costs in the wastewater tariffs, current replacement (market) values of assets should be taken into account, instead of the historical asset values which underestimate invested amounts and depreciation costs considerably. This fact is due to long useful (economic) lives and high local inflation rates encountered in the past. Thus, valuation of assets is required before the calculation of depreciation costs, which is generally not the case in Turkish municipalities.

The Regulation sets the principles for wastewater tariff setting in *Article 17* as follows:

- "Full cost recovery" and "polluter pays" principles shall be applied while setting the wastewater tariffs.
- Water consumption quantity measured by the water meters, and wastewater discharge quantity measured by the wastewater meters of private well owners or surface water users, who are not connected to the water distribution network but discharging into the municipal sewers, shall be considered while setting the wastewater tariffs (e.g. in case wastewater quantity cannot be measured, quantities fixed by the municipal administration shall be used as a basis of the discharged amount).
- Industrial wastewater tariffs shall be based on the pollution load and wastewater quantity discharged; non-industrial wastewater tariffs shall be based on water consumption measured or wastewater discharge quantities fixed by the municipal administration.

Minimum wastewater tariff for sustainability is 0.30 TL/m^3 (0.043 EUR/m^3 in October 2018, with an exchange rate of 7.0 TL/EUR). If calculated wastewater tariff is lower than this minimum amount, approval of the Ministry of Environment and Urbanization is required before charging. If this minimum amount shall not be sufficient to cover the cost of wastewater services to be provided in compliance with the existing legislation, then wastewater tariff at full cost recovery should be applied.

Thus, the application of "full cost recovery", "polluter pays" and "user pays" principles are also required in Article 17. Charging wastewater tariffs below the minimum level set in the Regulation is practically not possible since the approval of the Ministry of Environment and Urbanization is required. However, there is no automatic adjustment mechanism in the minimum wastewater tariff to be charged in the Regulation, at least based on local inflation. Consequently, while the minimum wastewater tariff of 0.30 TL/m^3 was equivalent to 0.13 EUR/m^3 in 2011, it reduced down to 0.043 EUR/m^3 as of 2018 by 67%, as a result of the devaluation of TL against EUR. If the real value of the minimum wastewater tariff would be maintained, then it should be increased to 0.90 TL/m^3 as of 2018. For municipalities that are not calculating full costs and not charging full cost recovery tariffs, this means that cost coverage rate of wastewater services will deteriorate continuously, and thus, financial sustainability shall not be maintained.

Three types of wastewater tariffs are specified in *Article 19* of the Regulation: (1) Variable wastewater tariffs based on the pollution load and the quantity discharged; (2) Fixed wastewater tariffs based on customer categories, such as septic tank emptying (e.g. suction truck fees) or meter reading fees; (3) Connection fees when new customers are connected to the municipal sewerage system.

Municipalities and WSAs have the discretion of using one or a combination of a few wastewater tariff types listed in *Article 19* of the Regulation. Fixed and variable wastewater tariffs *(Type 1 and Type 2 in Article 19)* shall be billed regularly, however connection fees *(Type 3)* are charged once at the time of connection to the municipal sewerage system.

Municipalities and WSAs shall establish the necessary cost accounting systems to ensure and monitor that wastewater tariff revenues collected are used for the wastewater service provision costs exclusively, as stipulated in *Article 21* of the Regulation.

Wastewater tariffs shall be included in the water bills to be issued regularly as stated in *Article 22* of the Regulation. Generally, water bills are issued on a monthly basis, based on actual meter readings of all registered customers in Turkey.[12]

Dissemination of information to the public is regulated in *Article 23* of the Regulation with the objective of increasing public awareness.[13] "*Guidelines for the Calculation of Wastewater Tariffs*" are also prepared and announced by the Environmental Management Directorate of the Ministry of Environment and Urbanization, including the application of a detailed calculation methodology with specific examples.

[12]In some municipalities bi-monthly billing is also practiced.

[13]Before taking the decision of wastewater tariffs in the Municipal Council (General Assembly of WSAs) a report shall be prepared on actual service provision costs in current and previous years, planned investment program and rationale for proposed tariffs to inform the public and get their opinions and recommendations on wastewater tariff levels. Decisions taken on wastewater tariffs by the Municipal Council (General Assembly of WSAs) after getting the comments and recommendations of the public shall be disseminated to the public at large through mass media and official websites opinions and recommendations on wastewater tariff levels.

It can be concluded that even though there is no legislation to determine water tariffs, there exists a comprehensive regulation and calculation methodology for wastewater tariffs in Turkey. Municipalities have the full discretion of setting water and wastewater tariffs without requiring approval from any central governmental administration or Regulatory Commission. "User pays", "polluter pays" and "full cost recovery" principles must be applied while setting tariffs.

However, the following issues/constraints exist for the municipalities in general:

- Difficulties in calculating the actual (full) costs of service provision separately for water and wastewater services (i.e. total cost and unit cost per cubic meter) due to the lack of up-to-date, reliable and accurate data and cost accounting systems, resulting in setting tariffs without knowing the actual costs and by linking increases to inflation or by comparing tariffs charged in neighboring municipalities.
- Generally, cash O&M costs are taken into account in cost calculations since they can be tracked easily without giving due importance to non-cash cost allowances for depreciation costs, bad debts, and employee termination benefits. These costs have very high shares in the real cost composition of the water/wastewater sector, resulting in the underestimation of costs, and consequently, lower than required tariffs.[14]
- Lack of reliable asset registers and asset valuation results in underestimation of asset values and depreciation costs, and consequently, charging lower than the required tariffs.
- If tariffs charged do not cover the real O&M, investment and financing costs, determined by taking billing and collection efficiency into account, then financial sustainability (i.e. generating positive cumulative cash flows at all times) cannot be ensured, requiring subsidies from other municipal revenues and/or building up debts.[15]
- Water/wastewater tariffs are set without taking household income and affordability into account, generally by assuming that all population is poor and without considering selective subsidies to certain poor clusters of the population, which lowers revenue generation potential considerably and is inconsistent with the user pays and polluter pays principles.
- To be able to protect domestic customers and improve the revenue base simultaneously, commercial and institutional (public) tariffs are set considerably higher to cross-subsidize the households, which is against the polluter pays principle.

[14]Thus, cost recovery is perceived as recovery of cash O & M costs only without taking investment and financing costs into account.

[15]Implication of non-revenue water is a major issue in Turkey.

11.13 Legislation on Water Losses (Non-revenue Water) in Turkey

The most inefficient operating performance of the Turkish municipalities is the high level of non-revenue water (NRW). Even though in some limited number of metropolitan municipalities, NRW, as a percentage of the water produced, is below 30% (e.g. 25% in Istanbul and Izmit). NRW is generally over 50% in many municipalities in Turkey (e.g. 63% in Kahramanmaras, 58% in Erzurum, 57% in Sanliurfa).

Having observed this major problem of Turkish municipalities (WSAs), the former Ministry of Forestry and Water Works prepared the "Regulation for the Reduction of Water Losses in the Water Supply and Distribution Systems", which was published in the Official Gazette No. 28994 dated May 08, 2014. The main principles of the Regulation are given in Annex 11.6. As per this Regulation, annual water balance, which is based on IWA definition, must be prepared and disseminated by all municipalities. Annual water balance of a typical WSA is shown in Annex 11.7 (OSKI 2015).

To be able to comply with this very ambitious regulation, municipalities have to reduce their physical losses in distribution networks and house connections by replacing/renewing/repairing pipes, which requires high investment costs, and decrease their administrative losses by improving managerial procedures to detect illegal and unauthorized consumption.

Additionally, the establishment of a separate "Water Loss Reduction Unit" is required according to the "Regulation of Technical Methods for the Control of Water Losses in the Water Supply and Distribution Systems", prepared by the former Ministry of Forestry and Water Works and published in the Official Gazette No. 29418 dated July 16, 2015. As per Article 37.4 of this Regulation, a specialized unit must be established in Municipalities (WSAs) to prevent, monitor and control the physical and administrative water losses in the water supply and distribution systems, which will also be responsible for ensuring coordination and definition of tasks and responsibilities of related departments. This Water Loss Reduction Unit (WLRU) must have the necessary equipment for leak detection, should identify and report water losses, and develop measures for loss reduction.

Acknowledgment The first author gratefully acknowledges the support of Istanbul University/Institute of Marine Sciences and Management in her office work and also Research Assistant MSc. Duygu Ulker for data processing and formatting this chapter.

Furthermore, the authors extend their thanks to the Springer Institution and to the Chief Editor Prof. Dr. Nilgun Harmancioglu for inviting them to prepare this chapter entitled "Urban and Industrial Water Uses".

Annexes

Annex 11.1 Treatment Types, Capacities and Discharge Points of the Wastewater Treatment Plants in Istanbul Metropolitan Area (Burak and Demir 2016)

No	Treatment plant	Treatment type	Capacity m^3/day	Population capacity	Discharge point
1	Yenikapi 1988	Primary-physical treatment	873.000	3.160.000	Treated wastewater is discharged to 64 m in depth, to the lower layer flows to Black Sea
2	Baltalimani 1997	Primary-physical treatment	625.000	3.000.000	Treated wastewater is discharged to 70 m in depth undercurrent in the strait 350 m away from the coast.
3	Buyukçekmece 1998	Primary-physical treatment	155.120	620.000	Treated wastewater is discharged to the 40 m depth of the Marmara Sea.
4	Uskudar 1992	Primary-physical treatment	77.760	350.000	Treated wastewater is discharged into the 47 m in depth undercurrent of the Bosphorus, flows through to Black Sea.
5	Kadikoy 2003	Primary-physical treatment	1.420.000	3.000.000	Treated wastewater is discharged into the 51,5 m in depth undercurrent of the Bosphorus, flows through to Black Sea.
6	Kuçukçekmece 2003	Primary-physical treatment	350.000	1.400.000	Treated wastewater is discharged to the 37,81 m depth of Marmara Sea.
7	Kuçuksu 2004	Primary-physical treatment	640.000	1.400.000	Treated wastewater is discharged to the 67 m in depth undercurrent of the Bosphorus, flows through to the Blacksea
8	Atakoy 2010	Advanced biological treatment	600.000	45.000	Treated wastewater is discharged into the Marmara Sea by Ayamama Creek.
9	Tuzla 1. Plant 1998	Advanced biological treatment	150.000	1.000.000	Treated wastewater is discharged into the 46 m depth of Marmara Sea.
	2. Plant 2009		100.000	500.000	

(continued)

No	Treatment plant		Treatment type	Capacity m^3/day	Population capacity	Discharge point
10	Pasakoy	1. Plant 2000	Advanced biological treatment	125.000	500.000	Treated wastewater is discharged into the Riva Creek by 6 km length tube and transmitted to Blacksea by this way.
		2. Plant 2009		125.000	500.000	
11	Terkos 2000		Advanced biological treatment	2.000	7.000	Treated wastewater is discharged into the Kuçukçekmece Lake Basin.
12	Bahçesehir 2004		Biological treatment	7.400		Treated waste water is discharged into Kuçukçekmece Lake Basin.
13	Pasabahçe 2009		Primary-physical treatment	570.000	2.000.000	Treated wastewater is discharged into undercurrent of Bosphorus.
14	Ambarli (2012)		Advanced biological treatment	400.000	1.600.000	Treated wastewater is discharged into the Marmara Sea.

Annex 11.2 Summary of EU and Turkish Drinking Water Treatment Parameters

Parameter	Parameter value		Unit
	EU drinking water directive	Turkish regulation No. 25730	
Microbiological parameters			
Escherichia coli (E.coli)	0	0	(number/ 100 ml)
Enterococci	0	0	(number/ 100 ml)
Chemical parameters			
Antimony	5.0	5.0	μg/l
Arsenic	10	10	μg/l
Benzene	1.0	1.0	μg/l
Benzo(a)pyrene	0.010	0.010	μg/l
Boron	1.0	1.0	mg/l
Bromate	10	25	μg/l
Cadmium	5.0	5.0	μg/l
Chromium	50	50	μg/l
Copper	2.0	2.0	mg/l
Cyanide	50	50	μg/l

(continued)

Parameter	Parameter value		Unit
	EU drinking water directive	Turkish regulation No. 25730	
1.2-dichloroethane	3.0	3.0	μg/l
Fluoride	1.5	1.5	mg/l
Lead	10	25	μg/l
Mercury	1.0	1.0	μg/l
Nickel	20	20	μg/l
Nitrate	50	50	mg/l
Nitrite	0.50	0.50	mg/l
Pesticides-individual	0.10	0.10	μg/l
Pesticides – Total	0.50	0.50	μg/l
Polycyclic Aromatic Hydrocarbons	0.10	0.10	μg/l
Selenium	10	10	μg/l
Tetrachloroethene and Trichloroethene	10	10	μg/l
Trihalomethanes – Total	100	150	μg/l
Indicator parameters			
Aluminium	200	200	μg/l
Ammonium	0.50	0.50	mg/l
Chloride	250	250	mg/l
Clostridium perfringens (including spores)	0	0	(number/ 100 ml)
Colour	Acceptable	–	–
Conductivity	2500	2500	μS/ cm at 20 °C
Hydrogen Ion Concentration	>6.5 and <9.5	>6.5 and <9.5	pH units
Iron	200	200	μg/l
Manganese	50	50	μg/l
Odor	Acceptable	–	–
Oxidisability	5.0	5.0	mg/l O2
Sulphate	250	250	mg/l
Sodium	200	200	mg/l
Taste	Acceptable	–	–
Colony count 22 Deg. C	No abnormal change	–	
Coliform bacteria	0	0	number/ 100 ml
Total organic carbon (TOC)	No abnormal change	–	mg/l
Turbidity	Acceptable (not exceeding 1.0 NTU for surface water treatment)	–	NTU
Free Residual Chlorine	–	0.5	mg/l
Radioactivity parameters			
Tritium	100		Bq/l
Total Indicative Dose	0.10		mSv/year

Annex 11.3 Wastewater Infrastructure of Organized Industrial Zones

Organized Industrial Zone (OIZ)	Basin	City	Waste water treatment	Discharging point
Tuzla	Marmara	Istanbul	Yes	
IMES OIZ	Marmara	Istanbul	Yes	
Dudullu	Marmara	Istanbul	Yes	
Ikitelli	Marmara	Istanbul	Yes	
Beylikduzu	Marmara	Istanbul	Yes	
Birlik	Marmara	Istanbul	Yes	
Istanbul Deri OIZ	Marmara	Istanbul	Yes	
Istanbul Asian Side OIS	Marmara	Istanbul	Yes	
Kocaeli OIZ – 8 pcs	Marmara	Kocaeli	Yes	
Canakkale OIZ	Marmara	Canakkale	No	
Biga OIZ	Marmara	Canakkale	No	
Bursa OIZ	Marmara	Bursa	Yes	
Balikesir OIZ – 2pcs	Marmara	Balikesir	Yes	
Yalova OIZ – 2 pcs	Marmara	Yalova	No	
BTSO	Susurluk	Bursa	Yes	Ayvali Creek
DOSAB	Susurluk	Bursa	Yes	Nilufer Creek
NOSAB	Susurluk	Bursa	Yes	Ayvali Creek
Gursu	Susurluk	Bursa	Yes	Delicay by DSI Cenup channel
Kestel	Susurluk	Bursa	Yes	Nilufer Creek
Bursa Deri	Susurluk	Bursa	Yes	–
Hasanaga	Susurluk	Bursa	No	Susurluk Creek
MKP	Susurluk	Bursa	No	–
Balikesir	Susurluk	Balikesir	No	Simav Creek
Balikesir II	Susurluk	Balikesir	No	–
Aliaga OIZ	Kuzey Ege	Izmir	Urban WWT	
Manisa	Gediz	Manisa	Yes	
IZBAŞ	Gediz	Manisa	Yes	
IAOIS	Gediz	Manisa	Yes	
KOSBI	Gediz	Manisa	Yes	
Manisa Turgutlu	Gediz	Manisa	No	
Manisa Salihli	Gediz	Manisa	Yes	Hayatli Creek
Salihli Leather OIZ	Gediz	Manisa	Yes	
Akhisar	Gediz	Manisa	No	

(continued)

Organized Industrial Zone (OIZ)	Basin	City	Waste water treatment	Discharging point
Kula Dericileri	Gediz	Manisa	Yes	
Usak OIZ	Gediz	Usak	Yes	
Torbali OIZ	Kuçuk Menderes	Izmir	No	
Buca (Ege Giyim)	Kuçuk Menderes	Izmir	Pretreatment	IZSU sewerage
ITOB Tekeli	Kuçuk Menderes	Izmir	Yes	
Tire	Kuçuk Menderes	Izmir	Yes	
Odemis (Under construction)	Kuçuk Menderes	Izmir	–	
Pancar (Under construction)	Kuçuk Menderes	Izmir	–	
Aydin OIZ	Buyuk Menderes	Aydin	Capacity is not enough	
Astim OIZ	Buyuk Menderes	Aydin	Under start-up	
Denizli OIZ	Buyuk Menderes	Denizli	Yes	
Usak Karma OIZ	Buyuk Menderes	Usak	Yes	
Karahalli OIZ	Buyuk Menderes	Usak	No	
Sandikli OIZ	Buyuk Menderes	Afyon	No	
Dinar OIZ	Buyuk Menderes	Afyon	No	
Kumluca Food Specialization	Bati Akdeniz	Antalya	Yes	
Antalya OIZ	Antalya	Antalya	Yes	Antalya municipality sewerage
Bucak OIZ	Antalya	Antalya	No	
Isparta Deri OIZ	Antalya	Isparta	Yes	
Isparta Suleyman Demirel OIZ	Burdur Lakes	Isparta	Yes	Burdur Lake
Burdur OIZ	Burdur Lakes	Burdur	No	Burdur municipality sewerage
Afyon OIZ	Akarcay	Afyon	No	
Bolvadin OIZ	Akarcay	Afyon	No	
Iscehisar Marble OIZ	Akarcay	Afyon	Pretreatment	
Suhut (Under Construction)	Akarcay	Afyon	No	

(continued)

Organized Industrial Zone (OIZ)	Basin	City	Waste water treatment	Discharging point
Sinanpasa (Under Construction)	Akarcay	Afyon	No	
Aksehir OIZ	Akarcay	Afyon	No	
Emirdag	Sakarya	Afyon	No	
Polatli	Sakarya	Ankara	Pretreatment	Gulveren creek
Eskisehir	Sakarya	Eskisehir	Yes	Porsuk creek
Sivrihisar	Sakarya	Eskisehir	–	–
Beylikova Besi İh.	Sakarya	Eskisehir	–	–
Kutahya	Sakarya	Kutahya	Pretreatment	Sewerage
Kutahya Merkez II	Sakarya	Kutahya	–	
Ostim	Sakarya	Ankara	Pretreatment	Sewerage
Ivedik	Sakarya	Ankara	Pretreatment	Sewerage
ASO I	Sakarya	Ankara	Pretreatment	Sewerage
ASO II	Sakarya	Ankara	No	Ankara Creek
Baskent	Sakarya	Ankara	Pretreatment	Ankara Creek
Cubuk Hay. Ih.	Sakarya	Ankara	No	–
Dokumculer	Sakarya	Ankara	No	Municipality sewerage
Beypazari	Sakarya	Ankara	No	–
Bilecik I	Sakarya	Bilecik	Yes	–
Bilecik II	Sakarya	Bilecik	No	Karasu
Bozoyuk	Sakarya	Bilecik	Pretreatment	Municipality sewerage
Osmaneli	Sakarya	Bilecik	No	–
Pazaryeri	Sakarya	Bilecik	Pretreatment	–
Sogut	Sakarya	Bilecik	No	–
Inegol	Sakarya	Bursa	Yes	Kalburt D.
Inegol Mob. Ih.	Sakarya	Bursa	–	
Yenisehir	Sakarya	Bursa	Yes	Goksu creek
Sakarya I	Sakarya	Sakarya	Pretreatment	Municipality sewerage
Sakarya II	Sakarya	Sakarya	Yes	Dinsiz creek
Sakarya III	Sakarya	Sakarya	Pretreatment	–
Karasu	Sakarya	Sakarya	–	–
Ferizli	Sakarya	Sakarya	–	–
Kaynarca	Sakarya	Sakarya	–	–
Bolu OIZ	Bati Karadeniz	Bolu	No	Bolu sewerage-Buyuksu creek
Gerede OIZ	West Blacksea	Bolu	Partly pretreatment	Municipality stabilization ponds
Gerede Leather OIZ	West Blacksea	Duzce	Partly	Ulusu creek

(continued)

Organized Industrial Zone (OIZ)	Basin	City	Waste water treatment	Discharging point
Duzce OIZ	West Blacksea	Duzce	Pretreatment	Duzce wwt
Duzce II. OIZ	West Blacksea	Duzce	Pretreatment	Duzce wwt
Karabuk OIZ	West Blacksea	Duzce	Pretreatment	Karabuk wwt
Zonguldak Caycuma OIZ	West Blacksea	Karabuk	No	
Zonguldak Eregli OIZ	West Blacksea	Zonguldak	No	
Amasya Centre OIZ	Yesilirmak	Amasya	In the planning stage	
Merzifon OIZ	Yesilirmak	Amasya	In the planning stage	
Corum OIZ	Yesilirmak	Corum	Yes	Corum Wastewater Treatment
Samsun Kavak OIZ	Yesilirmak	Samsun	In the planning stage	
Samsun Centre OIZ	Yesilirmak	Samsun	In the tender phase	
Tokat Centre OIZ	Yesilirmak	Tokat	Yes	Tokat wwt
Erbaa OIZ	Yesilirmak	Tokat	Yes	Erba wwt
Turhal OIZ	Yesilirmak	Tokat	No	
Suluova OIZ	Yesilirmak	Amasya	No	
Suluova Breeding OIZ	Yesilirmak	Amasya	No	
Samsun Food OIZ	Yesilirmak	Samsun	No	
Samsun Basin OIZ	Yesilirmak	Samsun	No	
Niksar OIZ	Yesilirmak	Tokat	No	
Zile OIZ	Yesilirmak	Tokat	No	
Sungurlu OIZ	Yesilirmak	Corum	No	
Kayseri OIZ	Kizilirmak	Kayseri	Under construction	Kayseri wwt
Kirikkale OIZ	Kizilirmak	Kirikkale	Yes	
Korgun OIZ	Kizilirmak	Camkiri	Yes	
Kirsehir OIZ	Kizilirmak	Kirsehir	No	Kirsehir wwt
Kastamonu OIZ	Kizilirmak	Kastamonuu	No	
Sivas OIZ	Kizilirmak	Sivas	No	Sivas wwt
Yozgat OIZ	Kizilirmak	Yozgat	No	Yozgat wwt
Bafra OIZ	Kizilirmak	Samsun	No	Bafra wwt
Sabanozu OIZ	Kizilirmak	Cankiri	No	Şabanozu wwt
Konya No1 OIZ	Konya Closed Basin	Konya	No	

(continued)

Organized Industrial Zone (OIZ)	Basin	City	Waste water treatment	Discharging point
Konya OIZ	Konya Closed Basin	Konya	Yes	
Eregli OIZ	Konya Closed Basin	Konya	Not operational	
Beysehir OIZ	Konya Closed Basin	Konya	No	
Nigde OIZ	Konya Closed Basin	Nigde	Yes	
Bor OIZ	Konya Closed Basin	Nigde	Under construction	
Aksaray OIZ	Konya Closed Basin	Aksaray	No	
Karman OIZ	Konya Closed Basin	Karaman	No	
Mersin Tarsus OIZ	East Mediterranean	Mersin	Yes	
Silifke OIZ	East Mediterranean	Mersin	Partly pretreatment	Silifke Municipality wwt
Anamur OIZ (Planning)	East Mediterranean	Mersin	–	
Adana Haci Sabanci	Seyhan	Adana	Yes	Ceyhan creek-Out of basin
Kahramanmaras OIZ	Ceyhan	Adana	Project approval phase	–
Osmaniye OIZ	Ceyhan	Osmaniye	Yes	Burnaz Kaynak Grubu Protected Area
Kadirli OIZ	Ceyhan	Osmaniye	No	Municipality Sewerage
Giresun OIZ	East Blacksea	Giresun	No	Giresun Municipality sewerage
Gumushane OIZ	East Blacksea	Giresun	No	Harsit creek
Ordu OIZ	East Blacksea	Ordu	No	Ordu municipality sewerage
Fatsa OIZ	East Blacksea	Ordu	No	Fatsa Municipality sewerage
Arsin OIZ	East Blacksea	Trabzon	Under construction / Biological	Rizvan creek
Besikduzu OIZ	East Blacksea	Trabzon	No	Land
Van OIZ	Van Lake	Van	Yes/Biological	Morali creek

Annex 11.4 Existing Main Turkish Acts and Legislation Governing the Tasks, Responsibilities and Operations of WSAs

Title of the act	Act no.
Act on Metropolitan Municipalities	5216
Act on Municipalities	5393
Act on the Establishment of İstanbul Water and Sewerage Administration (ISKI)	2560
Act on Municipal Revenues	2464
Act on Protection of General Public Health	1593
Turkish Commercial Code	6762
Act on Waters	831
Act on Environment	2872
Act on Ground Waters	167
Act on Public Financial Management and Control	5018
Act on Public Tendering	4734
Act on Public Tender Contracts	4735
Act on Right of Information	4982
Act on Civil Servants	657
Act on Work	4587
Act on Social Security	506
Act on Retirement Pension Fund	5434
Act on Social Security Administration (SGK)	5510
Act on Expropriation	2942
Act on Debts	818
Turkish Penal Code	5237
Act on Bankruptcy and Enforcement	2004
Act on Motor Vehicles	237
Act on Income Tax	193
Act on Value Added Tax (VAT)	3065
Act on Daily Allowances (Per Diems)	6245
Act on Stamp Tax	488
Act on Notification of Information	7201

Annex 11.5 The Duties and Responsibilities of the WSA Related to Water Services Given in Article 2 of the Act No. 2560

In order to supply water for domestic and industrial consumption, the WSA has to:

(i) undertake all the responsibilities in the distribution, research and project design, as well as in the construction, transfer and operation of facilities both in operation or under construction, and in the maintenance, renovation and upgrading of these facilities;

(ii) to conduct research and project design activities to provide the facilities necessary for collecting wastewater and storm water, their transportation and safe disposal from settlement areas to designated sites; to undertake the construction of these facilities as well as taking over the operation of existing ones, their maintenance, repair and upgrading responsibilities;

(iii) to prevent the establishment of facilities or the carrying out of the activities that will lead to the contamination or pollution of water reservoirs, lakes, coastal areas or other water resources within its responsibility area resulting from wastewater or industrial discharge; to carry out all necessary technical, administrative and legal measures to prevent contamination or loss of water resources;

(iv) to take over and carry out various responsibilities given to the District Municipalities by various Acts and regulations pertaining to the provision of water and wastewater services and to exercise its authority in implementing these services;

(v) to purchase and lease all kinds of assets or to sell any outmoded vehicles or equipment; to establish or operate either individually or jointly with other public and private firms the necessary facilities or to participate in their operations already established or to be established for this purpose;

(vi) to expropriate or to exercise the right-to-use all types of immovable assets required in carrying out of its services; and

(vii) to raise loans and credits from international financing institutions and lenders to upgrade and develop its facilities and operations and to utilize modern technology for its services with the approval of the Ministry of Finance.

Annex 11.6 Main Principles of the Regulation

(6.1) Main Principles for the Management of Water Supply and Distribution Systems:

(a) Continuous measurement of water quantity and water flow at each source that feeds the water distribution system;

(b) Continuous measurement of water pressure at critical points of the water distribution system;

(c) Digitizing the maps of the water supply and distribution system and development of the Geographic Information System (GIS) data base;

(d) Installation of monitoring and control systems (e.g. SCADA);

(e) Establishment of main pressure and DMA zones.

(6.2) Main Principles for the Reduction of Water Losses in Water Supply and Distribution Systems:

(a) Determination of Annual Water Balance

1. Determination of System Input;
2. Determination of Authorized Consumption;
3. Determination of Physical and Administrative Water Losses;
4. Determination of Non-Revenue Water (NRW)

(b) Prevention of Water Losses

1. Prevention of Unauthorized Consumption;
2. Provision of Optimum Operating Balance with effective pressure management in the network;
3. Repair and maintenance at locations where physical leaks are detected;
4. Periodic/preventive maintenance and renewal of networks;
5. Development of technical and administrative capacity to be able to detect physical losses/leaks.

Article 9 (Reduction of the Water Losses) of the Regulation stipulates that water losses must be reduced:

- In metropolitan municipalities and provincial central districts (which applies to 30 metropolitan municipalities and 51 provincial districts in Turkey) below 30% within 5 years (2019) and 25% in the following 4 years (2023);
- In all other (non-metropolitan) municipalities below 30% within 9 years (2023) and 25% in the following 5 years (2028).

Article 10 (Responsibility of Information Dissemination) of the Regulation stipulates the following:

- Municipalities (WSAs) should prepare an Annual Water Losses Report (template is presented in an annex; Standard Water Balance Form is exactly the Water Balance of IWA Definition – a typical example is in the Annex) and send to the Ministry of Forestry and Water Works in written form by the end of February each year.
- Municipalities (WSAs) are obliged to provide all information and documents to the representatives of the Ministry of Forestry and Water Works during audits on site to verify the validity of information in the Annual Water Losses Report.
- Municipalities (WSAs) are obliged to place the Annual Water Losses Report in their official websites for a period of 1 year after submission to the Ministry of Forestry and Water Works.

Annex 11.7 Annual Water Balance – A Typical Example (WSA)/Turkey

Unit	System input	Authorized consumption	Billed authorized consumption	Billed metered consumption	Revenue water
m^3/y	41,718,351	27,774,269	22,768,067	22,768,067	22,768,067
%	100.00	66.58	54.58	54.58	54.58
l/ ca-d	153.8	102.4	84.0	Billed unmetered consumption	84.0
m^3/y				0	
%				0.00	
			Unbilled authorized consumption	Unbilled metered consumption	Non-revenue water
m^3/y			5,006,202	0	18,950,284
%			12.00	0.00	45.42
l/ ca-d			18.5	0.0	69.9
				Unbilled unmetered consumption	
m^3/y				5,006,202	
%				12.00	
l/ca-d				18.5	
		Water losses	Apparent losses	Unauthorized consumption	
m^3/y		13,944,082	4,604,250	2,871,327	
%		33.42	11.04	6.88	
l/ca-d		51.4	17.0	10.6	
				Meter faults	
m^3/y				1,732,924	
%				4.15	
l/ ca-d				6.4	
			Real losses	Pipe leakages	
m^3/y			9,339,832	9,339,832	
%			22.39	22.39	
l/ ca-d			34.4	34.4	
				House conn. leakages	
m^3/y				0	
				WTP/water reservoirs overflows	
m^3/y				0	

References

Act No 5302 (2005) Special Provincial Administrations law (Il ozel idaresi kanunu). 22 February 2005, http://www.resmigazete.gov.tr/eskiler/2005/03/20050304-1.htm

Act No 5491 (2006) Act amending the act on environment (Cevre kanununda degisiklik yapilmasina dair kanun). 26 April 2006, http://www.resmigazete.gov.tr/eskiler/2006/05/20060513-1.htm

Alegre H, Baptista JM, Cabrera Jr, Cubillo F Duarte P, Hirner, Merkel W, Renato, P (2006) Performance indicators for water supply services. IWA Publishing, London, ISBN: 1843390515

Beler-Baykal B, Oructut N (2017) Grey water as reliable/renewable source of water for cities of the future and an appraisal based on Istanbul, Turkey. Paper presented at the IWA Cities of the Future Conference: Embrace the Water, Gothenburg, Sweden, June 12–14, 2017

Blinda M (2012) More efficient water use in the Mediterranean. (Plan Bleu Papers no 14), Plan Bleu, Valbonne

Blue Flag map (2018.) https://fee.carto.com/viz/6ed46582-5aed-11e5-9391-0e9d821ea90d/embed_map?wmode=opaque. Accessed 02 July 2018

Blue Plan (2008) Strategy Mediterranean for sustainable development: methodological note of water efficiency index calculation. Blue Plan Regional Activities Centre, Sophia Antipolis

Brettel M, Friederichsen N, Keller M, Rosenberg M (2014) How virtualization, decentralization and network building change the manufacturing landscape: an industry 4.0 perspective. Int J Mech, Ind Sci Eng 8:37–44

Burak S (2007) Mediterranean strategy for sustainable development: monitoring progress and promotion of water demand management policies, National Report of Turkey. Water demand management, progress and policies, Zaragoza, Spain, 19–21 March 2007, UNEP/MAP/Blue Plan

Burak S (2011) Water use efficiency and economic approach. National Study Turkey Final Version Report. Plan Bleu, UNEP/MAP Regional Activity Centre

Burak S, Demir V (2016) Land-based pollution in the Sea of Marmara and assessment of pollution abatement strategies vol 42. The Sea of Marmara Marine Biodiversity, Fisheries, Conservation And Governance. Turkish Marine Research Foundation (TUDAV), Istanbul

Burak S, Margat J (2016) Water management in the Mediterranean region: concepts and policies. Water Resour Manag:1–19. https://doi.org/10.1007/s11269-016-1389-4

Burak S, Mat H (2010) Municipal water demand and efficiency analysis: case studies in Turkey. Water Policy 12:695–706. https://doi.org/10.2166/wp.2009.209

Burak S, Dogan E, Gazioglu C (2004) Impact of urbanization and tourism on coastal environment. Ocean Coast Manag 47:515–527. https://doi.org/10.1016/j.ocecoaman.2004.07.007

Burak S, Bilge A, Dernek D (2017) A network model simulation proposal for river basin management plans (RBMPs) in Turkey. Paper presented at the 10th World Congress on Water Resources and Environment “Panta Rhei”, Athens, Greece, 5–9 July 2017

Degrémont (1991) Water treatment handbook, vol 2, 6th edn. Lavoisie Publishing Inc, Paris

Demir O, Yildiz M, Sercan U, Arzum CS (2017) Atiksularin geri kazanilmasi ve yeniden kullanilmasi. Harran Universitesi Muhendislik Dergisi 2:1–14

Envest Planners (2004) Public finance assessment, EU technical assistance for environmental heavy-cost investment planning in Turkey. Ministry of Environment, Ankara, Turkey

European Commission (2012) Commission staff working document Turkey 2012 progress report. European Commission, Brussels

Farooq S, Ansari ZI (1981) Philosophy of water reuse in Islamic perspective. Desalination 39:273–281. https://doi.org/10.1016/S0011-9164(00)86131-7

German Water Partnership (2016) Water 4.0. German Water Partnership, Berlin

Giresunlu E, Beler-Baykal B (2015) An appraisal of microbiological indicators for grey water reuse as a priority issue. Paper presented at the CEST 2015 Proceedings of the 14th International Conference on Environmental Science and Technology, Rhodes, Greece, September 3–5, 2015

Giresunlu E, Beler-Baykal B (2016) A case study of the conversion of grey water to a flush water source in a Turkish student residence hall. Water Sci Technol: Water Supply: ws:2016078. https://doi.org/10.2166/ws.2016.078
Hyder Consulting Ltd (1997) Feasibility study for Cesme, Alacatı and surrounding area. Southwest Environmental Project, Final Report, The World Bank, Ministry of Tourism, Turkey, May 1997
IME (Institut Méditerranéen De L'eau) (2017) Interreg Mediterranean Projet Europeen Co-Evolve; Etude sur la disponibilité et la qualité de l'eau en Méditerranée vol (No du marché 17/M0163), Marseille, France
IPEMED (2018) Reuse of treated waste water in the Mediterranean and impacts on territories. 8th World Water Forum, Brasilia-Brazil
Official Gazette 18132 (1983) Environment law (Cevre kanunu). Act No 2872, 09 August 1983, http://www.mevzuat.gov.tr/MevzuatMetin/1.5.2872.doc
Official Gazette No 29418 (2015) Regulation of Technical Methods for the Control of Water Losses in the Water Supply and Distribution Systems (Icme suyu temin ve dagitim sistemlerindeki su kayiplarinin kontrolu yonetmeligi teknik usuller tebligi). Accessed 16 July 2015
Official Gazette No. 17523 (1981) ISKI Act (Istanbul Su ve Kanalizasyon Idaresi Genel Mudurlugu kurulus ve gorevleri hakkinda kanun). 23 November 1981. http://www.iski.gov.tr/web/assets/SayfalarDocs/Mevzuat%20ve%20Y%C3%B6netmelikler/%C4%B0SK%C4%B0_Kanunu.pdf
Official Gazette No. 24021 (2000) Organized Industrial Zones (OIZs) in Turkey are established as per Law No. 4562 on Organized Industrial Zones approved by the National Assembly on April 12th, 2000 and issued in dated April 15th, 2000. http://www.mevzuat.gov.tr/MevzuatMetin/1.5.4562.pdf
Official Gazette No. 25531 (2004) Metropolitan municipality law. Act No 5216, 23 July 2004, http://www.resmigazete.gov.tr/eskiler/2004/07/20040723.htm
Official Gazette No. 25730 (2005) Water Intended for Human Consumption. Act No 7510, 17 February 2005, http://www.fao.org/faolex/results/details/en/c/LEX-FAOC151039
Official Gazette No. 25874 (2005) Municipal law. Act No 5393, 13 July 2005, http://www.enofis.com.tr/mevzuat/kanunlar/Belediye%20kanunu.pdf
Official Gazette No. 28489 (2004) Establishment of new Metropolitan Municipalities (On dort ilde Buyuksehir Belediyesi ve yirmi yedi ilce kurulmasi ile bazi kanun ve kanun hukmunde kararnamelerde degisiklik yapilmasina dair kanun). Act No 6360, 06 December 2012, http://www.mevzuat.gov.tr/MevzuatMetin/1.5.6360.pdf
OSKI (Ordu Su ve Kanalizasyon Idaresi) (2015) Su kayip/kacak raporu. OSKI Genel Mudurlugu, 2015
Ozturk İ, Calli B, Arikan O, Mahmut A (2015) Atiksu aritma çamurlarinin islenmesi ve bertarafi. Turkiye Belediyeler Birligi, Ankara
Republic of Turkey Ministry of Environment and Forestry (2006) EU integrated environmental approximation strategy (2007–2023). Republic of Turkey Ministry of Environment and Forestry, Ankara
The Council of the European Union (2008) Decision on accession partnership with Turkey. 4–18
The World Bank (1998) Turkey – Cesme-Alacati water supply and sewerage project, world development sources, 1998–2. The World Bank, Washington, DC
The World Bank (2016) Republic of Turkey: sustainable urban water supply and sanitation, reaching compliance with the European Union's water framework directive in a sustainable way – challenges and opportunities for Turkey's water supply and sanitation sector. The World Bank, Washington, DC
TS 266 (2005) Water intended for human consumption. TS 266, Ankara
TUBITAK MAM (2010–2013) Havza koruma eylem planlarinin hazirlanmasi projesi. TUBITAK MAM, Gebze
TURKSTAT (2017a) Municipal waste water statistics, 24875, 22 November 2017
TURKSTAT (2017b) Municipal wastewater indicators, 1994–2016
TURKSTAT (2017c) Municipal water statistics, 24874, 21 November 2017

TURKSTAT (2017d) Organized industrial zones water and wastewater statistics 2012–2016, 24872, 06 November 2017
TURKSTAT (2017e) Water abstraction for municipal water supply network, 1994–2016
TURKSTAT (2018) Sectoral water and wastewater statistics 2016, 27672, 16 February 2018
UNEP/MAP (2012) State of the Mediterranean marine and coastal environment. UNEP/MAP, Athens
UNEP/MAP (2015) Marine litter assessment in the Mediterranean. UNEP/MAP, Athens
UNEP/MAP/BLUE PLAN (2007) Water demand management, progress and policies: Proceedings of the 3rd Regional Workshop on Water and Sustainable Development in the Mediterranean. Zaragoza, Spain vol MAP Technical Reports Series No. 168, 2nd edn., Athens

Chapter 12
Transboundary River Basins

Unal Ozis, Nilgun B. Harmancioglu, and Yalcin Ozdemir

Abstract Transboundary water courses provide roughly 70 km^3/year or 40% of the gross surface water potential originating in Turkey. The Euphrates-Tigris Basin represents about four fifths of this figure; the rest is contributed by the basins Orontes, Kura-Araks, Chorokhi, Maritza, and a few other quite small basins. Turkey is the upstream riparian in the Euphrates-Tigris, Kura-Araks, Chorokhi, and the small basins, and a downstream riparian in the Orontes and Maritza basins. The total water potential of the Euphrates-Tigris Basin exceeds 90 km^3/y, where Turkey provides, in rough figures, 60%, Iraq 25%, Iran 10% (excluding Kharkeh), and Syria 5% of it. In Turkey, the average water potential of Euphrates is around 32 km^3/y, and that in Tigris around 24 km^3/y, including tributaries flowing directly to downstream countries. Ultimately, as a long-term average, half of the Euphrates-Tigris water potential originating from Turkey, about 40% in Euphrates and 65% in Tigris, will continue to flow towards the downstream countries. However, the amount of water in any allocation agreement should be set according to different levels of probabilities of discharges because of the significant stochastic variation of discharges, even after regulation by the huge reservoirs in Turkey. The water potential of the transboundary Euphrates-Tigris Basin is quoted with large differences according to various sources; hence, the determination of the accurate water potential is an essential prerequisite for any allocation among riparian states and eventual diversions to other middle-eastern countries. The water potential of the Euphrates Subbasin appears to be insufficient in Iraq so that the excess water of the Tigris Subbasin should be transferred to Euphrates to satisfy the irrigation needs along the Euphrates banks in Iraq.

Keywords Turkey · Euphrates · Tigris · Orontes · Kura-Araks · Chorokhi · Maritza · Transboundary river basin · Water power · Irrigation

U. Ozis · N. B. Harmancioglu (retired) (✉) · Y. Ozdemir
Faculty of Engineering, Civil Engineering Department, Dokuz Eylul University, Izmir, Turkey
e-mail: nilgun.harmancioglu@deu.edu.tr; yalcin.ozdemir@deu.edu.tr

N. B. Harmancioglu, D. Altinbilek (eds.), *Water Resources of Turkey*, World Water Resources 2, https://doi.org/10.1007/978-3-030-11729-0_12

12.1 Introduction

More than 200 watercourses in the world are of 'transboundary' and/or 'boundary forming' nature; they cover almost half of the continents. The 1997 UN-Convention used the unfortunate, misleading term 'international' for these watercourses, although they have been called 'transboundary' watercourses for several decades of preceding drafts and discussions. The term 'multinational' watercourse could have been a rational compromise if the expression 'national' were to be maintained in the terminology.

Most of the transboundary and/or boundary-forming watercourses often cause conflicts of interest among the riparian countries and around 300 treaties between various states have been issued for the use of these watercourses (Bilen 1994; Biswas 1994; Kolars 1994; Wolf 1994; Kibaroglu et al. 2011).

Turkey's transboundary river basins (Fig. 12.1), including border-crossing tributaries, are:

(a) Maritza (Meric) Basin, and the adjacent small Velika (Kocadere) and Rezovo (Mutludere) creeks;
(b) Chorokhi (Coruh) Basin and the adjacent Sarp Creek;
(c) Kura-Araks (Kura-Aras) Basin and the small Baradost Creek to the east of Yuksekova;
(d) Orontes (Asi) Basin and the adjacent small Qweik (Balik) creek;
(e) Euphrates-Tigris (Firat-Dicle) Basin.

Turkey is the upstream riparian in Euphrates-Tigris, Kura-Araks, Chorokhi, and the small basins, and a downstream riparian in Orontes and Maritza basins (Ozis 1997; Ozis et al. 1997, 2001, 2004a, 2013a, b; Ozis and Ozdemir 2009, 2010; Ozdemir et al. 2013).

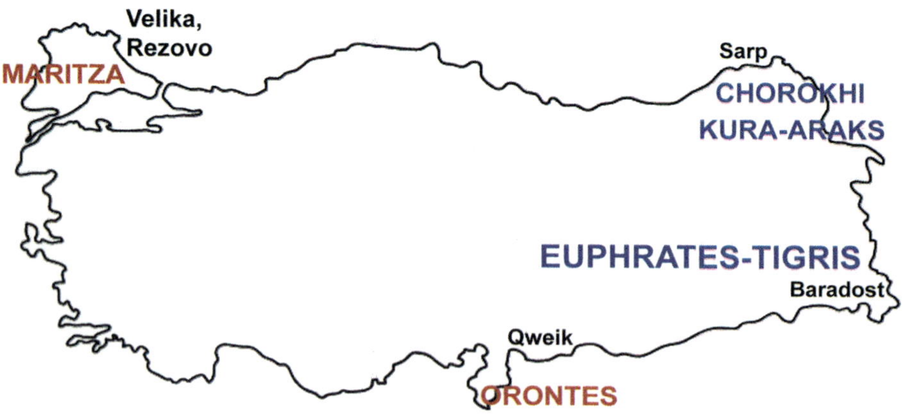

Fig. 12.1 Overview of transboundary river basins in Turkey. (U. Ozis & Y. Ozdemir)

Transboundary water courses cover an area of 250,000 km^2 or roughly one third of the land surface of Turkey. They provide about 70 km^3/y or 40% of the gross surface water potential originating in Turkey; the Euphrates-Tigris Basin represents about four fifths of this figure.

The development of land and water resources in Turkey worries downstream countries with regard to an anticipated decrease in quantity and deterioration in quality of the water. These are especially related to the implementation of the Southeastern Anatolia Project (Turkish acronym: GAP) in the Euphrates-Tigris Basin, where Turkey is the upstream riparian in both main subbasins.

The worries of the downstream countries are based on:

(a) evaporation losses from reservoir surfaces created by large dams (although these dams are indispensable to beneficially regulate the highly varying discharges of the Euphrates-Tigris Basin and to control floods and sedimentation, thus serving also to downstream riparian countries);
(b) the diversion to and consumption by large irrigation systems in the Basin (which appear to be basically equitable and reasonable uses);
(c) substantial urban and industrial water supply requirements in the region (with significant population increase and vital need for socio-economical development of the country).

Water scarcity in the Middle-East (presumably to be still worsened by anticipated negative effects of climate change processes), as well as water allocation claims of the riparian states, raise issues of conflicts rather than cooperation with regard to the development of the Euphrates-Tigris Basin (Ozis and Ozdemir 2009).

Besides the reactions from downstream riparian states, Turkey is confronted with several pressures from other foreign sources. International geopolitical and energy issues of states and organizations outside the region urge Turkey to restrict her uses and allocate large amounts of water to middle-eastern countries.

Turkey voted against the 1997 UN Convention that required several permissions and restrictions, as they would slowdown the implementation of her projects towards the urgent needs for the social and economical development of the southeastern regions. On the other hand, Turkey states that her projects make also ‘optimal use’ of the available land and water resources and also conform to the ‘equitable and reasonable use’ and ‘causing no significant harm’ principles.

Dams in Turkey provide significant benefits also to downstream countries, such as sediment retention, flood mitigation, and temporarily low flow augmentation. Nevertheless, water allocation disputes among riparians on one hand, water diversion issues from Tigris to Euphrates and eventually from Euphrates to the neighbouring Jordan and Orontes Basins on the other hand, place the Euphrates-Tigris Basin in the foreground of international interests.

12.2 Maritza (Meric) Basin, and the Adjacent Small Velika (Kocadere) and Rezovo (Mutludere) Creeks

12.2.1 Geographical Position of the Maritza Basin

Maritza is one of the leading river basins of the Balkans, flowing in Bulgaria, Greece and Turkey. It covers the large part of Thrace, the north-west region of Turkey.

The river Maritza (Meric), its southern tributary Arda (Arda), its northern tributary Tundza (Tunca) originate in Bulgaria. The river Maritza (called Evros in Greece) forms a short stretch between Svilengrad and Kapikule near Edirne. It forms the boundary between Bulgaria and Greece, than between Greece and Turkey. It enters Turkey for about 20 km at the Karaagac district of Edirne, then forms again the boundary between Turkey and Greece and discharges to the Aegean Sea near Enez.

The northern tributary Tundza, with a drainage area of around 7900 km^2 in Bulgaria, forms first the boundary between Bulgaria and Turkey. It then enters Turkey and joins Maritza near Edirne. The southern tributary Arda crosses the border to flow in Geece and joins Maritza near Edirne, shortly after the confluence of Maritza and Tundza, where Maritza forms the boundary between Greece and Turkey. The southeastern tributary of Maritza, called Ergene, flows entirely in Turkey and joins Maritza roughly 35 km before reaching the Aegean Sea (Figs. 12.2, 12.3, 12.4 and 12.5). The drainage area of the Maritza Basin is about 34,065 km^2 in Bulgaria, 3685 km^2 in Greece, and 14,850 km^2 in Turkey so that the total is roughly 52,600 km^2.

Fig. 12.2 Maritza (Meric) Basin in Bulgaria, Greece, Turkey; and Adjacent Two Small Creeks in Turkey and Bulgaria. (U. Ozis & Y. Ozdemir)

Fig. 12.3 Tundza (Tunca) River and the 136 m long Ekmekcizade Bridge (1615) in Edirne. (Photo by U. Ozis)

Fig. 12.4 Maritza (Meric) River and the 263 m long Mecidiye Bridge (1847) in Edirne. (Photo by U. Ozis)

On the north-eastern slopes of the Istranca mountains adjacent to the Maritza Basin, the Rezovo (Mutludere) creek, which is among the creeks flowing to the Blacksea, originates in Turkey. It forms then the boundary between Turkey and Bulgaria. The Velika (Kocadere) creek originates in Turkey and flows to Bulgaria. These creeks with a water potential of 0.1 km^3/y, originating from a roughly 400 km^2 drainage area in Turkey, can be used for water supply diversions of Kirklareli-Istanbul area and may affect Bulgaria.

Fig. 12.5 Ergene River and the 1360 m long Uzunkopru Bridge (1443) at Uzunkopru. (Photo by U. Ozis)

12.2.2 Water Potential of the Maritza Basin

The water potential of Maritza originating in Turkey is around 1.2 km^3/y contributed by the tributary Ergene, 0.4 km^3/y by the tributary Tundza, and 0.2 km^3/y by the creeks on the east bank of Maritza, thus totalling 1.8 km^3/y.

In the upper-riparian Bulgaria, the water potential of Maritza and the tributary Arda is around 5.1 km^3/y and the tributary Tundza 0.6 km^3/y. The additional water potential of Maritza and Arda from Greece is 0.5 km^3/y. Hence, the total water potential of the entire Maritza Basin, including Turkey, is in the order of 8 km^3/y (Ozis 1997; Sen 2002; Ozis et al. 2006, 2013a, b; Kibaroglu 2008a; Kramer and Schellig 2011).

12.2.3 Water Resources Development in the Maritza Basin

Seven dams on the upper valleys of Maritza's small tributaries were constructed in Bulgaria with a total reservoir volume of roughly 1.2 km^3. Two dams, Koprinka and the low Jrebchevo, exist in the upper part of Tundza, with a total reservoir volume of about 0.5 km^3. A cascade of three dams on Arda create a total reservoir volume of about 1.0 km^3. The total reservoir volume of 2.7 km^3 in Bulgaria proved to be insufficient to control and mitigate the floods of especially Maritza and Tundza. These dams in Bulgaria, some equipped with power plants, regulate the flow for irrigation purposes along the Plovdiv (Filibe) plain to the north of the Rodop mountains.

Greece constructed a low dam on Arda, serving as an afterbay to regulate the outflow of cascading hydroelectric plants in Bulgaria and to supply water for irrigation of roughly 50,000 ha of land between Arda and the western bank of Maritza.

Turkey has 1,200,000 ha agricultural land in the Maritza Basin, often called Meric-Ergene Basin and anticipates to irrigate at least one third of it by surface water (DSI 1995; Kibaroglu 2008a, Malkarali et al. 2008). Turkey constructed seven dams of modest sizes and more than one hundred small reservoirs in this basin to cover part of the irrigation needs.

12.2.4 Water-Related Issues in the Maritza Basin

The water related problems of the Maritza Basin, especially in Turkey, are twofold. On one hand, the water potential of the river basin is not sufficient for the irrigation requirements of the entire basin; on the other hand, the existing schemes are not capable of flood mitigation, especially in the Edirne area where Tundza and Arda tributaries confluence with the main river Maritza.

Turkey and Greece have constructed several stretches of levees with 170 km total length along both banks of the lower Maritza River. Flood mitigation is a very important issue, especially for Turkey (Eroglu 2006; Gunduz 2006; Dmitrov et al. 2008; Malkarali et al. 2008; Kramer and Schellig 2011).

The insufficiency of Turkey's water potential and the inconveniences in its areal distribution necessitate to harness the water of the main river. However, diversions in Bulgaria and Greece often cause water shortages in the river. Turkey even 'bought' water released from the reservoirs of Bulgaria's dams during the drought of 1993.

Water quality problems, especially due to industrial and agricultural fertilizer pollution, is another important issue in the basin (Samsunlu et al. 1996).

Turkey and Bulgaria made several partial agreements on cooperation in the Maritza Basin (Kibaroglu 2008a; Kibaroglu et al. 2011). The construction of the long-discussed Suakacagi dam on Tundza in Turkey, with its reservoir extending in Bulgaria, will bring some remedy to the flood and drought problems of Turkey in this basin. Last but not the least, changing formations of the sandy Maritza delta deserve special attention.

Bulgaria and Greece are members of the European Union and have to act in conformity with the EU Water Framework Directive, whereas Turkey is waiting since half a century to be an EU member (Bosnjakovic 2000; TMMOB 2006; Dalkilic and Harmancioglu 2008; Bilen 2009; Sumer and Muluk 2011). The EU often refers to the needs and demands of the downstream riparians Syria and Iraq in the Euphrates-Tigris Basin but does not make any comments on the needs of the downstream riparian Turkey in the Maritza Basin!

The lower-riparian Turkey has to manage the situation in the Maritza Basin within a broader frame, considering her sensitive relation with Bulgaria and Greece, the importance of the river and its tributary for Bulgaria, and the use of Arda waters in Greece. Besides the implementation of the transboundary Suakacagi dam on Tundza, the regulation of excess waters by off-stream reservoirs in Turkey, which are unregulated by the upstream dams, may partly contribute to the physical solution of the problem.

12.3 Chorokhi (Coruh) Basin and the Adjacent Small Sarp Creek

12.3.1 Geographical Position of the Chorokhi Basin

The Chorokhi (Coruh) Basin originates in the north-eastern region of Turkey. It is located to a great extent in Turkey with a drainage area of 19,872 km^2. Chorokhi, together with its small tributary Machakhela (Cakal) creek, then crosses the border to Georgia. Chorokhi receives there its last major tributary Ajaristskali and discharges into Blacksea near Batumi (Fig. 12.6). The drainage area in Georgia is 2479 km^2 so that the total area of the Chorokhi Basin is 22,351 km^2. These figures include the

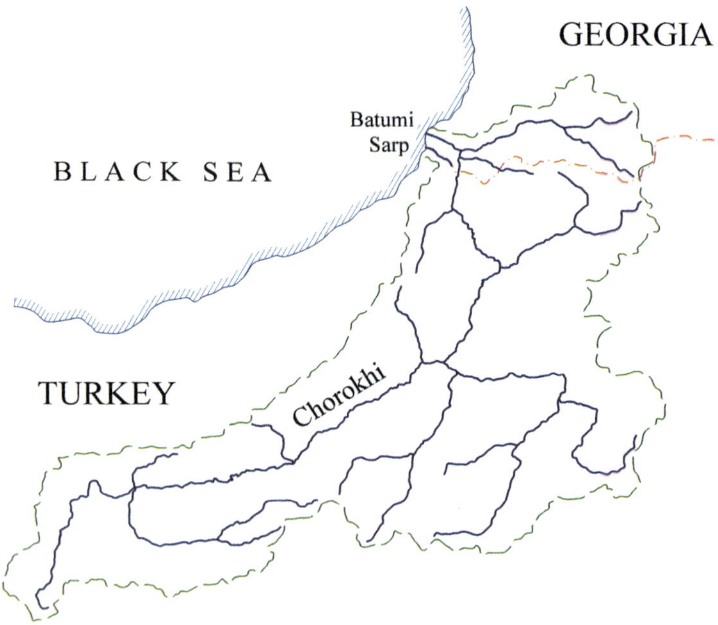

Fig. 12.6 Chorokhi (Coruh) basin in Turkey and Georgia; and the adjacent Sarp creek along the boundary. (U. Ozis & Y. Ozdemir)

Fig. 12.7 Chorokhi (Coruh) river valley. (Photo by Y. Ozdemir)

drainage area of the border-crossing Machakhela creek, 181 km^2 in Turkey and 360 km^2 in Georgia.

The small Sarp Creek flowing to the Blacksea at the north-eastern corner of Turkey, adjacent to Chorokhi Basin, forms the border between Turkey and Georgia.

12.3.2 Water Potential of the Chorokhi Basin

The water potential of the Chorokhi Basin in Turkey (Fig. 12.7) is 6.1 km^3/y (Ozis 1997; Ozis et al. 2006; Klaphake and Scheumann 2011; Ucar and Gurer 2013). This is close to one tenth of the total domestic transboundary water potential of Turkey. The additional water potential originating from Georgia appears to be in the order of 2.7 km^3/y so that the total water potential of the Chorokhi Basin is around 8.8 km^3/y. The water potential of the Sarp creek, originating from Turkey, is in the order of 0.1 km^3/y.

12.3.3 Water Resources Development in the Chorokhi Basin

Turkey foresees to irrigate 160,000 hectares of land and is constructing water power schemes generating a total of 12 billion kWh/y electrical energy in Chorokhi Basin,

Fig. 12.8 The 249 m high (above foundation) Deriner dam on Chorokhi (Coruh) river. (Photo by Y. Ozdemir)

two of them on tributaries and ten of them forming a cascade of dams with almost 1500 m head difference along the main river (EIE 1985; DSI 1995; TMMOB 2006; Sucu and Dinc 2008; Sarac and Eciroglu 2008). These include some of the highest concrete arch dams of the world, like the 270 m high Yusufeli (under construction), 249 m high Deriner (Fig. 12.8), and the 190 m high Artvin (Fig. 12.9) dams (heights above foundation), supplying hydroelectric plants of 540, 670, 332 MW capacity, respectively. Upstream of the existing modest Atshesi power plant, Georgia plans to construct a cascade of three larger high-head power plants (Shuakhevi, Keremkheti, Khertvisi) in the Ajaristskali subbasin.

12.3.4 Water-Related Issues in the Chorokhi Basin

The irrigation of 160,000 ha of land, compared to the water potential of Chorokhi, will not cause a handicap for the downstream riparian Georgia; moreover, flow regulation through the dam cascades will augment the low flows, trap the sediments, and mitigate the floods also for the benefit of Georgia.

However, Georgia worries about the reduction of sediments since this may cause the shrinkage of the delta through the erosion of the Blacksea, so that the effect of

Fig. 12.9 The 190 m high (above foundation) Artvin dam on Chorokhi (Coruh) river. (Photo by Y. Ozdemir)

sediment reduction on the formation of the delta near Batumi should be studied carefully (Ozis and Ozdemir 2009; Kibaroglu et al. 2011; Klaphake and Scheumann 2011; Ozis et al. 2013a, b).

Furthermore, both countries face pollution problems in Lower Chorokhi river and its tributaries, requiring appropriate measures. It should be noted that Georgia adopted the principles of the European Water Framework Directive.

12.4 Kura-Araks (Kura-Aras) Basin and the Small Baradost Creek

12.4.1 Geographical Position of the Kura-Araks Basin

The Kura-Araks (Kura-Aras) Basin consists of two main subbasins: Kura and Araks (Aras), both originating in the north-eastern region in Turkey. The total drainage area of the Kura-Araks Basin in Turkey is 28,479 km^2, with 4887 km^2 in Upper Kura, 20,408 km^2 in Upper Araks (including Kars and Arpacay tributaries), 2350 km^2 in Sarisu, and 832 km^2 in Kotur tributaries.

Kura originates in Turkey, crosses the border to Georgia, receives some creeks from Armenia, then continues to Azerbaidjan, joins Araks near Sabirabad, and flows into the Caspian Sea. The drainage area of the Kura Subbasin is 4887 km^2 in Turkey,

Fig. 12.10 Kura-Araks (Kura-Aras) Basin in Turkey, Georgia, Armenia, Iran, Azerbaidjan. (U. Ozis & Y. Ozdemir)

roughly 33,000 km^2 in Georgia, 7200 km^2 in Armenia, and 41,000 km^2 in Azerbaidjan, so that the entire Kura drainage area is about 86,000 km^2.

Araks originates in Turkey whereas the tributary Arpacay originates in Armenia and joins the tributary Kars to form the boundary between Turkey and Armenia. Arpacay joins Araks at the boundary. Araks forms then the boundary between Turkey and Armenia, for a short stretch between Turkey and Azerbaidjan (Nakhcevan). Next, Araks forms the boundary between the Islamic Republic of Iran (thereafter will be written as 'Iran' only) and Azerbaidjan (Nakhchevan). It also forms in succession the boundary between Iran and Armenia, again Iran and Azerbaidjan, and finally joins Kura in Azerbaidjan (Figs. 12.10 and 12.11).

The drainage area of the Araks Subbasin is about 23,592 km^2 in Turkey (including 2350 km^2 of Sarisu and 832 km^2 of Kotur tributaries, crossing the border to Iran), roughly 22,600 km^2 in Armenia, 39,000 km^2 in Iran, and 19,000 km^2 in Azerbaidjan (including 5500 km^2 of Nackhcivan), thus totalling around 104,000 km^2. The entire basin's drainage area, lying east of the Blacksea and discharging into the Caspian Sea, is in the order of 190,000 km^2 (Figs. 12.10,12.11 and 12.12).

The Baradost Creek, adjacent to the Tigris Subbasin, originates near Yuksekova and has a water potential of about 0.1 km^3/y in Turkey. It crosses the border to Iran and flows through to the Urmia (Orumiyeh) lake.

Fig. 12.11 Araks (Aras) River and the 130 m long Cobandede Bridge (1297). (Photo by U. Ozis)

Fig. 12.12 The Mount Ararat (Buyuk Agri Dagi) (peak el. 5165 m) between Aras River and its tributary Sarisu, east of Dogubayazit in Turkey. (Photo by U. Ozis)

12.4.2 Water Potential of the Kura-Araks Basin

The flow originating from Turkey in the Kura-Araks Basin is about 0.9 km^3/y in Upper Kura, 1.6 in Upper Araks, 0.6 in Kars, 0.1 in Sarisu, and 0.1 in Kotur tributaries, hence totalling 3.3 km^3/y (Ozis 1997; Ozis et al. 2006, 2013a, b; Baran and Cakmakoglu 2001; Klaphake and Kramer 2011).

The additional water potential of Kura (called Mtkvari in Georgia) is roughly 11 km^3/y in Georgia, 2 km^3/y in Armenia, and 7 km^3/y in Azerbaidjan, amounting to a total of 21 km^3/y, including Turkey. The additional water potential of Araks is roughly 4 km^3/y in Armenia, 5 km^3/y in Iran, and 3 km^3/y in Azerbaidjan, thus totalling 14 km^3/y, including Turkey. The total water potential of the Kura-Araks Basin is thus in the order of 35 km^3/y, again including Turkey.

12.4.3 Water Resources Development in the Kura-Araks Basin

Turkey plans to generate 2.3 billion kWh/y electrical energy and irrigate up to 480.000 hectares of land in the Kura-Araks Basin (DSI 1995). Its water potential in Turkey will barely meet the entire irrigation demand even when areal distribution differences of this potential can be solved.

The Arpacay dam at the boundary-forming stretch, providing flow regulation for the Serdarabat weir to irrigate the Igdir plain, was constructed in the 1980's on the basis of the 1927 treaty between Turkey and USSR, which stipulated equal use of Arpacay waters. Turkey has constructed several dams and hydroelectric plants for energy, together with some for irrigation purposes, like Kayabeyi on Kura, Cildir downstream of the lake, and Kars on the same named tributary.

Georgia and Azerbaidjan increased the capacity of Lake Jandari to create a storage volume of 25 km^3. Several dams for various purposes, mainly irrigation, are constructed in the basin. Among these, the dams Mingechevir with an active storage of 4.7 km^3, Shamkir with an active storage of 1.4 km^3 built in the Kura Subbasin in Azerbaidjan, and the dam Aras with an active storage of 1.2 km^3 between Iran and Nakhcevan in the Araks Subbasin, are noteworthy.

It is reported that 2–2.5 million hectares of land are actually under irrigation in the Kura-Araks Basin, with roughly 1 million ha in Azerbaidjan, 0.5 million ha in Iran, 0.3 million ha in Georgia, 0.2 million ha in Armenia, and 0.2 million ha in Turkey.

12.4.4 Water-related Issues in the Kura-Araks Basin

The riparians of the Kura-Araks Basin concluded several bilateral agreements concerning the waters of the basin; they are, however, far from establishing a joint integral basin development plan (Kibaroglu et al. 2011; Klaphake and Kramer 2011). Return flows from irrigation and pollution from domestic and industrial uses, enhanced with the scarcity of water in dry years, are of great importance for the basin. Even the release of 1.8 m^3/s from Turkey to Iran in the Sarisu tributary could not be realized during some dry periods.

The extension of irrigated areas in Turkey will affect the availability of water in downstream riparians. The positive effect of Turkey's dams on flood mitigation, discharge regulation, and sediment retention appears to be limited for the downstream riparians.

The Kura-Araks Basin bears, beyond water resources, great importance for Turkey with regard to political and economical relations with Georgia, Armenia, Iran, and especially Azerbaidjan, so the development of the basin will be affected by multilateral approaches in the region.

12.5 Orontes (Asi) Basin and the Adjacent Small Qweik (Balik) Creek

12.5.1 Geographical Position of the Orontes Basin

The river Orontes (Asi) originates near Baalbek in Lebanon's Bekaa valley, crosses the border to flow northwards to Syria and forms later the boundary between Syria and Turkey for around 25 km. It enters then into Turkey, makes a turn near the Amik Plain, and discharges to the Mediterranean Sea. The north-eastern tributary Afrin originates in Turkey, crosses the border to Syria, and enters back to Turkey near Reyhanli. Afrin joins Karasu near the Amik Plain. The northern tributary Karasu originates also in Turkey, forms the boundary between Turkey and Syria for around 30 km, then flows again in Turkey, and finally joins Orontes near the Amik Plain (Fig. 12.13).

The figures for the drainage area of the entire Orontes Basin vary from around 21,000–26,000 km^2, the latest being 24,745 km^2, determined through digital elevation maps and the geographical information system. The drainage area of Orontes in Turkey is about 6000 km^2, including roughly 1000 km^2 of the Afrin tributary's upper creek. It is 2000 km^2 in the upper riparian Lebanon, and 17,000 km^2 in Syria. Hence, the total drainage area of the Orontes Basin is in the order of 25.000 km^2 (Ozis and Ozdemir 2009; Ozis et al. 2013a, b; Karatas 2016; Selek 2016).

The Qweik (Balik) Creek between the Orontes Basin and the neighboring Euphrates Subbasin, ending in desert land near Aleppo in Syria, has a potential of about 0.2 km^3/y which originates from around 1000 km^2 in Turkey (Ozis 1997). Although a treaty dating back to 1921 depicts the equal use of Qweik waters for water supply of Aleppo, the city is actually supplied from the reservoir of the Tabqa (At-Thawra) dam on Euphrates.

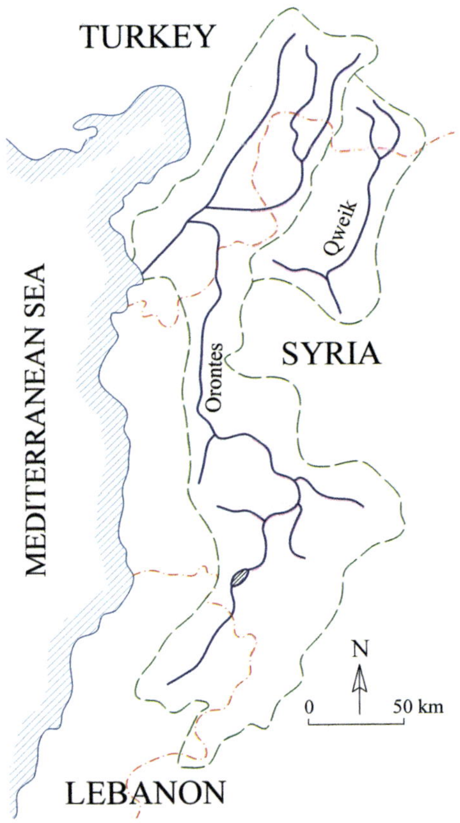

Fig. 12.13 Orontes (Asi) Basin in Lebanon, Syria, Turkey, and the Adjacent Small Qweik (Balik) Creek. (U. Ozis & Y. Ozdemir)

12.5.2 Water Potential of the Orontes Basin

The water potential of the Orontes Basin in Turkey is in the order of 1.2 km^3/y, including 0.2 km^3/y from the Afrin tributary's upper creeks. From Lebanon originates 0.4 km^3/y of the potential, and 1.2 km^3/y originates from Syria so that the entire water potential of Orontes is roughly 2.8 km^3/y (Baran et al. 1997, 2006; Ozis 1997; Ozis et al. 2006; Maden 2011a, b; Scheumann et al. 2011b; Karatas 2016; Selek 2016).

Furthermore, groundwater is actually an important source for irrigation and domestic water purposes in Turkey and especially in Syria. The total safe groundwater potential of the entire basin is in the order of 0.7–0.8 km^3/y (Karatas 2016).

12.5.3 Water Resources Development in the Orontes Basin

Turkey intends to irrigate roughly 200,000 ha of land in Orontes Basin (DSI 1995; Odemis et al. 2016; Selek 2016). Turkey has constructed some modest dams and several small reservoirs in this basin to cover part of the irrigation needs.

The Tahtakopru Dam on the northern Karasu tributary, formerly serving to irrigate 13,000 ha, is recently heightened by 11 m to irrigate additional 34,000 ha of land and to supply water to three hydroelectric plants producing 135 GWh/y. The 105 m high, recently constructed Buyuk Karacay Dam on the western tributary Karacay, supplies water to Antakya and its vicinity.

The Reyhanli Dam on the eastern Afrin tributary, with a remarkable crest length of 9.3 km, is anticipated to enter into service in 2019 and will serve to irrigate 60,000 ha land. The 42 m high Yarseli dam on the small Beyazcay tributary, westwards of the boundary forming stretch of the Orontes, serves to irrigate 7300 ha land.

Lebanon has the dam Assi to control the upper part of the Orontes river, irrigating around 20,000 ha of land with the contribution of karst springs.

Syria first restored the historical Katinah dam, upstream of Homs, and then constructed the Rastan dam between Homs and Hamah and the Mhardeh dam downstream of Hamah to create an active storage of 0.5 km^3/y. Roughly three dozens of additional small dams and reservoirs on side creeks of the main river increase the total active storage in the Orontes Basin in Syria to roughly 1.1 km^3.

Besides certain domestic and industrial uses, Syria has extensive irrigation activities in the Orontes Basin, like the Ghab and some other projects near Homs and Hamah, resulting in a total area of irrigated land of more than 200,000 ha. Part of the irrigation and urban water demand is actually supplied by groundwater.

12.5.4 Water-Related Issues in the Orontes Basin

Turkey cannot irrigate 200,000 ha of land in Orontes Basin through the use of only the domestic water potential. On the other hand, Syria's irrigation activities in the Orontes Basin significantly affect the discharges, on one hand leaving virtually quite limited amounts of water to enter Turkey, especially in dry seasons, and on the other hand, releasing flood discharges in wet seasons (Odemis et al. 2016; Turhan and Kibaroglu 2016). Moreover, there are several problems related to water quality (Kibaroglu and Jaubert 2016).

Turkey and Syria agreed in 2009 to jointly construct a 'Friendship' dam on the boundary forming stretch of the river between the two countries. This project would irrigate 8000 ha of land, protect 6000 ha of land against floods, and generate 13 GWh/y hydroelectric energy. The construction began in 2011; however, the upheaval in Syria, followed by issues of territorial integrity, refrained its realization for several years (Kibaroglu and Scheumann 2013; Kibaroglu and Sumer 2016; Selek 2016; Scheumann and Shamaly 2016).

The Hatay province of Turkey, located largely in Orontes Basin, is designated as being part of Syria on many Syrian maps, and Syria systematically refused to discuss water issues related to the Orontes Basin. The case of the Friendship Dam constitutes also a 'de facto' recognition of Turkey's rights related to the Hatay province.

The reciprocal positions of the two countries in Orontes and Euphrates, although at different scales, necessitates the inclusion of the Orontes Basin in discussions about the Euphrates-Tigris Basin in particular, and of middle-eastern water problems in general (TMMOB 2006; Ozis and Ozdemir 2009; Kibaroglu et al. 2011; Ozis et al. 2013a, b; Kibaroglu and Sumer 2016).

12.6 Euphrates-Tigris (Firat-Dicle) Basin

12.6.1 Geographical Position of the Euphrates-Tigris Basin

12.6.1.1 The Euphrates-Tigris Basin in the Middle-East

Euphrates and Tigris are the two principal branches of the transboundary river basin, joining each other 70 km north of Bassorah in Iraq, forming the Shatt-al-Arab, and discharging 100 km thereafter into the Gulf. A large closed basin in Iraq, the Thartar Subbasin, is artificially linked with Tigris and Euphrates.

Turkey is the upstream riparian of both the Euphrates and the Tigris. Iran is the upstream riparian of certain eastern tributaries of Tigris. Saudi Arabia is the upstream riparian of some virtually non-contributing creeks southwest of Euphrates. Syria is largely the upper downstream riparian of Euphrates and, for a short stretch, the downstream (boundary) riparian to Tigris. Iraq is the main downstream riparian of Tigris and Euphrates.

The Karun Basin in Iran, including the Dez Subbasin, discharges to the Gulf at the eastern edge of the Shatt-al-Arab delta near Khoramshahr and Abadan. It is geographically conceived also as a subbasin of the Shatt-al-Arab by certain references; in the present text, however, the river Karun will not be considered as a subbasin of the Euphrates-Tigris Basin.

The drainage area of the Euphrates Subbasin is 121,560 km^2 in Turkey, 87,300 km^2 in Syria, non-contributing 58,000 km^2 in Saudi Arabia, and 182,300 km^2 in Iraq; thus, the total is 450,000 km^2 at the beginning of Shatt-al-Arab.

The drainage area of the Tigris Subbasin is 57,615 km^2 in Turkey, 850 km^2 in Syria, 39,400 km^2 in Iran (excluding Karkheh), and 146,150 km^2 in Iraq (without Thartar); thus the total is 244,015 km^2 up to the beginning of Shatt-al-Arab. The Shatt-al Arab flows across three groups of marshlands: (a) Qurna, close to the confluence of Euphrates and Tigris; (b) Hamar to the west of Shatt; (c) Havize to the east of Shatt. The area of these marshlands largely vary according to the seasons and to stochastic variations in hydraulicity.

Fig. 12.14 The Euphrates-Tigris Basin in Turkey, Syria, Saudi Arabia, Iran (the tributary Kharkeh excluded), Iraq. (U. Ozis & Y. Ozdemir)

The entire drainage area of the Euphrates-Tigris Basin (Figs. 12.14, 12.15 and 12.16), including the closed Thartar Basin and the final part along Shatt-al-Arab down to the Gulf, is roughly 700,000 km^2.

Fig. 12.15 The Euphrates River near Karkamis. (Photo by U. Ozis)

Fig. 12.16 The Tigris River near Diyarbakir with the 180 m long Dicle-Bridge of roman origin. (Photo by A. Alkan)

12.6.1.2 The Euphrates Subbasin

Euphrates originates as the tributaries Karasu and Murat from the north-east. Through its course, Murat receives the tributaries Peri and Munzur, largely fed from karstic springs further down. Murat joins Karasu close to the Keban dam site, and both discharge into the Keban reservoir.

The Euphrates River receives, in succession, the tributary Tohma from the north-west between Keban and Karakaya dam sites, the tributary Kahta from the west between Karakaya and Ataturk dam sites, the tributary Goksu from the west between Ataturk and Birecik dam sites, and the tributary Nizip from the west between Birecik and Karkamis dam sites, the latter located close to the boundary between Turkey and Syria.

Some smaller tributaries of Euphrates in Turkey cross the border to Syria, e.g. Sajour (Sacir) to the west of Euphrates, Balikh (Culap) to the east of Euphrates, and three creeks (Circip, Zerkan, Cagcag) farther east, forming then the tributary Khabur in Syria. The tributary Sajour from the north-west joins Euphrates to the south of Jerablus. Euphrates receives the tributary Balikh from the north near Rakka and the tributary Khabour from the north-east near Buseyra. The Euphrates crosses the Syrian-Iraqi border near Abou Kemal and joins Tigris at Qurna to the northwest of Bassorah.

12.6.1.3 The Tigris Subbasin

Tigris originates from Lake Hazar to the southeast of Elazig, close to the tributary Murat of Euphrates and upper levels of the Keban reservoir on Euphrates. Tigris receives, from the north, the tributaries Batman, Garzan and Bitlis along its course towards east before turning to the south.

On the way to the Turkish-Syrian border, Tigris receives the tributary Botan from the east and forms for around 30 km the boundary between Turkey and Syria to the south of Cizre. Then, for about 7 km, it establishes the boundary between Syria and Iraq until it reaches Faish Khabur.

The eastern tributary of Tigris, Khabour (Habur) and its tributary Hezil, both originate in Turkey. Khabour crosses the border to Iraq and continues until its confluence with Hezil. Hezil forms the Turkish-Iraqi border for around 30 km until its confluence with Khabour. Khabour then continues as the Turkish-Iraqi border and joins Tigris near the Turkish-Iraqi-Syrian border junction.

The Greater Zap (Buyuk Zap) and its tributary Shamadinan (Semdinli) originate in Turkey. Then, they cross the border, and Shamadinan joins the Greater Zap in Iraq. The Greater Zap flows into Tigris southwards of Mossul. The Lesser Zap (Kucuk Zap) originates in Iran, crosses the border to Iraq, and joins Tigris westwards of Kerkouk (Kerkuk). Another eastern tributary, Adheim (Al-Uzaym), originates in Iraq and joins Tigris between Samarra and Baghdad. The tributary Diyala originates in Iran, crosses the border to Iraq, and joins Tigris to the south of Baghdad.

The Karkheh River, with all its tributaries, flows almost entirely in Iran and crosses the border to Iraq at a quite low elevation near Amarah to join Tigris. Tigris continues to flow in Iraq to join Euphrates at Qurna north of Bassorah.

12.6.2 *Water Potential of the Euphrates-Tigris Basin*

12.6.2.1 Water Potential of the Euphrates Subbasin

The average water potential of Euphrates and Tigris are cited with large differences in numerous publications. The figures cited below are based on the authors' own investigations, based on various data and approaches (Ozdemir 1998; Ozis et al. 1998, 1999a, b, 2000; Ozdemir and Ozis 2000; Ozdemir et al. 2002).

In these studies, observations of monthly discharges at key stream-gauging stations on Euphrates and Tigris in Turkey are analyzed and missing data estimated through correlation analyses. Long-term average discharges (m^3/s) are evaluated for the period of 1940–1989, as natural discharges without modification by large reservoirs and major withdrawals.

Similar studies are carried out for the discharges at the southern tributaries, which reach the main watercourse in the Turkish territory, or which cross the border separately towards the downstream riparian (Ozdemir and Ozis 2000). Furthermore, the unit average discharge ($l/sec/km^2$) values of these tributaries were instrumental in estimating the contribution of drainage areas uncovered by stream-gauging stations, or where reliable figures were not available neither in Turkey nor in other riparian states (Ozdemir 1998).

The average discharge of the main watercourse near Karkamis at the Syrian border is around 30.5 km^3/y. The small tributaries of Euphrates, crossing the Syrian border, contribute an additional 1.5 km^3/y. Hence, the total water potential of the Euphrates Subbasin in Turkey is roughly 32 km^3/y.

The contribution of the interim drainage area in Syria between the Turkish border and the Iraqi border is about 4 km^3/y. It should be noted that about 40% of this potential originate from several springs with significant discharges, located close to the Turkish border. Thus, the water potential of Euphrates near Abu-Kemal in Syria at the Iraqi border amounts to roughly 36 km^3/y.

The contribution of the interim drainage area in Iraq, partly through small creeks from Iraq to Syria on the left bank, partly downstream of the Syrian border and upstream of the Hit stream-gauging station, is in the order of 1 km^3/y. The contribution of the interim drainage area from Hit to the junction with Tigris near Al Qurna to form the Shatt-al-Arab can be neglected. The drainage area of the rather theoretical south-west right bank tributaries in Saudi Arabia has virtually no contribution. Hence, the total water potential of the Euphrates Subbasin is around 37 km^3/y.

12.6.2.2 Water Potential of the Tigris Subbasin

The average discharge of the main watercourse near Cizre at the Syrian border is around 17 km^3/y (Baran et al. 1995). The contribution of the interim drainage area upstream of the Iraqi border to the Hezil-Habur tributary basin from Turkey is 2 km^3/y, and that of the interim drainage area upstream of the Iraqi border to the tributary Greater Zap in Turkey is 5 m^3/sec. Thus, the total water potential of the Tigris Subbasin in Turkey amounts to 24 km^3/y (Ozis et al. 1999a). The contribution of the small watershed in Syria, on the right bank of the border-forming Tigris, is negligible.

The upper drainage areas of certain eastern tributaries of Tigris in Iran contribute 2 km^3/y to Lesser Zap, 6 km^3/y to Diyala-Hamreen, and 2 km^3/y to southern subbasins. The total water potential of the Tigris Subbasin in Iran is then roughly 10 km^3/y.

The interim drainage area in Iraq between the Turkish border and Mossul, including the Hezil-Habur tributary, contributes about 2 km^3/y. The contribution of the interim drainage area of the tributary Greater Zap in Iraq is somewhat more than 5 km^3/y and that of the Lesser Zap less than 5 km^3/y, so that the total is in the order of 10 km^3/y. The contribution of the tributary Adheim and that of the remaining interim drainage area between Mossul and Bagdad in Iraq is 3 km^3/y. Hence, the water potential of Tigris at Bagdad amounts to 41 km^3/y (Ozis et al. 1997).

The closed Thartar Basin between Euphrates and Tigris is linked to both water courses and serves actually as an interim reservoir for flood mitigation. The basin has virtually no direct contribution to the Euphrates-Tigris Basin.

The water potential of the interim drainage area of the tributary Diyala-Hamreen and the remaining interim drainage area of southern subbasins, excluding Karkheh, between the Iranian border and Shatt-al-Arab is about 8 km^3/y. Thus, the contribution from Iraq to the water potential of Tigris equals 23 km^3/y. Hence, the total water potential of the Tigris Subbasin at Al-Qurna amounts to roughly 57 km^3/y.

12.6.2.3 Total Water Potential of the Euphrates-Tigris Basin

The total water potential of the Euphrates sub-Basin is 37 km^3/y; of this total, Turkey provides 32 km^3/y, which is about 85% of this potential flowing from less than 30% of the drainage area. The water potential of the Euphrates subbasin is around 4 km^3/y in Syria and 1 km^3/y in Iraq; practically, no contribution comes from Saudi Arabia.

The total water potential of the Tigris Subbasin is 57 km^3/y; Turkey provides about 40% of this potential or 24 km^3/y coming from less than 25% of the drainage area. The water potential of the Tigris Subbasin is negligible in Syria; it is in the order of 10 km^3/y in Iran (excluding Karkheh) and 23 km^3/y in Iraq.

Hence, the average total water potential of the Euphrates-Tigris Basin is in the order of 94 km^3/y or roughly 3000 m^3/s. Turkey provides about 60% of the total water potential of Euphrates-Tigris Basin originating from 25% of the entire drainage area.

12.6.3 Water Resources Development in Euphrates-Tigris Basin

12.6.3.1 Water Resources Development in Turkey

The Euphrates-Tigris Basin represents more than 25% of irrigable agricultural lands and more than 40% of the hydroelectric potential of Turkey. The development of land and water resources of this basin is considered as the driving force for socio-economic development of the region.

The ‘Southeastern Anatolia Project‘(Turkish acronym: GAP) encompasses the Lower Euphrates and the Western and Central Tigris regions of the basin (Harmancioglu and Ozis 1978, 1983; DSI 1980; Ozis 1982, 1983, 1994, 1997; Harmancioglu 1986; Uskay 1987; Ozbek 1989; Congar 1994; Unver 1994, 1997a, b; Altinbilek 1997, 2004; Altinbilek and Akcakoca 1997; Akuzum et al. 1997; Bagis 1997; Avci and Yanik 1997; Bayazit and Avci 1997; Kulga and Cakmak 1997; Turkman 1998; Ozis et al. 2013a, b; Tortajada 2000; Biswas and Tortajada 2001; Aydogdu and Yenigun 2008; Ozis and Ozdemir 2009; Baskan 2011; Sen 2011; Tigrek and Kibaroglu 2011; Topcu 2011; Ozdemir et al. 2013; Kibaroglu and Gursoy 2015; Harmancioglu and Cetinkaya 2016).

In the Upper Euphrates Basin in Turkey, covering Karasu, Murat, Peri, Munzur rivers and their tributaries, more than one hundred dams, mostly for irrigation and/or energy production, are proposed. Some of them are already in operation.

The total area to be irrigated in Upper Euphrates is 0.45 million hectares; the anticipated total energy generation is 19 TWh/y (billion kWh/y). The 207 m high rock-fill Keban dam has an active storage capacity up to 25 km^3 and generates 6 TWh/y energy. It is the key reservoir for the regulation of Euphrates discharges (Fig. 12.17) (DSI 1995).

A total of 53 dams in Lower Euphrates and 15 dams in Western and Central Tigris in Turkey will regulate the flows for flood control, irrigation and/or energy production, as well as for certain urban and industrial water supply schemes.

Around 1.7 million hectares of agricultural land will be irrigated in the context of GAP, two-thirds in Lower Euphrates and one third in Western and Central Tigris. 18 hydroelectric power plants with 20 TWh/y in Lower Euphrates and 12 plants with 8 TWh/y in Western and Central Tigris in Turkey are planned. Outside the scope of the Southeastern Anatolia Project, 22 dams and 30 hydroelectric schemes with 9 TWh/y are planned on Eastern Tigris tributaries in Turkey.

Fig. 12.17 The Keban Dam on Euphrates. (Photo by Y. Ozdemir)

Fig. 12.18 The Karakaya dam on Euphrates. (Photo by U. Ozis)

Several dams, power plants, and irrigation schemes are already in operation. Among these are the rock-fill Ataturk Dam with 85 million m^3 embankment, creating a reservoir of 48 km^3 and generating up to 9 TWh/y (Harmancioglu and Ozis 1981; Harmancioglu 1986; Ozis et al. 1990, 1992; Ozis and Harmancioglu 1994); the Sanliurfa twin tunnels, each with 26.4 km length, conveying 330 m^3/sec (Kurt 1992; Tanriverdi 1992); and the 173 m high arched-gravity Karakaya Dam generating 8 TWh/y (Ozis and Ozel 1989). These schemes are particularly noteworthy among those on Lower Euphrates (Figs. 12.18, 12.19, 12.20 and 12.21).

Fig. 12.19 The Ataturk dam on Euphrates. (Photo by U. Ozis)

Keban – Karakaya – Ataturk – Birecik – Karkamis dams form a cascade of reservoirs on the Euphrates main river down to the border with Syria. Among the major dams of the Upper Tigris, the Kralkizi and Dicle dams on Tigris, Batman Dam on Batman, and Alkumru Dam on Botan are in operation; and Ilisu Dam on Tigris is under construction.

12.6.3.2 Water Resources Development in Syria

Three dams are located on the mainstream Euphrates in Syria: Teshreen with the maximal reservoir level approaching the Turkish border; At-Thawra (Tabqa) as the key dam for irrigation, energy production and urban water supply to Aleppo; and Al-Baath to regulate the discharge of the former dam. Turkey's proposal to jointly set up a high dam (Yusufpasa), using the head of Teshreen dam in Syria and Karkamis dam in Turkey, as a more beneficial plan to both countries, has not been received favorably by her downstream neighbor. Three dams for irrigation (Saab, Taaf, Shuhey) are located on Khabur and two tributaries in Syria, originating as the tributaries Circip to the west and Zerkan and Cagcag to the east, both of which lie to the east of Euphrates in Turkey (Kolars and Mitchel 1991; Karadamur and Hadid 1992; Wakil 1993; Bilen 1996, 1997; Kout 2008).

Another proposal by Turkey to heighten and shift the location of the Cizre Dam towards the end of the Turkish-Syrian border formed by Tigris has also not been received favorably by her neighbor. The purpose of the proposal was to divert part of Tigris waters to supplement Syria's irrigation needs in the Khabur region. Turkey constructed then the lower Cizre Dam entirely within the country. Syria is actually anticipating to pump 1.25 km^3/y water from Tigris, where the river forms the boundary between Syria and Iraq for about 7 km.

Syria plans to irrigate 0.8 million ha of land by Euphrates and 0.15 million ha by Tigris. However various factors, especially soil quality, appears to limit it to 0.3–0.4 million ha in Euphrates.

Fig. 12.20 The Sanliurfa twin tunnel, fed from the Ataturk dam's reservoir. (Photo by U. Ozis)

12.6.3.3 Water Resources Development in Iran

The hydroelectric potential of the upper stretches of certain eastern tributaries of Tigris in Iran, such as the Upper Lesser Zap and Upper Diyala, can eventually be harnessed by high-head diversion plants, diverting water either by weirs or partly regulated by dams. There is no accurate information about such hydroelectric schemes; however, their operation would not cause serious problems as long as the diverted discharges flow back to the same basin.

Fig. 12.21 The Sanliurfa main irrigation canal. (Photo by U. Ozis)

The Karkheh river, with all its tributaries, flows almost entirely in Iran and crosses the border to Iraq at a quite low elevation near Amarah to join Tigris. The Karkheh Subbasin is important for hydropower and agricultural production. Iran planned a series of six dams in Karkheh Basin, namely Tang Maashoureh on the tributary Kashkan, Sazbon and Seymareh on the tributary Seymareh, and Pa Alam and Karkheh on Karkheh. The earth-fill dam Karkheh and the concrete arch dam Seymare are in operation.

12.6.3.4 Water Resources Development in Iraq

One major dam, the Haditha Dam upstream of the Hit stream-gauging station, is located on Euphrates in Iraq. The scheme incorporates also a hydroelectric power plant. This is followed by the Ramadi weir, Habbaniyah and Abu-Dibbis (Razaza) off-stream reservoirs, and Dibban, Warrar, Hindiyah and Nassiriyah weirs (barrages), all serving irrigation purposes (Hadithi 1978; Bilen 1997; Altinbilek 2004; Chen et al. 2011a; Ohara et al. 2011; Onucyildiz et al. 2016). Significant water schemes are apparently neither possible nor anticipated on the ephemeral dry creeks at the south-west regions of the Lower Euphrates in Saudi Arabia.

The Mossul (formerly Saddam) Dam and Fattah and Samarra weirs are located on the main river Tigris in Iraq. Duhok Dam on Duhok, Bekme Dam on Greater Zap, Dokan and Dibbis dams on Lesser Zap, Adheim Dam on Adheim (Al-Uzaym), Derbendikhan and Hamrin dams, and Diyala Weir on Diyala are dams and weirs located on eastern tributaries of Tigris in Iraq. Some dams are equipped with power plants and are capable of generating a total of 10 TWh/y energy. Kut, Dibban and Gharraf weirs, all supplying water to irrigation systems, are located on Lower Tigris in Iraq, southwards of Baghdad (Bilen 1997; Chen et al. 2011a; Ansari and Knutsson 2011; Ohara et al. 2011; Onucyildiz et al. 2016).

There is no accurate information on hydroelectric schemes harnessing the hydroelectric potential of the upper stretches of certain eastern tributaries of Tigris in Iraq. High-head diversion power plants, with water diverted either by weirs or partly regulated by dams, can be implemented in these regions.

The Thartar closed basin in northwestern Iraq is used to store excess flood waters of Tigris, forming the Lake Thartar. It is also linked with Euphrates and may be used, among other options, to transfer water from Tigris to Euphrates for irrigation along its banks (Kolars and Michell 1991; Ozis 1994; Bilen 1997; Ozis et al. 2004a, b, 2013a, b; Ozis and Ozdemir 2009, 2010; Ohara et al. 2011). The link between the two canals can preferably be directly established, bypassing the turbid waters of the Lake Thartar and avoiding excessive evaporation losses.

A long canal, called also the 'Third River', was built between Euphrates and Tigris in southern Iraq, to provide an efficient collection of the drainage systems. The marshlands of Shatt-al-Arab's delta have been significantly reduced thereafter (Karpuzcu et al. 2009; Chen et al. 2011b).

Iraq anticipates to irrigate about 2.0 million ha of land along Tigris and even more than 2.0 million ha along Euphrates, without substantiating the details of many projects (Bilen 1997; Altinbilek 2004; Ohara et al. 2011). The figures cited for the total anticipated irrigation areas in Iraq reach even 5.8 million ha in some studies (Kucukmehmetoglu and Guldmann 2010; Kucukmehmetoglu and Geyman 2013). Turkey uses the same land classification criteria as the U.S. Bureau of Reclamation in evaluating irrigable areas, whereas it is not often the case in Syrian and Iraqi evaluations.

12.6.4 *Water-Related Issues in the Euphrates-Tigris Basin*

12.6.4.1 Discrepancies on Water Potential Estimates

Water scarcity in the Middle-East in general, and water allocation claims of the riparian states in particular, make the Euphrates-Tigris Basin one of the foremost conflict centers of the world. This conflict is enhanced by international geopolitical and energy issues stemming from states and organizations outside the region.

In various publications dealing with the middle-eastern water conflict, the figures for the average total water potential of Euphrates varied from 29 to 37 km^3/y and those of Tigris from 42 to 58 km^3/y (Ozis and Ozdemir 2009, 2010; Ozdemir et al. 2013; Ozis et al. 2013a, b).

The differences between 37 and 29 km^3/y (up to 8 km^3/y) for Euphrates on one hand, and between 58 and 42 km^3/y (up to 16 km^3/y) for Tigris on the other hand, are due to classified observations, lack of information, data bias, and disinformation. These discrepancies should definitely be clarified, and the accurate long-term water potential of the Euphrates-Tigris Basin has to be determined by contribution of all parties involved before entering the discussions on any water allocation agreement. This corresponds basically to the first stage of the 'three-stage plan' proposed by Turkey to her neighbors since 1980's, relating to development of the Euphrates-Tigris water and land resources.

12.6.4.2 Discrepancies on Estimates of Irrigation Demands

In addition to several urban and industrial requirements, Turkey anticipates, besides hydroelectric energy generation, to irrigate around 1.6 million ha of agricultural land in Euphrates and 0.65 million ha in Tigris subbasins.

Turkey will use as a long-term average, including evaporation losses from reservoirs, 19 km^3/y (600 m^3/sec) out of the 32 km^3/y water originating in the country in the Euphrates Subbasin. This corresponds to about 50% of the average total Euphrates water potential of 37 km^3/y. The ultimate long-term average discharge to be released to Syria from Euphrates and tributaries will thus be in the order of 13 km^3/y (400 m^3/sec).

In the Tigris Subbasin, Turkey will use as a long-term average, including evaporation losses from reservoirs, 8 km^3/y (250 m^3/sec) out of the 24 km^3/y water originating in Turkey. This corresponds to about less than 15% of the average total Tigris potential of 57 km^3/y. The ultimate long-term average discharge to be released to Syria and Iraq from Tigris and tributaries will thus be in the order of 16 km^3/y (500 m^3/sec).

The ratio of the long-term average water use in Turkey (including evaporation losses from reservoirs) will be in the order of less than 30% of the long-term average total water potential of the Euphrates-Tigris Basin. This is quite an equitable and reasonable water use for the upper riparian Turkey, with 56 km^3/y or 60% of the total potential originating in the country. Furthermore, the reservoirs created by dams in Turkey will also provide significant benefits to downstream countries, such as sediment retention, flood mitigation, and temporarily the augmentation of low flows (Ozis et al. 2004a, b, 2013a, b; Ozis and Ozdemir 2009, 2010). With regard to water quality, the salinity in Euphrates will be in the order of 700 ppm after all irrigation activities in Turkey are completed, thus not causing significant harm to Syria (Bilen 1997).

Syria anticipates, besides certain urban and industrial water requirements, to irrigate 0.8 million ha of land from Euphrates, eventually adding up 0.15 million ha land from Tigris. However, various factors, especially the soil quality, appear to limit the irrigation to half of these figures (Kolars and Mitchel 1991; Wakil 1993; Bilen 1997; Kibaroglu and Unver 2000; Altinbilek 2004; Salman and Mualla 2008; Oztan and Axelrod 2011). If Syria irrigates the anticipated 0.8 million ha, a very small part of the Euphrates water will flow to Iraq. A diversion of surplus Tigris discharges into Khabur Subbasin in Syria may also be investigated upstream of eventual major diversions in Iraq.

Syria claims that the 500 m^3/s provisional allocation on the basis of the 1987 protocol should be accepted as her definite share. On the other hand, studies in

Turkey show that the long-term average release can be in the order of 400 m^3/s and should be staged according to probability levels of discharges.

Iraq anticipates to irrigate about 2.0 ha of land along Tigris and even more than 2.0 million ha along Euphrates without substantiating the details of many projects (Bilen 1997; Altinbilek 2004; Ohara et al. 2011). The figures cited for total anticipated irrigation areas in Iraq reach even 5.8 million ha in some studies (Kucukmehmetoglu and Guldmann 2010; Kucukmehmetoglu and Geyman 2013). Iraq claims that Turkey should release even 700 m^3/s from Euphrates.

Turkey uses the same land classification criteria as the U.S. Bureau of Reclamation in evaluating irrigable areas, whereas it is not often the case in Syrian and Iraqi evaluations. The application of the second stage of the 'three-stage-plan' proposed by Turkey, depicting a joint determination of irrigation areas and agricultural water needs of each country in accordance with the same criteria, can help to clarify this very critical issue.

12.6.4.3 Attempts for Cooperation

Turkey and Iraq had signed in 1946 a protocol, setting certain conditions related to the development of Tigris and Euphrates. However, this had not been applied by the Iraqi side and became obsolete in the meantime. Turkey proposed since 1980's a 'three-stage plan' to Syria and Iraq for the ultimate integral development plan of the Euphrates-Tigris Basin in an optimal, equitable and reasonable context (Tekeli 1990; Turan 1993; Bilen 1997; Altinbilek 2004).

The first stage depicts the determination of the water potential of the basin. The second stage consists of the determination of irrigation areas and water needs of each country. The third stage foresees the elaboration of a master plan for the optimal integral development of the basin. The 'three-stage-plan' is basically a quite reasonable operational tool with regard to resolution of conflicts and creation of cooperation. However, it has not been welcomed by downstream riparian states, and Turkey was not able to reach a consensus.

Turkey and Syria signed an economic cooperation protocol in 1987, covering a provisional clause (Art. 6). It states that Turkey will release 500 m^3/s as monthly average flow at the Syrian border during the filling of Ataturk reservoir and until the three riparian states reach an agreement on the final allocation of the basin's waters. Deficiencies in any month will be compensated the next month. Syria and Iraq agreed in 1990 that 42% of the water released from Turkey will be used by Syria and 58% by Iraq (Bilen 1997).

Impoundment of the Ataturk reservoir necessitated the closure of the bottom outlet for about 1 month due to technical reasons. Downstream riparian states were informed on this matter 50 days in advance, and the release was increased up to 820 m^3/s from 23 November 1989 on in order to compensate for the necessary cut. When the impoundment began on 13 January 1990, a flood of hostile allegations that 'Turks have cut the water of Euphrates' spread over most of the press in the Arab world and various other countries. The cries of 'water wars' did not even cease after resuming the release of 500 m^3/s on 12 February 1990 (Bilen 1997; Ozis et al. 2004a, b; Karpuzcu et al. 2009).

Syria considers 13 km^3/y of Euphrates waters as her 'share'. The provisional 500 m^3/sec depicted in the 1987 protocol between Turkey and Syria correspond to about slightly less than 16 km^3/y. On the other hand, Syria agreed in 2000 to release 58% of the Euphrates flows coming from Turkey further to Iraq, so she will let flow 9 km^3/y to Iraq and use 7 km^3/y in Syria. Since the part of the Euphrates waters originating in Syria is about 4 km^3/y, Syria requires an increase of at least 2 km^3/y additional water, either from Euphrates or through diversion from Tigris.

Iraq considers Tigris as totally her own and refuses to deal with both rivers as a single watercourse basin. Iraq claims that Turkey should increase the 500 m^3/s discharge to 700 m^3/s, thus to about 22 km^3/y, in order to receive 13 km^3/y in Euphrates through Syria. It should be kept in mind, however, that the flow of 500 m^3/s is provisional, and the long-term average allocation can only be in the order of 400 m^3/s (Ozis and Ozdemir 2009, 2010; Ozis et al. 2013a, b). Moreover, any such allocation should be timely staged according to probability levels of river discharges because of the highly varying stochastic nature of the discharges in spite of the huge reservoirs in Turkey.

The development of the Euphrates-Tigris Basin in Turkey is, in fact, a means of cooperation rather than a cause of conflict in the region (Bagis 1994; Kibaroglu 1995, 1997, 2000, 2002; Kibaroglu and Unver 2000; Kibaroglu et al. 2005, 2011; Kibaroglu and Baskan 2011). The agricultural and industrial products will serve to satisfy the needs of many middle-eastern countries. The generation of surplus secondary hydroelectric energy in wet years will reduce oil and other fossil primary resources consumption at many locations. In very dry years, Turkey may release additional water against indemnity for some temporarily refrained domestic irrigation activities (Ozis and Ozdemir 2009, 2010; Ozis et al. 2013a, b).

12.6.5 Hydro-political Attitudes From Outside the Basin

The impact of the water scarcity in the Jordan (Seria) River Basin led to a flood of publications from 1980's onwards, covering also the Euphrates-Tigris Basin (Karpuzcu et al. 2009; Kavvas et al. 2011; Kibaroglu and Scheumann 2011, 2013; Kibaroglu et al. 2011; Kramer and Kibaroglu 2011; Ohara et al. 2011; Scheumann et al. 2011a, b; Turan 2011; Williams 2011; Ozger et al. 2013; Kankal and Uzlu 2014). Some of them consider the Euphrates-Tigris Basin as the emergency resource to alleviate the water shortages in the Jordan Basin. The relatively small Orontes River Basin, adjacent to the Euphrates subbasin, has a potential of 2.5–3 km^3/y. Lebanon, Syria and Turkey are the riparian states. Orontes should also be taken into account when dealing with water allocations in the Middle-East because Turkey is rather a downstream riparian in this Basin.

In 1997, the United Nations adopted the "Convention on the law of the non-navigational uses of international rivers", foreseeing the 'equitable and reasonable use' principle, 'without causing significant harm' to other riparian states (Cano 1989; UN 1997; UN/ECE 1997; Wouters 2000; ASCE 2004; WRPM 2007). The accepting votes were far from being a solid absolute majority although only three countries (Burundi, China, Turkey) voted against it.

The definition of these watercourses as 'international' instead of 'transboundary' was an unfortunately misleading term; 'multinational' watercourse would have been a rational compromise, if the expression 'national' were to be maintained in the terminology. Various clauses and restrictions of this convention can be abused by some riparian states. The inclusion of regional economic integration organizations as 'parties' may cause troubles.

The publishing of an article (Peters 2006) in the U.S. Armed Forces Journal, with a map showing most of the Euphrates-Tigris, Kura-Araks, Chorokhi Basins in Turkey as territories of a future 'Free Kurdistan' state, can neither be explained simply as the sole opinion of that article's retired colonel author, nor the provision is compatible by the more than half a century long alliance between U.S.A. and Turkey in NATO and other matters.

Terrorist activities all over the country, focused mainly in the southeastern regions, were intensified when Turkey began to develop her water and land resources in the Euphrates-Tigris Basin. This is especially true with regard to the implementation of the Southern Anatolia Project and the related irrigation schemes, which constitute the driving force in the social and economic development of that region. More than 30,000 Turkish citizens of diverse ethnic origins, either civilians or belonging to security forces, have lost their lives by terrorist attacks for the last four decades. In some foreign countries, including several members of the European Union and NATO, such terrorist activities in Turkey were claimed to be realized by rebels or freedom fighters.

Furthermore, declarations by the E.U., proposing joint management of the Euphrates-Tigris Basin, upset Turkey by evoking reminders of the 1920 Sèvres treaty, which became obsolete by the 1923 Lausanne treaty. Some groups in the E. U. and several international or national non-governmental organizations are trying to hinder dam and hydropower constructions, as experienced on some large projects in the Euphrates-Tigris Basin. The case of the Ilisu Dam on Tigris is a prominent example (Bilen 2003).

12.6.6 *Impacts of Discharge Regulation by Dams in Turkey*

The discharges of the Euphrates-Tigris Basin in Turkey show significant seasonal variations as well as year-to-year fluctuations, although they are largely fed, besides rain, from snowmelt and karst spring discharges (EIE 1955–1993; DSI 1961–1995; Baran et al. 1987; Bilen 1996, 1997; Ozis et al. 1999a, 2000).

The discharges of Euphrates at Dutluca gaging station, just downstream of the Ataturk Dam, which represents about five sixths of the Euphrates Subbasin flows originating in Turkey, showed an average value of about 870 m^3/s during 1937–1980 water years with the annual averages varying from 460 to 1500 m^3/s. Monthly mean discharges, rising up to 4360 m^3/s in May 1969, were usually about 300 to 500 m^3/s during the 8 months between July and February. These values were well below the average and much less during months like August and September, even decreasing to around 170 m^3/s (Ozis et al. 2000).

In view of the above fluctuations, large reservoirs have to be created by dams on the two main rivers and their tributaries in order to regulate the discharges for water power, irrigation, flood control (DSI 1980, 1995; Fistikoglu et al. 2008). The total active storage capacity in Turkey will ultimately reach the order of 60–70 km^3 in Euphrates, equaling almost twice the average annual flow volume in the country. It will be 20 km^3 in Tigris, being in the order of the average annual flow volume in Turkey. The firm energy of the hydroelectric plant at the foot of the Tabqa dam in Syria have been increased annually by several hundreds of GWh for the last 40 years and more, benefiting primarily from the regulation at the Keban reservoir in Turkey.

The topography in downstream countries does not permit the creation of such large storages, especially in Euphrates. In addition, the unit evaporation rate from water surface is significantly higher than that in Turkey. The total active storage in downstream countries will be around 15 km^3 in Euphrates (mainly in Syria), and 20 km^3 in Tigris (basically in Iraq).

The annual average evaporation from reservoir surfaces in Turkey will be about 5 km^3/y in Euphrates and 1.5 km^3/y in Tigris; that in downstream countries about 2 km^3/y in Euphrates and 6 km^3/y in Tigris (Ozis 1994, 1997; Ozis et al. 2004a, b, 2006, 2013a, b).

Reservoirs created in Turkey provide significant benefits to downstream countries, such as the retention of sediments, mitigation of floods, and temporary augmentation of low flows. Hence, the evaporation from these reservoirs should not be considered as an unnecessary loss of water and should not be solely debited to Turkey.

After completion of all anticipated schemes on waters originating in Turkey, around 13 km^3/y or 40% in Euphrates and 16 km^3/y or 65% in Tigris will continue to flow downstream, as long-term average values. However, it should be kept in mind that any water allocation has to be timely staged according to probability levels of discharges because of the highly varying stochastic nature of the discharges, even after the regulation by the huge reservoirs in Turkey (Figs. 12.22 and 12.23).

It should be kept in mind that climate changes may affect the discharges of the basin in the future (Kibaroglu 2008b; Yenigun et al. 2008; Ozdogan 2011; Bozkurt and Sen 2013; Ay et al. 2017) (more information can be found in Chap. 14 of this book).

Fig. 12.22 The Euphrates river downstream of Ataturk dam. (Photo by U. Ozis)

Fig. 12.23 The Tigris river at Hasankeyf (with remains of the historical bridge from twelfth century). (Photo by U. Ozis)

12.6.7 Equitable and Reasonable Use of the Euphrates-Tigris Water Potential

The water balance for a year of average hydraulicity is given in Table 12.1 for Euphrates and in Table 12.2 for Tigris Subbasins for an average total water potential of 94 km^3/y for the Euphrates-Tigris Basin (37 km^3/y from Euphrates and 57 km^3/y from Tigris) (Ozis et al. 2013b). These water balances are based on the ultimate development of the Euphrates-Tigris Basin under equitable and reasonable

Table 12.1 Approximate water balance for average hydraulicity in Euphrates subbasin

Water balance of the Euphrates subbasin in a year of average flow (in km^3/y)	
Natural flow in Turkey at Karkamis	+30.5
Natural flow of tributaries at Syrian border	+1.5
Total natural flow originating from Turkey	**+32.0**
Evaporation from reservoirs in upper Euphrates	−1.0
Evaporation from the Keban reservoir	−1.0
Irrigation in upper Euphrates (0.45 Mio.ha)	−4.0
Water supply in upper Euphrates	−0.5
Return flow in upper Euphrates	+1.0
Evaporation from reservoirs in lower Euphrates	−3.0
Irrigation in lower Euphrates (1.15 Mio.ha)	−12.5
Water supply in lower Euphrates	−1.5
Return flow in lower Euphrates (to main river)	+1.0
Return flow in lower Euphrates (to tributaries))	+2.5
Outflow from Turkey to Syria	**+13.0**
Flow originating in Syria	+4.0
Evaporation from reservoirs in Syria	−2.0
Irrigation in Syria (0.7 Mio.ha)	−8.5
Water supply in Syria	−1.5
Return flow in Syria	+2.5
Outflow from Syria to Iraq	**+7.5**
Flow originating in Saudi Arabia	0.0
Flow originating in Iraq	+1.0
Evaporation from reservoirs in Iraq	−1.5
Transfer from Tigris to Euphrates in Iraq	+13.0
Irrigation in Iraq (1.5 Mio.ha)	−21.0
Water supply in Iraq	−1.5
Return flow in Iraq	+5.5
Outflow from Euphrates to Shatt-al-Arab and Gulf delta	**+3.0**

considerations. Eventual significant water diversions in Iran from the eastern Tigris tributaries is not taken into account; the Karkeh Subbasin in Iran is excluded; the Karun Basin is considered as a separate basin entirely in Iran, and is therefore also excluded.

The water potential of the Euphrates Subbasin is not sufficient to satisfy the irrigation needs along its banks in Iraq. On the other hand, Tigris offers an excess of water potential in the order of 10–15 km^3/y, which should be diverted to Euphrates and beyond in order to relieve the water shortages in the Middle-East. This transfer may occur via modification of the existing Tigris-Thartar-Euphrates canal system or by alternative options.

Turkey has been accused to act according to the absolute sovereignty doctrine with regard to the development of Euphrates-Tigris Basin. However, the water

Table 12.2 Approximate water balance for average hydraulicity in Tigris subbasin

Water balance of the Tigris subbasin in a year of average flow (in km^3/y)	
Natural flow in Turkey at Cizre	+17.5
Natural flow of Greater Zap at Iraqi border	+4.5
Natural flow of other tributaries at Iraqi border	+2.0
Total natural flow originating from Turkey	**+24.0**
Evaporation from reservoirs in Western Tigris	−1.0
Evaporation from reservoirs in eastern Tigris	−0.5
Irrigation in Central Tigris (0.65 Mio.ha)	−7.5
Water supply in Central Tigris	−1.0
Return flow in Central Tigris (to the main river)	+0.5
Return flow in Central Tigris (to the tributaries)	+1.5
Outflow from Turkey to Syria and Iraq	**+16.0**
Flow originating in Syria	0.0
Irrigation in Syria (0.1 Mio.ha)	−1.0
Outflow from Turkey and Syria to Iraq	**+15.0**
Flow originating in Iran and flowing to Iraq	+10.0
Flow originating in Iraq	+23.0
Evaporation from reservoirs in Iraq	−4.5
Transfer from Tigris to Euphrates in Iraq	−13.0
Irrigation in Iraq (2.0 Mio.ha)	−28.0
Water supply in Iraq	−4.0
Return flows in Iraq	+8.5
Outflow from Tigris to Shatt-al-Arab and Gulf delta	**+7.0**

claimed by Turkish schemes is well in the range of equitable and reasonable use principles reflected in Articles 5 and 6 of the 1997 UN Convention (UN 1997; Wouters 2000).

This is especially the case when the following points are taken into consideration: (a) the economical feasibility of the projects prepared for the State Hydraulic Works (DSI), (b) the application of modern water-saving irrigation techniques, (c) the activities of the Southeastern Anatolia Project Regional Development Administration (GAPBKI) in the frame of the gigantic plan for the social and economic development of that region of Turkey (Unver and Voron 1993; Unver et al. 1993; Akmandor et al. 1994; Unver 1994, 1997a, b; Ozis 1994; Disisleri Bakanligi 1996; Karakaya 1996; Kutan 1996; Yakis 1996; Altinbilek 1997, 2004; Altinbilek and Akcakoca 1997; Akuzum et al. 1997; Bagis 1997; Bayazit and Avci 1997; Bilen 1997, 2009; Kulga and Cakmak 1997; Yilmaz and Kadilar 1997; Alpaslan and Harmancioglu 2001; Ozis et al. 2004a, b, 2013a, b; TMMOB 2006; Ozis and Ozdemir 2009, 2010; Yildiz and Ozbay 2011).

In fact, detailed analyses to be carried outside the scope of this chapter would show that the relevant factors cited in paragraphs ‘a’, ‘b’, ‘f’ of Article 6 favor definitely the developments in Turkey. As of paragraph ‘c’, the entire population of

Turkey depend heavily on these developments; as of paragraph ‘d’, positive effects significantly overweigh the adverse ones, leading to cooperation rather than to conflict; as of paragraph ‘e’ the potential uses of the watercourse are the most optimal; and as of paragraph ‘g’ the alternatives, wherever available, are far from being reasonable.

The development of the Euphrates-Tigris Basin in the upstream riparian Turkey, using only half of the 60% of Basin’s water potential originating in the country, appears well equitable and reasonable under consideration of all relevant factors.

12.7 Conclusions

Development projects in transboundary watercourses of Turkey, including the Southeastern Anatolian Project (GAP), basically comply with UN’s principles of equitable and reasonable use. Moreover, dams in Turkey provide also significant benefits to downstream countries, such as flood mitigation, sediment retention, and temporary low flow augmentation.

The average water potential originating in Turkey corresponds to about 60% of the entire Euphrates-Tigris water potential. Ultimately, as a long-term average, half of the potential originating in Turkey, about 40% in Euphrates and 65% in Tigris, will continue to flow towards downstream countries.

However, any water allocation has to be timely staged according to probability levels of discharges because of the significant periodicity and the highly varying stochastic nature of the discharges even after the regulation by the huge reservoirs in Turkey.

The water potential of the Euphrates is not sufficient to cover the large irrigation requirements along its banks in Iraq. Therefore, the surplus water of Tigris has to be transferred to Euphrates in Iraq, eventually allowing the transfer of a few km^3/y to adjacent Orontes and Jordan river basins of the region.

Acknowledgments The authors are grateful for the support and encouragement on matters of transboundary watercourses and especially on the Southeastern Anatolia Project, to the management and members of: Dokuz Eylul and Ege Universities, State Hydraulic Works, Electrical Works Survey Authority, Turkish Electricity Production Corporation, Southeastern Anatolia Project Regional Development Administration, United Nations Environment Program, Ata Construction Company, Turkey’s Chamber of Civil Engineers, and last but not the least, to their colleagues and former students whose names often appear in the list of references of their papers.

The first and third authors extend their thanks to the Springer Institution and to the chief editors Prof. Dr. Nilgun Harmancioglu and Prof. Dr. Dogan Altinbilek for inviting them to prepare this chapter of the book “Water resources of Turkey”.

References

Akmandor N, Pazarci H, Koni H (1994) Orta Dogu ulkelerinde su sorunu. TESAV Toplumsal, Ekonomik, Siyasal Arastirmalar Vakfi, Ankara

Akuzum T, Kodal S, Cakmak B (1997) Irrigation management in GAP. Int J Water Resour Dev 4:547–560

Alpaslan A, Harmancioglu N (2001) Turkiye'de ve Dunya'da su politikalari. III. Ulusal Hidroloji Kongresi, Dokuz Eylul Universitesi, Izmir, pp 111–118

Altinbilek D (1997) Water and land resources development in southeastern Turkey. Int J Water Resour Dev 3:311–332

Altinbilek D (2004) Development and management of the Euphrates-Tigris basin. Int J Water Resour Dev 1:15–33

Altinbilek D, Akcakoca H (1997) Innovative approaches in water resources development in the southeastern Anatolia project. Int J Water Resour Dev 4:485–503

Ansari NA, Knutsson S (2011) Toward prudent management of water resources in Iraq. J Adv Sci Eng Res 1:53–67

ASCE (2004) The trans-boundary sharing of water in times of water scarcity. American Society of Civil Engineers, Environmental and Water Resources Institute, Reston

Avci I, Yanik B (1997) Sinirasan ve sinir olusturan su kaynaklarimiz: potansiyel, su talepleri ve sorunlari. In: Su Kongresi ve Sergisi '97, Makina Muhendisleri Odasi ve 15 Diger Meslek Odasi, Istanbul, pp 7–15

Ay M, Karaca OF, Yildiz AK (2017) Trend analysis of streamflow in the Euphrates-Tigris basin. In: Second IWA regional symposium – The past, present and future of world's water resources, Proceedings CD, International Water Association, Cesme, pp 104–110

Aydogdu MH, Yenigun K (2008) Turkiye'nin en buyuk entegre su kaynaklari projesi GAP'ta sulama ve drenaj sistemlerinin ozet degerlendirmesi. In: TMMOB 2.Su politikalari kongresi, bildiriler kitabi, vol. 1, Insaat Muhendisleri Odasi, Ankara, pp 159–174

Bagis AI (ed) (1994) Water as an element of cooperation and development in the Middle East. Hacettepe University and Friedrich-Naumann-Foundation, Ankara

Bagis AI (1997) Turkey's hydropolitics of the Euphrates-Tigris basin. Int J Water Resour Dev 4:567–581

Baran T, Cakmakoglu O (2001) Aras havzasi su potansiyeli. In: III. Ulusal Hidroloji Kongresi, Dokuz Eylul Universitesi, Izmir, pp 629–632

Baran T, Harmancioglu N, Ozis U (1987) Turkiye'nin akarsu havzalarinda karst pinar katkilari. In: IX. Teknik Kongre bildiriler kitabi, vol. 2, Insaat Muhendisleri Odasi, Ankara, pp 299–311

Baran T, Demirci N, Ozis U (1995) Dicle havzasinin Cizre akim gozlem istasyonunda su potansiyeli. In: Insaat Muhendisliginde Gelismeler, II. Teknik Kongre bildiriler kitabi, Bogazici Universitesi, Istanbul, pp 233–242

Baran T, Sayar E, Ozis U (1997) Asi havzasi su potansiyeli. In: Insaat Muhendisliginde Gelismeler III. Teknik Kongresi bildiriler kitabi, vol. 3, Ortadogu Teknik Universitesi, Ankara, pp 667–678

Baran T, Ozis U, Ozdemir Y (2006) Sinir-asan Asi havzasi su potansiyeli ve yararlanilmasi. In: TMMOB Su politikalari kongresi, bildiriler kitabi, vol. 2, Insaat Muhendisleri Odasi, Ankara, pp 571–581

Baskan A (2011) Liberalization of Turkey's hydroelectricity sector. In: Kibaroglu A, Scheumann W, Kramer A (eds) Turkey's water policy, national frameworks and international cooperation. Springer, Berlin, pp 83–91

Bayazit M, Avci I (1997) Water resources of Turkey: potential, planning, development and management. Int J Water Resour Dev 4:443–452

Bilen O (1994) Prospects for technical cooperation in the Euphrates-Tigris basin. In: Biswas AK (ed) International waters of the Middle East from Euphrates-Tigris to Nile. Oxford University Press, Oxford, pp 95–116

Bilen O (1996) (2.pr. 2000a) Ortadogu su sorunlari ve Turkiye. TESAV Toplumsal, Ekonomik, Siyasal Arastirmalar Vakfi, Ankara

Bilen O (1997) (2.pr. 2000b) Turkey and water issues in the Middle East. Southeastern Anatolia Project Regional Development Administration, Ankara
Bilen O (2003) Cevre emperyalizmi ve Ilisu baraji ornegi. Avrasya-Bir Vakfi, Ankara
Bilen O (2009) Turkiye'nin su gundemi: su yonetimi ve AB su politikalari. Devlet Su Isleri, Ankara
Biswas AK (ed) (1994) International waters of the Middle East from Euphrates-Tigris to Nile. Oxford University Press, Bombay
Biswas AK, Tortajada C (2001) Development and large dams: a global perspective. Int J Water Resour Dev 1:9–21
Bosnjakovic B (2000) Regulation of international watercourses under the UN/ECE regional agreements. Water Int 4:544–553
Bozkurt D, Sen OL (2013) Climate change impacts in the Euphrates-Tigris basin based on different model and scenario simulations. J Hydrol 480:149–161
Cano JG (1989) The development of the law of international water resources and the work of the international law commission. Water Int 3:167–171
Chen ZQR, Kavvas ML, Ohara N, Anderson ML, Yoon J (2011a) Coupled regional hydroclimate model and its application to the Tigris-Euphrates basin. J Hydrol Eng 12:1059–1070
Chen ZQR, Kavvas ML, Ohara N, Anderson ML, Yoon J (2011b) Impact of water resources utilization on the hydrology of Mesopotamian marhlands. J Hydrol Eng 12:1083–1092
Congar B (1994) Harran-Ceylanpinar ovalarinda yillik yeraltisuyu bilancolarinin GAP projesi uygulanmasi ile degisimi. In: 2. Erguvanli Muhendislik Jeolojisi Kollokyumu, Istanbul Teknik Universitesi, Istanbul, pp 1–7
Dalkilic Y, Harmancioglu N (2008) Avrupa Birligi su cerceve direktifinin Turkiye'de uygulama olanaklari. In: TMMOB 2. Su politikalari kongresi, bildiriler kitabi, vol. 2, Insaat Muhendisleri Odasi, Ankara, pp 415–424
Disisleri Bakanligi (1996) Orta-Dogu'da su sorunu. T.C. Disisleri Bakanligi, Bolgesel ve Sinirasan Sular Dairesi, Ankara
Dmitrov D, Balabanova S, Koshinchanov G (2008) Floods information and forecasting services. In: Taskin konferansi, Devlet Su Isleri XI. Bolge Mudurlugu, Edirne, pp 38–46
DSI (1961 to 1995) Rasat yilliklari (1959 ila 1989). Devlet Su Isleri, Etut ve Plan Dairesi, Ankara
DSI (1995) Haritali istatistik bulteni. Devlet Su Isleri, Ankara
DSI (Uskay S, Taskapilioglu E, Gurkan N, Palali Y, Toprakoglu H, Abidoglu A, Koknel B) (1980) Guneydogu Anadolu Projesi, Devlet Su Isleri, Etud ve Plan Dairesi, Ankara
EIE (1955 to 1993) Akim neticeleri (1936 ila 1989). Elektrik Isleri Etud Idaresi, Ankara
EIE (1985) Coruh nehri havzasi enerji imkanlari. Elektrik Isleri Etud Idaresi, Ankara
Eroglu V (2006) Taskinlarin havza bazinda yonetimi. In: 1. Ulusal taskin sempozyumu, Devlet Su Isleri, Ankara, pp 3–15
Fistikoglu O, Ozdemir Y, Ozis U (2008) Firat-Dicle havzasinda akislarin barajla duzenlenmesi. Su ve enerji konferansi, bildiriler kitabi, vol. 2, Devlet Su Isleri XXVI. Bolge Mudurlugu, Artvin, pp 33–42
Gunduz DN (2006) Meric-Arda-Tunca nehirleri taskin yonetimi calismalari. In: 1. Ulusal taskin sempozyumu, Devlet Su Isleri, Ankara, pp 123–131
Hadithi AHA (1978) Optimal utilization of the water resources of the Euphrates river in Iraq. Dissertation. University of Arizona
Harmancioglu N (1986) Impact of economic factors on development of lower Euphrates Basin in Turkey. In: Water Power 85 – International conference on Hydropower, 1985 Las Vegas, American Society of Civil Engineers
Harmancioglu N, Cetinkaya CP (2016) Euphrates and Tigris basins. In: Singh VP (ed) Handbook of applied hydrology, 2nd edn. Mcgraw Hill, New York, Chapter 99, pp 99.1–99.4
Harmancioglu N, Ozis U (1978) Enerji bedellerinin degisimine Asagi Firat planlamasinin duyarligi. DSI Teknik Bulteni 42:3–12
Harmancioglu N, Ozis U (1981) Enerji buhrani ve Ataturk Baraji. Ege Universitesi, Insaat Fakultesi Dergisi, Ataturk'un 100.dogum yili ozel sayisi: 451–474

Harmancioglu N, Ozis U (1983) Dynamics of water resources development: lower Euphrates case, Turkey. In: Bayazit M (ed) Application of system analysis on water resources development, NATO Workshop, 1983 Istanbul, Technical University, Istanbul, Rep. 8
Kankal M, Uzlu E (2014) The status of transboundary rivers of Turkey. Water Resour 41:649–665
Karadamur SE, Hadid B (1992) Hydraulic structures on Euphrates river. In: International conference on protection and development of the Nile and other major rivers, Proceedings, vol. 2/2, pp 8.1–8.14
Karakaya B (1996) The controversy over water in the Middle East and the world. In: Water conflict in the Middle East. Aydinlar Ocagi, Ankara, pp 11–32
Karatas A (2016) Sustainable water management in Hatay. In: Kibaroglu A, Jaubert R (eds) Water resources management in the lower Asi-Orontes river basin. MEF University, Istanbul, pp 111–123
Karpuzcu M, Gurol MD, Bayar S (eds) (2009) Transboundary waters and Turkey. Institute of Technology, Gebze
Kavvas ML, Chen ZQ, Anderson ML, Ohara N, Joon JY, Xiang F (2011) A study of water balances over the Tigris-Euphrates watershed. Phys Chem World 36:197–203
Kibaroglu A (1995) Prospects for cooperation in the Euphrates-Tigris river basin. In: Turkish Review of Middle East Studies, Annual 1994/95, Foundation for Middle East and Balkan studies (OBIV), Istanbul, pp 137–155
Kibaroglu A (1997) Firat-Dicle havzasinda su sorununa kapsamli bir bakis. In: Sinir asan sularimiz, Celal Bayar Universitesi Yuksek Ogrenim Vakfi, Manisa, pp 3–23
Kibaroglu A (2000) Institutions for effective use of the Euphrates-Tigris waters. In: IWRA Xth World Water Congress, International Water Resources Association, Melbourne, Sec J-L, Paper 6
Kibaroglu A (2002) Building a regime for the waters of the Euphrates-Tigris river basin. Kluwer Law International, The Hague
Kibaroglu A (2008a) Meric nehri havzasi sinir asan su politikalari. In: Taskin konferansi, Devlet Su Isleri XI. Bolge Mudurlugu, Edirne, pp 1–11
Kibaroglu A (2008b) Kuresel iklim degisikliginin sinirasan su kaynaklari politikasina etkileri. In: TMMOB 2.Su politikalari kongresi, bildiriler kitabi, vol. 2, Insaat Muhendisleri Odasi, Ankara, pp 347–355
Kibaroglu A, Baskan A (2011) Turkey's water policy framework. In: Kibaroglu A, Scheumann W, Kramer A (eds) Turkey's water policy, national frameworks and international cooperation. Springer, Berlin, pp 8–25
Kibaroglu A, Gursoy SI (2015) Water development and poverty reduction. Water Int 5–6:824–838
Kibaroglu A, Jaubert R (eds) (2016) Water resources management in the lower Asi-Orontes river basin. MEF University, Istanbul
Kibaroglu A, Scheumann W (2011) Euphrates-Tigris rivers system political rapprochement and transboundary water cooperation. In: Kibaroglu A, Scheumann W, Kramer A (eds) Turkey's water policy, national frameworks and international cooperation. Springer, Berlin, pp 277–299
Kibaroglu A, Scheumann W (2013) Evolution of transboundary politics in the Euphrates-Tigris river systems: new perspectives and political challenges. Glob Gov 19:279–305
Kibaroglu A, Sumer V (2016) Transboundary water relations in the Asi Basin: the case of Syrian-turkish relations. In: Kibaroglu A, Jaubert R (eds) Water resources management in the lower Asi-Orontes river basin. MEF University, Istanbul, pp 139–150
Kibaroglu A, Unver O (2000) An institutional framework for facilitating cooperation in the Euphrates-Tigris river basin. Int Negot 2:311–330
Kibaroglu A; Klaphake A, Kramer A, Scheumann W, Carius A (2005) Cooperation on Turkey's transboundary waters. Federal Ministry for the Environment, Nature Conservation and Nuclear Safety, Adelphi Research GmbH, Berlin
Kibaroglu A, Scheumann W, Kramer A (eds) (2011) Turkey's water policy: national frameworks and international cooperation. Springer, Berlin

Klaphake A, Kramer A (2011) Kura-Aras river basin: burgeoning transboundary water issues. In: Kibaroglu A, Scheumann W, Kramer A (eds) Turkey's water policy, national frameworks and international cooperation. Springer, Berlin, pp 263–275

Klaphake A, Scheumann W (2011) Coruh river basin: hydropower development and transboundary cooperation. In: Kibaroglu A, Scheumann W, Kramer A (eds) Turkey's water policy, national frameworks and international cooperation. Springer, Berlin, pp 251–261

Kolars JE (1994) Problems of international river management: the case of the Euphrates. In: Biswas AK (ed) International waters of the Middle East from Euphrates-Tigris to Nile. University Press, Oxford, pp 44–94

Kolars JE, Mitchell WA (1991) The Euphrates river and the Southeast Anatolia development project. Southern Illinois University, Carbondale

Kout W (2008) Integrated water resources management in Syria. In: Sengupta M, Salwari R (eds) Taal 2007 – the 12th world Lake conference. Indian Ministry of Environment and Forests, Jaipur, pp 2300–2314

Kramer A, Kibaroglu A (2011) Turkey's position towards international water law. In: Kibaroglu A, Scheumann W, Kramer A (eds) Turkey's water policy, national frameworks and international cooperation. Springer, Berlin, pp 215–228

Kramer A, Schellig A (2011) Meric river basin: transboundary water cooperation at the border between the EU and Turkey. In: Kibaroglu A, Scheumann W, Kramer A (eds) Turkey's water policy, national frameworks and international cooperation. Springer, Berlin, pp 229–249

Kucukmehmetoglu M, Geymen A (2013) The significance and impacts of large investments: the case of the Euphrates & Tigris river basin. In: International perspective on water resources & the environment, Dokuz Eylul University & American Society of Civil Engineers, Izmir, Paper 120

Kucukmehmetoglu M, Guldmann JM (2010) Multiobjective allocation of transboundary water resources: case of the Euphrates and Tigris. J Water Resour Plan Manag 1:95–105

Kulga D, Cakmak C (1997) Toward sustainable water management in the southeastern Anatolia project (GAP). Int J Water Resour Dev 4:541–546

Kurt C (1992) Sanliurfa tunnel system. In: Broch E, Lysne D (eds) Hydropower'92, 1992 Lillehammer. Balkema, Rotterdam, pp 137–142

Kutan R (1996) Water disputes in middle-eastern countries. In: Water conflict in the Middle East. Aydinlar Ocagi, Ankara, pp 33–49

Maden TE (2011a) Turkiye-Suriye iliskilerinde suyun rolu. Ortadogu-Analiz 35:33–40

Maden TE (2011b) Türkiye-Suriye iliskilerinde Asi nehri. Ortadogu-Analiz, 31/32:40–49

Malkarali S, Korkmaz MS, Sezen N (2008) Meric nehri taskini ve taskinlar ile ilgili gelistirilen uluslararasi projeler. In: Taskin konferansi, Devlet Su Isleri XI. Bolge Mudurlugu, Edirne, pp 12–37

Odemis B, Evrendilek F, Kazgoz D (2016) Sustainable agricultural water management in the Asi Basin. In: Kibaroglu A, Jaubert R (eds) Water resources management in the lower Asi-Orontes river basin. MEF University, Istanbul, pp 97–109

Ohara N, Kavvas ML, Anderson ML, Chen ZQR, Yoon J (2011) Water balance study for the Tigris-Euphrates river basin. J Hydrol Eng 12:1071–1082

Onucyildiz M, Abdulmohsin MA, Buyukkaracigan N (2016) Firat-Dicle Havzasi ve Irak su yapilari. Selcuk Universitesi – Sosyal ve Teknik Arastirmalar Dergisi 12:118–151

Ozbek T (1989) Das Sudost-Anatolien-Projekt-GAP. In: V.Internationales Weiterbildungsprogramm, vol.16, Technische Universität, Berlin, pp 307–317

Ozdemir Y (1998) Water potential of the Euphrates-Tigris basin in the Middle East. M.S.-Thesis, Dokuz Eylul University, Izmir

Ozdemir Y, Ozis U (2000) Turkiye'de Asagi Firat'in ve Dicle'nin guney kollarinin uzun sureli akislari. Insaat Muhendisleri Odasi – Teknik Dergi, 1:2075–2100 & Long-term discharges of Lower Euphrates' and Tigris' southern tributaries in Turkey. Chamber of Civil Engineers, Digest-2000:621–625

Ozdemir Y, Ozis U, Baran T, Demirci N, Fistikoglu O, Canga R (2002) Firat-Dicle havzasinin Turkiye, Suriye, Irak, Iran'daki su potansiyeli. Turkiye Muhendislik Haberleri, 420-421-422:27-34

Ozdemir Y, Ozis U, Baran T (2013) Water resources development in the Euphrates-Tigris basin. In: International perspective on water resources & the environment. Dokuz Eylul University & American Society of Civil Engineers, Izmir, Paper 153

Ozdogan M (2011) Climate change impacts on snow water availability in the Euphrates-Tigris basin. Hydrol Earth Syst Sci 9:2789–2803

Ozger M, Abdollahzadehmoradi Y, Singh VP (2013) Scaling properties of Euphrates and Tigris basin daily streamflow data. In: International perspective on water resources & the environment, Dokuz Eylul University & American Society of Civil Engineers, Izmir, Paper 237

Ozis U (1982) Aménagement de la basse Euphrate en Turquie. Travaux 565:68–72. & Ausbauplan des unteren Euphrat-Beckens in der Türkei. Die Wasserwirtschaft 5: 207–210. & The development plan for the lower Euphrates basin in Turkey. Nat Resour Dev 16:73–82

Ozis U (1983) Ausbau des westlichen Tigris-Beckens in der Türkei. Die Wasserwirtschaft 4:109–112. & Development plan of the Western Tigris basin. Int J Water Resour Dev 4:343–352

Ozis U (1994) La gestion des besoins et d'approvisionnements en eau dans le bassin Euphrate-Tigre. In: IWRA VIII. World congress on water resources, 1994 Cairo. vol. 2(T5–S2), International Water Resources Association, Carbondale, pp 1.1–1.12

Ozis U (1997) Sinir-asan sular ve Turkiye. In: Sinir asan sularimiz. Celal Bayar Universitesi Yuksek Ogrenim Vakfi, Manisa, pp 69–97

Ozis U, Harmancioglu N (1994) Muhendislik ekonomisi acisindan Ataturk barajinin planlanmasi. In: DSI 40'inci Kurulus Yili, Su ve Toprak Kaynaklarinin Gelistirilmesi Konferansi Bildirileri, vol 3. Devlet Su Isleri, Ankara, pp 1269–1278

Ozis U, Ozdemir Y (2009) Turkey's transboundary watercourses and the Euphrates-Tigris basin. In: Karpuzcu M, Gurol MD, Bayar S (eds) Transboundary waters and Turkey. Institute of Technology, Gebze, pp 21–57. (p 23–59 in the turkish version)

Ozis U, Ozdemir Y (2010) Sinirasan akarsu olarak Firat-Dicle havzasi. In: Izmir Su sempozyumu (2009) bildiriler kitabi, Insaat Muhendisleri Odasi, Izmir, pp 153–163

Ozis U, Ozel I (1989) Karakaya dam and power plant. Int Water Power & Dam Constr 7:20–24

Ozis U, Basmaci E, Harmancioglu N (1990) Ataturk nears completion. Int Water Power & Dam Constr 9:12–16

Ozis U, Basmaci E, Harmancioglu N (1992) Ataturk hydro-power scheme on Euphrates. In: Königsberger A, Matthias HB, Varga T (eds) 7. International Seminar – Water Energy Plants, Technical University, Vienna, pp 15–29 (pp 17–31 in the German version)

Ozis U, Ozdemir Y, Baran T (1997) Dicle ve kollarının Iran ve Irak'taki su potansiyelinin belirlenmesi. In: Türkiye Insaat Muhendisligi 14.Teknik Kongresi, Insaat Muhendisleri Odasi, Izmir, pp 565–580

Ozis U, Ozdemir Y, Baran T, Demirci N, Fistikoglu O, Canga R, Ozen (Saf) B, Tariyan S (1998) Mezopotamya su potansiyelinin kestirimine bir yaklasim. In: II. Ulusal Hidroloji Kongresi, Teknik Universite, Istanbul, pp 9–20

Ozis U, Baran T, Demirci N (1999a) Turkiye'de Dicle havzasindaki baslica akis gozlem istasyonlarinin aylik akislari. Insaat Muhendisleri Odasi – Teknik Dergi, 4: 2029–2045. & Monthly discharges at principal streamgaging stations of Tigris basin in Turkey. Chamber of Civil Engineers, Digest '99, pp 613–616

Ozis U, Turkman F; Baran T, Ozdemir Y (1999b) Guneydogu Anadolu Projesi ve hidropolitik yonleri. In: Izmir Su Kongresi, Muhendislik ve Diger Meslek Odalari, Izmir, pp 443–462. & Yapi Dunyasi, 50:37–47

Ozis U, Fistikoglu O, Canga R (2000) Keban, Karakaya, Ataturk baraj yerlerinde aylik akislar. Insaat Muhendisleri Odasi – Teknik Dergi, 1:2101–2119. & Monthly discharges at Keban, Karakaya, Ataturk dam sites. Chamber of Civil Engineers, Digest 2000, pp 627–630

Ozis U, Baran T, Ozdemir Y, Fistikoglu O (2001) Turkiye acisindan sinir-asan sular. Meteoroloji Muhendisligi 1:34–39

Ozis U, Ozdemir Y, Dalkilic Y, Turkman F, Baran T (2004a) Development of the Euphrates-Tigris basin. In: Harmancioglu N, Fistikoglu O, Dalkilic Y, Gul A (eds) EWRA symposium on water resources management: risks and challenges for the 21st century. European Water Ressources Association and Dokuz Eylul University–SUMER, Izmir, pp 85–95

Ozis U, Ozdemir Y, Dalkilic Y, Turkman F, Baran T (2004b) Su siyaseti acisindan sinir-asan Firat-Dicle havzasi. Silâhli Kuvvetler Dergisi 382:16–33

Ozis U, Turkman F, Baran T, Ozdemir Y, Dalkilic Y (2006) Su hukuku. In: Ozis U, Turkman F, Baran T (eds) Su kaynaklarinin gelistirilmesi. Dokuz Eylul Universitesi, Izmir, pp 3.1–3.36

Ozis U, Ozdemir Y, Baran T (2013a) Cooperative issues in the trans-boundary Euphrates-Tigris basin. In: International perspective on water resources & the environment, Dokuz Eylul University & American Society of Civil Engineers, Izmir, Paper 154

Ozis U, Ozdemir Y, Baran T (2013b) Das grenzüberschreitende Euphrat-Tigris Becken. WasserWirtschaft 7(8):74–78

Oztan M, Axelrod M (2011) Sustainable transboundary groundwater management under shifting political scenarios: the Ceylanpinar aquifer and Turkey-Syria relations. Water Int 5:671–685

Peters R (2006) Blood borders – How a better Middle East would look. Armed Forces Journal (June 2006)

Salman M, Mualla W (2008) Water demand management in Syria: centralized and decentralized views. Water Policy 6:549–562

Samsunlu A, Maktav D, Kapdasli S (1996) Transboundary water issues between Greece, Bulgaria and Turkey: the case of Meric/Evros river. In: Ganoulis J, Duckstein L, Litherathy P, Bogardi I (eds) Transboundary resources management – institutional and engineering approaches. Springer, Berlin, pp 439–449

Sarac M, Eciroglu O (2008) Coruh havzasi ve EIE. In: Su ve enerji konferansi, bildiriler kitabi, vol.1, Devlet Su Isleri XXVI. Bolge Mudurlugu, Artvin, pp 108–117

Scheumann W, Shamaly O (2016) The Turkish-Syrian friendship dam in the Orontes river: benefits for all? In: Kibaroglu A, Jaubert R (eds) Water resources management in the lower Asi-Orontes river basin. MEF University, Istanbul, pp 125–129

Scheumann W, Baumann V, Mueller AL, Mutschler D, Steiner S, Walenta T (2011a) Environmental impact assessment in turkish dam planning. In: Kibaroglu A, Scheumann W, Kramer A (eds) Turkey's water policy, national frameworks and international cooperation. Springer, Berlin, pp 139–159

Scheumann W, Sagsen I, Tereci E (2011b) Orontes river basin: downstream challenges and prospects for cooperation. In: Kibaroglu A, Scheumann W, Kramer A (eds) Turkey's water policy, national frameworks and international cooperation. Springer, Berlin, pp 301–312

Selek B (2016) Development and management of water resources of Turkey with specific reference to the Asi basin. In: Kibaroglu A, Jaubert R (eds) Water resources management in the lower Asi-Orontes river basin. MEF University, Istanbul, pp 31–42

Sen Z (ed) (2002) Sinir asan sularimiz. Su Vakfi, Istanbul

Sen Z (2011) Water for energy. In: Kibaroglu A, Scheumann W, Kramer A (eds) Turkey's water policy, national frameworks and international cooperation. Springer, Berlin, pp 69–81

Sucu S, Dinc T (2008) Coruh havzasi projeleri. In: TMMOB 2.Su politikalari kongresi, bildiriler kitabi, vol. 1, Insaat Muhendisleri Odasi, Ankara, pp 33–38

Sumer V, Muluk C (2011) Challenges for Turkey to implement the EU water framework directive. In: Kibaroglu A, Scheumann W, Kramer A (eds) Turkey's water policy, national frameworks and international cooperation. Springer, Berlin, pp 43–67

Tanriverdi I (1992) Sanliurfa tunelleri. Devlet Su Isleri, Ankara

Tekeli S (1990) Turkey seeks reconciliation for the water issue induced by the southeastern Anatolia project (GAP). Water Int 4:206–216

Tigrek S, Kibaroglu A (2011) Strategic role of water resources in Turkey. In: Kibaroglu A, Scheumann W, Kramer A (eds) Turkey's water policy, national frameworks and international cooperation. Springer, Berlin, pp 27–42

TMMOB (2006) Su politikalari kongresi bildiriler kitabi. Insaat Muhendisleri Odasi, Ankara

Topcu S (2011) Water for agriculture: a major and inefficient consumer. In: Kibaroglu A, Scheumann W, Kramer A (eds) Turkey's water policy, national frameworks and international cooperation. Springer, Berlin, pp 93–115

Tortajada C (2000) Evaluation of actual impacts of the Ataturk dam. Int J Water Resour Dev 4:453–464

Turan I (1993) Turkey and the Middle East: problems and solutions. Water Int 1:23–29

Turan I (2011) The water dimension in turkish foreign policy. In: Kibaroglu A, Scheumann W, Kramer A (eds) Turkey's water policy, national frameworks and international cooperation. Springer, Berlin, pp 179–195

Turhan E, Kibaroglu A (2016) Political economy and political ecology of the turkish section of Asi River basin. In: Kibaroglu A, Jaubert R (eds) Water resources management in the lower Asi-Orontes river basin. MEF University, Istanbul, pp 19–30

Turkman F (1998) The development plan of the transboundary Euphrates-Tigris basin and environmental considerations. In: Loucks DP (ed) Restoration of degraded rivers: challenges, issues and experiences. Kluwer, The Hague, pp 263–276

Ucar I, Gurer I (2013) Temperature and precipitation variation at Coruh river basin due to climate change. In: International perspective on water resources & the environment, Dokuz Eylul University & American Society of Civil Engineers, Izmir, Paper 605

UN (1997) Convention on the law of the non-navigational uses of international watercourses. United Nations, New-York

UN/ECE (1997) Meeting of the parties to the convention on the protection and use of transboundary watercourses and international lakes. United Nations, Economic Commission for Europe, Genève

Unver O (1994) Southeastern Anatolia project (GAP) of Turkey: water resources development within the context of integrated regional socioeconomic development. In: IWRA VIII.World congress on water resources, 1994 Cairo. vol. 2(T5-S2), International Water Resources Association, Carbondale, pp 5.1–5.12

Unver O (1997a) South-Eastern Anatolia integrated development project (GAP), Turkey: an overview of issues of sustainability. Water Resour Dev 2:187–208

Unver O (1997b) Southeastern Anatolia project (GAP). Water Resour Dev 4:453–483

Unver O, Voron B (1993) Improvement of canal regulation techniques: the southeastern Anatolia project -GAP. Water Int 3:157–165

Unver O, Voron B, Akuzum T (1993) Improvement of field water distribution and irrigation techniques: the southeastern Anatolia project – GAP. Water Int 3:166–178

Uskay S (1987) Guneydogu Anadolu Projesi DSI calismalari. In: IX.Teknik Kongre Bildiriler Kitabi, vol.2, Insaat Muhendisleri Odasi, Ankara, pp 7–19

Wakil M (1993) Analysis of future water needs for different sectors in Syria. Water Int 1:18–22

Williams PA (2011) Turkey's water diplomacy, a theoretical discussion. In: Kibaroglu A, Scheumann W, Kramer A (eds) Turkey's water policy, national frameworks and international cooperation. Springer, Berlin, pp 197–214

Wolf AT (1994) A hydropolitical history of the Nile, Jordan and Euphrates river basins. In: Biswas AK (ed) International waters of the Middle East from Euphrates-Tigris to Nile. University Press, Oxford, pp 5–43

Wouters P (2000) National and international water law – achieving equitable and sustainable use of water resources. Water Int 4:499–512

WRPM (Draper S, ed) (2007) Special issue: transboundary water sharing. J Water Resour Plan Manag 5:377–471

Yakis Y (1996) Turkey, the Middle East and their water problem. In: Water conflict in the Middle East. Aydinlar Ocagi, Ankara, pp 59–67

Yenigun K, Gumus V, Bulut H (2008) Trends in streamflow of the Euphrates basin. Water Manag 4:189–198

Yildiz D, Ozbay O (2011) Su Firat'in suyu. Truva, Istanbul

Yilmaz S, Kadilar R (1997) Siniri asan sularimiz. In: Sinir asan sularimiz. Celal Bayar Universitesi Yuksek Ogrenim Vakfi, Manisa, pp 27–66

Chapter 13
River Basin Management

Bülent Selek and Zeliha Selek

Abstract Sustainable river basin management can be achieved through the conservation-utilization balance between sustainable use of water and conservation of its quality. Turkey, as a European Union (EU) candidate country, is conducting studies to ensure compliance with EU norms and requirements on river basin management. River Basin Management Plans (RBMPs) are prepared in accordance with the EU Water Frame Directive (WFD) for 25 river basins in the country to achieve "good status" by 2036 in all water environments by implementing the required measures. In this context, previously completed Basin Protection Action Plans (BPAPs) are transformed into RBMPs. Important projects related to river basin management, taking into consideration both water quality and quantity, were carried out in Turkey in the past decade. In particular, the establishment of the basin institutional structure and delegation in the country has been an important step in terms of water basin management. The stream network forming river basins in Turkey have a more complicated structure than those in other EU countries, and this makes basin management more difficult. The problems encountered in water resource management in Turkey have originated from institutional structures, water management based on administrative boundaries, insufficient databases, poor monitoring and surveillance, and insufficient sanctions and policies because the national Draft Water Law has not been finalized yet.

Keywords Basin management · Water framework directive · Institutional structure · River basins of Turkey · River basin management plans

B. Selek (✉)
Ministry of Agriculture and Forestry, Ankara, Turkey

Z. Selek
Technology Faculty, Gazi University, Ankara, Turkey
e-mail: zelihaselek@gazi.edu.tr

N. B. Harmancioglu, D. Altinbilek (eds.), *Water Resources of Turkey*, World Water Resources 2, https://doi.org/10.1007/978-3-030-11729-0_13

13.1 Introduction

Water is essential for life and is important for both social and economic development. The relationship between water and socio-economic development was clearly defined in the United Nations (UN) Millennium Development Goals (MDGs), agreed upon in 2000, by drawing attention to targets on water and sanitation to enhance the overall aim of ensuring environmental sustainability (UNESCO-IHP 2014). On the other hand, Sustainable Development Goals (SDGs), built on MDGs, aim to eliminate poverty, protect our planet and enforce peace and prosperity for all people (UN 2015). The SDGs cover 17 objectives which will be achieved by 2030, and three of them (Goal 6: Clean Water and Sanitation, Goal 13: Climate Action, Goal14: Life Below Water) are directly related to water.

Socio-economic development requires sustainable management of water. Rockström et al. (2009) claimed that the world lost its natural sustainability due to human interventions on natural resources. While water is at the core of sustainable development and production, it is also a basic human right. Therefore, as people struggle to have more water, there should be an equitable access to adequate quantity and quality of the resource for everyone. Zeitoun (2011) suggested a tool, "The Global Web of National Water Security", which demonstrates the strong nexus and interdependencies between water security and other critical areas. In this regard, Integrated Water Resources Management (IWRM) has been considered as the best way to achieve an efficient and sustainable management of water resources. Along the same line, the European Union (EU) Water Framework Directive (WFD) foresees a new era for European water management, focusing on understanding and integrating all aspects of the water environment to effectively and sustainably manage them (Teodosiu et al. 2003).

Turkey's Tenth Development Plan aims to reduce disparities between its regions and recognizes sustainable management of the country's water resources as an essential condition for its economic development. Water-related activities are systematically managed through central planning. In this framework, Ministry of Environment and Urbanization (MoEU) and the abolished Ministry of Forestry and Water Affairs (MoFWA) (This ministry was combined with the Ministry of Food, Agriculture and Livestock with the presidential decree issued in June, 2018, and named as the Ministry of Agriculture and Forestry) have prepared national environment related strategies in the fields of water, waste and climate-change supported by management plans, including:

- Strategic Plan of MoEU (2013–2017);
- Strategic Plan of MoFWA (2013–2017);
- Groundwater Management Action Plan (2013–2024);
- National Climate Change Strategy (2010–2020);

- National Climate Change Adaptation Strategy and Action Plan (2011–2023);
- National Basin Management Strategy (2014–2023);
- National Biodiversity Strategy and Action Plan (2007);
- Natura 2000 Implementation Strategy of Turkey (2011);
- National Climate Change Action Plan (2011–2023);
- River Basin Protection Action Plans (2009–2013);
- Waste Water Treatment Action Plan (2015–2023);
- Integrated Urban Development Strategy and Action Plan (2010–2023);
- Turkish Industrial Strategy Document (2011–2014);
- National Disaster Management Strategy Document;
- National Disaster Response Plan.

The National Basin Management Strategy (NBMS) of Turkey, as led by the abolished MoFWA, is prepared to inform stakeholders of longer-term investment programs in watershed rehabilitation and water management. It also serves to ensure that such investments meet key objectives, including livelihood support and income generation, conservation and sustainability of natural resources, reduced vulnerability to climate change, and fiscal efficiency. NBMS is also consistent with the EU environmental and water management standards, which is a critical component of a strong integrated natural resource management policy framework. It also identifies a strategy that prioritizes the needs of the country and strengthens Turkey's sustainable development agenda.

River Basin Protection Action Plans, for all 25 river basins across Turkey, were completed through a protocol between the General Directorate of Water Management (GDWM) and Marmara Research Centre of the Scientific and Technological Research Council of Turkey (TUBITAK- MAM). These plans include the identification of:

- existing point and non-point pollution sources;
- existing condition of treatment facilities;
- existing water quality classifications as per Water Pollution Control Regulations of Turkey; and
- proposed infrastructure investment program to improve water quality standards in respective basins.

Both the European Union and Turkey adopt the basin scale management approach for water resources. Basin scale management, which is the basic principle of the EU Water Framework Directive (WFD), enables the protection and the sustainable use of water resources (RIBAMAP 2018). Turkey, as a candidate country to EU, is preparing River Basin Management Plans (RBMPs) in concordance with WFD for 25 river basins to achieve their "good status" by 2036 by implementing the required measures.

13.2 Basin Management in Turkey

13.2.1 Current Situation

Turkey is divided into 25 hydrological basins (Fig. 13.1), and the average annual flow from these basins is 181 billon m^3. According to the data of the Directorate General for State Hydraulic Works, about one third of these flows belong to the Euphrates-Tigris Basin, which is located to the east of the country. In terms of drainage area, the Euphrates-Tigris Basin is followed by Kizilirmak and Sakarya basins. In addition, in terms of average annual flow rates, the Euphrates-Tigris again ranks the first, followed by the Eastern Mediterranean, the Eastern Black Sea, and Antalya Basins (Table 13.1).

It is essential that the water potential of a basin is primarily assessed within the basin. However, the rates and the temporal distributions of rainfall in Turkey vary highly from region to region. While the annual rainfall is 2500 mm in the Eastern Black Sea Region, it decreases down to 320 mm per year in the Central Anatolian Region, especially around Konya. Where the rainfall is low, droughts occur and they affect almost every sector. Among the most common consequences of droughts are the slowdown in regional growth, decreases in farmers' incomes, difficulties in supplying basic nutrients, serious losses in industries where agricultural production is directly connected, and consequently unemployment. To eliminate such adverse effects, it is necessary to invest in water resources, to be frugal in the use of existing

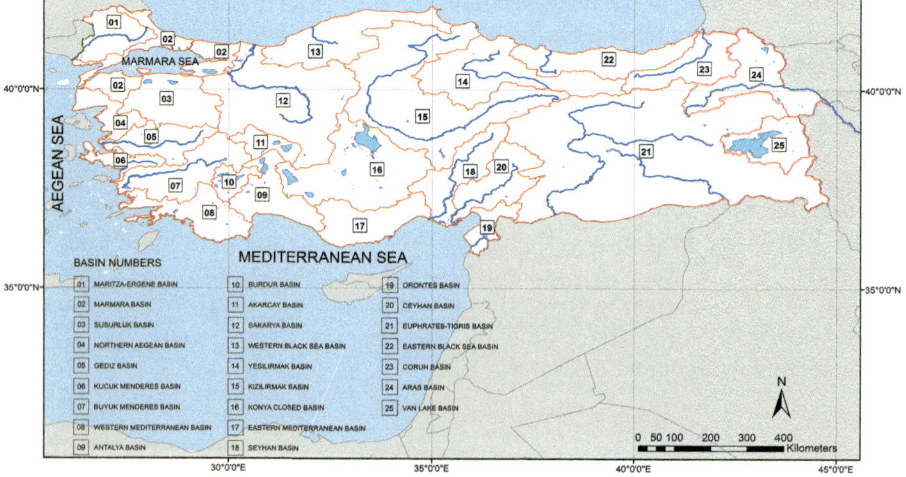

Fig. 13.1 River basins in Turkey

Table 13.1 General information about the river basins in Turkey

Names of river basins	Precipitation area (km^2)	Precipitation area %	Annual average flow (km^3)	Annual average flow (%)	Average annual yield (l/s/km^2)
(01) Meric-Ergene Basin	14,560	1.9	1.842	1.02	4.0
(02) Marmara Basin	24,100	3.1	7.540	4.17	9.9
(03) Susurluk Basin	22,399	2.9	4.226	2.34	6.0
(04) Northern Aegean Basin	10,003	1.3	1501	0.83	4.8
(05) Gediz Basin	18,000	2.3	1.545	0.85	2.7
(06) Kucuk Menderes Basin	6907	0.9	0.528	0.29	2.4
(07) Buyuk Menderes Basin	24,976	3.2	2.969	1.64	3.8
(08) Western Mediterranean Basin	20,953	2.7	6.969	3.85	10.5
(09) Antalya Basin	19,577	2.5	13.083	7.24	21.2
(10) Burdur Lake Basin	6374	0.8	0.256	0.14	1.3
(11) Akarcay Basin	7605	1.0	0.326	0.18	1.4
(12) Sakarya Basin	58,160	7.5	5.158	2.85	2.8
(13) Western Black Sea Basin	29,598	3.8	9.914	5.48	10.6
(14) Yesilirmak Basin	36,114	4.6	6.582	3.64	5.8
(15) Kizilirmak Basin	78,180	10.0	6.120	3.39	2.5
(16) Konya Closed Basin	53,850	6.9	2.647	1.46	1.6
(17) Eastern Mediterranean Basin	22,048	2.8	8.240	4.56	11.8
(18) Seyhan Basin	20,450	2.6	6.786	3.75	10.5
(19) Asi Basin	7796	1.0	1.813	1.00	7.4
(20) Ceyhan Basin	21,982	2.8	7.372	4.08	10.6
(21) Firat-Dicle Basin	184,918	23.7	55.419	30.65	9.5
(22) Eastern Black Sea Basin	24,077	3.1	16.461	9.11	21.7
(23) Coruh Basin	19,872	2.6	7.047	3.90	11.2
(24) Aras Basin	27,548	3.5	4.182	2.31	4.8
(25) Van Lake Basin	19,405	2.5	2.263	1.25	3.7
Total	**779,452**	**100.0**	**180.789**	**100.00**	**7.3**

resources, and to transfer water between basins only in compulsory situations. Thus, the ecological, social, and demographic conditions of Turkish river basins and the use of basin resources may show significant differences among the basins, depending on the geographical region they are in. Also, the horizontal and vertical distributions of the basin drainage areas are other important factors differentiating rivers basins and their employment in Turkey.

In recent years, with the contribution of scientific institutions and non-governmental organizations, awareness and support have increased in society on the importance of natural resources and the values provided by sustainable management of their assets. Furthermore, programs and practices have been significantly increased for rehabilitation of damaged areas, afforestation, soil conservation, and protection of biodiversity.

The precipitation distribution of Turkey shows great differences among seasons and regions. Water demands in some river basins seem to have exceeded the existing water potential. Besides water quantity, there are also large differences country-wide in terms of water quality. In the western part of the country where high population and intensive industrialization exist, the status of water quality is worse than the other basins.

According to international standards, the current technically exploitable and economically feasible amount of renewable water is approximately 112 billion m^3 per year (DSI 2018) and 1385 m^3 per person per year according to 2017 population (TUIK 2018). Thus, Turkey is now regarded as a country facing "water stress". Most of Turkey's land is located in the semi-arid climate zone, and in some regions, rainfall is limited to 5 or 6 months per year. Considering also the impacts of climate change, water management has become a very important issue for Turkey.

The increase of water demand due to rapid population growth, the lack of appropriate water resource availability, over-exploitation of water resources, resulting from increasing industrial and agricultural activities, decreases in groundwater resources, and problems caused by water pollution increase the importance of basin-scale water resources management.

In view of the above problems, significant steps have been taken towards studies on water management for sustainable development of water resources in terms of water quality/quantity and socio-economic development. As a candidate country to the European Union, Turkey has begun to align its legislation with the European Union legislation. The increasing diversity of stresses on water resources has necessitated that the management of river basins is realized by an integrated approach.

Following from the above, Turkey is re-developing its water management policy, taking into account her own needs and international standards. Along this line, Basin Protection Action Plans (BPAPs) were prepared between 2009–2013 by the abolished MoFWA, which covered the implementation of actions for urban and industrial water management, urban wastewater management, and control of diffuse pollution. Implementation of these actions have been followed up since 2013 by the former MoFWA, and now it is under the responsibility of Ministry of Agriculture and Forestry (MoAF). On the other hand, preparation of River Basin Management Plans for four basins (Konya Closed, Susurluk, Buyuk Menderes, Meric-Ergene) in accordance with the EU Water Framework Directive was started by GDWM under the abolished Ministry of Forestry and Water Affairs in 2014.The studies were developed within the context of an EU project named "Technical Assistance for the Conversion of River Basin Action Plans into River Basin Management Plans" (RIBAMAP 2014). These four plans are completed but not yet been put into practice. It is expected that all BPAPs will be converted into River Basin Management Plans by 2023. GDWM has also started to prepare Flood Basin Management Plans, Drought Basin Management Plans, and Sectoral Water Allocation Plans since 2012. On the other hand, master plans for each of the 25 basins, used in the preparation of RBMPs, have been prepared by the General Directorate of State Hydraulic Works (DSI).

Management and protection of water resources, as well as provision of water services to the users, are the tasks of the State. These activities are carried out by a number of public institutions and establishments in Turkey.

Water management studies in the above plans cover the following subjects:

- management of water sources (determination of their status, allocation, efficiency of water resources systems, effects of climate change, etc.)
- water use (drinking, domestic-commercial-industrial demands, irrigation, etc.)
- planning of efficient water resources systems (supply and distribution, energy-focused water structures, etc.)
- sustainability of the current and future allocation of water resources
- monitoring of water quantity and quality
- preservation of the quality of water bodies
- minimizing the environmental impact of water use on the natural environment.

There are various institutions in Turkey assigned with tasks and responsibilities regarding water management. These institutions act in accordance with the duties entitled to themselves via relevant legislation. As there are significant overlaps in the tasks and responsibilities of these institutions, water management becomes very complicated. The following sub-sections summarize these institutions, their tasks, and responsibilities.

13.2.2 Competent Authorities of Water Management in Turkey

In conjunction with the transition to a new government system in Turkey, some ministries were combined in order to reduce the number of ministries. In this context, the Ministry of Agriculture and Forestry is established through the Presidential Decree dated 10th July, 2018 by merging the former Ministry of Forestry and Water Affairs and the former Ministry of Food, Agriculture and Livestock. The names of the competent authorities related to water management issues directly or indirectly are the following:

Public Institutions and Organizations

- Ministry of Agriculture and Forestry: Directorate General for Combating Desertification and Erosion; Directorate General for Forestry; DSI; GDWM; Directorate General for Nature Conservation and National Parks; Directorate General for Meteorology; Directorate of Information Technologies; Strategy Development Directorate; Turkish Water Institute. Directorate General for Agricultural Reform; Directorate General for Crop Production; Directorate General for Agricultural Researches and Policies, Directorate General for Fisheries and Aquaculture, Directorate of Educational Publications and Publications Department, Department of GIS.

- Ministry of Environment and Urbanization: Directorate General for Spatial Planning; Directorate General for Environmental Impact Assessment, Permits and Inspection; Directorate General for Environmental Management; Directorate General for Preservation of Natural Heritage; Directorate General for Bank of Provinces (ILBANK); Directorate General for Infrastructure Services; Directorate General for Local Administrations.
- Ministry of Energy and Natural Resources: Directorate General for Mining Operations,
- Ministry of Culture and Tourism,
- Ministry of National Education and Ministry of Health,
- Ministry of Treasury and Finance,
- Ministry of Development,
- Ministry of Interior: Local Administrations: Governorates, District Governorships, Special Provincial Administrations, Municipalities, other units
- Regional Development Administration,
- Disaster and Emergency Management Authority (AFAD),
- Development Agencies,

Other Stakeholders

Participation of Non-governmental Organizations (NGOs), universities, private sectors, etc. to basin management studies is important for exchange of information and expertise. The following stakeholders also play a role in the management procedure:

- NGOs related to land and water resources, biodiversity and rural development, associations, etc.),
- professional organizations,
- irrigation unions,
- rural communities living in the basins,
- urban communities,
- scientific and educational institutions (Scientific and Technical Research Council of Turkey-TUBITAK, Universities, Research Institutes, etc.),
- related private sector institutions and organizations,

Water service providers are also stakeholders in the water sector. The main water service providers in Turkey are:

- General Directorate and Regional Directorates of State Hydraulic Works (DSI),
- General Directorate for the Bank of Provinces (ILBANK),
- Water and Sewerage Administrations of Metropolitan Municipalities (There are 30 Metropolitan Municipalities in Turkey),
- Water and Sewerage Directorates of Municipalities,
- Special Provincial Administrations.

Coordination and cooperation among the above mentioned institutions are necessary for preparation of river basin management plans, realization of water related investments, and consideration of other legal, administrative, and technical issues.

13.2.3 *Institutional Organization of Basin Management in Turkey*

A number of governmental and non-governmental organizations have direct and indirect interest in the development and conservation of water resources in Turkey (Alpaslan et al. 2007) The Turkish administrative system related water resources management has three levels: the national, the provincial, and the local level (Kibaroglu et al. 2011). The organization of basin management in Turkey comprises Water Management Coordination Committee and Basin Management Central Committee at the national level; 26 Basin Management Committees and 81 Provincial Water Management Coordination Committees at the basin and provincial basin level (Fig. 13.2). A Communiqué on the Establishment, Duties and Work Procedures and Principles of Basin Management Committees was also issued in the Official Gazette dated 20.05.2015 and numbered 29361. The purpose of this Communiqué is to regulate the necessary procedures and principles related to the establishment and operation of the Basin Management Central Committee, the Basin Management Committees, and the Provincial Water Management Coordination Committees. It serves to ensure coordination among institutions and monitor the implementation of basin protection action plans, plans for basin, flood and drought management, sectoral water allocation, and basin protection plans for drinking water basins.

The By-law on the Preparation, Implementation and Follow-up of Basin Management Plans, published in the Official Gazette no 30224, dated the 28th of October 2017, also refers to the four tier central and basin management organization, which covers the tasks and responsibilities of the committees established.

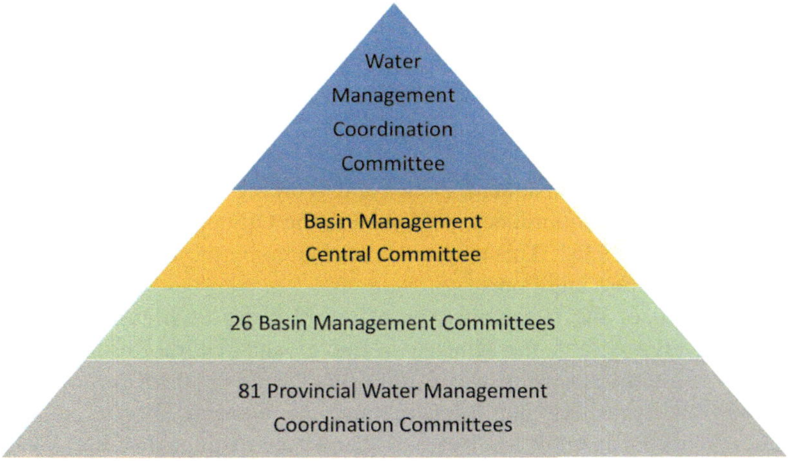

Fig. 13.2 Institutional organization of basin management in Turkey

Structure and responsibilities of basin management organizations are as follows:

13.2.3.1 Water Management Coordination Committee (WMCC)

Water Management Coordination Committee (WMCC) was established on the basis of the circular issued by the Prime Minister's Office numbered 2012/17 (Official Gazette dated 20 March 2012, No: 28239). The Committee serves for coordination and cooperation on water issues, using a holistic approach developed by the supreme level participation of related institutions. The WMCC is chaired by the Ministry of Agriculture and Forestry.

The Committee consists of the minister or the deputy minister of MoAF and the highest level representatives of MoEU, the Ministry of Interior, the Ministry of Foreign Affairs, the Ministry of Health, the Ministry of Industry, the Ministry of Energy and Natural Resources, the Ministry of Culture and Tourism, the Ministry of Development, and the relevant general directorates of MoAF.

The tasks and responsibilities of WMCC are:

- to determine the necessary measures for protection of water resources, using an integrated river basin management approach;
- to ensure inter-sectoral coordination and cooperation among institutions for efficient water management and to enhance water investments;
- to develop strategies, plans and policies for achieving the objectives mentioned in relevant national and international documents;
- to evaluate the issues that need to be covered by public institutions and organizations in the context of basin management plans;
- to ensure coordination and cooperation between relevant institutions.

13.2.3.2 Basin Management Central Committee (BMCC)

The Basin Management Central Committee (BMCC) is established by the Communiqué on the Establishment, Duties and Working Principles and Procedures of Basin Management Committees, published in the Official Gazette dated 20 May 2015 and numbered 29361. This committee is chaired by the Undersecretary of the Ministry of Agriculture and Forestry.

It is composed of the representatives of the Ministry of Industry, MoEU, the Ministry of Foreign Affairs, the Ministry of Energy and Natural Resources, MoAF, the Ministry of Interior, the Ministry of Culture and Tourism, the Ministry of Health, the Ministry of Transport and Infrastructure, the General Directorate of State Hydraulics Work-DSI, GDWM, Head of the Water Institute (SUEN), ILBANK (Iller Bankasi that supports municipal investments), and the Head of AFAD (Directorate of Hazards and Emergency Management).

The tasks and responsibilities of the BMCC are as follows:

- following-up and enhancing practices to ensure that short, medium and long term applications specified in basin protection action plans are implemented;
- ensuring coordination among institutions for the preparation and implementation of basin management plans and their follow-up;
- ensuring coordination among institutions for the preparation of flood and drought management plans and following up the implementation of plans upon their completion;
- ensuring coordination among institutions within the scope of the National Basin Management Strategy;
- coordinating and assessing developments in the special provisions determined for drinking water basins;
- discussing the issues raised by the Basin Management Committees and developing solutions;
- forwarding the non-resolved issues to the WMCC;
- ensuring the implementation of decisions taken by the WMCC at basin scale.

13.2.3.3 Basin Management Committees (BMCs)

The basin management committees had formerly been established with the Communiqué on Basin Management Committees (BMCs) on 18 June 2013 in the Official Gazette no 28681. The establishment of Basin Management Committees in 25 basins, in which all stakeholders participate, is completed. The Communiqué published on 18 June 2013 was repealed with another Communiqué on the Establishment, Duties and Working Principles and Procedures of Basin Management Committees. This Communique was published in the Official Gazette dated 20 May 2015 and numbered 29361. The BMCs formed prior to the effective date of this communiqué have been restructured in accordance with the provisions of the new communiqué.

The members of the BMCs are governors or deputy governors of other provinces, general directors of water and sewerage administrations in metropolitan municipalities, mayors of municipalities, representative of GDWM, representative of the Regional Directorate of DSI in the responsible coordinating city, representative of the Ministry of Foreign Affairs and SUEN in case of transboundary basins, as well as representatives of universities, organized industrial zones, and NGOs. The committees meet under the chairmanship of the Coordinator Governor.

Their tasks and responsibilities are the following:

- supporting basin scale studies and flood/drought management plans;
- monitoring and evaluating the implementation of the basin protection action plans, basin, flood and drought management plans and to inform the relevant institutions and organizations of the decisions taken by the committee;

- evaluating the results of the audits and sanctions prepared by the relevant institutions or organizations, concerning the reports formed as a result of the meeting of Provincial Water Management Coordination Committees;
- submission of the evaluation reports prepared by Provincial Water Management Coordination Committees;
- following the studies for the protection of drinking and utility water resources and ensuring that the required special provisions are applied;
- ensuring accession to the knowledge and active participation of people in the process of preparing, reviewing and updating of basin, flood and drought management plans;
- recording monitoring results related to water quality and quantity in a common database to be formed by GDWM, sharing this information with relevant basin units, and evaluating basin, flood and drought management plans on the basis of processed and reported monitoring results;
- ensuring the access of stakeholders to information on the preparation, review and updating of basin management plans under the coordination of the BMCC.

13.2.3.4 Provincial Water Management Coordination Committees (PWMCCs)

The Provincial Water Management Coordination Committees (PWMCCs) are composed of the Director General of Water and Sewerage Administration in Metropolitan Municipalities, Mayor of Municipalities, Chairperson of Provincial General Council or General Secretary of Special Provincial Administration, the highest level representatives of MoAF, the Ministry of Energy and Natural Resources, the General Directorate of DSI, the General Directorate of Forestry, the General Directorate of Meteorology, the General Directorate of Highways, high level representatives of the Development Agency and ILBANK in the province, the Director of Provincial Environment and Urbanization, the Director of Provincial Agriculture and Forestry, the Director of Provincial Industry, the Director of Provincial Culture and Tourism, the Director of Provincial Public Health, the Director of Provincial Disaster and Emergency, Chairman of the Chamber of Industry and Commerce, and representatives of Irrigation Unions. This committee gathers under the chairmanship of the Governor or the Deputy Governor of the Province.

The duties of the committee are as follows:

- providing necessary contributions for the basin, flood and drought management plans to be prepared by the Ministry of Agriculture and Forestry,
- ensuring the application of river basin protection action plans, and basin, flood and drought management plans on a provincial basis,
- following the studies for the protection of drinking and utility water resources and ensuring the implementation of the prepared special provisions at provincial scale.

13.2.4 The Importance of EU Water Framework Directive for Basin Management

Due to the increasing rates of urbanization and industrialization in Turkey, the demand for effective environmental management of river basins and environmental infrastructure investments will continue to rise. Recent data indicate that approximately 92.5% of Turkey's population lives in urban areas of cities and municipalities (TUIK 2018). Investments to implement the EU environmental acquis are also expected to place an increasing burden on Turkey's public sector finances over the next two decades. Moreover, the demand is increasing for quality urban dwellings, more environmentally sustainable urban planning, and more water resources of good quality. In order to cope with the varying levels of local and regional changes and pressures, Turkey's efforts for harmonization with the EU WFD mainly consist of pilot projects and changes in the current legislation, focusing on different aspects of water quality and quantity.

The Water Framework Directive (WFD) of the European Union (EU) is one of the most comprehensive water policy documents of the EU. In October 2000, the "Directive 2000/60/EC of the European Parliament and of the Council of 23 October 2000, establishing a framework for Community action in the field of water policy" (Water Framework Directive or WFD), was adopted and came into force in December 2000. The introduction of the EU Water Framework Directive 2000/60/EC (WFD) aimed to initiate a new era for European water management, focusing on understanding and integrating all aspects of the water environment for effective and sustainable water management (Teodosiu et al. 2003). The purpose of the directive is to establish a framework for the protection of inland surface waters (rivers and lakes), transitional waters (estuaries), coastal waters, and groundwater. It will ensure that all aquatic ecosystems with regard to their water demands, terrestrial ecosystems, and wetlands meet the "good status" condition by 2015, which is a strict deadline. Member States that avail themselves of an extension beyond 2015 are required to achieve all WFD environmental objectives by the end of the second and third management cycles, which extend from 2015 to 2021 and 2021 to 2027, respectively (European Commission 2012).

Many water policy actions were designed by the European Commission to help nations' waters reach a good ecological status. Turkey, who has the status of an accession country to the EU, has to adopt this comprehensive policy, as well as other EU directives directly relevant to basin management.

13.3 River Basin Management Projects in Turkey

As a candidate country to the European Union, Turkey has begun to align its legislation with the European Union legislation. Due to the increasing diversity of stresses on water resources, the management of river basins must be realized by an integrated approach, considering both water quantity and quality in a holistic manner. Following these requirements, Turkey is re-developing its water management policy, considering both her own needs and international standards. Effective management of water resources is essential for the realization of sustainable development because these resources are essential for life. Yet, they decrease day by day and face the threat of pollution. In this context, the preparation of River Basin Management Plans in accordance with the WFD was started in 2014 to be implemented in the future years.

The first step for river basin management planning in Turkey started with the "Draft Buyuk Menderes River Basin Management Plan", which was prepared as an output of the Twinning Project "Capacity Building Support to the Water Sector in Turkey" (2007–2010) between Turkey, Slovakia, the United Kingdom, and the Netherlands (EMWIS 2018). The Draft National Implementation Plan (DNIP) for the Water Framework Directive was prepared by this Twinning Project. It was intended via this project to provide a basis for decision-making in water management. The Draft WFD National Implementation Plan, prepared in 2010, brings together the work required to complete the transposition of the Water Framework Directive and the relevant environmental *acquis*. The Plan fills the gap in the implementation of the WFD in Turkey through the realization of required actions. The DNIP addresses planning and implementation of measures for all the water bodies in Turkey.

Basin Protection Action Plans (BPAPs) were prepared for all 25 basins between 2009 and 2013. BPAPs are considered as a predecessor for the WFD RBMPs (River Basin Management Plans in accordance with Water Framework Directive-WFD). These serve as a good basis for RBMPs, giving a characterization of the basin, analyzing the principal sources of pollution (point and diffuse), and proposing required measures. However, the BPAPs are not prepared solely by considering the WFD and its requirements for the River Basin Management Plans, so there is a gap between BPAPs and RBMPs. Preparation of BPAPs for 11 basins (Marmara, Susurluk, Kuzey Ege, Kucuk Menderes, Buyuk Menderes, Burdur, Konya Kapali, Ceyhan, Seyhan, Kizilirmak, and Yesilirmak) were completed in 2010. River Basin Protection Action Plans for 14 Basins (Antalya, Dogu Akdeniz, Bati Karadeniz, Firat -Dicle, Dogu Karadeniz, Bati Akdeniz, Coruh, Aras, Asi, Meric-Ergene, Van, Akarcay, Gediz, and Sakarya) were started in December 2011 and are now completed.

The first River Basin Management Plan of Turkey was initiated in 2014 within the scope of the Technical Assistance Project "Conversion of River Basin Protection Action Plans to River Basin Management Plans Project" (RIBAMAP 2014), which includes four basins (Buyuk Menderes, Meric – Ergene, Susurluk, and Konya Closed Basins). These plans are completed in 2018. By the year 2023, 25 Basin Protection Action Plans will be converted to River Basin Management Plans (RBMPs), which are expected to achieve "good status" in Turkish water resources by that year.

The RBMPs prepared under the supervision of the abolished Ministry of Forestry and Water Affairs are to be approved by the Water Management Coordination

Committee (WMCC). Program of measures implemented by the competent authorities, which are central institutions and authorities in relevant basins, will be followed up by the GDWM. Basin Management Committees and Provincial Water Management Coordination Committees will play a crucial role for the follow-up of measures and their reporting to GDWM on the basis of the By-law on the Preparation, Implementation and Follow-up of Basin Management Plans, published in the Official Gazette no 30224 dated 28 October, 2017.

Important projects related to water management were carried out in Turkey in the past decade. More detailed information on some important projects related to basin management projects in Turkey is given in the following subsections.

13.3.1 Basin Master Plans

Basin master plans prepared by General Directorate of DSI are an important basis for all basin management studies. Comprising such elements as data collection and evaluation, investigation, and technical, economic and environmental studies, these master plans, which will contribute to social and economic development in a basin, are prepared by evaluating water and soil resources potential with a holistic approach.

Purpose of the basin master plan studies are as follows;

- investigation of basin water potential, water quality, soil resources, water and land use and water needs,
- determination of the basin water budget by also considering water rights,
- identification of sectoral water needs,
- determination of water demand and the associated sectoral priorities, considering the available water potential,
- examination of technical, economic and environmental feasibility of development projects
- recommendation of optimum operation methods and policies for the planned facilities, that will enable the rational use of resources for various purposes

Basin master plans in 23 basins are completed, and the studies in the remaining two basins will be completed by the end of 2018.

13.3.2 Basin Protection Action Plans (BPAPs)

A picture of Turkish river basins was provided with the previously prepared BPAPs, through which urgent measures were developed and put into practice for areas under pressure. In BPAPs, all pollution sources in the basins were examined, and the measures to be taken at basin scale were determined to improve water quality in the basins. At present, the implementation of BPAPs, completed in 25 basins in 2013, is being executed. The implementation phase covers 15 actions, such as the prevention of water pollution, installation of water structures, and afforestation. These actions are to be completed by 2023. As an example of the results of such actions, the number of urban wastewater treatment plants increased from 503 as of 2013 to 859 in 2018; 437 plants are still to be established. The number of solid waste regular storage facilities increased from 64 as of 2013 to 88 in 2018, with 38 yet to be built. At present, conversion of BPAPs into RBMPs has been continued by the Ministry of Agriculture and Forestry.

13.3.3 River Basin Management Plans (RBMPs)

River Basin Management Plans are important for conservation of water resources and sustainable development. These plans are a requirement of the EU WFD, as well as one of the closing criteria of the Environment and Climate Change Chapter that Turkey has to abide with.

In RBMPs, pressures and effects on water bodies are determined, monitoring activities are carried out to determine the quantity and quality status of the water, followed by detailed economic analyses and modeling activities. Finally, the necessary measures, costs and institutions to realize these measures are determined for the purpose of achieving good water status.

RBMPs, which are the highest levels of basin scale planning, are completed in four basins (Buyuk Menderes, Susurluk, Konya and Meric Ergene) in the context of RIBAMAP (2014) project. They are continued in seven basins (Gediz, Kucuk Menderes, Burdur, Northern Aegean, Western Mediterranean, Akarcay and Yesilirmak). RBMPs of 25 basins are planned to be completed by until 2023.

According to conclusions of the Technical Assistance Project "Conversion of River Basin Protection Action Plans to River Basin Management Plans Project" (RIBAMAP), the current situation of the surface water bodies in terms of water quality and quantity in four basins is identified as in the following (RIBAMAP 2018):

- In Meric-Ergene Basin, 4 out of the 120 surface water bodies (3.3%) are in good status.
- In the Buyuk Menderes Basin, 23 out of the 134 surface water bodies (16.9%), are in good status.
- In Konya Closed Basin, 12 out of the 92 surface water bodies (13%) are in good status.
- In the Susurluk Basin, 16 out of the 156 surface water bodies (10.2%), are in good status.

For the groundwater bodies, the current situation is as in the following:

- In Meric-Ergene Basin, it has been determined that none of the 12 groundwater bodies are in good water condition at the present.

- In the Buyuk Menderes Basin, 17 out of the 38 groundwater bodies (44.7%), are in good status.
- In Konya Closed Basin, 3 out of the 18 groundwater bodies (16.7%), are in good status.
- In the Susurluk Basin, 7 out of the 22 groundwater bodies (31.8%), are in good status.

The results of the RIBAMAP project show that many water bodies in four basins are not in good status. In the project, a total of 4483 measures have been identified in the four basins to ensure that all water bodies have access to good water status and that those in good status are protected. The total investment cost of these measures for four basins is around 1.5 billion €.

13.3.4 Flood Management Plans

After the preliminary assessment of flood risks, flood management plans are prepared by developing flood hazard and flood risk maps at each basin in accordance with the EU flood directive. The plans are completed in four basins by now and are continued in 19 basins. The flood management plans for the remaining basins will be finished by 2023.

13.3.5 Drought Management Plans

Drought management plans are prepared by analyzing drought indices, water budget studies, drought risk maps, and sectoral vulnerability analyses in each basin. In these plans, the transition from the crisis management approach to the risk management approach is adopted. The plans are finished in five basins by now and are continued in ten basins. The remaining basins will have their plans completed by 2023.

13.3.6 Basin Based Sectoral Water Allocation Plans

The current situation and future sectoral water demands (drinking, environmental need, agriculture, trade, energy, tourism, mining, and recreation etc.) are determined for water potential projections at basin scale (for the climatic conditions of normal, mild, medium, severe and very severe arid periods). Next, sectoral water allocation plans are being prepared by taking into account sectoral priorities and socio-economic factors. Water allocation plans are finished in only one basin (Seyhan) by today and are ongoing in four basins (Akarcay, Gediz, Konya, K. Menderes). They are to be completed by 2023 for the remaining basins.

13.3.7 National Water Information System (NWIS)

The purpose of the National Water Information System (NWIS) is to collect data related to water resources of the country through a unique system to be shared with relevant institutions and organizations. Through NWIS, it is aimed to provide support for the management of water related data on a GIS basis and to support the formation of national water policy. The system was established by the end of 2017, and efforts for developing its functionality and sustainability in disseminating sound information are currently continued.

13.4 Challenges in Basin Management in Turkey

Turkey has begun the process of alignment with the WFD but faces various challenges related to its implementation, including the requirement to fund a massive national program of water supply and sanitation, covering not only cities and towns, but also rural areas. The implementation process also raises a number of technical challenges in Turkey, having one of the largest surface areas among the other EU member countries (Moroglu and Yazgan 2008). The stream network forming river basins in Turkey have a more complicated structure than in other EU countries, and this makes basin management more difficult. Taking into consideration the cultural and socio-economic status of Turkey, a unique system of watershed management is foreseen and efforts to this end are continued. The goal is not only to achieve public health targets but also to ensure ecological sustainability. In addition, as WFD includes important recommendations related to transboundary waters, Turkey will need to review her policies and management plans in such basins.

The tasks and responsibilities of Turkish authorities do not have clear-cut boundaries. Often, different institutions fail to acknowledge that they try to run similar projects. Therefore, especially water service providers, include identical plans in their investment programs for the same locations. For example, the population criteria in Law No: 1053 on the Supply of Drinking, Utility and Industrial Water to the Municipal Settlements (Official Gazette July 16, 1968, No: 12951, amended by Law no 5625 dated 18 April 2007) was revoked by the Law No: 5625, according to which the General Directorate of State Hydraulic Works is authorized to provide water supply to all cities with municipalities. ILBANK is also authorized on the same matter and this leads to a conflict. However, municipalities demand the construction of facilities by DSI due to the fact that the reimbursement conditions of DSI's expenses are better than ILBANK.

In Turkey, the increase in the number of metropolitan municipalities to 30 within a period of 35 years is in line with the trend all over the world. Provision of water and waste water services at required levels has evolved into a complex business. Accordingly, institutional, human and economic capacities of water service providers are becoming more and more significant, requiring larger institutions in to be formed. Thus, it is recommended that province-wide water and sewerage administrations are also established for the remaining 51 provinces similar to those of metropolitan cities.

On the other hand, horizontal and vertical coordination networks should be established between ministries. Conflicts between institutions can be prevented by the WMCC, the BMCC, the BMCs and the PWMCCs, by ensuring effective participation of water service providers in the management process.

In Turkey, the problems in water resource management concern administrative boundaries, institutional structures, databases, monitoring and control points, laws, sanctions and policies. These problems can be summarized as:

- administrative activities cannot be carried out as required since administrative boundaries and basin boundaries (covering the natural drainage areas) do not overlap,
- lack of integration among water-related plans with other sectoral plans,
- authorities and responsibilities of existing institutions are not explicitly stated in legal statements,
- institutional coordination is poor,
- the main authority at local scale is not identified,
- there is no sufficient database on water resources and basins,
- lack of data prevents effective management and planning,
- lack of adequate data on water resources prevents the evaluation of management processes,
- water ecosystems are not well known,
- numerous laws and regulations create conflicts in jurisdiction,
- sanctions are not dissuasive,
- policies on water resources management are short term, whereas they have long-term implications.

Insufficient coordination among institutions is a major weakness in the realization of integrated river basin management in Turkey. Different organizations conduct their activities on water resources according to their own plans. This often leads to unnecessary duplications in basin management studies, causing waste of time and waste of money for the same water resource. To avoid duplications of plans, sufficient coordination should be ensured between institutions and municipalities.

It is also necessary to strengthen legal and institutional capacities for sustainable management of basins and to ensure coordination and cooperation among institutions and stakeholders. The following issues can be recommended to strengthen legal and institutional capacities:

- developing a basin classification system by compromising with relevant institutions, which will form the basis for basin studies, and demonstrating boundaries and areas of basin, sub-basins and micro-basins,
- developing basin management policies,

- monitoring of implementation and carrying out institutional arrangements at national and basin levels to ensure that the assessment is carried out in coordination with relevant institutions and authorized representatives of the stakeholders;
- prioritization of basins according to the pressures and impacts on water resources, thus, carrying out basin investments and activities by relevant institutions and organizations according to appropriate priorities;
- establishment of a "basin information system" by using geographical information system (GIS) to include information also on investments, related pressures, and impacts;
- establishment of effective database systems;
- water quality management of basins with the "polluter pays principle";
- effective participation of basin committees in basin management;
- production of long-term policies;
- definition of tasks and responsibilities of institutions at basin scale.

The current Draft Water Law should enter into force as soon as possible to ensure holistic management of water resources, to prevent forthcoming problems and to set the essential legal basis for all relevant water management issues.

For effective water management at basin scale, the current situation where the basin boundaries do not overlap with the administrative boundaries should be changed in a way to establish a single responsible administration for the whole basin. It should also be noted that the tasks regarding water and waste water should be undertaken by the same institution. The overlaps of tasks regarding monitoring should especially be eliminated. Detailed guidelines should be prepared for BMCs to prepare the RBMPs. Sufficient capacity building and training activities should be organized at basin scale in order to disseminate the issues regarding both the theory and practice of the WFD, which is an important basin management tool.

13.5 Concluding Remarks

As a European Union candidate country, Turkey has taken important steps in recent years towards implementing the EU Water Framework Directive. Institutional arrangements have been made for water basin management, preparation of basin-based plans has been started, and some of them have been completed and put into practice. Some concluding remarks can be pointed out as the required steps for effective basin management as in the following:

- holistic management of water resources at basin scale, planning, protection, rehabilitation, monitoring, supervision and implementation should be executed by one responsible institution wherever possible; otherwise coordination among institutions should be ensured by clear tasks and responsibilities enacted by legislation;

- effective participation of basin committees in management;
- production of long-term policies;
- definition of tasks and responsibilities of the institutions in the frame of basin boundaries.
- production of sufficient data at basin level;
- establishment of horizontal and vertical coordination networks between relevant ministries.
- Investigating alternative finance models to meet the investment costs of the measures identified in RBMPs to achieve good water status in basins.

References

Alpaslan AH, Atac A, Yesil N (2007) River Basin Management Plans in Turkey during the Accession Period to European Union, International River Basin Management Congress, DSI, 22–24 March 2007, Antalya

DSI (2018) http://en.dsi.gov.tr/land-water-resources. Accessed Sept 2018

EMWIS (2018) http://www.emwis.org/initiatives/fol060732/capacity-building-support-water-sector-turkey. Accessed Sept 2018

European Commission (2012) Commission staff working document, European overview (1/2) accompanying the document: "report from the commission to the European Parliament and the council on the implementation of the water framework directive (2000/60/EC) river basin management plans". COM (2012) 670 final

Kibaroglu A, Sheumann W, Kramer A (2011) Turkey's water policy, National Framework and International Cooperation, Springer, https://doi.org/10.1007/978-3-642-19636-2

Moroglu M, Yazgan M (2008) Implementation of EU water framework directive in Turkey. Desalination 226(1–3):271–278. https://doi.org/10.1016/j.desal.2007.01.245

RIBAMAP (2014) http://ribamap.ormansu.gov.tr/index. Accessed Sept 2018

RIBAMAP (2018) Final draft report of technical assistance project "Conversion of River Basin Protection Action Plans to River Basin Management Plans Project", GDWM

Rockström J, Steffen W, Noone K, Persson Å, Chapin FS, Lambin E, Lenton TM, Scheffer M, Folke C, Schellnhuber H, Nykvist B, De Wit CA, Hughes T, van der Leeuw S, Rodhe H, Sörlin S, Snyder PK, Costanza R, Svedin U, Falkenmark M, Karlberg L, Corell RW, Fabry VJ, Hansen J, Walker B, Liverman D, Richardson K, Crutzen P, Foley J (2009) Planetary boundaries: exploring the safe operating space for humanity. Ecol Soc 14(2):32. [online] URL: http://www.ecologyandsociety.org/vol14/iss2/art32/

Teodosiu C, Barjoveanu G, Teleman D (2003) Sustainable water resources management 1. River basin management and the EC water framework directive. Environ Eng Manage J (EEMJ) 2 (4):377–394

TUIK (2018) http://www.tuik.gov.tr/UstMenu.do?metod=temelist. Accessed Sept 2018

UN (2015) https://www.un.org/sustainabledevelopment/sustainable-development-goals/. Accessed Sept 2018

UNESCO-IHP (2014) Water in the post-2015 development agenda and sustainable development goals – discussion paper, 21th session of the intergovernmental Council of the International Hydrological Programme of UNESCO (18–20 June 2014, UNESCO Headquarters, Paris, France)

Zeitoun M (2011) The global web of national water security, (online) Available at: http://onlinelibrary.wiley.com/doi/10.1111/j.1758-5899.2011.00097.x/pd. Accessed May 2018

Chapter 14
Impacts of Climate Change on Precipitation Climatology and Variability in Turkey

Murat Turkes, M. Tufan Turp, Nazan An, Tugba Ozturk, and M. Levent Kurnaz

Abstract In this chapter, changes in seasonal precipitation climatology, extreme weather conditions, and aridity conditions of Turkey are evaluated for the period of 2021–2050 with respect to the reference period of 1971–2000 by using regional climate model simulations. Projections of future climate conditions are modeled by forcing Regional Climate Model, RegCM4.4 of the International Centre for Theoretical Physics (ICTP) with MPI-ESM-MR global climate model of the Max Planck Institute for Meteorology. The outputs of MPI-ESM-MR are used to generate 10 km resolution data by the double nesting method under both RCP4.5 and RCP8.5 emission scenarios. The seasonal time-scale performance of RegCM4.4 in reproducing the observed climatology over Turkey is tested by using the output of the global climate model. The projection results show a strong decrease in precipitation for almost all parts of the domain according to the output of the regional model. The intensity of drought conditions is projected to increase. According to the projection results, more arid conditions are expected in the region for the near future. Therefore, drier than present climate conditions are projected to occur more intensely over Turkey.

M. Turkes
Center for Climate Change and Policy Studies, Bogazici University, Istanbul, Turkey
e-mail: murat.turkes@boun.edu.tr

M. T. Turp (✉) · N. An
Center for Climate Change and Policy Studies, Bogazici University, Istanbul, Turkey

Institute of Environmental Sciences, Bogazici University, Istanbul, Turkey
e-mail: tufan.turp@boun.edu.tr; nazan.an@boun.edu.tr

T. Ozturk
Center for Climate Change and Policy Studies, Bogazici University, Istanbul, Turkey

Department of Physics, Isik University, Istanbul, Turkey
e-mail: tugba.ozturk@isikun.edu.tr

M. L. Kurnaz
Center for Climate Change and Policy Studies, Bogazici University, Istanbul, Turkey

Department of Physics, Bogazici University, Istanbul, Turkey
e-mail: levent.kurnaz@boun.edu.tr

N. B. Harmancioglu, D. Altinbilek (eds.), *Water Resources of Turkey*, World Water Resources 2, https://doi.org/10.1007/978-3-030-11729-0_14

Keywords Climate change impacts · Climate modeling · Aridity in Turkey · Precipitation extremes

14.1 Introduction

Water resources are gradually shrinking due to the decrease in precipitation and the increase in evapotranspiration that stem from increasing global temperatures. These dry conditions will limit the sustainability of regional water resources in the near, mid, and long term. Therefore, in this section, precipitation climatology in Turkey is analyzed in order to see especially the potential impact of climate change on total precipitation amounts, aridity conditions, and precipitation extremes.

For the future period of 2070–2099 with respect to 1960–1989, temperature increases and precipitation declines are predicted in the Mediterranean Region (Gibelin and Déqué 2003). The decrease in average precipitation is associated with a significant decrease in soil moisture and is expected to have an impact on water resources in the Mediterranean Basin (Gibelin and Déqué 2003). It has been observed that the cyclones throughout the Mediterranean region affect the climate and the weather conditions of the region. The cyclones cause changes in fundamental climate parameters such as temperature and precipitation, and extreme weather events such as storms and floods. Extreme weather events are unpredictable because the reasons causing these events are very diverse (Lionello et al. 2006a). Observations of precipitation between October and March in the Mediterranean indicate that precipitation has increased from the mid-nineteenth century to the mid-twentieth century and reached its maximum levels in the 1960s. It started decreasing from that date forward and declining rapidly towards the end of the twentieth century (Xoplaki et al. 2004). In the winter months, there is a positive trend of average precipitation in the northern Mediterranean, while a negative trend is observed in the southern part. In other seasons, a decrease is broadly foreseen, especially in the summer. However, in the case of extreme precipitation and arid periods, not only temporal shifts occur, but also the probability and frequency of their occurrence increase, leading to the possibility of floods and droughts (Gao et al. 2006). In the Mediterranean region, a decrease in winter precipitation and a temperature increase of 0.75 °C has been observed especially in the last century. In the same region, temperature increases of 4–7 °C and changes in the order of −40% to +20% in precipitation are expected in the next century (Lionello et al. 2006b). Annual maximum and minimum temperatures have been observed for the summer and winter seasons in the future periods of 2030–2059 and 2070–2099 with respect to the reference period of 1979–1999. The greatest increase in maximum temperature is seen in southern Europe. Average precipitation is decreasing in the Mediterranean Basin and southern Europe in winter, spring and summer seasons. In addition, an increase is observed in the precipitation extremes in autumn, winter and spring seasons throughout the region (Goubanova and Li 2007).

Except for the Northern Mediterranean (such as the Alps), there is a significant decrease in precipitation, and the highest increase in temperature occurs during the summer months. This increase between the years, especially in summer, also causes extremely high temperature events in summer (Giorgi and Lionello 2008). Another study examining the precipitation trends in the Mediterranean region between 1951 and 2000 indicates a significant decrease during the winter season between October and March. Besides, there is an increase in precipitation in the northern and western Mediterranean during the winter, and a decrease in precipitation in the spring and autumn. However, negative changes in all precipitation periods are noteworthy in the Southern and Eastern Mediterranean regions (Jacobeit et al. 2007). When the future precipitation regime for the winter is examined by dividing the Mediterranean Region into north and south, an increase in precipitation in the northern parts of the Mediterranean Basin and a decrease in precipitation in the southern part are anticipated. These results indicate that climate change is not distributed homogeneously in the Mediterranean (Lionello and Giorgi 2007). Average annual temperature increases in Europe are projected to be higher than the world average, and it is expected that the maximum seasonal warming will be in northern Europe in winter and in the Mediterranean region in summer. It is very likely that the majority of the Mediterranean will have a decrease in precipitation and wet days. Most probably, the precipitation and evaporation trends in the Mediterranean Region indicate that hydrological droughts will increase especially in the summer months. Besides, it is highly likely that there will be a decline in the number of snowy days and snowfall in Europe in general (Calbó 2009).

Under the A2 SRES scenario in the years 2071–2100 compared to the 1971–2000 period, a precipitation decrease of 1.75 mm/day in the winter in the southern part of Turkey is expected. Similar decreases are also foreseen for spring and summer seasons as well. The B1 SRES scenario, which is the optimistic case like RCP4.5, also has trends similar to the A2 scenario, which is the pessimistic case like RCP8.5, but it is predicted that these trends will be less severe (Altinsoy et al. 2011).

In particular, a significant increase in the minimum nightly temperatures in Turkey, especially in the Mediterranean Basin in the spring and summer seasons, are observed. When the period 1901–2005 is examined, it can be seen that the amount of precipitation increased by 1% in the top half of the northern hemisphere, while it decreased by 3% in the Mediterranean Basin in 10 years. In general, it has been determined that there is a remarkable decrease in the number of frosty and snowy days, especially after 1990s, and an increase in the number of warm days and nights, with the lowest nighttime and the highest daytime temperatures in the Eastern Mediterranean. Long-term declining trends in precipitation since 1970s and arid conditions have been particularly influential in most of the Mediterranean region. Turkey is one of the regions most affected by this drought. In the future projection of 2071–2100, an increase in precipitation is expected at 0.5–2 mm/day in the whole Mediterranean Basin in winter months and an increase between 2.5 and 5 °C in winter temperatures. A rise in summer temperatures of 1 °C is expected and further increase is predicted in the spring and autumn seasons (Turkes 2012a).

Between 1960 and 2010, it is observed that the precipitation trends decline in the Mediterranean Region. It is more likely to happen in maximum daily precipitation. According to the RCP4.5 and RCP8.5 scenarios, it is predicted that precipitation decreases in all seasons in the eastern Mediterranean region (Akcakaya et al. 2013). In the Mediterranean Region, based on the years 1970–2000 for the future period of 2070–2100, the increases in temperature are the highest in summer and the lowest in winter. Besides, a decrease in precipitation is foreseen in almost all seasons in the whole Mediterranean (Ozturk et al. 2015). According to the RCP4.5 scenario, precipitation decreases by 20% between 2041 and 2070 for the eastern Mediterranean, and depending on the RCP8.5 scenario, increases in the western Mediterranean during the rainy season and decreases in the eastern Mediterranean between 2016 and 2040. Reductions in precipitation in the Mediterranean are predicted between 2041–2070 and 2071–2099 during the summer (Demircan et al. 2017). Between the years 1979 and 2011, especially one of the most important causes of extreme precipitation in the south of Turkey is the upper air turbulence moving west to east and strong low pressure systems over the eastern Mediterranean basin. Extreme rainfall variability during the year occurs more frequently at the end of autumn and the beginning of winter. Extreme events usually last for 1 day, and there are no extreme weather events lasting 4 days (Lolis and Turkes 2017).

When precipitation periods in Turkey are analyzed, maximum precipitation is seen in December–January, while minimum precipitation is observed in July–August (Kadioglu et al. 1999). Considering the long-term annual average and monthly total precipitation in Turkey, a downward trend in precipitation is observed overall Turkey. Between 1° and 6 °C temperature rise and decline in precipitation except for winter months is expected in the 2016–2099 future period. A decrease in precipitation in autumn throughout Turkey is expected except for some local regions (Demircan et al. 2017). An increase in temperature between 5° and 7 °C for the summer in almost all of Turkey and a decrease of 40% in precipitation across the southern and western sections in the future period of 2071–2100 are predicted. Due to rising temperatures and decreasing precipitation, drought events are expected to increase in severity, frequency and duration. However, it is anticipated that, in the hot seasons, the amount of water in the rivers that are fed with melted snow will experience a decrease as a result of the increased temperature. Also, it is expected that the agriculture sector will be affected due to the decrease in water amount and increase in the number of arid days due to increasing temperature and decreasing precipitation (Sen et al. 2012; Sen 2013).

According to the past data of 1860–2005 and the 2005–2100 RCP4.5 scenario, the Mediterranean Region has been found to be warm and arid. Studies conducted with future projections have also shown that the drought and warming in the Mediterranean region will take place at a greater rate than in the past until the end of the twenty-first century (Mariotti et al. 2015). Furthermore, from the mid-twentieth century, the increase in average temperature in the entire Mediterranean Basin is greater than the increase in the global mean temperature, and a decrease in annual mean precipitation is observed. Since the 1960s, the intensity, duration, and number of heatwaves on the eastern Mediterranean have increased at least five times

(Ruti et al. 2016). Moreover, aerosols and greenhouse gases change the water cycle and affect the drought of the Mediterranean by affecting the cloud properties and the radiative balance in the Basin. Accordingly, from the early twentieth century, the precipitation in the Mediterranean tends to decrease (Tang et al. 2018). Gediz Basin, located in the western Mediterranean, is examined for the future period of 2017–2099, related with the 1985–2005 scenario; accordingly an increase of 0.8 °C of the annual mean temperature under RCP4.5 (optimistic scenario) and 2.0 °C under RCP8.5 (pessimistic scenario), and also a decrease of 35 mm/year of precipitation under RCP4.5 and a decrease of 90 mm/year under RCP8.5 are observed (Gorguner et al. 2017).

14.2 Data and Model Description

Regional climate model, RegCM, has been effectively used as a dynamical downscaling tool and applied to several domains (i.e. the Mediterranean, Africa, North America, Central America, South America, East Asia, Central Asia, South Asia, Europe) for regional climate change and climate variability studies over the last two decades (Almazroui 2012, 2016; Almazroui et al. 2016; Chen et al. 2003; Coppola et al. 2014; Gao et al. 2002; Giorgi et al. 2004a, b, 2012, 2014; Gu et al. 2012; Ozturk et al. 2011, 2012, 2017, 2018; Mariotti et al. 2014; Turp et al. 2014; Sylla et al. 2016). For purposes of this chapter, the regional climate change simulations are conducted using the version 4.4 of RegCM, which is a hydrostatic regional climate model developed by the Abdus Salam International Centre for Theoretical Physics (ICTP) (Giorgi et al. 2012; Pal et al. 2007). The dynamic structure of RegCM4.4 is the hydrostatic version of the Pennsylvania State University's National Atmospheric Research Center (National Center for Atmospheric Research (NCAR) of the Pennsylvania State University), which is called MM5 (the Mesoscale model) (Grell et al. 1994). SUBBATS (Giorgi et al. 2003a, b), which is the sub-grid scaled version of the previous scheme of BATS1E (Biosphere-Atmosphere Transfer Scheme) (Dickinson et al. 1993), is used for the surface. Community Land Model (CLM) version 3.5 is also included in the dynamic structure of the code as an option (Oleson et al. 2008; Tawfik and Steiner 2011). Radiative transfer in RegCM4.4 is modeled by using NCAR Community Climate Model, version CCM3 radiation package (Kiehl et al. 1996). It is modeled through the solar radiation transfer δ-Eddington approach (Kiehl et al. 1996). The part of the cloud radiation uses three parameters including the amount of cloudiness, cloud liquid water content, and effective droplet radius. In the model, planetary boundary layer PBL scheme, based on the non-local diffusion concept developed by Holtslag et al. (1990), is used. Convective rainfall patterns of the model are calculated by choosing one of the three schemes, namely the modified-Kuo scheme (Anthes 1977; Anthes et al. 1987), Grell scheme (Grell 1993), and the MIT-Emanuel scheme (Emanuel 1991; Emanuel and Živkovic-Rothman 1999). In this chapter, BATS (Biosphere and Atmosphere Transfer Scheme) (Dickinson et al. 1993) is used as the land-surface scheme, and Grell scheme is used as a cumulus convection scheme with the Fritsch-Chappell type closure (Fritsch and Chappell 1980).

Projected changes in mean precipitation climatology, precipitation extremes and aridity conditions during the period of 2021–2050 (near-term) with respect to the reference period of 1971–2000 are investigated for Turkey, using regional climate model simulations. The outputs of MPI-ESM-MR (global circulation model of the Max Planck Institute for Meteorology) (Taylor et al. 2012) are dynamically downscaled to 10 km resolution by using the double nesting method under both RCP4.5 and RCP8.5 emission scenarios (Van Vuuren et al. 2011). RCP4.5, which is a medium-low emission scenario, stabilizes after 2100 at 4.5 $W \cdot m^{-2}$ without overshoot pathway, while RCP8.5 is the highest of the four reaches at 8.5 $W \cdot m^{-2}$ in 2100 on a rising trajectory.

14.3 Projected Changes in Precipitation Climatology

14.3.1 Model Biases in Seasonal Temperature and Precipitation Climatology

In this chapter, the regional climate model's performance is first investigated for simulation of the observed climate of the region by comparing model results with the CRU observational dataset. The Climatic Research Unit (CRU) dataset of the University of East Anglia comprises a 0.5° grid resolution monthly mean climatology of multiple variables for global land surface, excluding Antarctica (Mitchell and Jones 2005). Model's skill in simulating observed climatology is spatially evaluated for seasonal temperature and precipitation values. For this purpose, the regional climate model is run with the forcing data of MPI-ESM-MR global climate model for the period of 1971–2000 to investigate model biases for four seasons, which are taken as December-January-February (DJF, winter), March-April-May (MAM, spring), June-July-August (JJA, summer), and September-October-November (SON, autumn).

Seasonal temperature values obtained from the RegCM4.4 driven by MPI-ESM-MR global datasets are presented in Fig. 14.1. The seasonal surface temperature values of observational gridded CRU dataset are also represented in Fig. 14.2 for comparison. The outputs obtained from the regional climate model show detailed temperature distribution with respect to the CRU data due to higher resolution (10 km). Results of the regional climate model show that the climatology of the Turkey domain is reproduced for all seasons. Outputs of the regional climate model produce colder temperature values than the results of the observational dataset, especially during the winter season around mountainous and higher plateaus terrains of the domain, like the south-east part of Turkey. This bias could be due to bias in measurements at the stations because the climatological and meteorological stations are very likely constructed in the valleys of the mountainous parts of the region. In

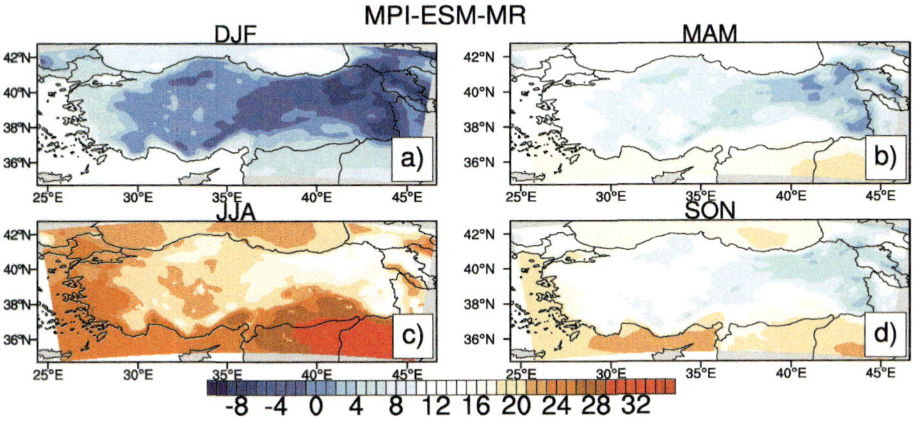

Fig. 14.1 Mean air temperatures (°C) obtained from RegCM4.4 with MPI-ESM-MR dataset for the period of 1971–2000: (**a**) Winter, (**b**) Spring, (**c**) Summer, and (**d**) Autumn season

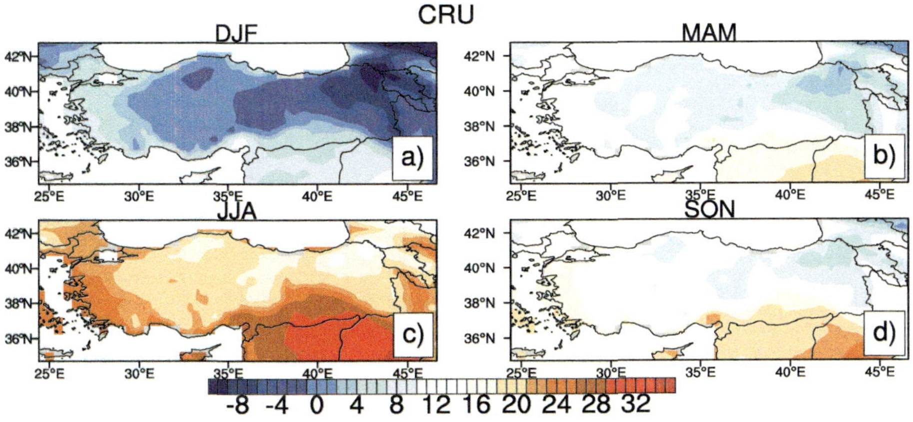

Fig. 14.2 Mean air temperatures (°C) obtained from CRU dataset for the period of 1971–2000: (**a**) Winter, (**b**) Spring, (**c**) Summer, and (**d**) Autumn season

other parts of the domain, the regional climate model produces similar results with the observational dataset.

Seasonal precipitation values obtained from the RegCM4.4, driven by MPI-ESM-MR global datasets and the observational gridded CRU dataset, are represented in Figs. 14.3 and 14.4, respectively. Results show that the regional climate model overestimates precipitation over almost all part of the domain, especially for colder seasons when Turkey has wet conditions. The reason for this bias in estimation can be the fact that RegCM overestimates air pressure and water vapor values along with lower wind speeds compared to the driving datasets (Almazroui et al. 2015). Model outputs agree with observations for the summer season. Therefore, the overall performance of the regional climate model, RegCM, is reasonable.

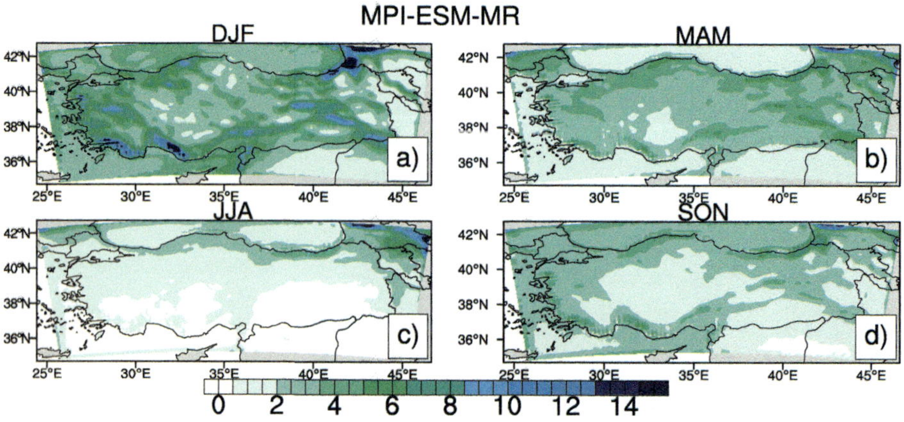

Fig. 14.3 Average precipitation amounts (mm/day) obtained from RegCM4.4 with MPI-ESM-MR dataset for the period of 1971–2000: (**a**) Winter, (**b**) Spring, (**c**) Summer, and (**d**) Autumn season

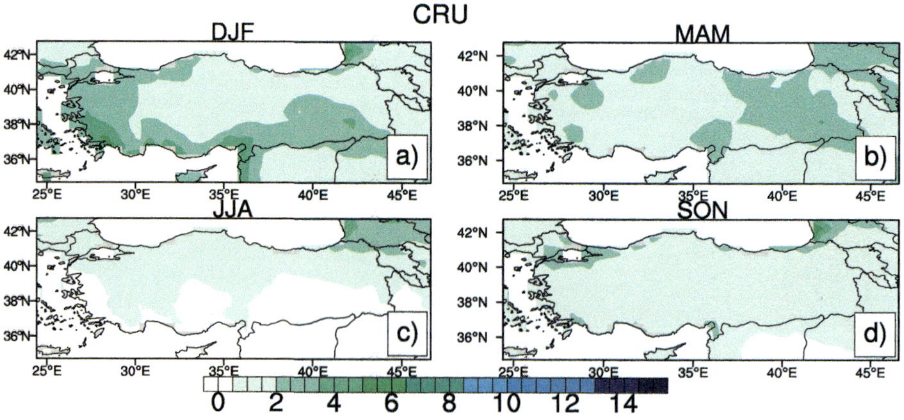

Fig. 14.4 Average precipitation amounts (mm/day) obtained from CRU dataset for the period of 1971–2000: (**a**) Winter, (**b**) Spring, (**c**) Summer, and (**d**) Autumn season

14.3.2 Changes in Precipitation Climatology

The projected changes in precipitation for the future period of 2021–2050 with respect to the present period of 1971–2000, based on the IPCC's RCP4.5 and RCP8.5 emission scenarios, are presented in Figs. 14.5 and 14.6. According to the model results, there is no marked change in the amount of precipitation in the all parts of the domain for dry seasons, which are summer and autumn. Results show much decrease in precipitation amounts with respect to the present period, especially over the south-west part of Turkey for the winter season. On the other hand, there will be an increase in precipitation over the northern part of Turkey.

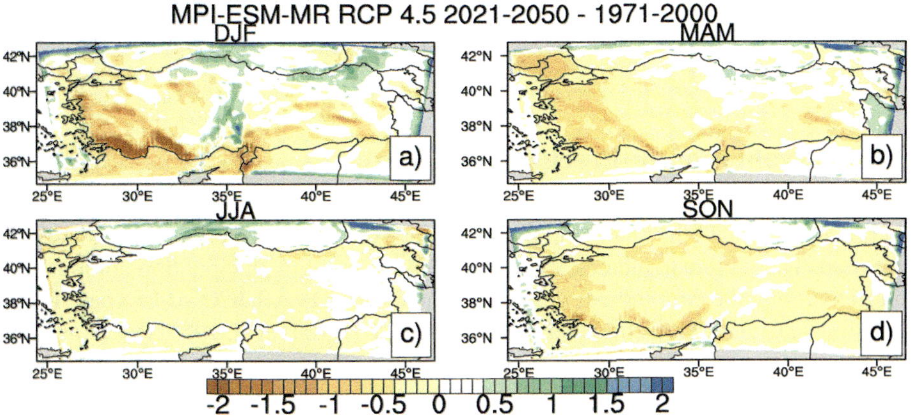

Fig. 14.5 Projected changes in precipitation (mm/day) using the Regional Climate Model RegCM, which is forced by the Global Climate Model MPI-ESM-MR with RCP4.5 emission scenario for the period of 2021–2050, with respect to the reference period of 1971–2000: (**a**) Winter, (**b**) Spring, (**c**) Summer, and (**d**) Autumn season

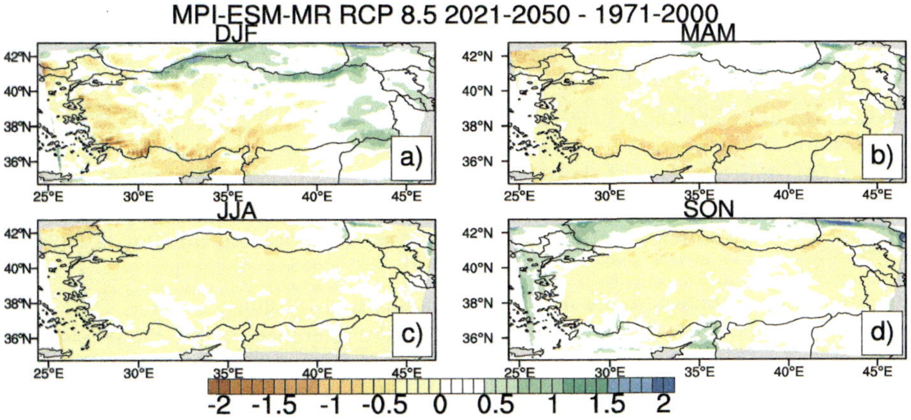

Fig. 14.6 Projected changes in precipitation (mm/day) using the Regional Climate Model RegCM, which is forced by the Global Climate Model MPI-ESM-MR with RCP8.5 emission scenario for the period of 2021–2050, with respect to the reference period of 1971–2000: (**a**) Winter, (**b**) Spring, (**c**) Summer, and (**d**) Autumn season

The regional climate model simulates drier conditions for most part of Turkey, which is already arid and semi-arid. Model outputs of both RCP4.5 and RCP8.5 emission scenarios show alike results for future projections of precipitation. According to the atlas of global and regional climate projections of the IPCC Fifth Assessment Report (AR5), most parts of the South Europe/Mediterranean domain experience a decrease in average precipitation (IPCC 2013). Decrease in precipitation will be significant over the southern-southwestern parts of Turkey and Cyprus in the winter season for the period of 2081–2100, with respect to the period of 1986–

2005 (IPCC 2013). The regional climate model, RegCM, gives more or less similar results with the mentioned in AR5 over Turkey even for the near future.

14.4 Evaluation of Precipitation Extremes

In order to evaluate the change in precipitation extremes, 9 different climate indices are used. These climate indices, except the daily maximum precipitation amount, are chosen among the core indices defined by the CCl/CLIVAR/JCOMM Expert Team on Climate Change Detection and Indices (ETCCDI) (Karl et al. 1999; Peterson et al. 2001; Zhang et al. 2005). Firstly, all of the indices are computed using the daily precipitation output of RegCM for both the reference period (1971–2000) and the near-future period (2021–2050). Subsequently, annual and/or seasonal changes in multi-year averages are illustrated.

The definitions of the 9 climate indices for precipitation extremes are as follows:

Consecutive dry days index: number of the dry periods where the daily precipitation amount is less than 1 mm for at least consecutive 5 days

Consecutive wet days index: number of the wet periods where the daily precipitation amount is more than 1 mm for at least consecutive 5 days

Wet days: number of days where the daily precipitation amount is at least 1 mm

Heavy_10_mm: number of days where the daily precipitation amount is at least 10 mm

Heavy_20_mm: number of days where the daily precipitation amount is at least 20 mm

Heavy_25_mm: number of days where the daily precipitation amount is at least 25 mm

Maximum precipitation: daily maximum precipitation amount

Wet days with respect to the 90th percentile of the reference period: the percentage of wet days where the daily precipitation amount for the future period is greater than the 90th percentile of the daily precipitation amount for the reference period

Wet days with respect to the 95th percentile of the reference period: the percentage of wet days where the daily precipitation amount for the future period is greater than the 95th percentile of the daily precipitation amount for the reference period.

When the change in the number of wet days is examined (Fig. 14.7), it is expected that the decreases in the western and southern regions of the country will reach an average of 20 days in the near future according to the RCP4.5 scenario. Also increases by at least 10 days in the Black Sea coastline are expected. According to the RCP8.5 scenario, a reduction in the number of wet days is foreseen over almost all the country. It is expected that this decline will be more around the Middle Taurus Mountains.

When the changes in daily extreme precipitation are analyzed under both scenarios, it is projected that the number of heavy (daily precipitation amount $\geq$ 10 mm) and very heavy (daily precipitation amount $\geq$ 20 mm and 25 mm) precipitation days

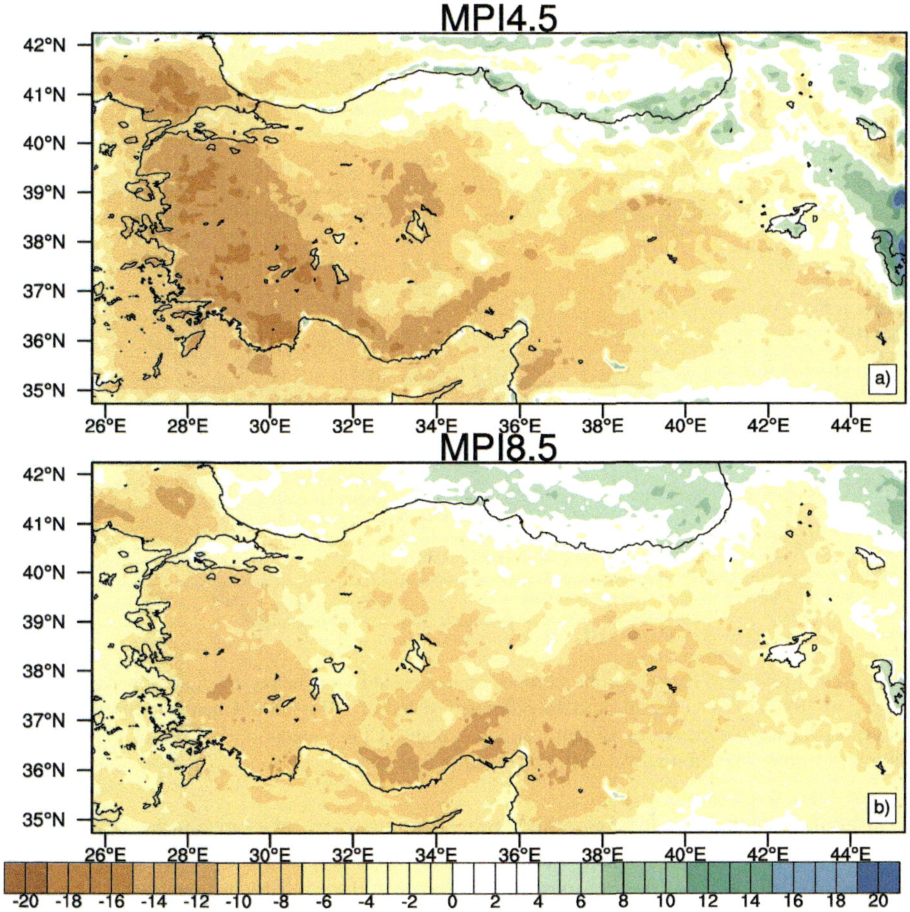

Fig. 14.7 Annual average changes in wet days under (**a**) RCP4.5 scenario and (**b**) RCP8.5 scenario for the period of 2021–2050 with respect to the reference period of 1971–2000

in the future will decrease except at the Black Sea coasts (Fig. 14.8). This reduction is expected to be greater in the RCP4.5 scenario. The number of days when the daily total precipitation amount is greater than or equal to 10 mm, especially in the southern and western Turkey, will decrease more than that in other areas, while it will increase in the Black Sea region. A similar change is anticipated in the number of days when the total daily precipitation amount is greater than or equal to 20 mm and 25 mm, but this change is to be less than that seen on 10 mm and above wet days.

According to the RCP4.5 scenario, consecutive dry days in the future will increase in the west of the country, while in other regions, it is expected to show a slight decrease in general (Fig. 14.9a). According to the RCP8.5 scenario, an increase towards southern Marmara and the southwestern and inner Anatolian regions; a fair amount of decrease in other places are assumed (Fig. 14.9b). When

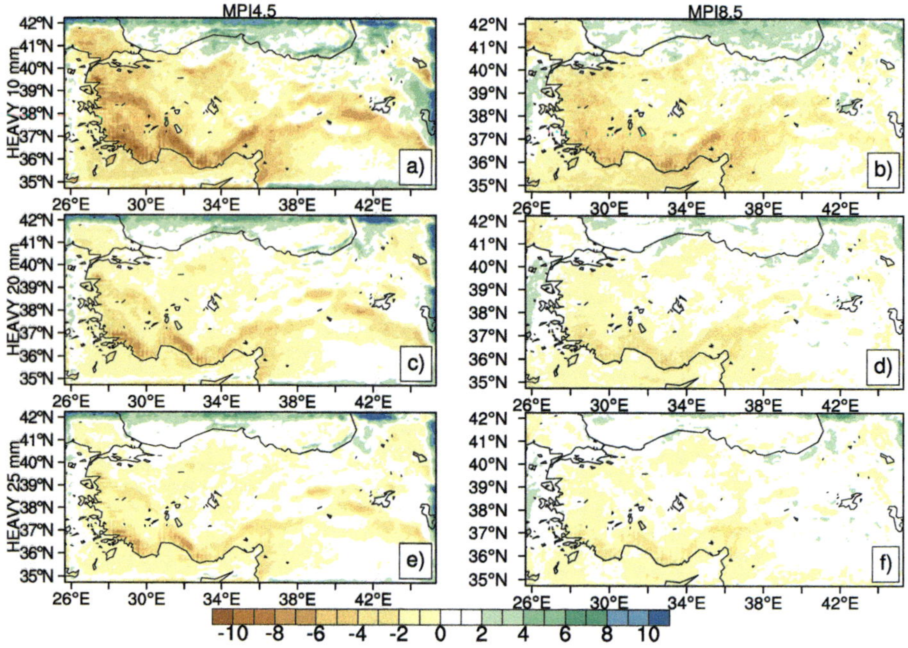

Fig. 14.8 Projected changes in the number of heavy and very heavy precipitation days for the period of 2021–2050 with respect to the reference period of 1971–2000. Change in (**a**) 10 mm wet days, (**c**) 20 mm wet days, (**e**) 25 mm wet days under RCP4.5 scenario, (**b**) 10 mm wet days, (**d**) 20 mm wet days, (**f**) 25 mm wet days under RCP8.5 scenario

consecutive wet days are examined, the opposite view of consecutive dry days is encountered (Fig. 14.9). The consecutive wet days under the RCP4.5 scenario will decrease in the west of the country and increase in the middle and eastern Black Sea, as well as in northeastern parts of the country (Fig.14.9c). On the contrary, under the RCP8.5 scenario, a decrease in the consecutive wet days will be expected in the whole country, which will be more especially over the Antalya-Karaman-Mersin region (Fig. 14.9d). According to the RCP8.5 scenario, the increase in the vicinity of Sinop and Kastamonu provinces contrasts with the change seen throughout the country.

It is predicted that the annual maximum daily precipitation amounts will decrease by 5 mm/day in the entire country according to the RCP4.5 scenario (Fig. 14.10a). This decline is expected to be greater across the Thrace, coastal Aegean and Mediterranean regions. According to the RCP8.5 scenario, the average daily maximum precipitation is expected to increase slightly along the Black Sea coastline, while a decrease of up to 2 mm/day is expected in the country (Fig. 14.10b). It is predicted that the amount expected to be realized according to the RCP4.5 scenario will be relatively higher than the RCP8.5 scenario.

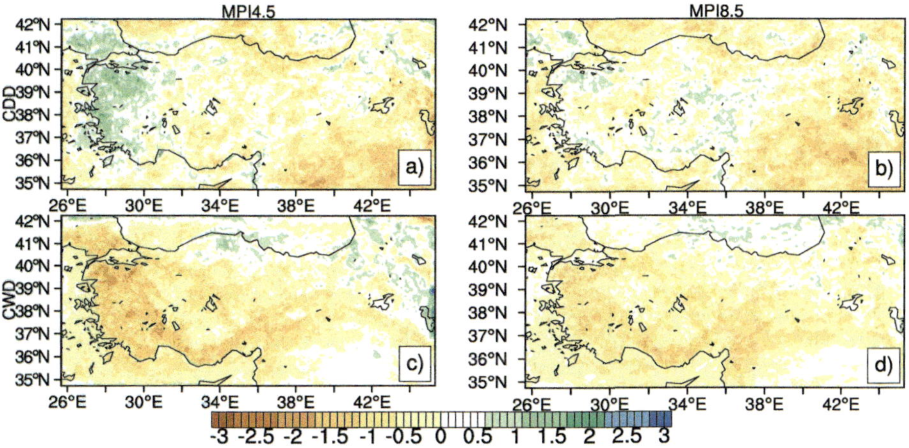

Fig. 14.9 Projected annual average changes in consecutive dry days for the period of 2021–2050 with respect to the reference period of 1971–2000: (**a**) RCP4.5 scenario, (**b**) RCP8.5 scenario; and annual average change in consecutive wet days for the period of 2021–2050 with respect to the reference period: (**c**) RCP4.5 scenario, (**d**) RCP8.5 scenario

Figure 14.11 shows the annual average percentage of days when the daily precipitation amount in the near-future is above the 90th percentile of the reference period. Under the RCP4.5 scenario, 5% of the winter season (about 5 days) in Turkey will have heavier precipitation conditions than those happened in the past. This percentage rate will reach 20% (18 days) in the summer season. RCP8.5 scenario-based projection gives similar results with the RCP4.5 scenario-based projection results for the winter season, while it projects more days with heavy precipitation (25% – around 23 days) than those seen in the RCP4.5 scenario.

Figure 14.12 shows the seasonal average percentage of days when the daily precipitation amount in the near-future is above the 95th percentile of the reference period. Under the RCP4.5 scenario, 5% of the winter season (about 5 days) in Turkey will have heavier precipitation conditions than those happened in the past. This percentage rate will reach 10% (9 days) in the summer season. RCP8.5 scenario-based projection gives similar results with the RCP4.5 scenario-based projection results for the winter season, while it projects more days with heavy precipitation (15% – around 13 days) than those seen in the RCP4.5 scenario.

14.5 Evaluation of Aridity Conditions

Increase in aridity conditions with climate change and/or land degradation depending on human activities is a very crucial global issue in terms of future risks due to food security and health conditions. Decreases in total precipitation

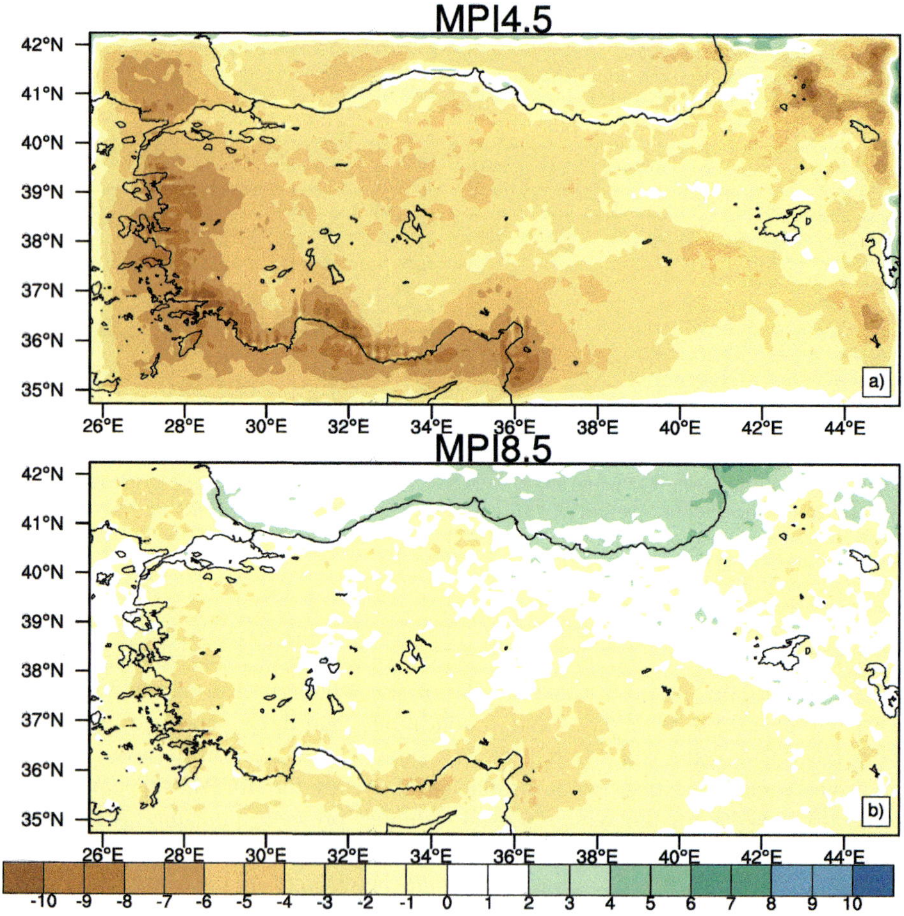

Fig. 14.10 Projected annual changes in daily maximum precipitation (mm/day) for the period of 2021–2050 with respect to the reference period of 1971–2000 under (a) RCP4.5 scenario, (b) RCP8.5 scenario

over the years, repeated dry seasons, and increasing severity and duration of drought events accelerate the transition to desertification (Turkes 2012b, c).

Aridity can be described briefly as climatological drought (Turkes 1998, 2012b, c). The areas where the climatologically long-term dry conditions (i.e. climatological drought) is seen throughout the year are called arid or arid areas (Turkes 2012a).

Aridity Index (AI) can be considered as a bioclimatic index since it takes into account both the physical (precipitation and evaporation) and the biological processes (plant transpiration). Moreover, this index is one of the most suitable indicators to examine the processes of desertification (Middleton and Thomas 1997; Arora 2002; Salvati et al. 2009; Turkes 2013).

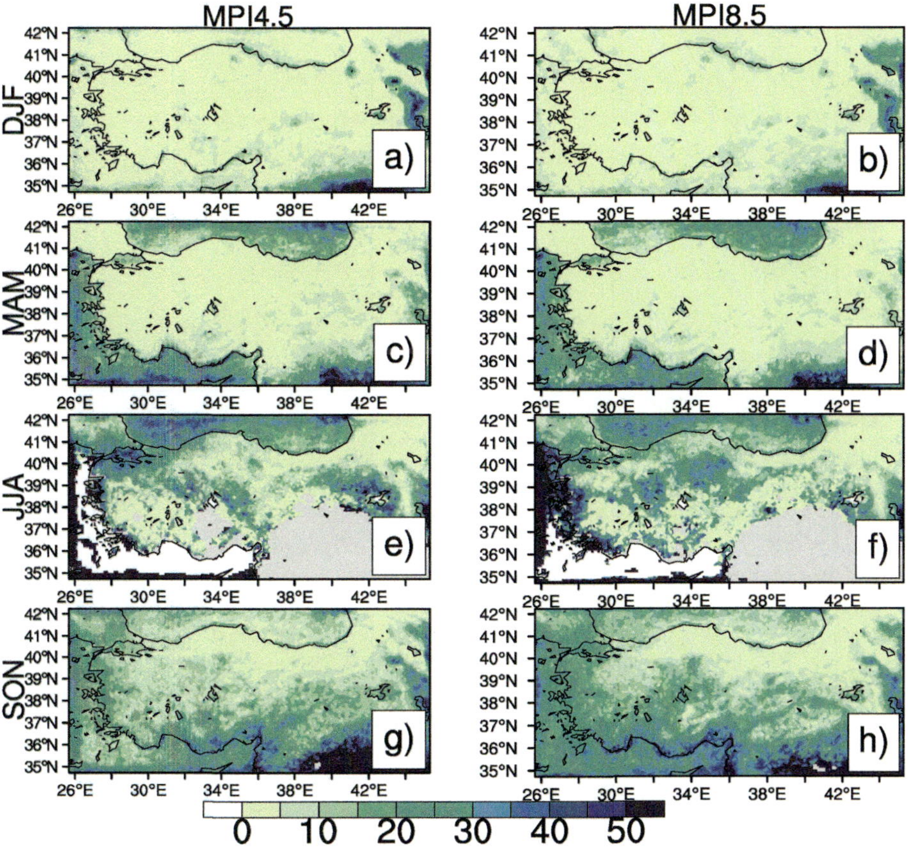

Fig. 14.11 Projected seasonal average percentage of days when the daily precipitation amount in the period of 2021–2050 is above the 90th percentile of the period of 1971–2000: (**a**) Winter (DJF), (**c**) Spring (MAM), (**e**) Summer (JJA), (**g**) Autumn (SON) under the RCP4.5 Scenario and (**b**) Winter (DJF), (**d**) Spring (MAM), (**f**) Summer (JJA), (**h**) Autumn (SON) under the RCP8.5

AI is defined as the ratio of total precipitation (P) to the potential evapotranspiration (PET) by the United Nations Environmental Programme (UNEP) (Turkes 1998, 1999):

$$AI = \frac{P}{PET} \tag{14.1}$$

Total precipitation amounts are directly provided by the RegCM's output, while the potential evapotranspiration values are empirically estimated by using the formula of Romanenko (1961). Romanenko's PET formula with its requirement for only two variables (i.e. mean air temperature (T_{mean}) and relative humidity (RH)) is

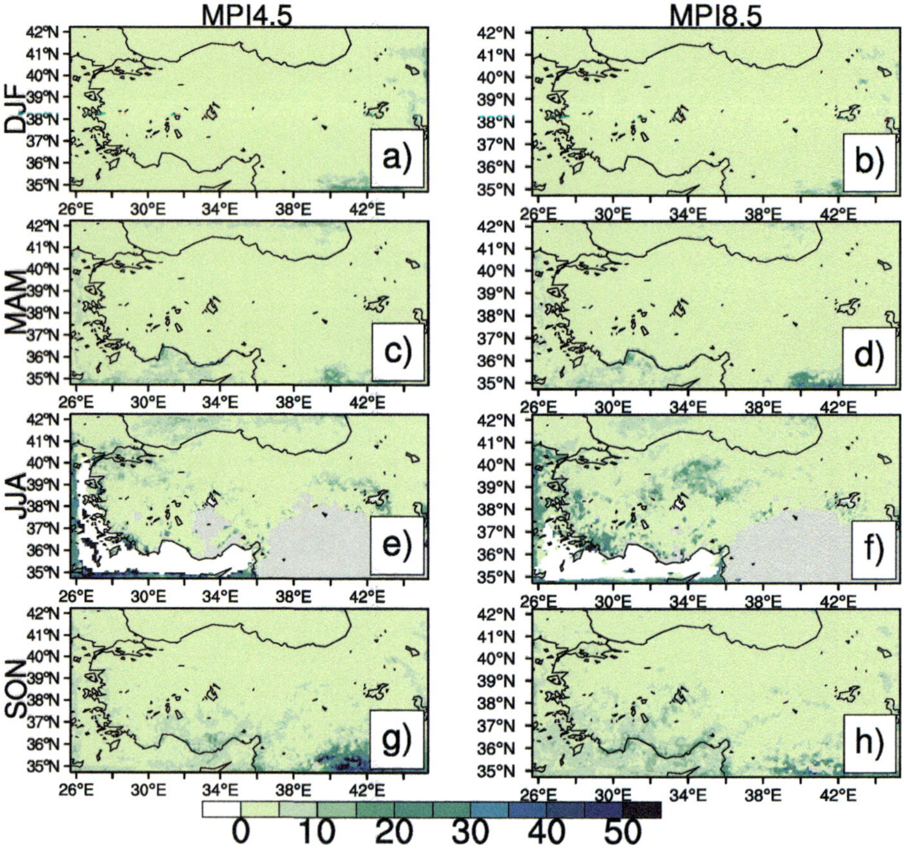

Fig. 14.12 Projected seasonal average percentage of days when the daily precipitation amount in the period of 2021–2050 is above the 95th percentile of the period of 1971–2000: (**a**) Winter (DJF), (**c**) Spring (MAM), (**e**) Summer (JJA), (**g**) Autumn (SON) under the RCP4.5 Scenario and (**b**) Winter (DJF), (**d**) Spring (MAM), (**f**) Summer (JJA), (**h**) Autumn (SON) under the RCP8.5

more practical than the other methods in the literature. Romanenko's PET formula is as follows (Romanenko 1961; Sahin 2012):

$$PET = 0.0018\ (25 + T_{mean})^2\ (100 - RH) \tag{14.2}$$

Considering the AI classification modified by Turkes (2013), Turkey can be divided into seven categories: arid, semi-arid, dry sub-humid, moist sub-humid, semi-humid, humid & very humid, hyper humid (Table 14.1).

According to the map of the aridity index drawn for the reference period (1971–2000), except for a very small portion of arid lands (0.06%), there are no hyperarid or drylands experiencing the desert-like conditions in Turkey (Fig. 14.13). However, as previously stated by Turkes (2010, 2012b, c, 2013), in Turkey, there are semi-arid,

Table 14.1 Modified version of aridity index classification for the climate of Turkey (Turkes 2013)

Aridity index	Classification
< 0.20	Arid
0.20–0.50	Semi-arid
0.50–0.65	Dry sub-humid
0.65–0.80	Moist sub-humid
0.80–1.00	Semi-humid
1.00–2.00	Humid, very humid
> 2.00	Hyper-humid

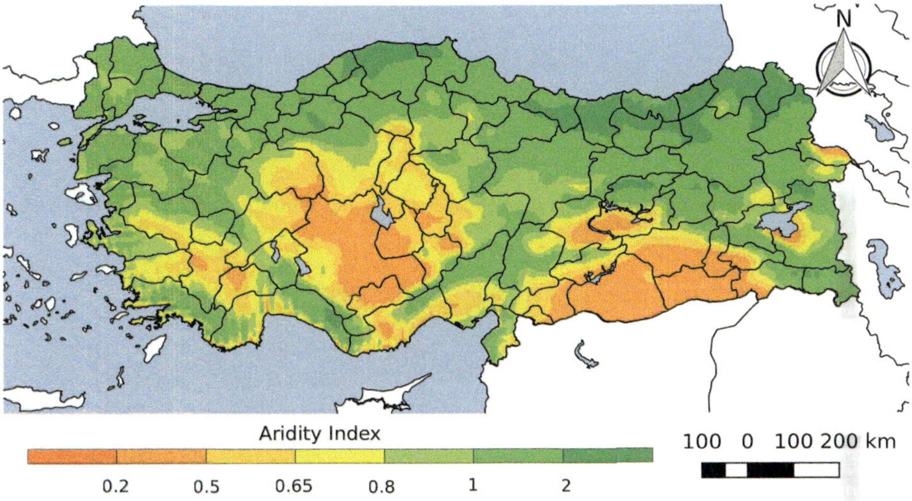

Fig. 14.13 Geographical distribution of AI for the reference period (1971–2000)

dry subhumid, and moist subhumid regions, which are prone to desertification or can be impacted by the desertification at different severity levels. Konya Plain in the Central Anatolia Region, Sanliurfa, Mardin, the eastern part of Gaziantep, southern parts of Diyarbakir and Batman, the western part of Sirnak in the Southeastern Anatolia Region, some parts of Malatya and Elazig in the valley of the upper Euphrates river, a small area in Denizli and Burdur in the Southwestern Anatolia Region, and Igdir province in the Eastern Anatolia Region have semi-arid climate conditions. The Northeastern Black Sea Region and some parts in the western Black Sea Region are hyper-humid areas in Turkey for the period of 1971–2000. Except for these arid and hyper-humid regions, the majority of the areas in Turkey are classified as semi-humid, humid, and very humid.

When the projection results are evaluated for the near future, Turkey's future is expected to go towards more arid conditions in both scenarios (Figs. 14.14 and 14.15). According to the RCP4.5 scenario, it is seen that areas where semi-arid conditions prevail in the past will expand slightly in the near-future (Table 14.2). That is, it is projected that semi-arid and dry sub-humid conditions will dominate in

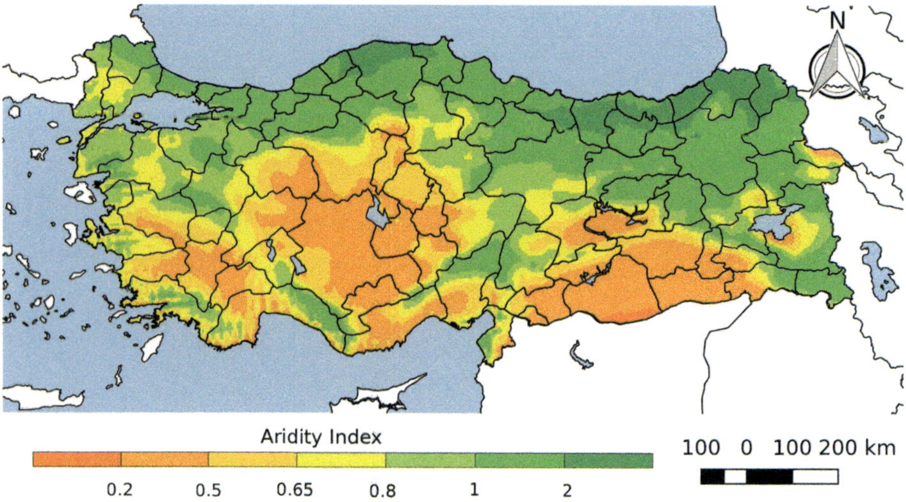

Fig. 14.14 Geographical distribution of projected AI for the future period (2021–2050) under RCP4.5 scenario

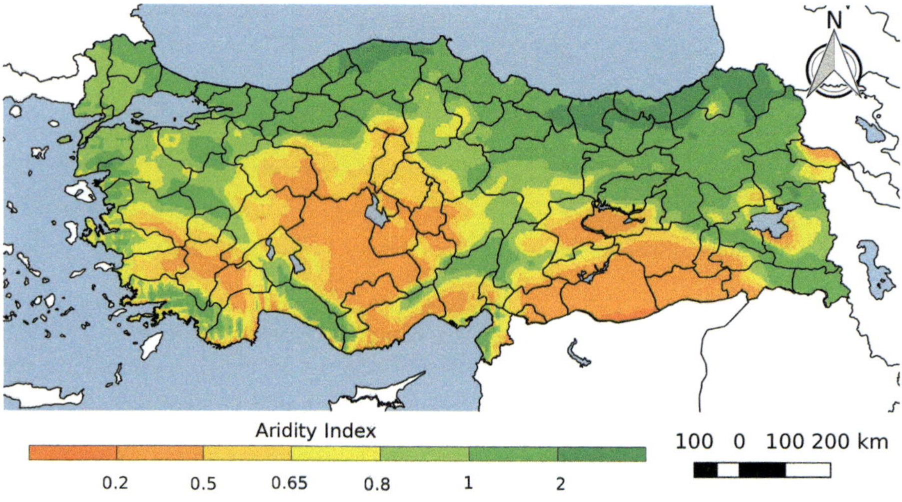

Fig. 14.15 Geographical distribution of projected AI for the future period (2021–2050) under RCP8.5 scenario

the future period (2021–2050) over the Central Anatolia Region, the Southeastern Anatolia Region, and the inner Aegean Region (Fig. 14.14). It is also expected that moist sub-humid conditions will prevail throughout Thrace's Edirne and Tekirdag provinces. Similar results are obtained according to the RCP8.5 scenario, as those in the RCP4.5 scenario, but the severity of the aridity will be slightly less in the 2021–2050 period under the RCP8.5 scenario (Fig. 14.15). For example, according to the

Table 14.2 Comparison of arid and humid areas in Turkey from past to near-future

Aridity classification	Area (%) for reference period (1971–2000)	Area (%) for future period (2021–2050) under RCP4.5 scenario	Area (%) for future period (2021–2050) under RCP8.5 scenario
Arid	0.06	0.29	0.26
Semi-arid	12.86	21.58	19.50
Dry sub-humid	11.77	14.44	14.07
Moist sub-humid	11.32	13.36	12.51
Semi-humid	16.71	16.67	18.24
Humid, very humid	40.44	28.67	30.63
Hyper-humid	6.85	4.98	4.79

RCP4.5 scenario, the conditions that are moist sub-humid in Edirne and Tekirdag provinces become humid and very humid under the RCP8.5 scenario.

According to the classification of aridity index, calculated based on the outputs obtained from the climate model, it is projected that the humid and very humid areas prevalent in the 1971–2000 period will substantially diminish in the future under both scenarios (i.e. RCP4.5 and RCP8.5), whereas especially the semi-arid areas will be more extended (Table 14.2). In other words, with climate change, it is anticipated that the areas prone to desertification will be widened, while the areas in which desertification does not occur, especially in the Black Sea Region, will be lessened. However, it is expected that the semi-humid regions that are inclined to desertification in deteriorating conditions will remain almost the same due to the RCP4.5 scenario and will increase a little according to the RCP8.5 scenario.

14.6 Conclusions

In this chapter, impacts of climate change on mean precipitation climatology, precipitation extremes and aridity conditions for Turkey are projected using the regional climate model RegCM4.4. Regional climate model is driven by MPI-ESM-MR global climate model for the period of 2021–2050. The MPI-ESM-MR global climate model is chosen as this model represents the climatology of the Eastern Mediterranean Basin more accurately than comparable global models. First, the ability of the RegCM4.4 regional climate model to reproduce the observed climatology is evaluated for four climatological seasons for the period of 1971–2000 by comparing outputs of regional model with the CRU dataset. Results show that the RegCM regional model's capacity in reproducing the observed temperature and precipitation climatology is reasonable. The CRU dataset is created using meteorological data from observation stations. Prior to year 2000, these stations were not wide spread and were also prone to human reading errors. There are few stations from Turkey, which are included in the CRU dataset. Therefore, a perfect correspondence between the model outputs and the CRU dataset should not be expected. Even under these circumstances, the model outputs give reliable results when compared to the CRU dataset. Also, it must be mentioned that the main difference between the climate model outputs and the observation datasets come from the physics used in the global climate models. Regional climate models used to downscale the climate variables decrease the differences between the model outputs and the observations.

The MPI-ESM-MR global model with RCP4.5 and RCP8.5 emission scenario outputs is used as forcing data for the regional climate model RegCM4.4 for the near future (2021–2050) in order to project changes in precipitation climatology, extremes and aridity conditions. According to model results of future projections, there will be a decrease in precipitation amounts over almost all parts of Turkey. Drier conditions will likely occur over the most part of Turkey, which is already arid and semi-arid.

In the near future (2021–2050), most of the drier conditions will be observed in the western and southern parts of Turkey and especially during the winter season. This is to be expected as the western and southern parts of the country normally get most of the precipitation in the winter season. As rain is infrequent in the summer months, a relative decrease in the precipitation amounts will not be readily visible in the model outputs. It should also be remarked that the northern and the northwestern part of the country does not lose much of the precipitation in any season, and some slight increase in the winter season can be observed along the Black Sea coastal region. Even though the uncertainty in the climate models can be considered slightly high, there might be some good news for the Central Anatolia region for the winter months, especially in the RCP4.5 emission scenario.

For Turkey, the difference between the precipitation outputs of RCP4.5 and RCP8.5 scenarios for the near future (2021–2050) is clearly observable. Common sense suggests that precipitation should decrease more in the worse scenario (RCP8.5) than the better scenario (RCP4.5). However, the model outputs show that the opposite can be true. In Turkey, higher temperatures lead to more evaporation, and, in turn, this leads to more precipitation. In the far future of the climate models (2071–2100), this behavior will change as the atmospheric pressure systems will shift, causing a more severe drying in our region.

The used models produce outputs every 3 min. But, to be able to compare the data with observations, average (temperature) or sum (precipitation) of the data obtained from the models is needed. Therefore, important information regarding the strength of the precipitation events is lost. The situation for Turkey is not the loss in the total amount of precipitation but the change in the distribution of the precipitation pattern. The distribution pattern of the precipitation events changes in such a way that both extreme precipitation events and also prolonged droughts in the region occur. This change in precipitation patterns is not unique to Turkey, but it is certainly a result of global climate change.

Therefore, the projected drought (decrease in precipitation) conditions will very likely affect Turkey, which already has a mostly arid and semiarid climate and environment, and this makes Turkey extremely vulnerable to climate change, particularly to increased droughts. As the rainstorms also come as extreme events, the water holding capacity of the upper layers of the soil is negatively affected. Heavy rains after long drought events also lead to soil erosion in the region.

Acknowledgements This research has been supported by Bogazici University Research Fund Grant Number 13561. We also thank Zekican Demiralay and Basak Bilgin from Bogazici University for their valuable contributions.

References

Akcakaya A, Eskioglu O, Atay H, Demir O (2013) Yeni senaryolar ile Turkiye icin iklim degisikligi projeksiyonlari. Meteoroloji Genel Mudurlugu, Ankara

Almazroui M (2012) Dynamical downscaling of rainfall and temperature over the Arabian Peninsula using RegCM4. Clim Res 52:49–62. https://doi.org/10.3354/cr01073

Almazroui M (2016) RegCM4 in climate simulation over CORDEX-MENA/Arab domain: selection of suitable domain, convection and land-surface schemes. Int J Climatol 36:236–251. https://doi.org/10.1002/joc.4340

Almazroui M, Islam MN, Al-Khalaf AK, Saeed F (2015) Best convective para-metrization scheme within RegCM4 to downscale CMIP5 multi-model data for the CORDEX-MENA/Arab domain. Theor Appl Climatol 124:807–823

Almazroui M, Islam MN, Al-Khalaf AK, Saeed F, Dambul R, Rahman MA (2016) Simulation of temperature and precipitation climatology for the CORDEX-MENA/ Arab domain using RegCM4. Arab J Geosci 9(1):1–13

Altinsoy H, Ozturk T, Turkes M, Kurnaz ML (2011) Projections of future air temperature and precipitation changes in the Mediterranean Basin by using the global climate model. In: Proceedings of the national geographical congress with international participation, Turkish Geographical Society, Istanbul University, 7–10 September 2011

Anthes RA (1977) A cumulus parameterization scheme utilizing a one-dimensional cloud model. Mon Weather Rev 105(3):270–286

Anthes RA, Hsie EY, Kuo YH (1987) Description of the Penn State/NCAR mesoscale model version 4 (MM4). NCAR, Boulder

Arora VK (2002) The use of the aridity index to assess climate change effect on annual runoff. J Hydrol 265(1–4):164–177

Calbó J (2009) Possible climate change scenarios with specific reference to Mediterranean regions. In: Water scarcity in the Mediterranean. Springer, Heidelberg, pp 1–13

Chen M, Pollard D, Barron EJ (2003) Comparison of future climate change over North America simulated by two regional models. J Geophys Res-Atmos 108(D12):4348. https://doi.org/10.1029/2002JD002738

Coppola E, Giorgi F, Raffaele F, Fuentes-Franco R, Giuliani G, Llopart-Pereira M, Mamgain A, Mariotti L, Diro GT, Torma C (2014) Present and future climatologies in the phase I CREMA experiment. Clim Chang 125(1):23–38

Demircan M, Gurkan H, Eskioglu O, Arabaci H, Coskun M (2017) Climate change projections for Turkey: three models and two scenarios. Turk J Water Sci Manag 1(1):22–43

Dickinson RE, Kennedy PJ, Henderson-Sellers A (1993) Biosphere-Atmosphere Transfer Scheme (BATS) version 1e as coupled to the NCAR community climate model. National Center for Atmospheric Research, Climate and Global Dynamics Division

Emanuel KA (1991) A scheme for representing cumulus convection in large-scale models. J Atmos Sci 48(21):2313–2329
Emanuel KA, Živkovic-Rothman M (1999) Development and evaluation of a convection scheme for use in climate models. J Atmos Sci 56(11):1766–1782
Fritsch JM, Chappell CF (1980) Numerical prediction of convectively driven mesoscale pressure systems. Part I: convective parameterization. J Atmos Sci 37(8):1722–1733
Gao X, Zhao Z, Giorgi F (2002) Changes of extreme events in regional climate simulations over East Asia. Adv Atmos Sci 19(5):927–942
Gao X, Pal JS, Giorgi F (2006) Projected changes in mean and extreme precipitation over the Mediterranean region from a high resolution double nested RCM simulation. Geophys Res Lett 33(3):L03706
Gibelin AL, Déqué M (2003) Anthropogenic climate change over the Mediterranean region simulated by a global variable resolution model. Clim Dyn 20(4):327–339
Giorgi F, Lionello P (2008) Climate change projections for the Mediterranean region. Glob Planet Chang 63(2–3):90–104
Giorgi F, Bi X, Qian Y (2003a) Indirect vs. direct effects of anthropogenic sulfate on the climate of East Asia as simulated with a regional coupled climate-chemistry/aerosol model. Clim Chang 58 (3):345–376
Giorgi F, Francisco R, Pal J (2003b) Effects of a subgrid-scale topography and land use scheme on the simulation of surface climate and hydrology. Part I: effects of temperature and water vapor disaggregation. J Hydrometeorol 4(2):317–333
Giorgi F, Bi X, Pal J (2004a) Mean interannual variability and trends in a regional climate change experiment over Europe. I. Present-day climate (1961–1990). Clim Dyn 22(6–7):733–756
Giorgi F, Bi X, Pal J (2004b) Mean interannual variability and trends in a regional climate change experiment over Europe. II: climate change scenarios (2071–2100). Clim Dyn 23(7–8):839–858
Giorgi F, Coppola E, Solmon F, Mariotti L, Sylla MB, Bi X, Elguindi N, Diro GT, Nair V, Giuliani G, Turuncoglu UU, Cozzini S, Güttler I, O'Brien TA, Tawfik AB, Shalaby A, Zakey AS, Steiner AL, Stordal F, Sloan LC, Brankovic C (2012) RegCM4: model description and preliminary tests over multiple CORDEX domains. Clim Res 52:7–29
Giorgi F, Coppola E, Raffaele F, Diro GT, Fuentes-Franco R, Giuliani G, Mamgain A, Llopart MP, Mariotti L, Torma C (2014) Changes in extremes and hydroclimatic regimes in the CREMA ensemble projections. Clim Chang 125(1):39–51
Gorguner M, Ishida K, Kavvas ML (2017) Dynamically downscaled CMIP5 climate projections over a Mediterranean-climate watershed in western Turkey. In: Dunn CN, Van Weele B (eds) Proceedings of the world environmental and water resources congress 2017: water, wastewater, and stormwater; urban watershed management; and municipal water infrastructure, Sacramento
Goubanova K, Li L (2007) Extremes in temperature and precipitation around the Mediterranean basin in an ensemble of future climate scenario simulations. Glob Planet Chang 57(1–2):27–42
Grell GA (1993) Prognostic evaluation of assumptions used by cumulus parameterizations. Mon Weather Rev 121(3):764–787
Grell GA, Dudhia J, Stauffer DR (1994) A description of the fifth-generation Penn State/NCAR mesoscale model (MM5). National Center for Atmospheric Research. Mesoscale and Micro-scale Meteorology Division, Boulder
Gu H, Wang G, Yu Z, Mei R (2012) Assessing future climate changes and extreme indicators in east and South Asia using the RegCM4 regional climate model. Clim Chang 114(2):301–317
Holtslag AAM, De Bruijn EIF, Pan HL (1990) A high resolution air mass transformation model for short-range weather forecasting. Mon Weather Rev 118(8):1561–1575
IPCC (2013) Van Oldenborgh, GJ, Collins M, Arblaster J, Christensen JH, Marotzke J, Power SB, Rummukainen M, Zhou T (eds) Annex I: atlas of global and regional climate projections. In: Stocker TF, Qin D, Plattner GK, Tignor M, Allen SK, Boschung J, Nauels A, Xia Y, Bex V, Midgley PM. (eds) Climate change 2013: the physical science basis. Contribution of working group I to the Fifth Assessment Report of the Intergovernmental Panel on Climate Change. Cambridge University Press, Cambridge, United Kingdom and New York, NY, USA

Jacobeit J, Dunkeloh A, Hertig E (2007) Mediterranean rainfall changes and their couses. In: Lozán JL, Grassl H, Hupfer P, Menzel L, Schönwiese CD (eds) Global change: enough water for all? Wissenschaftliche Auswertungen, Hamburg, pp 195–199
Kadioglu M, Ozturk N, Erdun H, Sen Z (1999) On the precipitation climatology of Turkey by harmonic analysis. Int J Climatol 19(15):1717–1728
Karl TR, Nicholls N, Ghazi A (1999) CLIVAR/GCOS/WMO workshop on indices and indicators for climate extremes: workshop summary. Clim Chang 42:3–7
Kiehl JT, Hack JJ, Bonan GB, Boville BA, Briegleb BP, Williamson DL, Rasch PJ (1996) Description of the NCAR Community Climate Model (CCM3). Climate and Global Dynamics Div – National Center for Atmospheric Research, Boulder
Lionello P, Giorgi F (2007) Winter precipitation and cyclones in the Mediterranean region: future climate scenarios in a regional simulation. Adv Geosci 12:153–158
Lionello P, Bhend J, Buzzi A, Della-Marta PM, Krichak SO, Jansà A, Maheras P, Sanna A, Trigo IF, Trigo R (2006a) Cyclones in the Mediterranean region: climatology and effects on the environment. In: Lionello P, Malanotte-Rizzoli P, Boscolo R (eds) Developments in earth and environmental sciences, vol 4. Elsevier, Amsterdam, pp 325–372
Lionello P, Malanotte-Rizzoli P, Boscolo R, Alpert P, Artale V, Li L, Luterbacher J, May W, Trigo R, Tsimplis M, Ulbrich U, Xoplaki E (2006b) The Mediterranean climate: an overview of the main characteristics and issues. In: Lionello P, Malanotte-Rizzoli P, Boscolo R (eds) Developments in earth and environmental sciences, vol 4. Elsevier, Amsterdam, pp 1–26
Lolis CJ, Turkes M (2017) Atmospheric circulation characteristics favouring extreme precipitation in Turkey. Clim Res 71(2):139–153
Mariotti L, Diallo I, Coppola E, Giorgi F (2014) Seasonal and intraseasonal changes of African monsoon climates in 21st century CORDEX projections. Clim Chang 125(1):53–65
Mariotti A, Pan Y, Zeng N, Alessandri A (2015) Long-term climate change in the Mediterranean region in the midst of decadal variability. Clim Dyn 44(5–6):1437–1456
Middleton NJ, Thomas DSG (eds) (1997) World atlas of desertification. Edward Arnold, London
Mitchell TD, Jones PD (2005) An improved method of constructing a database of monthly climate observations and associated high-resolution grids. Int J Climatol 25(6):693–712
Oleson KW, Niu GY, Yang ZL, Lawrence DM, Thornton PE, Lawrence PJ, Stöckli R, Dickinson RE, Bonan GB, Levis S, Dai A, Qian T (2008) Improvements to the community land model and their impact on the hydrological cycle. J Geophys Res Biogeo, 113(G1). https://doi.org/10.1029/2007JG000563
Ozturk T, Turkes M, Kurnaz ML (2011) RegCM4.3.5 iklim modeli benzetimleri kullanilarak Turkiye'nin gelecek hava sicakligi ve yagis klimatolojilerindeki degişikliklerin cozumlenmesi. Ege Cografya Dergisi 20:1): 17–1): 27
Ozturk T, Altinsoy H, Turkes M, Kurnaz ML (2012) Simulation of temperature and precipitation climatology for Central Asia CORDEX domain by using RegCM 4.0. Clim Res 52:63–76
Ozturk T, Ceber ZP, Turkes M, Kurnaz ML (2015) Projections of climate change in the Mediterranean Basin by using downscaled global climate model outputs. Int J Climatol 35:4276–4292. https://doi.org/10.1002/joc.4285
Ozturk T, Turp MT, Turkes M, Kurnaz ML (2017) Projected changes in temperature and precipitation climatology of Central Asia CORDEX region 8 by using RegCM4.3.5. Atmos Res 183:296–307. https://doi.org/10.1016/j.atmosres.2016.09.008
Ozturk T, Turp MT, Turkes M, Kurnaz ML (2018) Future projections of temperature and precipitation climatology for CORDEX-MENA domain using RegCM4.4. Atmos Res 206:87–107. https://doi.org/10.1016/j.atmosres.2018.02.009
Pal JS, Giorgi F, Bi X, Elguindi N, Solmon F, Gao X, Rauscher SA, Francisco R, Zakey A, Winter J, Ashfaq M, Syed FS, Bell JL, Diffenbaugh NS, Karmacharya J, Konaré A, Martinez D, Da Rocha RP, Sloan LC, Steiner AL (2007) Regional climate modeling for the developing world: the ICTP RegCM3 and RegCNET. Bull Am Meteorol Soc 88(9):1395–1409

Peterson TC, Folland C, Gruza G, Hogg W, Mokssit A, Plummer N (2001) Report on the activities of the working group on climate change detection and related rapporteurs 1998–2001. WMO, Geneve, pp 1–143
Romanenko VA (1961) Computation of the autumn soil moisture using a universal relationship for a large area. In: Proceedings of Ukrainian Hydrometeorological Research Institute, Kiev, 1961
Ruti PM, Somot S, Giorgi F, Dubois C, Flaounas E, Obermann A, Dell'Aquila A, Pisacane G, Harzallah A, Lombardi E, Ahrens B, Akhtar N, Alias A, Arsouze T, Aznar R, Bastin S, Bartholy J, Béranger K, Beuvier J, Bouffies-Cloché S, Brauch J, Cabos W, Calmanti S, Calve JC, Carillo A, Conte D, Coppola E, Djurdjevic V, Drobinski P, Elizalde-Arellano A, Gaertner M, Galàn P, Gallardo C, Gualdi S, Goncalves M, Jorba O, Jordà G, L'Heveder B, Lebeaupin-Brossier C, Li L, Liguori G, Lionello P, Macià̀s D, Nabat P, Önol B, Raikovic B, Ramage K, Sevault F, Sannino G, Struglia V, Sanna A, Torma C, Vervatis V (2016) MED-CORDEX initiative for Mediterranean climate studies. Bull Am Meteorol Soc 97(7):1187–1208
Sahin S (2012) An aridity index defined by precipitation and specific humidity. J Hydrol 444:199–208
Salvati L, Scarascia MEV, Zitti M, Ferrara A, Urbano V, Sciortino M, Giupponi C (2009) The integrated assessment of land degradation. Ital J Agron 4(3):77–90
Sen OL (2013) A holistic view of climate change and its impacts in Turkey. Istanbul Policy Center, Istanbul
Sen B, Topcu S, Turkes M, Sen B, Warner JF (2012) Projecting climate change, drought conditions and crop productivity in Turkey. Clim Res 52:175–191
Sylla MB, Pal JS, Wang GL, Lawrence PJ (2016) Impact of land cover characterization on regional climate modeling over West Africa. Clim Dyn 46(1–2):637–650
Tang T, Shindell D, Samset BH, Boucher O, Forster PM, Hodnebrog Ø, Myhre G, Sillman J, Voulgarakis A, Andrews T, Faluvegi G, Fläschner D, Iversen T, Kasoar M, Kharin V, Kirkevåg A, Lamarque JF, Olivié D, Richardson T, Stjern CW, Takemura T (2018) Dynamical response of Mediterranean precipitation to greenhouse gases and aerosols. Atmos Chem Phys 18 (11):8439–8452
Tawfik AB, Steiner AL (2011) The role of soil ice in land-atmosphere coupling over the United States: a soil moisture–precipitation winter feedback mechanism. J Geophys Res Atmos 116 (D2):D02113
Taylor KE, Stouffer RJ, Meehl GA (2012) An overview of CMIP5 and the experiment design. Bull Am Meteorol Soc 93(4):485–498
Turkes M (1998) Iklimsel degisebilirlik acisindan Turkiye'de collesmeye egilimli alanlar. In: DMI/ ITU II. Hidrometeoroloji Sempozyumu Bildiri Kitabi, Meteoroloji Genel Mudurlugu, Ankara
Turkes M (1999) Vulnerability of Turkey to desertification with respect to precipitation and aridity conditions. Turk J Eng Environ Sci 23(5):363–380
Turkes M (2010) Klimatoloji ve meteoroloji. Kriter Publisher, Istanbul
Turkes M (2012a) Turkiye'de gozlenen ve ongorulen iklim degisikligi, kuraklik ve collesme. Ankara Universitesi Cevrebilimleri Dergisi 4(2):1–32
Turkes M (2012b) Kuresel iklim degisikligi ve collesme. In: Ozgen N (ed) Gunumuz Dunya Sorunlari – Disiplinlerarasi Bir Yaklasim. Egiten Kitap, Ankara, pp 1–42
Turkes M (2012c) Kuraklik, collesme ve Birlesmis Milletler Collesme ile Savasim Sozlesmesi'nin ayrintili bir cozumlemesi. Marmara Avrupa Arastirmalari Dergisi 20:7–56
Turkes M (2013) Iklim verileri kullanilarak Turkiye'nin collesme haritasi dokumani hazirlanmasi raporu. T.C. Orman ve Su Isleri Bakanliği Collesme ve Erozyonla Mucadele Genel Mudurlugu, Ankara
Turp MT, Ozturk T, Turkes M, Kurnaz ML (2014) RegCM4.3.5 Bolgesel iklim modelini kullanarak Turkiye ve cevresi bolgelerin yakin gelecekteki hava sicakligi ve yagis klimatolojileri icin ongorulen degisikliklerin incelenmesi. Ege Cografya Dergisi 23(1):1–24

Van Vuuren DP, Edmonds J, Kainuma M, Riahi K, Thomson A, Hibbard K, Hurtt GC, Kram T, Krey V, Lamarque JF, Masui T, Meinshausen M, Nakicenovic N, Smith SJ, Rose SK (2011) The representative concentration pathways: an overview. Clim Chang 109:5–31

Xoplaki E, González-Rouco JF, Luterbacher JU, Wanner H (2004) Wet season Mediterranean precipitation variability: influence of large-scale dynamics and trends. Clim Dyn 23(1):63–78

Zhang X, Hegerl G, Zwiers FW, Kenyon J (2005) Avoiding inhomogeneity in percentile-based indices of temperature extremes. J Clim 18(11):1641–1651

Chapter 15
Legal and Institutional Foundations of Turkey's Domestic and Transboundary Water Policy

Aysegul Kibaroglu

Abstract Turkey's water policy and management is a culmination of various laws and regulations governed by a range of national ministries and executive administrations. Over time, several changes were made in the existing legislation and institutions, which ended up with complex water management system in Turkey. Existing surface and groundwater laws have become insufficient in responding to the increasing water demand and diminishing water supply. On the other hand, neoliberal transformation of Turkish economy in the 1980s and the country's harmonization process with the European Union since the early 2000s have produced new primary and secondary water legislations in the domestic water, irrigation, hydropower and the environment sectors. In this context, this chapter, firstly, describes the principal water legislation in Turkey. Secondly, main water institutions are depicted with specific attention to the reorganization processes of various key ministries due to domestic and regional political changes. Finally, Turkey's transboundary water policy is delineated with its basic principles and prevailing practices.

Keywords Turkey · Water legislation · Water institutions · Transboundary water policy

15.1 Overview

Good water governance requires equitable and efficient legal and institutional arrangements, which should provide effective coordination and collaboration between all state institutions and the other stakeholders. Turkey's water policy and management is composed of several laws and by-laws, and is executed by a variety of national ministries and administrations. Water management legislation originates from the early years of the Republic such as the Village Law No. 442 (1924) and the

---

A. Kibaroglu (✉)
Faculty of Economics, Administrative and Social Sciences, Department of Political Science and International Relations, MEF University, Istanbul, Turkey
e-mail: aysegul.kibaroglu@mef.edu.tr

N. B. Harmancioglu, D. Altinbilek (eds.), *Water Resources of Turkey*, World Water Resources 2, https://doi.org/10.1007/978-3-030-11729-0_15

Law on Waters No. 831 (1926). Over time, several changes were made in the existing legislation and institutions, which ended up with complex water management system in Turkey.

The constitution, laws (code), decree-laws and by-laws constitute main categories of water legislation. On the other hand, various other legislation affects water resources management and allocation. It is possible to identify stipulations on water use, management and allocation in more than hundred different laws, decree-laws and by-laws. Thus, implementation and enforcement of water-related legislation becomes all the more difficult (Ozbay 2006).

The Constitution (Law No. 2709, Official Gazette dated 7.11.1982) establishes the basic principles, which govern natural resources. Accordingly, water is a public good under the State's trusteeship (Article 168). The authority to explore and manage water resources is vested in the State. The Constitution also introduces environmental rights as a human right in Article 56: "Everyone has the right to live in a healthy, balanced environment. It is the duty of the State and citizens to improve the natural environment, and to prevent environmental pollution." The Turkish Constitution assigns stewardship of the country's water resources to the public domain, where only public institutions may grant water use-rights to both public and private parties as long as they accommodate 'public benefit' or 'common wealth' (Kibaroglu and Baskan 2011).

On the other hand, the Civil Code (Law) (2001) deals with water in two separate categories: public water resources and water resources in the purview of private law and private ownership. This classification generates from the Civil Code, Article 715: "the assets under nobody's possession and the commodities at the service of the public shall be under the command and possession of the State." The Civil Code, Article 756 legalizes springs as a subject to private ownership. It stipulates, "any spring is an integral part of the land, the ownership of a spring may be allowed only together with the ownership of the land." Articles 715 and 756 should be assessed jointly, hence, "except for privately owned springs, surface and groundwater resources cannot be owned, but are subject to user rights which are granted for beneficial use only, such as domestic and agricultural use, fishing, hydropower generation, industry and mining, transportation and medicinal and thermal uses" (Kibaroglu and Baskan 2011). The Civil Code also contains clauses on water pollution. In cases of water resources get polluted Article 757 stipulates matters of compensation while Article 758 involves issues concerning their restoration (Kibaroglu et al. 2011).

Analysts argue that the Civil Code prioritizes the 'public' nature of water resources yet also respects historic rights and private ownership. As the question of springs demonstrated, some exceptions to the public ownership of water in the Code created problems, which are settled by courts afterwards. Furthermore, the Civil Code has suffered from the problems arising from the intensification of competition over limited water resources by a rapidly increasing population. As a consequence, the need for a clear legal delineation of the boundaries between the 'public' and 'private' spheres has gained prominence in the debates around water management (Sumer 2013).

15.2 Principal Water Legislation

15.2.1 Surface Water (Law No. 831, 1926) (Law No. 6200, 1954)

The Law on Waters, No. 831, which entered into force on 10 May 1926, pioneers water resources legislation in Turkey. It is a brief law with nine articles. Article 1 requires that municipalities are in charge of water supply to towns and cities, while in line with the Village Law (No. 442, 7 April 1924, Official Gazette No. 68), the Village Council of Elders takes care of water supply development in villages. Ministry of Health and Social Aid is considered to be the central authority for the implementation this law, and it is in charge of endorsing water infrastructure projects. The Law on Waters is still legally valid, however, its practical relevance has diminished in the meantime due to the enactment of new laws and institutional changes in the water sector (Sumer 2013).

Law No. 6200 (1953 Official Gazette No. 8592) authorizes the State Hydraulic Works (DSI, Turkish acronym) as the main state agency to develop surface and groundwater resources, to make optimal use of them and to develop them in such a way as to achieve optimum benefit (Article 1). DSI's regional directorates are organized along administrative boundaries, and Law No. 6200 does not explicitly recognize the 'river basin' as a unit of operation. Rather, its basic aim is to empower DSI with developing Turkey's freshwater resources. It emphasizes 'water quantity' development as against to 'water quality' improvement. In the 1950s, water resources development for economic and social development in Turkey was deemed crucial, thus, DSI was established as the main agency in charge of systematically developing water resources for the benefit of the country. However, in due course, the concerned ministries have adopted river basin management approach particularly when drafting a comprehensive 'new' water law (Sumer 2013).

15.2.2 Groundwater (Law No. 167, 1960)

According to Article 756 of the Civil Code and the Groundwater Law No. 167 (1960) groundwater resources are public waters, and, therefore, shall be under the command and possession of the State. Possessing a part of land does not convey the ownership of water under that land. Law No. 167 (1960) stipulates all issues concerning groundwater research, utilization, protection and registration. The Law put property rights into the public domain. User-rights are subject to licensing upon request (within the safe yield of an aquifer), and can neither be transferred nor sold (Kibaroglu and Baskan 2011).

Law No. 167 authorizes the DSI as the central public institution to manage groundwater resources. Compared to surface water management, whereby several institutions have had overlapping responsibilities, groundwater resources was originally single-handedly governed by DSI. However, this clear responsibility for groundwater management empowered by Law No. 167 did not last long and was modified through several subsequent legislation. DSI issues user licenses, monitors pollution but its mandate does not cover groundwater pollution control. Later, Law No. 3202, which instituted the General Directorate of Rural Services (GDRS), and the Law No. 2560 which introduced the Istanbul Water and Sewage Administration (ISKI, Turkish acronym) also began to operate within the domain of Law No. 167. Responsibilities of Water and Sewage Administrations, within the border of all metropolitan municipalities, are to take legal, technical and administrative measures for preventing groundwater pollution and decreasing quantity. Supplying potable water to rural communities by drilling groundwater wells is one of the main duties of the Special Provincial Administrations after the abolition of GDRS. This has resulted in administrative duplications, which, in turn, caused improper protection of groundwater resources. There are legal inconsistencies exacerbated by a lack of effective monitoring. In this regard, experts argue that the Law No. 167 should be revised by adding provisions against illegal well drilling (Sumer 2013).

15.2.3 Domestic Water (Law No. 1053, 1968) (Law 5625, 2007)

Law No. 1053 (1968) entrusts DSI with the provision of water supply for cities with more than 100,000 inhabitants, provided that the government authorizes DSI and that the concerned city council also approves. This Law has largely extended DSI's duties and is seen as complementary to the Law No. 6200. Article 10 of Law No. 1053 was amended in 2007. The Amended Law No. 5625 has annulled the city criterion (cities with a population over 100,000) and extended the duties of DSI. Thus, since 2007, DSI has been authorized for the domestic and industrial water supply of settlements all over Turkey, which have municipality administrations.

The Law No. 1053 specifies that, if deemed necessary, DSI shall give urgency to wastewater treatment plants in progress. Law No. 1053 characterizes how rapid urbanization in the late 1960s was incorporated in Turkey's water management. At the time, there was a need for extensive water infrastructure in urban areas due to boosted migration from rural to urban centers. DSI took the charge of building this infrastructure since the municipalities did not have sufficient financial and administrative capacities to carry on such projects (Sumer 2013).

15.2.4 Irrigation Water (Law No. 6172, 2011)

The Establishment Law 6200 entitles DSI to transfer operation and maintenance (O&M) of irrigation systems to irrigation management organizations, such as village administrations, municipalities, cooperatives, irrigation associations, and other private legal entities. 'Irrigation association' is a form of transfer considered innovative, where the irrigation scheme covers more than one local administrative unit, for example, a village or municipality. From the early 1960s, DSI had a program for such transfers relating to secondary and tertiary canals (Kibaroglu and Baskan 2011). Until 1993, however, DSI was able to transfer of O&M of irrigation systems amounting to only approximately 70,000 hectares to various types of irrigation management organizations. The process has gained momentum since 1993, and within the past two decades, the management of irrigation covering more than two million hectares has been handed over to local administrations or to irrigation associations (Kibaroglu et al. 2009).

Following the transfer of O&M, DSI maintains only the ownership of the resource infrastructure. The responsibility for the secondary and tertiary canals is transferred to the irrigation associations or irrigation cooperatives. A cooperative is different from irrigation association as it is owned and operated by its members who share its profits or benefits (Svendsen and Nott 2000).

The legal standing of the irrigation association should in principle be guaranteed by an enabling law, which authorizes its establishment and the transfer agreement between the state agency and the irrigation association (Kibaroglu et al. 2009). However, in Turkey, the accelerated transfer program progressed much faster than planned, and there was no opportunity to prepare an enabling law. The associations were established by reference to three laws: the Village Law (No. 442), the Local Government Law (No. 1580) and the Provincial Governance Law (No. 5442). The need for a new law that would determine the principles of irrigation association functioning was articulated by different agencies. In 2005, irrigation associations were brought under the jurisdiction of a new legislation pertaining to Local Administrative Unions (Law No. 5355, 26 May 2005). That legislation did not bring about major changes.

Irrigation associations finally gained public legal authority status following the legislation of the Irrigation Associations Law No. 6172, which entered into force 08.03.2011. With this law, many changes are brought to the structure and functioning of the irrigation associations. Accordingly, irrigation associations are set up by the local authorities in an irrigation zone and apply to DSI in order to sign the transfer agreement and protocol, which gives them the right to collect fees and assigns them the responsibility to distribute water and maintain the canals. According to the Law, irrigation associations are responsible for the operation, management, maintenance and repair (Ozerol 2013).

The Law No. 6172 changed the one farmer-one vote principle by increasing the weight of those farmers who own or rent tracts larger than the average in their irrigation associations. Currently, the number of votes in the election of councilors depends on the size of land a farmer owns or rents (for a period longer than 5 years) with a maximum of 5 votes per farmer.

The chairperson of the association is elected by the members of the irrigation association's assembly (parliament) for a 4-year term and is the head of the executive committee, which decides on matters related to the management of the associations.

Technical staffs are hired to operate the system. In line with the Law No. 6172, the associations are not allowed to spend more than 30% of their yearly budget on personnel expenditures.

The revenues of the association consist mostly of fees collected from the users. The fees depend on the crop that will be cultivated and are set by each association (Unver and Gupta 2002). Self-auditing mechanisms for the irrigation associations existed but were not widely used. Prior to the Law No. 6172, a group of councilors could be selected to audit the accounts, to question the chair, and to scrutinize the yearly activity report submitted to the council by the chair (Kadirbeyoglu and Ozertan 2015). The Law No. 6172 established an audit committee selected from among the councilors. However, the extent to which this committee can perform its duties depends on the power asymmetries in the local context. There are external checks and balances in the system as well in that it is the responsibility of the Governor's office to monitor the activities of the irrigation associations and to approve their fees and budgets. The Governor's office is in charge of establishing an audit commission to scrutinize the finances and administration of the associations.

On 19 April 2018, with a new Law No. 7139, major amendments were made in Law No. 6172 along with the amendments made in Law No. 6200. With this new law, chair of the irrigation association will be appointed by the Minister from among the civil servants upon the suggestion of DSI (Article 9). The irrigation association assembly will not elect the chair anymore. This is, in fact, against the main principle of Law No. 6172 which is based on decentralization and local management. It brings back central control of irrigation management by DSI and the Ministry. Moreover, responsibility for abolition of an irrigation association is taken away from the irrigation association assembly (Article 20). In determination of the fact that the irrigation association is not fulfilling its objectives, the association will be abolished with the approval of the Minister upon the suggestion of DSI (Camlibel 2018).

15.2.5 Hydropower (Law No. 4628, 2001) (Law No. 5346, 2005)

National energy sector, including hydroelectricity production, was deregulated in the 1980s. Before the 1980s, hydropower dams were constructed through public investments without the participation of the private sector (Baskan 2011). With the adoption of the Electricity Market Law No. 4628 in 2001, an independent public institution, namely the Energy Market Regulatory Authority (EMRA) was established with a responsibility of issuing licenses for production activities (including hydropower generation) in the electricity market.

For hydropower projects, in order for the private sector to get licenses a '*Water Use Rights Agreement*' should be signed between the DSI and the private entrepreneur. The "Law on the Utilization of Renewable Energy Resources for the Purpose of Generating Electrical Energy" (No. 5346), which entered into force in May 2005, entails "the guarantee of purchase" principle, which guarantees the purchase of a company's service by the government. This law provides a strong incentive for private investments.

According to the procedures, companies must first apply to DSI in order to sign Water Use Rights Agreements. In line with the "Regulation on the Procedure and Principle of Signing Water Use Right Agreement to Make Production in the Electricity Market" (2003), the agreement grants the production license to the private company. Private companies apply for those projects they select from EMRA's predetermined list of potential projects. The companies start to work after they get EMRA's approval (Baskan 2011).

Both the "Electricity Market Law" No. 4628 of 2001 and the Renewable Energy Law No. 5346 of 2005 enabled the Turkish government to speed up the development of hydropower potential by involving private investors and financial service institutions.

15.2.6 Environment (Law No. 2872, 1983)

The Environment Law of 1983 (No. 2872), which was revised on 26 April 2006, is envisaged as a framework law that stipulates main principles concerning environmental protection and pollution prevention in Turkey. It endorses the 'polluter pays principle' and handles environmental issues broadly. The aim of the law, which considers the environment as a whole, is not only to prevent and eliminate environmental pollution, but also to allow for the management of natural and historical values and land in such a way as to utilize and preserve its richness for future generations. According to its basic principles, citizens as well as the State bear responsibility for the protection of the environment. It emphasizes that every effort should be made to minimize and solve environmental problems in economic activities, in particular when determining production methods (Kibaroglu and Baskan 2011).

Since the Environment Law (1983) is a framework document, it was assumed that the relevant by-laws would be introduced in due course. The efforts towards European Union (EU) alignment have accelerated that legislative changes. Since the opening of the Chap. 27 on Environment in 2009, there have been two closing benchmarks for water management in Turkey: transposition of the Water Framework Directive (WFD) and the completion of the river basin management plans for 25 river basins in Turkey (Fehim 2012). Hence, numerous by-laws have been adopted particularly on environmental protection, water quality and river basin management (Ministry of Forestry and Water Affairs 2018).

15.2.6.1 By-Law on Environmental Impact Assessment (EIA)

This by-law was issued in 1993 and revised several times. The major revision was done in 2014. Accordingly, it is mandatory for all large-scale projects, including storage facilities (dam reservoirs) with a reservoir volume of 10 million cubic meters and more, and for run-off the river type hydropower projects constructions with capacities above 10 megawatt. A series of amendments were introduced to the 2014 By-Law in 2016, 2017 and 2018. The amendments inserted in 2017 require 'water transfer projects over 10 million cubic meter/year between basins' to go through an EIA process (Article 13).

Under this by-law as well as in accordance with the international agreements to which Turkey is a party, projects that would have impact on wetlands, lakes, protected areas and eco-systems, which are rich in biodiversity are subject to EIA. Turkey is party to the Barcelona Convention and the Ramsar Convention, which concern the protection of the Mediterranean Sea and the protection of wetlands, respectively. Turkey has not yet signed the United Nations Economic Commission for Europe (UNECE) Convention on Environmental Impact Assessment in a Transboundary Context (Espoo 1991), which refers to EIA in a transboundary context. However, as EU accession talks resumes, Turkey will have to consider signing the Espoo, Aarhus and other UNECE conventions (Kramer and Kibaroglu 2011).

15.2.6.2 By-Law on Water Pollution Control

The By-law on Water Pollution Control entered into force in 1988 (No. 19919), revised in 2004 (No. 25687), in 2008 (No. 26786) and in 2018 (No. 30332). It aims at both conserving the quality of water resources and water-dependent ecosystems, and protecting and improving water quality to meet human demands. This by-law is noteworthy in the sense that it is the first legislation specifically designed for the protection of water quality. It marks the rising significance of environmental concerns and water quality issues in Turkish water management policies. In 2004, it was amended by a new by-law, which had been envisaged in the framework for the National Program for the adoption of EU *acquis communautaire* (By-law on Water Pollution Control 2004). Through this new by-law, Turkey's basic legislation on water quality has mostly been aligned to that of the EU. Several authorities became responsible for different aspects of the prevention of water pollution: the Ministry of Environment and Forestry (MoEF), the Ministry of Health, the Ministry of Industry and Trade, the Ministry of Culture and Tourism, Provincial and District Governors, DSI, and the municipalities.

The By-law of 2004, required DSI and the MoEF to respectively prepare 'basin plans' and 'basin protection plans' (Article 5). The definitions of these plans are provided in Article 3. Moreover, the By-law (2004) defines emission limit values, i.e. "the maximum allowable discharge of pollutants into natural and artificial receiving water bodies." It divides inland surface waters into four categories, and defines water quality standards for each category of receiving water bodies (Orhan and Scheumann 2011). The By-law on Water Pollution Control also regulates "the permit system for direct (into receiving waters) and indirect dischargers (into municipal sewage systems)".

The third amendment of the By-law, dated 13 February 2008, enabled Turkey to harmonize its water pollution control legislation with the EU legislation. It empowered the MoEF provincial branches with the mandate of issuing discharge permits. Before this amendment, the governor was the responsible authority to take care of this matter. Various new concepts such as 'eutrophication', 'sensitive water zone', 'urban waste water', 'recreational areas' and 'bathing water' were introduced with this new version of the by-law. By-law (2008) encompasses a new legal obligation not to release untreated waste-water into receiving water bodies (Article 4). This introduces a more strict prohibition of the discharge of waste-water into drinking water bodies (Sumer 2013).

The third amendment of the By-Law (2004) enacted on 14.02.2018 (Official Gazette No. 30332). As the By-Law on Protection of Drinking Water Basins was prepared by the Ministry of Forestry and Water Affairs and entered into force on 28.10.2017 (Official Gazette No. 30224), the articles (16–20) on the protection of the drinking and domestic water resources were eliminated in order to prevent duplication in two related by-laws (Ministry of Environment and Urbanization 2018).

In addition to these pioneering by-laws, a series of other by-laws have entered into force on controlling water and environmental pollution stemming from dangerous substances, water intended for human consumption, protection of waters from agricultural nitrate pollution, urban waste water treatment, bathing water quality, groundwater, surface water and drinking water quality management and protection (Delipinar and Karpuzcu 2017). Among them, which plays a decisive role in river basin management is the By-Law on the Preparation of the River Basin Protection Plan and Management Plan of 2012 (Official Gazette No. 28444).

15.2.7 Draft Water Law (Bill)

The concerned ministries and experts frequently underlined the need for a comprehensive water law. One reason for this requirement is the 'outdated' nature of the existing laws such as the Law on Waters (1926), which do not respond adequately current water needs in the country. Significant changes in the economic and social conditions of the country started to challenge the implementation of existing laws. To illustrate, Groundwater Law No. 167 adopts the 'first come, first served' approach as regards to issuing use permits. This was not critical in early 1970s when the available water resources were matching the demand. However, over time, the available groundwater reserves became short to respond to the demands of the population (Sumer 2013).

Moreover, water user rights and ownership are not clearly defined in Turkey. According to the customary practice, assigned user rights enjoy the right of 'prior appropriation'. These user rights cannot be sold or transferred. User rights to water resources are subject to title deed registration (Kibaroglu and Baskan 2011). While private water is subject to specified legislation, there is no specific legislation governing public (surface) waters. This legal practice for 'user rights' was also valid for groundwater resources. With the adoption of the Law No. 167, groundwater resources were transferred from the 'private' to the 'public' domain. (Sumer 2013).

Principal laws such as the Law on Waters (1926) and the Groundwater Law No. 167 do not comprise a river basin management approach. However, the by-laws, which have been enacted after 2004, particularly within the context of the EU harmonization process, are clearly drafted with a river basin management approach. Therefore, while secondary legislation is updated, main laws remain largely intact. Adoption of a new water law, which would provide a framework compatible with basic principles of the EU WFD, was one of the commitments made by Turkey in the context of EU accession negotiations (Sumer 2013).

Against this background, the Ministry of Forestry and Water Affairs (MoFWA) prepared a new water bill in 2012. For feedback, the MoFWA sent the draft bill to various stakeholders, namely government institutions, metropolitan municipalities, NGOs, unions of chambers of engineers, universities, and association of irrigation associations and irrigation cooperatives.

The draft bill aims at eliminating the current situation of fragmented water management and to create an efficient governance scheme in which the MoFWA would be the single main responsible authority. It adopts the view that water resources should be conserved, improved, developed and used at basin level (Article 4). The draft bill envisions that MoFWA prepares a national water plan to meet social, economic and ecological needs, taking into consideration the current and future status of water resources in terms of quantity and quality (Article 6). Article 7 and Article 8 entail how basin management plans and flood control and flood management plans will be prepared either under coordination or supervision of the MoFWA, respectively. About the allocation of water resources, the bill brings forward the systems of 'water allocation certificates' and 'water allocation register' as well as 'basin water allocation plans.' MoFWA is authorized to prepare basin water allocation plans at the basin or sub-basin level through joint evaluation of surface and groundwater resources, and by taking into consideration water use priorities and all other needs (Article 12). Water allocations to citizens and legal entities shall be made by DSI, which will take the basin water allocation plans as the basis for allocation (Article 13). Water allocation certificates shall be issued for the allocated water resources and natural mineral waters, and this certificate shall be subject to a fee. Water shall be used in compliance with the water allocation certificate (Article 13/2). Water allocation register, which shall be publicly accessible, shall be kept by DSI (Article 15). Furthermore, the draft bill enshrines the principle of full-cost-recovery (Article 22/6). In this respect, it is a step forward in complying with some of the basic principles of the WFD.

However, there are shortcomings and loopholes as well. To illustrate, although the draft law deliberately targets the development of a participatory framework for water management, it falls short of explaining how this participation is going to be realized. Another issue is the definition of transboundary waters resources. Though 'transboundary waters' is mentioned in the definitions (Article 2). The draft law only stipulates bodies of water forming boundaries (Article 20). One of fundamental novelties that the draft law brings is the creation of new authorities for water pollution control and monitoring. Yet, it is not clear how the creation of these new authorities will act together with central institutions namely the DSI and the MoFWA.

Several public institutions (e.g. Union of Chambers of Turkish Engineers and Architects) and NGOs (e.g. Turkish Foundation for Combating Soil Erosion, for Reforestation and the Protection of Natural Habitats-TEMA, Turkish acronym) as well as experts and scientists published their criticisms on the draft water bill. However, it is not clear if and how the MoFWA revised the draft bill by taking into consideration these criticisms and suggestions. The draft water bill is still waiting to be discussed at the Parliament for further consultations and possible amendments.

15.3 Main Water Institutions

15.3.1 Ministry of Forestry and Water Affairs

An institutional structure of environmental protection and water management has emerged over the past three decades in Turkey, driven by an increased emphasis in domestic law, the expansion of activity in terms of bilateral and multilateral international agreements and the nation's efforts to meet EU criteria toward full membership. The Ministry of Environment, was established in 1991. Its main responsibilities included the implementation and enforcement of policies for the protection and conservation of the environment. However, the Ministry of Environment had limited resources and limited competence (Kibaroglu and Baskan 2011). Hence, with a governmental decree, the Ministry of Environment and Forestry (MoEF) was established in 2003, merging two central bodies: the Ministry of Environment and the Ministry of Forestry. MoEF had concerted aims of protecting and promoting of the environment, and ensuring the most appropriate and the most effective use and protection of the land and natural resources in rural and urban areas. MoEF played a key role in the EU harmonization process. It was regarded as the key bureaucratic establishment to assume overall coordination and responsibility for the approximation of the EU environmental *aquis*.

Later on, within the governmental reorganization process, MoEF was restructured and its duties and responsibilities were undertaken by two different ministries, namely the Ministry of Forestry and Water Affairs (MoFWA) and Ministry of Environment and Urbanization. The leading government body that has dealt with management and protection of water resources since 2011 has been the MoFWA. The main duties and the responsibilities of the Ministry comprise, among others, creating policies for sustainable protection and utilization of water resources; coordination of national water management; harmonization of the Turkish water legislation with the EU; protection, improvement and management of national and natural parks, protected wildlife reserves, wetlands and biological diversity; producing policies and strategies for the purposes of monitoring meteorological events and taking essential measures; crafting policies with regard to protection, improvement, managing, rehabilitation of the forestry, prevention of desertification and erosion, continuation of reforestation; in collaboration with the concerned government institutions monitoring and contributing to the international studies which are within the scope of the ministry.

After the June 24, 2018 elections, under the new government system of Turkey, new ministries were formed with a presidential decree (Hurriyet 2018). Thus, MoFWA and the Ministry of Food, Agriculture and Livestock merged and the new Ministry of Agriculture and Forestry was established. In the first statements of the new Minister, the importance of agriculture, forests and water resources were underlined (Ministry of Agriculture and Forestry 2018). Main tasks and responsibilities of the new ministry comprise, among others, food production safety, rural development, protection and efficient use of land, water and biodiversity. Water management stays among the main administrative units (directorate general) of the new ministry. "Water policy" will be handled together with other issues such as food, agriculture and forestry within the same ministry.

15.3.1.1 Directorate General of State Hydraulic Works (DSI)

Directorate General of State Hydraulic Works (DSI, Turkish acronym) is the primary executive state agency responsible for planning, design, construction, and operation of hydraulic structures in order to develop the nation's overall water resources in a sustainable manner. DSI, which was established by Law No. 6200 (18th December 1953), constructed a series of dams and hydroelectric power plants and built an extensive system of irrigation and drainage systems all over the country. It implemented large-scale projects for energy production, irrigation development, drinking water provision and flood control.

DSI was first established as a public agency under the Ministry of Public Works; later on functioned under the Ministry of Energy and Natural Resources, and has been affiliated to the Ministry of Forestry and Water Affairs since 2011. As a public agency, it is responsible for four major tasks: providing water supply for domestic and industrial use; taking necessary measures to prevent flood hazards from causing life and property losses; equipping all economically irrigable land with modern irrigation facilities; and developing technically viable capacity to generate hydroelectric energy. In order to achieve those tasks, DSI primarily constructs dam projects, which are at the center of the four tasks. Therefore, DSI is mainly known as a public agency developing dam projects (DSI 2012). DSI conducts water management and investments with 26 regional directorates, roughly settled up based on the river basin boundaries and dispersed throughout Turkey. They execute their work on behalf of the DSI General Directorate according to annual and 5-year development plans as well as investment programs (DSI 2012).

DSI carries out survey and planning for river basin development; collects data pertaining to the quality and quantity of the surface and groundwater resources; prepares master plans and feasibility reports to determine technically and economically optimal solutions of water resources projects in the river basin planning; where necessary, executes land expropriation as well as preparing resettlement action plans for people affected by dam constructions; and prepares environmental impact assessment reports (DSI 2012).

Since late 1980s, under economic liberalization programs, some of the tasks of DSI have been transferred to the private sector, irrigation associations and other state institutions (Kibaroglu et al. 2009). Thus, DSI prepares contract documents and implements bidding for the works to be contracted out to private sector entities; proposes inclusion of projects in the investment programs; supervises constructions; allocates water usage right to private sector for hydro power plants construction and operation; and transfers hydraulic structures to the agencies concerned: HEPPs to the electricity authority, water treatment plants to the municipalities, irrigation facilities to the irrigation organizations (DSI 2012).

15.3.1.2 Directorate General of Water Management

Directorate General of Water Management (DGWM), founded under the MoFWA, is a relatively new agency. It is responsible for developing policies for protecting and sustaining water resources, and coordinating and preparing river basin management plans (RBMPs) together with relevant stakeholders. Turkey is making concerted efforts to prepare RBMPs for 25 river basins aligned with the WFD, with the main goal of reconciling economic development and ecosystem maintenance.

Its main duties and responsibilities are determining water resources policies; providing coordination at national and international level; preparing the RBMPs, conducting the legislation studies on coordination of sectoral water allocation according to RBMPs; developing water quality standards and water quality monitoring systems for the whole country; developing policy and strategy related to flood control, preparing related legislation and flood management plans; preparing the National Water Database Information System; identifying and monitoring sensitive areas in terms of water pollution and nitrate; conducting studies on climate change effects on water resources, and in collaboration with concerned institutions, namely Ministry of Foreign Affairs, conducting studies on boundary and transboundary waters as well as following up international water conventions (Delipinar and Karpuzcu 2017).

15.3.1.3 Turkish Water Institute (SUEN)

The Turkish Water Institute (SUEN, Turkish acronym) was founded in 2011 as a national think tank under the MoFWA. SUEN is entrusted with the objective of conducting and supporting scientific research to strengthen Turkey's national and international water policy (Kibaroglu 2014). SUEN works in cooperation with national and international water-related institutions on issues such as sustainable water management, developing water policies, sustainable energy and capacity building for solving local and global water problems (SUEN 2018). SUEN's main tasks include conducting scientific studies to develop and support national and international water policies; following the activities, innovations and statistics of national and international water establishments; organizing national and international education programs; contributing to national and international fora, conferences, meetings, symposia, training programs and similar activities. Since its establishment, SUEN organized training programs to more than 500 people from 3 continents (more than 25 countries). The series of delegations from the concerned ministries and research institutes attends training courses organized by the SUEN with programs of lectures, covering the planning of water resources, water and wastewater treatment, water quality management and river basin planning (SUEN 2017).

15.3.2 Ministry of Environment and Urbanization

Ministry of Environment and Urbanization was established in 2011 (Law No. 6223, Official Gazette, No. 27984) along with the MoFWA. Its main duties and responsibilities encompass preparing environment legislation and monitoring its implementations; defining and implementing secondary legislation on waste water treatment; developing policies on environmental protection and prevention of environmental pollution; developing and approving environmental plans and ensuring their implementation; ensuring the implementation of the national EIA regulation; evaluation of environmental impacts of all facilities and industrial activities; giving environmental permissions and auditing; utilization of geographical information systems; developing plans and policies for mitigating the impacts of global climate change, and aligning with concerned institutions for monitoring and contributing to international activities related to environmental policy and legislation.

15.3.3 Ministry of Food, Agriculture and Livestock

The former Ministry of Agriculture and Rural Affairs was reorganized as the Ministry of Food, Agriculture and Livestock (MoFAL) in 2011 (Law No. 6223,

Official Gazette, No. 27958). It is the main state institution to develop policies on agriculture, food and livestock. It is responsible for making investigations and preparing projects to protect and improve soil, water, plant, animal and fisheries resources and products; controlling wastewater discharges into fish production areas and monitoring nitrates and pesticide parameters in surface and groundwater. It also has a supportive role for providing training to the farmer's associations. Rural Development Plan (2010–2013) has been prepared by the Ministry in order to develop the working and living conditions of rural population and to enable sustainable development.

Under the MoFAL, DG Agrarian Reform (TRGM, Turkish acronym) is in charge of carrying out activities that are critical for the development and practice of irrigated agriculture. The off-farm and on-farm investments to expand irrigated agriculture are undertaken by DSI and TRGM, respectively. DSI undertakes the large-scale irrigation investments, whereas TRGM carries out the land consolidation, on-farm development and land distribution activities.

Although they were abolished, DG Rural Services and DG Land and Water under the former Ministry of Agriculture and Rural Affairs are also relevant, since both organizations worked about water and land management, particularly in rural areas. DG Land and Water was abolished in 1984 and overtaken by DG Rural Services, which was established to consolidate all the governmental bodies that work about rural development. However, DG Rural Services was also abolished in 2005, mainly due to its high personnel costs. Most of the tasks as well as the personnel of DG Rural Services were overtaken by provincial administrations. This institutional restructuring in water and land management became disruptive and ended up with delays and failures in the efforts for land consolidation and increasing irrigation ratio and irrigation efficiency. Due to cuts in personnel and funding, land consolidation and on-farm development services slowed down after the DG Land and Water was abolished in 1984 and almost halted after the DG Rural Services was abolished in 2005. However, TRGM became legally authorized for the provision of on-farm development services with the amendment of the Soil Protection and Land Use Law in 2011. This change improves the authority of TRGM to implement on-farm development and creates an opportunity to align off-farm and on-farm irrigation development.

After the June 24, 2018 elections, under the new government system of Turkey, new ministries were formed with a presidential decree (Hurriyet 2018). Thus, MoFWA and MoFAL merged and the new Ministry of Agriculture and Forestry was established.

15.3.4 Ministry of Energy and Natural Resources

The Ministry's main tasks, among others, include evaluation of renewable energy resources and determination of policy and strategy for the purposes of increasing energy efficiency and productivity. Moreover, the Ministry had hosted the first agency responsible for stream flow measurement, hydropower planning and design, namely the Electrical Investigation Administration (Elektrik Isleri Etud Idaresi, EIE in Turkish acronym), from the year 1935 to the year 2011. EIE was the main agency responsible for hydrometric measurement (part of this responsibility lay with DSI). EIE conducted studies and surveys to explore the country's hydropower potential, executed engineering services and designed studies for dams and hydroelectric power plants. It also carried out studies for new and renewable energy sources (wind power, solar energy, etc.) to determine if they were feasible for producing electrical energy, and it oversaw and made hydrological plans for Build-Operate-Transfer (BOT) projects (World Bank 2006). Most of the tasks of EIE as relates to renewable energy is transferred to the General Directorate of Renewable Energy under the Ministry.

15.3.5 Ministry of Health

In the early days of the Republic, one of the most pressing issues was to improve public health, which explains why the Ministry of Health (MoH) formulated the Law on Waters in 1926. This led to investment in drinking water supply and the draining of swamps (Kibaroglu and Baskan 2011). Today, the Ministry is responsible for determining quality standards for drinking and domestic water, monitoring these standards and preparing legislation in these areas. It has control responsibilities on environmental protection and urban wastewater collection and treatment in terms of public health. In collaboration with the MoEF (and later on with MoFWA), MoH transposed the EU Drinking Water Directive to the Turkish legislation. MoH also extended contributions to the legislative drafting studies for bathing waters (Sumer 2016).

15.3.6 Ministry of Development

The Ministry of Development was formed in 2011 after the dissolution of the State Planning Organization (DPT in Turkish acronym). DPT, founded in 1960, was the strategic organization established to guide economic and social development through each of the Five Year Development Plans. Each Five Year Plan is a basic planning instrument, which defines investment priorities and the allocation of resources for public investment. To illustrate, the Tenth Development Plan (2014–2018) which was prepared by the Ministry of Development, along with experts from concerned ministries, underlines the main objective for land and water resources management as "preservation and development of quantity and quality of water and land resources, and development of a management structure that provides sustainable use of these resources, especially in the highly demanding agriculture sector". Moreover, as the main policy to achieve such an objective is determined as follows "shortcomings and uncertainties in the legislation on water management will be eliminated, duties, powers and responsibilities of institutions will be clarified, and collaboration and coordination among all institutions involved in water management will be enhanced".

15.3.7 *Ministry of Foreign Affairs*

Ministry of Foreign Affairs (MoFA) is the main state body to formulate and implement Turkey's transboundary water policy. Water has become an element of Turkish foreign policy, particularly in the 1980s when the Southeastern Anatolia Project (GAP, Turkish acronym) was initiated with an ambitious target of building of a series of large-scale dams and extensive irrigation systems in the Euphrates-Tigris river basin. With the increasing profile of the GAP in the international arena, a bureaucratic structure was systematically developed, which determined the principles, policies and practices concerning transboundary river basins (Kibaroglu 2014). In this context, a separate department/unit in charge of regional and transboundary waters was formed at MoFA, under the directorate general, which is responsible for issues pertaining to energy, water and the environment. In formulating transboundary water policy, MoFA acted essentially according to national interests defined by geographical and historical facts; bilateral and regional political relations as well as socio-economic development needs of the country (Rende 2002). When it comes to transboundary water issues, MoFA played a guiding and defining role and provided a reference point for other government institutions, namely MoFWA and its affiliated institution, DSI. Thus, since the foundation of the Republic, MoFA played the essential role in conducting transboundary waters negotiations and preparation of water treaties, agreements and protocols with Turkey's neighbors.

15.3.8 *Local Administrations*

Law No. 5393 on Municipalities (2005) assigns numerous powers and duties to municipalities, which are, for example, the construction of urban water supply and sewerage systems and wastewater treatment plants (Kibaroglu and Baskan 2011). Municipalities usually prefer to combine water and urban transport services as a means of obtaining revenue and cross-subsidizing public services. In the non-metropolitan areas, the primary concern of local government is usually water supply rather than wastewater disposal and treatment. However, separating water supply and sewerage services under different management lines preclude the possibility of an integrated approach (Cinar 2009).

Metropolitan areas have faced serious sewerage problems as a consequence of population increases from the 1980s onward. This has encouraged the establishment of new organizational models, which link water and wastewater management.

Starting with Istanbul and the establishment of ISKI in 1981, autonomous entities were created with the responsibility for planning, designing, constructing and operating water supply and sewerage services in metropolitan areas. In the beginning, ISKI was independent of the Istanbul Municipality, but after the reorganization of the municipality as a metropolitan administration in 1984, ISKI was subordinated to the Istanbul Metropolitan Municipality as a public entity with an independent budget. This water and sewerage administration model was extended to cover other metropolitan municipalities, such as Ankara in 1987 and Izmir in 1989. Today there are 30 water and sewerage administrations within metropolitan municipalities.

15.4 Transboundary Water Policy-Making

15.4.1 Main Principles

In order to understand transboundary water policy-making framework in Turkey, it is useful to delineate the main principles. Turkish authorities, namely the Ministry of Foreign Affairs (MoFA), have adopted certain principles in conducting transboundary water policy. In this context, MoFA states that "Turkey's policy regarding the use of transboundary rivers is based on the following principles" (Ministry of Foreign Affairs 2018):

- Water is a basic human need.
- Each riparian state of a transboundary river system has the sovereign right to make use of the water in its territory.
- Riparian states must make sure that their utilization of such waters does not give "significant harm" to others.
- Transboundary waters should be used in an equitable, reasonable and optimum manner.
- Equitable use does not mean the equal distribution of waters of a transboundary river among riparian states.
- Transboundary water disputes should be solved between the riparian countries, third party involvement for mediation should not be supported (Ministry of Forestry and Water Affairs 2013).
- The variable natural hydrological and meteorological conditions must be taken into account in the allocation and use of transboundary waters. These variable conditions make it necessary to share the risks of the droughts among all riparian countries. Thus, it is not possible to share waters through fixed quantities or quotas (Ministry of Forestry and Water Affairs 2013).
- Turkey is ready to share its experiences in building hydropower stations, dams and other water structures such as drinking water supply networks and irrigation systems as well as its potential in technology and human resources (Ministry of Forestry and Water Affairs 2013).

- The principle of sharing the benefits at basin level should be pursued (Ministry of Foreign Affairs 2018).

Turkey's upstream position particularly in the Euphrates-Tigris (ET) river basin as well as customary international water law principles influenced the Turkish authorities in formulating these principles (Kibaroglu 2014). According to the first principle, that is to say, "water is a basic human need," Turkey will always have the good intention to release as much of the available water as possible under the given hydrological and meteorological conditions. (Ministry of Foreign Affairs 1996).

First principle also reflects the official Turkish position that "needs" rather than "rights" enable the "equitable use" principle is operational. In this contention, by calculating the objective needs of domestic water users, major sectors of the economy - namely agriculture, as well as taking into consideration the demands of the riparian countries at an international level, it would be possible to turn the right to equitable use into tangible practices (Kibaroglu 2014).

Time and again, the MFA authorities emphasized that "Turkey views water as a catalyst for cooperation rather than a source of conflict." Traditionally, Turkey has also stressed the principle of "good neighborliness" which considers other riparians' interests in dealing with 'transboundary' or 'international' rivers. However, Turkish official discourse explicitly distinguishes between the terms "international rivers" and "transboundary rivers", and considers international rivers only to be those that constitute a border between two or more countries, such as the Meric river which forms the border between Turkey and Greece, and the Arpacay river (Aras basin) where it forms the border between Turkey and Armenia. In the Turkish official contention, while such boundary rivers are to be shared equally between the riparian countries, the water of transboundary rivers should be allocated equitably (Ministry of Foreign Affairs 1996).

Turkey adopts the main principles of customary international water law: "equitable utilization" and "no significant harm" principles. However, Turkey also underlines that "sovereign right to the use of water," as well as "equitable, reasonable and optimum use," rather than simply "equal use" are the defining principles of Turkish transboundary water policy. Turkey objects to the claim of the downstream countries that they should have the right of co-sovereignty on the waters of the upstream country or vice-versa (Kibaroglu 2014). Turkey's strong argument in claiming her sovereign rights over the portion of the transboundary rivers that are situated on her territory basically originates from the foundational principles of international law as codified in the UN Charter as "the principle of the sovereign equality of all its members" as well as "territorial integrity of the members." However, similar to numerous UN resolutions and international instruments, including treaties and transboundary water agreements, Turkey attempts to balance the issues arising out of the sovereignty paradox by putting forward a bunch of cooperation initiatives over its transboundary waters, such as concluding bilateral water allocation treaties, establishing joint water mechanisms, initiating joint projects such as joint dams, and joint technical trainings on water use and efficiency (Kibaroglu 2014).

Yet, Turkey's position regarding international water law was widely perceived as being reluctant, and the fact that Turkey voted against the UN Convention on the Law of the Non-navigational Uses of International Watercourses (1997) might support this view. However, Turkey acknowledges several basic principles of international customary water law. According to the Turkish position, the principle of equitable and reasonable utilization should serve as a guiding rule for the allocation of transboundary waters and the settlement of conflicts. Consequently, Turkey pleads for the limited territorial sovereignty doctrine but objects to the doctrine of co-sovereignty of the riparians, which would strengthen downstream interests (according to the Turkish position) in an asymmetrical manner (Kramer and Kibaroglu 2011). However, it seems that Turkey's reservations mainly stem from a reluctance to agree on far-reaching procedural rules (e.g., compulsory mechanisms for dispute settlement; detailed procedures for prior notification). This does not mean that Turkey rejects any transboundary cooperation. Interestingly, the historical bilateral agreements, which concerned riparians, include mechanisms for conflict resolution (Kibaroglu et al. 2011).

Turkey's state practice demonstrates that the concerned authorities had preferred direct negotiations as a basic diplomatic mechanism to settle disputes over transboundary water resources. Through her experience with the donor agencies during the construction of the Keban and Karakaya dams, Turkey developed a negative stance towards the possibility of third-party involvement in transboundary water issues (Gurun 1994). Turkey contended that the donors' intervention was solely in favor of protecting the rights of the downstream riparians and gave slight recognition of Turkish rights to develop and use the river system (Kibaroglu 2002). MoFA's stance as relates to the role of the third parties did not change much, however, with new institutions founded lately in water bureaucracy, such as the Turkish Water Institute, third parties' such as development agencies, international research institutes and think tanks have become partners in joint projects.

Turkish authorities adopted a progressive understanding in dealing with the transboundary water issues, namely the "benefit sharing" approach thanks to their increasing participation in the global water fora, such as the World Water Forum organized every 3 years by the World Water Council, the World Water Week gathered annually by the Stockholm International Water Institute, as well as through their growing understanding and perceptions about the evolving global water management paradigms.

Rather than sharing the waters through simple arithmetic, as suggested by Iraq and Syria, Turkey suggested sharing the benefits of water-based development projects and water structures by way of conducting joint inventory studies for water and land resources as a basis for a trilateral, final allocation agreement in the ET basin (Ministry of Foreign Affairs 1996). Turkish policy-makers argue that the "benefit-sharing approach" fits with Turkey's historical position and it provides opportunities for win-win solutions. In this respect, Turkey has come up with more concrete proposals, such as the joint dam development projects in the river basins, as initiatives for enhancing mutual benefits related to hydropower and irrigation. Examples of Turkish initiatives for joint development of scarce water resources driven by a pragmatic and workable approach to transboundary cooperation in these river basins include joint water storage projects, such as the Serdarabad regulator (already in operation) on the Aras river (Arpacay), the Suakacagi Dam (in planning and negotiation stage) on the Meric river (Tunca/Tunca), and the proposed Friendship Dam on the Orontes river (an item discussed in the Turkish-Syrian technical talks between 2009 and 2011).

15.4.2 Practices and Analysis

We can assess Turkey's transboundary water policy as the external consequences of the internal economic development strategy, which comprised production of agricultural commodities and achieving independency from energy imports (Kibaroglu et al. 2011). This national water development approach is complemented by a clear articulation of interest in transboundary water development. During the Cold War, Turkey pursued a unilateral water resources development strategy, which must be interpreted in the context of the often very tense relations with the other riparians (e.g. with Syria and Iraq) who, for their part, have for a long time followed a veto-strategy by trying to prevent Turkey from achieving its water resource development plans in the ET basin.

Most of the water diplomacy principles relate to the ET basin regarding its peculiarity among the other five major transboundary basins that Turkey shares with her neighbors. (Ministry of Foreign Affairs 2018). The principles adopted pertaining to the ET basin is the accumulation of years' of experience and practices Turkey developed during the endured diplomatic negotiations with her southern neighbors. Since 1970s, through bilateral and trilateral negotiations with Iraq and Syria, as well as paying visits to water and land resources development projects in downstream countries, Turkish authorities came to the conclusion that there would be increasing pressure on the Euphrates due to the magnitude of the planned irrigation projects of all three riparians in the Euphrates region. Thus, Turkish diplomats asserted that ET rivers constitute a single basin "due to the fact that they are linked by their natural course when merging at the Shatt-al-Arab in the Gulf, and also because of the Thartar Canal built to connect the two rivers inside Iraq." They also suggested that "all existing and future agricultural water uses need not necessarily be derived from the Euphrates" and that "irrigation water for areas fed by Euphrates may also be supplied from the Tigris" (Ministry of Foreign Affairs 1996). This principle, in a way, became an established and traditional principle, without any substantial revision since the mid-1990s.

Turkish decision-makers became overly cautious about hydro-meteorological conditions in the ET basin This is basically due to the fact that the 1987 Bilateral

Water Protocol put Turkey under obligation to release a certain amount (i.e., 500 m3/s) of water from the Euphrates River to Syria (Protocol 1987). This obligation urges Turkey to provide the promised amount of water even in dire conditions such as prolonged, severe droughts. Thus, a series of water diplomacy principles were adopted through that state practice. Turkey's stress on the principles of achieving "efficiency in water use," "prevention of pollution," and "basin wide data exchange" is the culmination of this understanding, which was developed over the years during the fruitless negotiation processes in the ET basin (Kibaroglu 2014).

Turkey's position on transboundary water issues is also characterized by proposals to jointly investigate water use and water needs in respective countries, instead of merely negotiating water rights. This paradigm shift is probably best illustrated by the Turkish offer to build joint dams with Georgia, Bulgaria and Syria that could serve the energy needs of both countries, and the proposed Three-Stage-Plan for the ET basin. The latter would contribute to water allocations that take into account water needs for agriculture, population, industrial water use, and basin-wide costs and benefits of different management options. Taking this Turkish proposal seriously, the offer could contribute to a sustainable water management strategy. However basin-wide and needs-based coordination is highly challenging in political terms, including open questions of distribution and institutionalisation. But in the long-term, the shift from water-rights negotiation to a needs-based approach is highly relevant in the context of water scarcity in international basins (Kibaroglu et al. 2005).

Finally, a new trend has developed in Turkish water diplomacy, which could be defined as "humanitarian water diplomacy." Turkey undertakes humanitarian responsibility by providing financial and technical assistance in the water sector with a specific focus in the Middle East and Africa. The main target of Turkish water aid is to ensure the provision of sustainable, safe drinking water as well as sanitation for vulnerable people living mainly in crisis areas without access to clean drinking water and in need of improved sanitation. Concerned institutions in Turkey, namely the AFAD, TIKA, DSI, SUEN, and the Municipalities and Water and Sewerage Administrations, carry out considerable aid programs individually and/or collectively in the water sector (AFAD 2017).

Turkey's water aid perspective envisages a model for an international water fund that focuses on urgent water related issues particularly in Africa and the Middle East. In line with an analysis of the leading international agencies (e.g. UNHCR, WHO, UNICEF), Turkey identified the following main water related priority/emergency issues, namely: water scarcity and famine, waterborne diseases, and the needs of refugees that require rapid global response (MFWA 2017). Through her firsthand experience, Turkey identified country cases in the Middle East and North Africa, namely Syria, South Sudan, Yemen, and Somalia who were also facing these emergency issues and in urgent need of water aid (Ministry of Forestry and Water Affairs 2017).

References

AFAD (2017) Humanitarian Aid for the Syrians in Turkey. https://www.afad.gov.tr/upload/Node/2373/files/Suriyeli_Siginmacilara_Yapilan_Yardimlar.pdf
Baskan A (2011) Liberalization of Turkey's hydroelectricity sector: institutional changes and the private companies. In: Kibaroglu A, Scheumann W, Kramer A (eds) Turkey's water policy: national frameworks and international cooperation. Springer, Berlin, pp 83–92
By-law on Water Pollution Control (2004) Official Gazette No. 25687, 31 December 2004
Camlibel N (2018) 7139 Sayılı Yeni Yasa ve Sulama Kooperatiflerinin ve Sulama Birliklerinin Yeni Yasal Durumu. Su Politikalari Dernegi, Ankara
Cinar T (2009) Privatization of urban water sewarage services in Turkey: some trends. Dev Pract 19 (3):362–363
Delipinar S, Karpuzcu M (2017) Policy, legislative and institutional assessments for integrated river basin management in Turkey. Environ Sci Pol 72:20–29
DSI (2012) Water and DSI, Ankara. http://www2.dsi.gov.tr/english/water-and-dsi/files/assets/basic-html/page8.html. Accessed 15 June 2018
Fehim B (2012) Basin management in Turkey. Directorate General for Water Management, Ministry of Forestry and Water Affairs, Ankara. http://www.suyonetimi.gov.tr/Libraries/su/HAVZA_IC_ING.sflb.ashx. Accessed 10 May 2018
Gurun K (1994) Akintiya Kurek Cekmek: Bir Buyukelcinin Anilari [Rowing against the flow: Memoirs of an ambassador]. Milliyet Yayinlari, Istanbul
Hurriyet (2018) President Erdogan announces ministers of Turkey's new cabinet, Ankara. http://www.hurriyetdailynews.com/president-erdogan-announces-ministers-of-turkeys-new-cabinet-134375. Accessed 09 July 2018
Kadirbeyoglu Z, Ozertan G (2015) Power in the governance of common-pool resources: a comparative analysis of irrigation management decentralization in Turkey. Environ Policy Gov 25 (3):157–171
Kibaroglu A (2002) Building a regime for the waters of the Euphrates-Tigris river basin. Kluwer Law Int, Amsterdam
Kibaroglu A, Klaphake A, Kramer A, Scheumann W, Carius A (2005) Cooperation on Turkey's transboundary waters, German Federal Ministry for Environment. Nature Conservation and Nuclear Safety, Berlin
Kibaroglu A, Baskan A, Alp S (2009) Neo-Liberal transitions in hydropower and irrigation water management in Turkey: main actors and opposition groups. In: Huitema D, Meijerink S (eds) Water policy entrepreneurs. A research companion to water transitions around the globe. Edward Elgar, Cheltenham, pp 287–303
Kibaroglu A, Scheumann W, Kramer A (eds) (2011) Turkey's water policy: national frameworks and international cooperation. Springer, Berlin
Kibaroglu A, Baskan A (2011) Turkey's water policy framework. In: Kibaroglu A, Scheumann W, Kramer A (eds) Turkey's water policy: national frameworks and international cooperation. Springer, Berlin, pp 3–25
Kibaroglu A (2014) An analysis of Turkey's water diplomacy and its evolving position vis-à-vis international water law. Water Int 40(1):153–167
Kramer A, Kibaroglu A (2011) Turkey's position toward international water law. In: Kibaroglu A, Scheumann W, Kramer A (eds) Turkey's water policy: national frameworks and international cooperation. Springer, Berlin, pp 215–229
Ministry of Agriculture and Forestry (2018.) https://www.tarim.gov.tr/Haber/1461/Bakan-Pakdemirli-Yurutulen-Calismalar-Ve-Projelerle-Ilgili-Brifing-Aldi. Accessed 11 July 2018
Ministry of Environment and Urbanization (2018.) http://cygm.csb.gov.tr/su-kirliligi-kontrolu-yonetmeliginde-degisiklik-yapilmasina-dair-yonetmelik-yururluge-girdi.-haber-223791. Accessed 10 June 2018
Ministry of Foreign Affairs (1996) Department of regional and transboundary waters water issues between Turkey-Syria and Iraq, Ankara

Ministry of Forestry and Water Affairs (2013) Havza Yonetimi ve Su Bilgi Sistemi Calisma Grubu Raporu, Ormancilik ve Su Surasi 21–23 March 2013, Ankara, pp 33–34
Ministry of Forestry and Water Affairs (2017) Turkey's water aid perspective: a vision for practical international water fund. Ministry of Forestry and Water Affairs, Ankara
Ministry of Foreign Affairs (2018) turkey's policy on water issues. http://www.mfa.gov.tr/turkey_s-policy-on-water-issues.en.mfa. Accessed 5 June 2018
Ministry of Forestry and Water Affairs (2018.) http://www.ormansu.gov.tr/mevzuat/y%C3%B6netmelik; http://apps.csb.gov.tr/mevzuat. Accessed 12 June 2018
Orhan G, Scheumann W (2011) Turkey's water policy for combating water pollution. In: Kibaroglu A, Scheumann W, Kramer A (eds) Turkey's water policy: national frameworks and international cooperation. Springer, Berlin, pp 117–139
Ozbay O (2006) Turkiye Su Mevzuatinin Gecirdigi Evreler Haber Bulten. Chamb Geol Eng 3:37–40
Ozerol G (2013) Institutions of farmer participation and environmental sustainability: a multi-level analysis from irrigation management in Harran Plain, Turkey. Int J Commons 7(1):73–91
Protocol on Matters Pertaining to Economic Cooperation Between the Republic of Turkey and the Syrian Arab Republic (1987) United Nations Treaty Series 87/12171, 17/7/1987
Rende M (2002) The Looming global water shortage and Turkey's water management in a transboundary context. NATO Parliamentary Assembly 48th Annual Session, Istanbul, 16–19 November 2002
Report on the Negotiations among Turkey, Syria, and Iraq on the Euphrates River (1975) State Hydraulic Works, Turkish Ministry of Energy and Natural Resources
SUEN (2017) https://suen.gov.tr/en/2017/03/31/irakli-uzmanlar-su-ve-atiksu-egitimiicin-suende. Accessed 10 June 2018
SUEN (2018) http://suen.ormansu.gov.tr/suen/AnaSayfa/AboutUs.aspx?sflang=en. Accessed 10 June 2018
Sumer V (2013) Legal context in water management policy in Turkey: evolution of complex framework in three stages. In: Kibaroglu A, Kirschner A, Mehring S, Wolfrum R (eds) Water law and cooperation in the Euphrates-Tigris region. Brill, Leiden, pp 229–257
Sumer V (2016) Water and politics in Turkey: structural change and EU accession. I.B. Tauris, London
Svendsen M, Nott G (2000) Irrigation management transfer in Turkey: process and outcomes. In: Groenfeldt, D, Svendsen, M (eds.), Case Studies in Participatory Irrigation Management, EDI Participatory Irrigation Management Services. 273 IBRD, Washington
Unver O, Gupta R (2002) Water management. Cross Cutting Issues. Middle East Technical University, Ankara
World Bank (2006) Irrigation and water resources with a focus on irrigation prioritization and management. Economic Sector Work, Working Paper. World Bank, Turkey

Chapter 16
Sustainability Issues in Water Management in the Context of Water Security

Nilgun B. Harmancioglu, Cem P. Cetinkaya, and Filiz Barbaros

Abstract This chapter basically complements Chap. 13 on river basin management in Turkey and elaborates further on the concepts of sustainability, integrated water resources management (IWRM), and water security. Sustainability and IWRM are long renowned concepts in Turkey, but their integration into developmental issues and water management has been rather slow. The same is true for the fundamental basis of water security although it is also a new concept in the world. In Turkey, the basic difficulty has been to perceive the link between sustainability, water resources, and development. Misconceptions on these linkages and drawbacks related to institutional and legal aspects of water management in the country have slowed down the development of sound policies for decision making in management. Studies on these issues long remained at academic level, and the early institutional responses were rather slow. The chapter discusses the above problems and presents examples of the few studies carried out on sustainability and water security, including the water-food-energy nexus.

Keywords Sustainability · Integrated water resources management (IWRM) · Water security · Water allocation · Water-food-energy nexus (WFE)

16.1 Perception of Basic Concepts

Chapter 1 of this book has mentioned that we live in an age of environmental (including water) crises at local, regional, national and global scales, which started practically as early as the 1970s. The international community reacted fast this

---

N. B. Harmancioglu (retired) (✉) · C. P. Cetinkaya · F. Barbaros
Faculty of Engineering, Civil Engineering Department, Dokuz Eylul University, Izmir, Turkey
e-mail: nilgun.harmancioglu@deu.edu.tr; cem.cetinkaya@deu.edu.tr; filiz.barbaros@deu.edu.tr

N. B. Harmancioglu, D. Altinbilek (eds.), *Water Resources of Turkey*, World Water Resources 2, https://doi.org/10.1007/978-3-030-11729-0_16

situation, and, through a series of large conventions and policy decisions thereof, developed some basic concepts to underlie environmental resources management. In particular, UN (United Nations) and its affiliated organizations have taken significant steps within the last 30 years towards setting these concepts and procedures to be followed (Harmancioglu 2017).

Although a number of developments were achieved up to the 1990s, the major event to delineate significant concepts in environmental (water) policy took place in 1992, i.e., the "UN Conference on Environment and Development" (UNCED Earth Summit) in Rio de Janeiro (UNCED 1992). The resulting document of this conference was the famous Agenda 21, which introduced the basic guidelines to be followed in resource management. Agenda 21 integrated environmental issues into economic and developmental decision making and defined "sustainability" as a management strategy. This reference document emphasizes that decisions must seek for sustainability in 3 dimensions: economic (efficiency), social (equity) and environmental (compatibility), sometimes with a fourth dimension relating to institutional issues. In essence, all new developments regarding environment and development today are still based on Agenda 21.

With respect to resource management, e.g. water resources management, Agenda 21 foresaw that the environment must be managed and protected as a *cohesive system*. Two main concepts accruing to this policy framework were *holistic approach to resource evaluation* and *integrated management of the environment (water) resources (IWRM)*. These concepts stemmed from the following two factors, as emphasized in Agenda 21 (Harmancioglu 2017; Harmancioglu et al. 1998):

(a) environment is recognized as *a "continuum"* of air, soil, and water components, which interact in a number of complex ways;
(b) integration of environmental issues into economic and developmental decision making, thus ensuring *"sustainability"* as a management strategy.

Several international key events were organized until recent times to elaborate environmental and, in particular, water management policies. After the conclusion of the International Decade for Action 'Water for Life', which prevailed between the years 2005 and 2015, a new agenda was defined as the "Post-2015 Development Agenda" in 2015. The formulation of this agenda is a process again developed by the UN to define the future global development framework. In 2015, the UN Millennium Goals (MDGs) completed their period of action and were converted to SDG's or Sustainable Development Goals to define the new agenda, which is now called as the Agenda 30 as it will be valid until 2030. SDG's emphasize socio-economic goals and their linkage to environmental issues by balancing the economic, social and environmental dimensions of sustainable development. A key concept defined within this framework is water security, which is the basic element of the Global Goal on Water (Harmancioglu 2017). "*Water security is … the capacity of a population to safeguard sustainable access to adequate quantities of acceptable quality water for sustaining livelihoods, human well-being, and socio-economic development, for ensuring protection against water-borne pollution and water-related disasters, and for preserving ecosystems in a climate of peace and political stability*" (UN-Water 2013). Water security has a central role in many security areas, such as human security, food security, energy security, economic security, and environmental security. It is recognized that these security areas are all linked to water security (UN-Water 2013, 2015; WWAP 2014).

Turkey was represented in the above international conventions and even organized some congresses herself, like the fifth World Water Forum in Istanbul in 2009 (WWAP 2009) and a few other international congresses in the following years. However, the perception of the policy development within the first 10 years has been rather slow and the significance of the basic concepts were not fully recognized. Thus, it took long for environmental and water communities to become aware of the new policies and related concepts as depicted in Agenda 21.

Sustainability was introduced as a new trend, more of a philosophical nature, into the Turkish environmental jargon after the release of Agenda 21 at the international platform in 1992. In the following years, even Local Agenda 21 programs were initiated at central levels, including municipalities. Thus, the term "sustainability" can be considered as a long renowned and popular concept in Turkey; yet, it is subject to many misconceptions regarding the basic policies it represents. Although the term has been used and used on every occasion, there has been a problem in the sense that every sector, every authority, or every individual evaluated sustainability through his/her own window so that everybody has his/her definition and understanding of it. The most general meaning attributed to the concept has been that the use of environmental resources in development activities should be restrained to a level at which long term and irreversible damages to the environment are prevented, particularly for the sake of future generations (Clark and Gardiner 1994; Harmancioglu et al. 2013). In simple terms, the concept of sustainability has been associated in the water sector with efforts on preventing long term adverse effects of development on the environment. Thus, in Turkey, the focus has been directed, almost solely, to environmental and water quality to be conserved. The idea has been also supported by adoption of the EU WFD (Water Framework Directive) in the Turkish legislation, governing environmental and water resources issues.

At present, the question still remains as to how sustainability, development, and environment (water) are linked together in tangible terms. As noted in the above, Agenda 21 and the following international documents thereafter emphasize that sustainability has to be considered in 3 dimensions: economic (efficiency), social (equity) and environmental (compatibility). Sustainability is achieved as the intersection of these factors, and this target intersection point can be tangibly derived by simultaneously considering all indicators relating to the three dimensions (OPTIMA 2007; Cetinkaya et al. 2008; Harmancioglu et al. 2013; Cetinkaya and Gunacti 2018). This property of the concept is still overlooked by the authorities in Turkey so that public projects do not achieve sustainability to the fullest extent. The three pillars of the sustainability concept are often assessed separately as economic, social, and environmental sustainability without merging them to arrive at realistic decisions on environmental and water management.

As described above for sustainability, similar criticisms are also valid for the policies of integrated water resources management (IWRM) and water security although the latter is a rather newer concept. These issues are elaborated in the following sections.

16.2 Practices of Integrated Water Resources Management

16.2.1 IWRM Problems in Turkey

Agenda 21 is the first international document to introduce the basic concept of integrated (holistic) water resources management (IWRM) along with sustainability. In fact, this concept is recognized as a must for achieving sustainability. As noted earlier, the concept of IWRM stems from two requirements for sustainability, i.e. that the environment must be considered as a cohesive whole and that environmental issues must be integrated with economic and social considerations in decision making for management.

Various international summits following Rio 1992 assessed whether the goals of Agenda 21 had been met or not. The results of these evaluations indicated that there are significant problems in the implementation of the basic guidelines and thus in achieving sustainable management of natural resources. Furthermore, it is concluded that the most important factor leading to the current water crisis is essentially the mismanagement of the resource (Harmancioglu 2017). Even in the recent SDGs, it is stated that "*Today it is widely recognized that an integrated approach to freshwater management offers the best means of reconciling competing demands with supply and a framework where effective operational actions can be taken. It is thus valuable for all countries at all stages of development*" (UN 2017). Besides the technical basis for IWRM, strong legal and institutional adjustments are required, and sound policy making statements must be made.

Just like sustainability, IWRM was received as a popular approach in Turkey but it was not fully recognized as to its requirements. It was supported by many environmental and water communities as a new trend; however, it meant, and to an extent still does, different things to different people. At first, IWRM was unfortunately mistaken for the management of water quality, stemming from the fact that the primary document of Agenda 21 focused on "environment and development", where the term "environment" was interpreted particularly as "water quality". This was also due to increasing water pollution problems in the country at the time being. Yet, this was an unfortunate approach as it overlooked the other major problem of water scarcity. Certainly, the basic question which complicated the perception of the concept was: integrate what and how?

The above questions were cleared to a significant extent at the beginning of the last decade through Turkey's efforts on adopting the EU WFD into Turkish legislation. Authorities recognized the importance of IWRM as a remedy to the increasing water problems in the country and started the preliminary preparations for IWRM projects. Yet, there is still a long way to go before the country can effectively fulfill the requirements of IWRM, particularly in incorporating economic and social dimensions into the environmental framework.

Summarized below are some of the important problems that Turkey experiences towards sustainable water management (Svendsen et al. 2005; Harmancioglu 2008; Michelsen and Harmancioglu 2009):

- Turkey has long suffered from lack of integrated planning. The first efforts to reverse this situation started with the development of "basin protection action plans" which are now being converted into basin management plans. Yet, there is no basin where an IWRM plan is practiced.
- Thanks to the EU WFD, the use of IWRM is now supported; however, it still may mean different things to different people. The development of a common/uniform language for IWRM and management terms is needed. The same holds true for the concept of "sustainability". Likewise, the difference between "planning" and "management" of water resources must be recognized.
- IWRM requires, above all, strong legal and institutional adjustments and the development of sound management policy statements. As it will be remembered from Chaps. 13 and 15 of this book, there are some problems associated with the legal framework in Turkey. In particular, no governing Water Law yet exists; after 5–6 years of efforts, it still remains in a draft form. Institutionally, some significant adjustments have been made at the central level, but there are some issues to be handled, as noted in the points below. These points also relate to policy making.
- Implementation of IWRM plans have a strong economic component. Like many other developing countries, Turkey also experiences financial constraints to develop the required information and conduct the analyses for the successful implementation of IWRM plans. Furthermore, coordination between institutions must be ensured so that they commit themselves to share the necessary funding for IWRM.
- Policy makers need to be educated and convinced to use the supporting science and analysis of IWRM.
- A critical problem that has to be solved is the disconnect between science, planning, policy and decision making. Capacity building, education, training and awareness raising in IWRM is strongly required in Turkey. This includes scientists, managers, and all other stakeholders including water users.
- Furthermore, the iterative nature of IWRM must also be recognized and adaptive management practices must be adopted. IWRM is an iterative process since it takes time, and the processes involved (environmental, social and economic) are dynamic. Thus, decision makers and managers must recognize that Turkey has a long way to go and that the IWRM process is not complete once the plans are made.
- To implement IWRM, goals and policies for management must be established first. It must also be recognized in Turkish institutions that a policy statement should be the first step towards IWRM. This step should cover the delineation of

objectives, constraints and instruments to be used in achieving the objectives of IWRM in a particular basin.

- IWRM requires the use of advanced techniques for decision making, and these techniques, including data, modeling, GIS, remote sensing, etc., have to be integrated, too. In Turkey, there are significant problems in data provision and management, but, fortunately, all water communities are aware of this fact. On the other hand, scientists, water resources managers and decision makers need to be provided with access and training on the use of sophisticated modeling tools. These tools are basically common at the academic level, but it will most probably take some more time for these tools to be applied at the authoritative level for decision making.
- One specific problem in integrated planning in Turkey is the consideration of groundwater resources and surface waters separately whereas they strongly interact. Since groundwater resources in many basins of the country are overexploited, improved control is needed to better manage them.

16.2.2 Examples of IWRM Plans

Probably the first IWRM plan attempted in Turkey is the Buyuk Menderes Basin integrated management plan. In 2002, the Government of the Netherlands provided support for the Turkish government towards implementation of the EU Water Framework Directive in Turkey. Within this context, a consortium led by Grontmij Consulting Engineers developed a project between January 2002 and November 2003, which was supported by MATRA Pre-Accession Program. One of the objectives of this project was to develop an Integrated River Basin Management Plan to be accomplished in the largest basin of the Aegean region. In addition to in-depth socioeconomic analyses, the project foresaw the use of DSS tools in management. The project was realized by a working group comprising the Provincial Directorate of the former Ministry of Environment and Forestry and the 21st Regional Directorate of DSI (Harmancioglu et al. 2005a). The Buyuk Menderes plan was in a way an introduction to WFD in Turkey and basically served for capacity building at authoritative levels.

The best example for integrated planning in Turkey is the Southeastern Anatolia Project (GAP) mentioned in many of the preceeding chapters of this book. GAP started with the establishment of individual structures for water resources development as early as the 70s, but it eventually adopted features of an integrated project. At present, GAP is a renowned large scale regional development project, covering various sectors such as power production, agriculture, rural-urban infrastructure, transportation, industry, tourism, housing, education, health etc. The major objective of this project is to support the socioeconomic development in the region, from which the whole country benefits. It is integrated in the sense that it has a basin administration and incorporates physical, social, economic and institutional features in an integrated framework.

Studies towards IWRM continued at the authoritative level through Basin Protection Action Plans (BPAPs) prepared between 2009 and 2013 by the former MoFWA (the Ministry of Foresty and Water Affairs), as mentioned in Chap. 13 of this book. In essence, BPAPs provide only an inventory of the status of river basins in the country. However, this can be considered as a good start since the first step towards IWRM is to assess the prevailing status of a basin before attempting to take any action. Next, an EU project on "Technical Assistance for the Conversion of River Basin Action Plans into River Basin Management Plans" was undertaken in 2014 to prepare River Basin Management Plans for Konya Closed, Susurluk, Buyuk Menderes, and Meric-Ergene basins. It is stated in Chap. 13 that "these four plans are completed but not yet been put into practice. It is expected that all BPAPs will be converted into River Basin Management Plans by 2023".

In contrast to studies at the central level, some detailed projects were realized as early as 2000 by some academic institutions. The authors of this chapter participated in such EU supported projects as SMART (SMART 2005), NOSTRUM-DSS (NOSTRUM-DSS 2007), and OPTIMA (OPTIMA 2007), where basin management studies were carried out in a water-scarce basin (Gediz basin along the Aegean coast) in Turkey. The same basin was investigated earlier for water management problems within an IWMI (International Water Management Institute) supported project on "Institutional Support Systems" (IWMI 2000). This project basically resulted in a basin status report, pinpointing all major problems in Gediz basin and, thus in Turkey, towards basin management (Svendsen et al. 2005). NOSTRUM-DSS was realized along the same line, and country-wide (including Turkey) reports were prepared on the basis on all relevant data (physical, environmental, social, economic and those relating to developmental activities such as agriculture, energy, construction, etc.). The basic objective of the project was to delineate the current and targeted decision making processes for water management (Harmancioglu et al. 2005a, b). The results of these projects indicated the following problems in the Gediz basin:

(a) No functional system for allocating water rights
(b) Deteriorating water quality resulting from urban and industrial wastewater, and agricultural return flows
(c) Competition for water (water allocation problems; increasing domestic and industrial water demand; changing patterns of demand)
(d) Water shortage (droughts + competing water uses)

This picture of the basin was valid for most of the Turkish river basins and, in a way, summarized the problems of water management in the country at the time being. It is interesting to note that these problems still remain unresolved, particularly due to the fact that IWRM practices have not yet been implemented.

The SMART project was also carried out for the Gediz basin, where the current situation and possible future changes in domestic, industrial and irrigation water demands and supply were estimated on the basis of prevailing trends. In particular, the project served to develop scenarios on the current (baseline scenario) and future (projected scenarios) for sectoral water demands and supplies in the basin (Table 16.1). An annual water budget simulation model called WaterWare (provided

Table 16.1 Scenarios Developed for the Gediz Basin (Harmancioglu et al. 2008)

Variables/driving forces	Baseline	BAU	Optimistic	Pessimistic
Birth control	Existing	Existing (partially successful)	Existing (successful)	Existing (unsuccessful)
Urban growth rate	1.5%/y	1.5%/y	1%/y	3%/y
Rural growth rate	−1% /y	−1% /y	−1% /y	−2% /y
Precipitation rate	700 mm/y	0%	0%	−10%
Groundwater supply	9 mm/y	0%	0%	−10%
Surface water supply	59 mm/y	0%	0%	−10%
Groundwater pollution	Class IV	Class IV	Class III	Class IV
Basin-out water supply (surface & ground)	0.2 mm/y	0.2 mm/y	0.4 mm/y	0.5 mm/y
Domestic water use (surface & ground)	7.4 mm/y	0%	0.5%/y	2.5%/y
Industrial water use (groundwater)	3 mm/y	0%	4%/y	8%/y
Irrigation water use	39 mm/y	0%/y	−40%/y	15%/y
Domestic water supply investments	sufficient	sufficient	sufficient	insufficient
Change in crop pattern	Cotton, grape, maize	Cotton, grape, maize	Grape, vegetable, maize	Cotton, grape
Irrigation m/o investments	insufficient	insufficient	sufficient	insufficient
Loss rate in irrigation system	30%	30%	10%	30%
Irrigated area	1070 km^2	0%	0%	0%
Industrial water use (surface)	0 mm	0 mm	4 mm	4 mm
Surface water quality	Class IV	Class IV	Class III	Class IV
Water exploitation awareness	Insufficient awareness	Insufficient awareness	Comprehensive awareness	Insufficient awareness

by Environmental Software Systems-ESS, Austria) was used to determine the performance of the existing river network system in terms of the available water. The analysis was mainly based on comparison of alternative water management scenarios (SUMER 2005). The project investigated methods and tools for long-term policy analysis and decision support for integrated coastal development. Yet, the emphasis was essentially on water resources and land use, and the resource balance between the coastal region and inland areas. A multi-sectoral integration of quantitative and qualitative analysis was used in the approach by combining advanced tools of systems engineering based on numerical simulation models, with methods of environmental, socio-economic and policy impact assessment using rule-based expert systems technology and interactive decision support methods (Cetinkaya et al. 2004; Harmancioglu et al. 2008).

OPTIMA is essentially the first project developed in Turkey for sustainable water management by merging the three pillars of sustainability mentioned earlier, i.e. environmental, social and economic. It is also the first study that develops an integrated framework for management so that it is considered important to summarize it within the context of this book. The work comprises an optimization system which considers water demand and supply, surface and groundwater, water technologies and efficiency of use, allocation strategies, costs and benefits. The system is composed of a dynamic simulation model with its associated databases and a water resources planning and optimization system. The project was applied through a web-based client-server to support distributed use and easy access for multi-criteria optimization and decision support (Cetinkaya et al. 2008; Cetinkaya and Harmancioglu 2008).

The dynamic simulation model was again WaterWare, which was used in the SMART project. The model describes water resources systems at basin scale, including the groundwater system for conjunctive use. It not only covers the physiographic and hydrological elements, but also aims to represent the institutional and regulatory framework as well as the socio-economic driving forces. The optimization component is based on heuristics and concepts of genetic programming, producing a realistic, detailed, dynamic and distributed representation of the river basin. For the Gediz basin, this component was used to identify sets of non-dominated pareto-optimal solutions in heavily constrained scenarios. These scenarios were actually the basis for an interactive discrete multi-criteria selection with the participation of stakeholders in Gediz. The multi-criteria approach herein covered global and sectoral demand and supply balances, reliability of supply, access, cost and benefits, including environmental and social aspects.

The basic scenarios for the basin were developed for a wet year (1982) and a typical dry year (1991) as the data for these years were complete. All data collected were on a daily basis. The study first developed an annual water balance water budget summary and some global criteria as supply-demand ratio, reliability, benefit cost ratio and economic efficiency. (Cetinkaya and Harmancioglu 2008). Then, some economic parameters (i.e. water costs, benefits and efficiencies) and sectoral budget are calculated through water balance simulations for the basin (Table 16.2).

Table 16.2 Sectoral Water Budget of Gediz Basin according to the 1991 Dry Year Scenario (Cetinkaya and Harmancioglu 2008)

Sector	Domestic	Irrigation	Industry	Light industry	Total
Demand (Mm^3)	38.51	559.21	63.07	31.54	692.32
Net Supply (Mm^3)	39.67	277.73	63.07	31.54	412
Cons. use (Mm^3)	23.1	265.17	28.38	15.77	332.42
Losses (Mm^3)	17.17	37.02	1.58	0.79	56.55
Shortfall (Mm^3)	0	290.04	0	0	290.04
Unallocated	1.16	8.55	0	0	9.72
Supp./Dem. (%)	100	48.13	100	100	58.15
Reliability (%)	100	54.52	100	100	82.69

For the optimization studies, first the Gediz basin stakeholders were asked to define their priorities and concerns over the basin. They identified water scarcity as one of the priority problems so that the reduction of sectoral water demands was determined as the objective to be optimized. Next, they also specified constraints and instruments to solve the major problem of water scarcity. The majority agreed on rehabilitation of the existing irrigation system and the use of more efficient irrigation technologies as the initial activities to be realized towards the optimization of basin management. Accordingly, the use of new instruments, i.e., water technologies that serve to reduce the demands and conveyance losses at irrigation and municipal demand nodes, were selected.

The optimization procedure was run again for the year of 1991, as the dynamic model simulations of WaterWare indicated that system performance was adversely affected by drought conditions in 1991. In this case, the rigid constraints specified by basin stakeholders were used, and about 10,000 runs in the optimization procedure were realized on the basis of these constraints. Optimization scenarios were evaluated through maximizing supply/demand ratio, reliability of supply and benefit/cost ratio criteria while keeping the direct and indirect water costs to a minimum. The reference point used in comparison of alternative solutions in the feasible set is the maximum or the minimum value for each criterion. The procedure produced 50 feasible solutions for the baseline 1991 scenario, which were then used for post optimal analysis with the Discrete Multi-Criteria optimization (DMC) tool of WaterWare. The results are presented in Table 16.3.

Similarly, Table 16.4 indicates a significant improvement in the annual net benefit which defines the expected income generated through the use of water. It can be followed from the above discussion that all these results are based on the consideration of environmental, social and economic as joint criteria for optimization.

Table 16.3 Comparison of Indicators for the Current Situation and after the Planned Investments (Cetinkaya and Harmancioglu 2008)

Indicator	Baseline 1991	Optimized 1991
Supply/demand	49.52%	61.8%
Reliability	80.53%	86.2%
Total Shortfall (Mm^3)	290.04	37.95
Total Unallocated (Mm^3)	9.72	4.01
Benefit/Cost	0.97	1.17
Economic Efficieny (€/m^3)	−0.01	0.05

Table 16.4 Total Net Benefit Values for Baseline and Optimized Scenarios and the Difference after Investments (Cetinkaya and Harmancioglu 2008)

	Indirect total net benefit (€/year)	Direct total net benefit (€/year)
Baseline	−1,730,000	72,180,000
Optimized	19,585,700	81,368,800
Difference	*21,315,700*	*9,188,800*

16.3 Water Security and Allocation

16.3.1 Preliminary Studies

Water security is a newer concept in Turkey, and it has been only 4–5 years since water authorities came to recognize it as a requirement. At the international level, the issue of water security was addressed as early as 2000, and in fact, the second World Water Forum in Hague resulted in a ministerial declaration on "Water Security in the 21st Century", which was one of the first documents to mention the concept (Harmancioglu 2017). However, the first working definition of water security was worked out within the framework of the Post-2015 Development Agenda (UN-Water 2017). This definition is presented earlier in Sect. 16.1 of this chapter. Essentially, the Post-2015 Development Agenda focused on the Global Goal for Water, using the definition of the water security concept, which "*will address multiple priority development areas under consideration: conflict and fragility; environmental sustainability; growth and employment; health, hunger, food and nutrition; inequities; energy; and WATER*" (UN-Water 2013; UNESCO 2013). Along the same line, the Global Goal for Water was the formulated as "*Securing Sustainable Water for All*" (UN-Water 2014). It is deemed that water security is essential to achievement of sustainability as it covers the three dimensions of sustainable development, i.e. social, economic and environmental. The currently valid Sustainable Development Goals (SDGs) are all based on this concept. It is also important to note that all nations should adopt IWRM as a required step towards achieving water security.

It may be followed from the above short discussion that the formulation of a working definition for water security and the following SDGs is a rather new process in the world. This also true for Turkey, and it will take some time for the concept to be fully understood and for actions to be taken in that regard. Yet, this time Turkey took a timely step to have the concept introduced to the relevant communities. The third Istanbul International Water Forum was organized in Istanbul in 2014, which focused on two basic themes: water security and legislation on water. The issue was followed after the forum by investigations on water allocation but no concrete results have been attained yet.

The Global Goal for Water identified 5 targets, 2 of which are related to water allocation that is recognized as a significant process to achieve water security. Within the last decades, equitable allocation of water has been in the agenda of all water communities as the issue even led to serious conflicts among water users. The three basic factors that necessitate equitable allocation are (Harmancioglu 2014; Le Quesne et al. 2007):

- *"Increasing complexity due to population growth, development pressures, and changing needs;*
- *Unequal distribution of water due to political changes, resource mismanagement, and climatic anomalies and thus water scarcity;*

- *Increased competition among uses and users, requiring more effective negotiation and allocation mechanisms".*

Often, construction of new storage and distribution facilities has been the traditional approach in meeting the increased water demands. This approach has been preferred in most of the developing countries such as Turkey. Recently, however, a more realistic stepwise procedure is recommended to include: situation assessment, scenario development, assessment of implications of different scenarios with respect to sustainability dimensions, i.e. environmental, economic (development) and social (equity) (Speed et al. 2013; Harmancioglu 2017). These steps are essentially the basic activities to be undertaken in IWRM.

In general, decisions on sectoral water allocation in Turkish river basins are made by DSI on an annual basis. The first priority in allocating water supplies among sectors is given to urban (domestic) water demands, followed by the environmental and agricultural sectors. Energy and industrial water demands attain the third priority. In preparation of annual allocation plans, DSI first estimates the annual demands of each sector and then realizes the allocation plan on the basis of the available water supplies for that year. This process is rather complex for highly populated basins where agricultural and energy production activities are also intense. The main complexity arises from the fact that no general rules or policies exist for water allocation in Turkey. To alleviate this situation, DSI and authorities aim to start developing water allocation plans in large basins.

There is only one activity towards water allocation in Turkey, which is the project on preparation of the Seyhan Basin Sectoral Water Allocation Plan (SYGM 2017). The project was initiated in 2017 by the General Directorate of Water Management (SYGM) of the former Ministry of Forestry and Water Affairs and is expected to be completed by 2023. The project was followed by an Action Plan prepared in the same year; however, no reports have been released yet as to the activities covered up to date.

A very recent study was carried out at the academic level, which again focused on Gediz Basin where irrigation is the major water consuming activity (Cetinkaya and Gunacti 2018). The work investigated how irrigation water was allocated to various crops, as previous droughts struck crop yields the most. Sixteen alternative allocation scenarios were developed to assess the crop yield of each alternative, which were represented by social, economic and environmental indicators as the basic criteria for selection of the most favorable plan. The selection was realized by Multi Criteria Analysis (MCA) methodology, using the Reference Point Approach (RPA), which was shown in the study as a powerful tool to rationally select among several possible alternatives in water allocation.

16.3.2 Water-Food-Energy Nexus

Water security assigns a central role to water in many other security areas that are linked to water, being the basic element of sustainable development. These areas are: human security (a good level of health and well-being), food security, energy security, economic security and environmental security as mentioned earlier. Sustainability requires that security is achieved in all these areas in relation with each other.

These areas share some common properties such as in the following (UN-Water 2013; WWAP 2014; UN-Water 2015; Harmancioglu 2017):

- Many people do not have access to them;
- Each security area is faced with a rapidly growing global demand;
- Resources are limited for each of them;
- All areas require healthy ecosystems;
- Each area has different regional availability and variations in supply and demand.

Food security and energy security "*generally mean reliable access to sufficient supplies of food or energy, to meet basic needs of individuals, societies, and nations, thus supporting lives, livelihoods and production*", and water security has a similar nature. This link between water, food and energy is referred to as the "Water-Food-Energy Security Nexus" (WFE) and is an important element of the Post-2015 Development Agenda (UN-Water 2013; WWAP 2014; UN-Water 2015; Harmancioglu 2017). This nexus actually underlines most of the Sustainable Development Goals defined so far and is thus a must for achievement of sustainability.

The major problem in Turkey, regarding the WFE Nexus, is that sustainability is sought in each area (water, food and energy) separately, overlooking their intricate relationship. Furthermore, each sector has its own associated problems since Turkey needs strong policy making and development in each area, considering also their strong relationship.

Along with water and energy problems, food security particularly experiences many difficulties which have emerged in Turkey within the last decade. Turkey relies on agriculture (including also animal husbandry) to cover her food demand; however, the agricultural sector currently reflects many deficiencies. First, food demand is continuously increasing due to rapid population growth. Second, migration from rural areas to provinces and metropolises has been significantly high within the last two to three decades so that, at present, the population residing in provinces is as high as 92.5% of the total population (TUIK 2018). This rate indicates that consumption in the highly populated provinces is also high.

Furthermore, the decrease in the rural population results in considerable losses of agricultural production. In recent years, Turkey started to import many of the essential agricultural products (including meat) from abroad. Furthermore, most important agricultural inputs are imported and any instable change in currency exchange rates directly affects the profitability of the sector. The very recent inflation crisis and increases in the prices of many agricultural inputs (e.g. fertilizers, insecticides, pesticides, fuel, equipment, seeds, etc.) have aggravated the situation. As an example, input distribution for wheat production in Western Anatolia consists of 56% fertilizers, 16% diesel fuel, and 22% seeds, which are mostly imported and are, therefore, directly affected by exchange rates (FATIMA D4.1.4 2018).

Considering the above economic constraints, most farmers have started to evaluate the alternative of quitting agricultural activities since agricultural profits on which they rely are in a declining trend. There are other related problems such that the government provides low subsidies or supports only for particular agricultural products through subsidies. Certainly, this reduces the competitive capacity of farmers performing at small or medium scales under free market conditions with respect to large scale farmers or food producers (FATIMA D4.2.3 2018). Consequently, farming will no longer be a profitable family business in the near future. Even now, the number of young farmers is decreasing as they prefer to migrate to large provinces for better chances of education and jobs. To avert this trend, the former Ministry of Food, Agriculture and Animal Husbandry started in 2018 a project to support young farmers, as depicted in the Official Gazette No: 30370 Article No: 2018/12. All the above problems indicate that strong policy making is required at the central level to reverse this situation.

16.3.3 Further Problems

It is worth mentioning some further problems that relate to water pricing in the irrigation sector in Turkey. Water prices are determined not on the basis of the amount of water used but on the basis of the irrigated area and the number of irrigation units per season to cover the general and O&M costs of the irrigation systems. The collection of fees for these costs and their payment schedules, which may vary among different irrigation regions, often create problems for farmers whose economic capacities are low. In general, when farmers run into difficulties due to unstable prices of various inputs (fertilizers, seeds, diesel fuel and energy), they refrain from paying fees for irrigation as interest rates and penalties on these fees are lower in comparison to other loans and payments. (FATIMA D4.2.3 2018). Thus, these problems cause significant financial constraints for irrigation associations and cooperatives that operate surface water irrigation systems in the Aegean, Mediterranean and the Black Sea basins since most of these systems have completed their economic life spans and reflect high O&M costs.

In Turkey, the price of irrigation water is approximately 0.05–0.10 EU/m^3, which is well below those in the European countries (FATIMA D1.2.3 2016). This low rate causes some farmers to use excess water in their irrigation systems, a situation which may look in contrast to the above mentioned financial constraints. However, some lower educated farmers mistakenly consider that they can improve the profitability of their agricultural products by using more water when faced with financial difficulties in paying for other agricultural inputs such as fertilizers (FATIMA D4.1.4 2018). Another reason for excessive water use is the effort to provide as much water as possible to saturate the soil. This is basically due to the low reliability of the distribution systems so that farmers prefer to use water to the fullest extent when it is made available to them. This practice is particularly observed in dry years when the amount of available irrigation water is restricted.

References

Cetinkaya CP, Gunacti MC (2018) Multi-Criteria Analysis of Water Allocation Scenarios in a Water Scarce Basin. Springer Water Resour Manag 32(8):2867–2884. https://doi.org/10.1007/s11269-018-1963-z

Cetinkaya CP, Harmancioglu N (2008) Optimization of water resources management schemes. River Flow 2008: Proc Int Conf Fluvial Hydraul 3:2313–2322

Cetinkaya CP, Barbaros F, Gundogdu H (2004) Determination of social and economical indicators for future water use scenarios in Gediz River Basin. EWRA Symposium on: Water Resources Management: Risks and Challenges for the 21st Century, Izmir, Proceedings V(I), pp 375–383

Cetinkaya Cem P, Fistikoglu O, Fedra K, Harmancioglu NB (2008) Optimization methods applied for sustainable management of water-scarce basins. IWA Publishing, J Hydroinformatics 10 (1):69–95. https://doi.org/10.2166/hydro.2007.011

Clark MJ, Gardiner J (1994) Strategies for handling uncertainty in integrated river basin planning. In Kirby C, White WR (eds) Integrated river basin development. Wiley, Part VIII, pp 437–445

FATIMA D1.2.3 (2016) Comparative analysis of externalities from current intensive and alternative sustainable crop production systems in pilot areas. Project Deliverable of "Farming Tools for external nutrient inputs and water Management – FATIMA" EU H2020 Project, Grant No: 633945

FATIMA D4.1.4 (2018) Summary report of on-farm audit results and impact. Baseline for promoting an "Energy efficiency plan" in agriculture. Project Deliverable of "Farming Tools for external nutrient inputs and water Management – FATIMA" EU H2020 Project, Grant No: 633945

FATIMA D4.2.3 (2018) Policy Analysis Report. Project Deliverable of "Farming Tools for external nutrient inputs and water Management – FATIMA" EU H2020 Project. Grant No 633945

Harmancioglu NB (2008) Management and mismanagement of water resources (in Turkish). KALDER, Symposium on "Izmir Water Resources: Their Quality and Management", Izmir, December 2008

Harmancioglu NB (2014) Efficient water allocation mechanisms in support of water security. Istanbul, 3. Istanbul Uluslararasi Su Forumu (3rd International Water Forum), SUEN, 27–29 Mayis, 2014

Harmancioglu NB (2017) Overview of water policy developments: pre- and post-2015 development agenda. Springer Water Resour Manag 31:3001–3021

Harmancioglu NB, Barbaros F, Cetinkaya CP (2013) Sustainability issues in water management. Springer Water Resour Manag 27(6):p1867–p1892

Harmancioglu N, Fedra K, Barbaros F (2008) Analysis for sustainability in management of water scarce basins: the case of the Gediz River Basin in Turkey. Desalination, IWA Publishing 226:175–182

Harmancioglu NB, Ozkul S, Baran T (2005a) Experiences of decision making processes for water management: Turkey Gediz River Basin. National Report Part II, NOSTRUM-DSS Project on "Network on Governance, Science and Technology for Sustainable Water Resource

Management in the Mediterranean. The Role of DSS Tools", supported by the European Commission FP6 Programme (2004–2007), p 36
Harmancioglu NB, Ozkul S, Baran T (2005b) Statistical information at country level: assessment of water management practices, use and needs, Turkey. National Report Part I, NOSTRUM-DSS Project on "Network on Governance, Science and Technology for Sustainable Water Resource Management in the Mediterranean. The Role of DSS Tools", supported by the European Commission FP6 Programme (2004–2007), 73 p
Harmancioglu NB, Singh VP, Alpaslan N (eds) (1998) Environmental data management. Kluwer Academic Publishers, Water Sci Technol Lib 27:298
IWMI (2000) Institutional support systems project – the Turkey activity. A project supported by the International Water Management Institute, Colombo, Sri Lanka (IWMI) and realized with cooperation of IWMI, DSI and Dokuz Eylul University, 1999–2000
Michelsen A, Harmancioglu NB (2009) Session Outcome document of Session 3.4.2 on strategic framework for effective and sustainable water resources management and protection. Final report of the 5th World Water Forum, Istanbul, p 48
NOSTRUM-DSS (2007) The INCO-MPC1 project on "Network on governance, science and technology for sustainable water resource management in the mediterranean. The role of DSS tools" (NOSTRUM-DSS), supported by the European Commission FP6 Programme (INCO-MPC-1-509158) (2004–2007) (www.feem-web.it/nostrum/fra/)
OPTIMA (2007) The INCO-MPC1 project on "Optimisation for sustainable water management – OPTIMA" supported by the European Commission FP6 Programme (INCO-MPC-1-509091) (2004–2007) (http://www.ess.co.at/OPTIMA)
Quesne TL, Pegram G, Heyden CVD (2007) Allocating scarce water: a primer on water allocation, water rights and water markets. WWF Water Security Series 1, WWF-UK
SMART (2005) The INCO-MED project on "Sustainable management of scarce resources in the coastal zone - SMART", supported by the European Commission FP5 Programme, Contract No. ICA3-CT-2002, 2002–2005 (http://www.ess.co.at/SMART)
Speed R, Li Y, Quesne TL, Pegram G, Zhiwei Z (2013) Basin water allocation planning. Principles, Procedures and Approaches for Basin Allocation Planning, UNESCO, Paris
SUMER (2005) Gediz case study deliverable D05.1. The INCO-MED project "Sustainable Management of Scarce Resources in the Coastal Zone - SMART" supported by the European Commission FP5 Programme (Contract number: ICA3-CT-2002) (2002–2005)
Svendsen M, Murray-Rust H, Harmancioglu NB, Alpaslan N (2005) Chapter 11: Governing closing basins: the case of the Gediz River in Turkey. In: Irrigation and River Basin Management, Options for Governance and Institutions (ed. by M. Svendsen), CABI Publishing, Cambridge, MA, pp 193–213
SYGM (2017) Project on preparation of the Seyhan Basin sectoral water allocation plan (Seyhan Havzasi Sektorel Su Tahsis Planı Hazirlanmasi Projesi). The Ministry of Forestry and Water Affairs, General Directorate of Water Management (SYGM), Ankara. 73 p
TUIK (2018.) http://www.tuik.gov.tr/HbGetirHTML.do?id=27587. Accessed Oct 2018
UN (1992) Agenda 21: Programme of Action for Sustainable Development, United Nations, New York, NY, USA
UN (2017.) https://sustainabledevelopment.un.org/index.php?menu=220. Accessed May 2017
UNCED (1992) United Nations Conference on Environment and Development (UNCED), Rio de Janeiro, 3–14 June 1992 (http://www.un.org/geninfo/bp/enviro.html)
UNESCO (2013) Free flow: reaching water security through cooperation. UNESCO Publishing, Paris
UN-Water (2013) Water security and the global water agenda. A UN-water analytical brief., http://www.unwater.org/topics/water-security/en/
UN-Water (2014) A post-2015 global goal for water: synthesis of key findings and recommendations from UN-Water, Executive Summary: Detailed Targets and Associated Indicators. Technical Advice. www.unwater.org

UN-Water (2015) Annual report. http://www.unwater.org/publications/un-water-annual-report-2015/
UN-Water (2017.) http://www.unwater.org/topics/water-security/en/. Accessed May 2017
WWAP (2009) The United Nations world water development report 3: water in a changing world. UNESCO/Earthscan, Paris/London
WWAP (2014) The United Nations world water development report 2014: water and energy. UNESCO, Paris

Chapter 17
Challenges for the Future

Dogan Altinbilek and Nilgun B. Harmancioglu

Abstract This chapter is basically a wrap-up of current problems regarding water resources and management in Turkey and discusses challenges for the future in the light of these problems. The most challenging issue appears to be water scarcity resulting from fast increases in population, urbanization, industrialization, agriculture, expansion of tourism, and increases in economic activities, climate change, and resource depletion. The chapter also discloses how authorities and water communities in the country react to these challenges and plan new targets for the future.

Keywords Water crisis · Water scarcity · Global Risks Reports · Climate change · Food security · Water allocation · Sustainability · Integrated Water Resources Management (IWRM) · Water security · Data · Future challenges

The preceding chapters of this book have discussed the current status of water resources and their development in Turkey, focusing also on associated problems and measures taken to resolve them. This chapter serves basically to refresh the list of major difficulties which still prevail and constitute challenges for future developments.

Turkey has taken significant strides since the foundation of the Republic in 1923 towards establishing a multifaceted framework for structural, institutional and legal aspects of water resources developments. Especially after the establishment of General Directorate of State Hydraulics Works (DSI) in 1954, Turkey has built many water resources systems and structures in Turkish river basins for purposes of

D. Altinbilek (✉)
Past Director General, State Hydraulics Work (DSI), Ankara, Turkey

Past Vice President, World Water Council, Marseille, France

Civil Engineering Department, Middle East Technical University, Ankara, Turkey
e-mail: hda@metu.edu.tr

N. B. Harmancioglu (retired)
Faculty of Engineering, Civil Engineering Department, Dokuz Eylul University, Izmir, Turkey
e-mail: nilgun.harmancioglu@deu.edu.tr

N. B. Harmancioglu, D. Altinbilek (eds.), *Water Resources of Turkey*, World Water Resources 2, https://doi.org/10.1007/978-3-030-11729-0_17

irrigation, domestic water supply, power generation, flood control, and other purposes. In particular, the last 3–4 decades have witnessed large scale and unique development plans and systems, including water transfer among basins and even transport to neighboring water scarce regions. In recent years, Turkey has hosted 3.5 million migrants from Syria, Iraq and Afganistan. Approximately 10% of the migrants live in special camps, and their water, sanitation and other basic needs are met by the Government of Turkey. Again Turkey is undertaking infrastructure activities in selected African countries to provide basic water needs and flood control.

On the other hand, new problems have emerged since the early 80s as Turkey gradually became exposed to global crises that also struck the majority of the countries in the world. The international community was fast to react to this situation and developed guidelines for environmental management, development of water resources and sustainability. Turkey followed these developments at a rather slower pace but was able catch up with the required procedures towards assessing and handling her problems. However, some overarching difficulties are still valid for most countries and for Turkey.

Water still continues to be of paramount importance in the world as it is linked to many sectors, i.e. environmental, socioeconomic, energy, food production, and the similar. World Economic Forum's Global Risks Reports have listed water crises as a one of the top-five risks in the world for the third consecutive year (World Economic Forum 2016, 2017, 2018). Turkey also experiences this crisis and is currently a country under water stress. In that regard, the most challenging issue appears to be water scarcity resulting from fast increases in population, increased competition for water, urbanization, industrialization, agriculture, expansion of tourism, increases in economic activities, climate change, and resource depletion.

Another major risk specified in the above Global Risks Reports is climate change. It already started to impact Turkey through changes in precipitation patterns, droughts and floods. Changes in the frequency and intensity of these events are being observed, and, as Chap. 14 of this book indicates, a significant decrease in precipitation for almost all parts of the country is expected in the future. The intensity of drought conditions is expected to increase so that more intense arid conditions are expected in the region. Chapter 3 of this book discloses that there will be 16% and 27% reductions in the water potential of Turkey by 2050 and 2075, respectively. While the existence of a large number of dams and small dams in Turkey helps adaptation to climate change, it also causes water losses due to evaporation from reservoir surfaces. However, there are limits to their effectiveness of water storage for adapting to future hydrological extremes in the water cycle. In that regard, the existing multiple water storage infrastructures in river basins must be managed in an optimal way with a systems approach considering water availability forecasts and demand projections. Another requirement is the evaluation of risks in water resources planning and the adoption of risk based design procedures particularly to mitigate climate induced risks.

Water scarcity coupled with climate change impacts leads to an increase in water demand, which is also due to increased agricultural use, growing population and cities, and rapid industrial development in Turkey. As Turkey's population heads to 95 million people by 2050, water demands for food, cities and industry are expected to increase significantly. Water management in the near future must consider links between the water scarcity, climate change, and food security.

In that regard, water allocation has become an important issue that deserves further attention and investigations since loosely-planned and controlled allocation among water users leads to conflicts. The fast population growth, increased competition for water, increasing (and newly emerging) demands, and hence, water scarcity all necessitate equitable allocation of the resources so that Turkey has to focus on developing allocation plans in all her basins. A unified coordinating mechanism must be attained for allocating water among irrigation, urban demand, industrial requirements and environmental protection.

The concept of IWRM (Integrated Water Resources Management) constitutes a basis for the solution of the above mentioned problems. It provides useful tools for climate change adaptation and mitigation for water management such as infrastructure, land use management, agriculture, water quality management, floods and droughts management and governance. Turkey has to elaborate the already started efforts to first solve the difficulties encountered in management. Among these is the recognition and full understanding of the need for sustainable development where water plays a central role. With regard to water scarcity, there also exists the need for efficient water use in agriculture as it is the most water consuming sector in Turkey. On the other hand, there is the recognition in Turkey that there are multiple dimensions to water resource management problems, i.e. different disciplines, different interests, different uses, ground and surface water, quantity and quality, and so on. This recognition provides opportunity to develop an integrated approach to basin water resource planning and management.

One of the requirements of IWRM is the availability of sound and reliable data upon which management decisions are made. Data provision is an important issue in Turkey since basin management and other water related activities are often hindered by data limitations. There is the need for accurate and up-to-date descriptive information and a national database on all aspects of water resources in basins, including water allocations, reservoir positions, groundwater elevations and quality, water quality conditions, available resources, etc. Furthermore, this information should be accessible to users, possibly at minimal costs depending on the type of data. In particular, groundwater quality monitoring is not widespread and the results are not publicly available, making it difficult to know if significant degradation of groundwater quality is occurring. To solve the data problems, the hydrometric network should be improved and reassessed at particular intervals in time. Furthermore, the concept of data management should also be recognized as a significant activity for provision of data (Harmancioglu et al. 1998).

As discussed in some of the preceding chapters of this book, a new national Water Law is definitely required to support decision making for water management. Studies on this issue started in the late 90s, and a draft law was finally achieved in 2010. However, the law still remains in its draft form and requires immediate attention of the authorities to lay legal groundwork for effective basin management and protection.

Turkey has developed links with various international institutions in the water sector, which provide access to international experience on basin governance problems. Furthermore, the country has the willingness to communicate and cooperate with these institutions. Along this line, there is a strong motive at the authoritative level in harmonizing standards, practices, and procedures with those of the EU. Adopting the EU WFD (Water Framework Directive) has provided significant benefits in Turkey's efforts towards preparation of basin management plans.

The above issues were discussed in-depth at the 2nd Council of Forestry and Water, which convened in 2017 with the participation of a wide audience, including public authorities, managers, universities, NGOs, and representatives of the private sector. The purpose of the Convention was to develop national policies and strategies towards sustainable management of water and forest resources and to delineate all pathways (structural, legal, resource management, application, monitoring and evaluation) that are needed to solve prevailing problems in the forestry and water sectors. For the water area, 51 concluding decisions were formulated in the form of Actions Plans to be put into practice (ORMANSU 2017). These decisions are actually targets which Turkey envisages as the required steps towards coping with future challenges in the foreseeable future. Among these, the following are worth noting as they address the major current and expected problems in water management:

- Preparation of a National Water Plan and a Water Security Plan to provide decision support tools for water authorities and managers; efforts towards this target are initiated in 2017 to be completed by 2023, the centennial of the foundation of the Turkish Republic;
- Completion of the National Water Information System and provision of its progressive continuity and service effectiveness to be effective as of 2018 and continue onward;
- Encouragement of activities which lead to more effective and efficient uses of water through: reduction of evaporation from reservoir surfaces; selection of crops that can resist drought conditions; reduction of losses in drinking water distribution systems; use of gray water; harnessing of rainwaters; use of water saving armatures; development of smart distribution systems for drinking water and irrigation waters; the application of the "Polluter pays" principle to manage save water; selection of industrial processes that reduce water consumption and do not cause pollution; encouragement of water saving means through education and dissemination of information (to arouse awareness in public); determination of water footprints and thus supporting water saving approaches;
- Provision of the reuse of irrigation return flows and treated wastewaters in irrigation;
- Continuation of water pollution prevention actions so as to support the application of more effective basin management plans; protection of drinking water basins; development of a more efficient management procedure based on a "basin-wide" approach;

- Preparation of a legal basis which will permit the management of water resources and irrigations by a single central authority;
- Improvement of the institutional capacities and performances of irrigation Water User Associations (several detailed proposals were considered to handle this issue);
- Develop the legal and institutional basis for flood insurance systems so as to reduce flood risks;
- Assessment of legislation on floods and precise specification of relevant authorities and responsibilities to better manage flood mitigation activities;
- Delineation of responsibilities, authorities and sanctions on a legal basis, regarding dam safety and security (an important issue for the aging dams in Turkey);
- Development of a National Drought Database; preparation of emergency action plans for drought periods; consideration of droughts within the framework of "disasters";
- Completion of hydrogeological investigations and surveys all over the country; determination of groundwater masses and their quality; implementation of measures to prevent groundwater pollution.

This chapter starts with the claim that "*Turkey has taken significant strides since the foundation of the Republic in 1923 towards establishing a multifaceted framework for structural, institutional and legal aspects of water resources developments*". This is not an exaggeration but a realistic summary of what Turkey has accomplished within the last 95 years, which is actually a short span of time in comparison with the practices of developed countries. This book presents several examples of how the country managed to develop her water resources potential in this period. Just to give one example, the hydroelectric energy production of Turkey was only 1 TWh/y in 1960 but increased to about 75 TWh/y in 2017, which is equal to half of the ecomically feasible hydroelectric potential of the country. The total installed capacity of hydroelectric power plants was 0.4 GW in 1960 and increased to 27.3 GW in 2017, thus nearly 70 times in 57 years. Certainly, establishing a multifaceted framework for water resources developments within such a short period of time implies that the problems encountered have also been multifaceted, particularly due to data limitations, institutional and legal inefficiencies in the beginning. However, early recognition of these problems has finally created the current status of Turkey with respect to water resources development, where the country not only handles her own problems but also provides support to neighbouring water scarce countries and to selected needy countries in Africa.

As of today, the global water crisis has inevitably struck Turkey as it adversely affects many parts of the world. The major factor leading to this crisis is water scarcity aggravated by impacts of climate change. Water scarcity has embittered a chain of threats to many related sectors such as environmental security, water security, food security, energy, agriculture, and socioeconomics, to name a few. Turkey has also recognized these and other emerging problems regarding sustainable water management and developed a fast response to the crises. As summarized in the above targets, which were defined at the 2017 Second Council on Forestry and Water, Turkey is firmly determined and committed to resolve these issues in concrete steps within the deadlines specified for each target.

References

Harmancioglu NB, Singh VP, Alpaslan N (eds) (1998) Environmental data management. Kluwer Academic Publishers, Water Science and Technology Library, vol 27, 298 p

ORMANSU (2017) 2. Ormancilik ve Su Surasi (2nd Council on Forestry and Water). TC Orman ve Su Isleri Bakanligi (Ministry of Forestry and Water), Afyon, 39 p. www.ormansu.gov.tr/duyuru/2017/10/10/2.ormancılık-ve-su-surası-kararları

World Economic Forum (2016) The global risks report 2016, 11th edn. Published by the World Economic Forum within the framework of The Global Competitiveness and Risks Team, Geneva, REF: 080116

World Economic Forum (2017) The global risks report 2017, 12th edn. Published by the World Economic Forum within the framework of The Global Competitiveness and Risks Team, Geneva, ISBN: 978-1-944835-07-1, REF: 050117

World Economic Forum (2018) The gobal risks report 2018, 13th edn. Published by the World Economic Forum within the framework of The Global Competitiveness and Risks Team, Geneva, ISBN: 978-1-944835-15-6, REF: 09012018

Index

N. B. Harmancioglu, D. Altinbilek (eds.), *Water Resources of Turkey*, World Water Resources 2, https://doi.org/10.1007/978-3-030-11729-0

D

E

F

G

H

9783030117283